I0041846

Marine Surfactants

This book explores the development of novel marine biosurfactants. The book also covers the utilization of marine surfactants for biological, biomedical, and environmental applications.

Marine Surfactants: Preparations and Applications aims to examine every aspect of marine-derived surfactants. The first part of the book discusses the isolation of marine surfactants from various organisms, including marine bacteria, algae, cyanobacteria, and so on. The editors also examine the cultivation of marine microorganisms and the harvesting of other natural biological resources from the sea. The next part of the book discusses the application of marine surfactants, including oil spill removal in the sea, bioremediation of polluted water and soil, treatments for breast cancer, restoration of marine environments, nanoparticle synthesis, and development of different kinds of emulsifiers. With contributions from world-renowned experts in the field, this book will be an essential resource in understanding and developing various marine-derived surfactants.

This book is intended for researchers and marine biotechnologists as well as medical practitioners working on a vast range of industrial and medical applications using marine materials. It would also be useful for students looking to understand the utilization of marine-derived surfactants.

Marine Surfactants

Preparations and Applications

Edited by Se-Kwon Kim and Kyung-Hoon Shin

CRC Press
Taylor & Francis Group
Boca Raton London New York

CRC Press is an imprint of the
Taylor & Francis Group, an **informa** business

First edition published 2023
by CRC Press
6000 Broken Sound Parkway NW, Suite 300, Boca Raton, FL 33487–2742

and by CRC Press
4 Park Square, Milton Park, Abingdon, Oxon, OX14 4RN

CRC Press is an imprint of Taylor & Francis Group, LLC

© 2023 selection and editorial matter, Se-Kwon Kim, Kyung-Hoon Shin; individual chapters, the contributors

Reasonable efforts have been made to publish reliable data and information, but the author and publisher cannot assume responsibility for the validity of all materials or the consequences of their use. The authors and publishers have attempted to trace the copyright holders of all material reproduced in this publication and apologize to copyright holders if permission to publish in this form has not been obtained. If any copyright material has not been acknowledged please write and let us know so we may rectify in any future reprint.

Except as permitted under U.S. Copyright Law, no part of this book may be reprinted, reproduced, transmitted, or utilized in any form by any electronic, mechanical, or other means, now known or hereafter invented, including photocopying, microfilming, and recording, or in any information storage or retrieval system, without written permission from the publishers.

For permission to photocopy or use material electronically from this work, access www. copyright.com or contact the Copyright Clearance Center, Inc. (CCC), 222 Rosewood Drive, Danvers, MA 01923, 978–750–8400. For works that are not available on CCC please contact mpkbookspermissions@tandf.co.uk

Trademark notice: Product or corporate names may be trademarks or registered trademarks and are used only for identification and explanation without intent to infringe.

ISBN: 978-1-032-30704-6 (hbk)
ISBN: 978-1-032-30966-8 (pbk)
ISBN: 978-1-003-30746-4 (ebk)

DOI: 10.1201/9781003307464

Typeset in Times New Roman
by Apex CoVantage, LLC

Contents

Preface

Around 70% of the ocean is covered by water, and the marine environment is an underexplored natural resource. Marine surfactants have gained much attention due to their interesting biological properties. Marine-derived species, including microbes, have produced fascinating surfactants that can be potentially used in several biological and biomedical applications.

In the current book, several contributors around the globe contributed to marine surfactant production, the culture method of microbes for surfactant production, and the essential bioprocess method to enlarge surfactant production. Purification of surfactants using acid precipitation, solvent extraction, liquid membrane extraction, foam fractionation, and membrane-based techniques are discussed in the introduction chapters. Different kinds of surfactant glycolipids, phospholipids and fatty acids, lipoproteins, and polymeric surfactants are provided. The diversity of marine biosurfactant-producing bacteria and their role in the degradation of petrochemical hydrocarbons are discussed. Finally, the production and characterization of biosurfactants to improve the effectiveness and bioavailability of insoluble antibiotics are discussed.

Bacteria producers of biosurfactants isolated from saline soil and caverns and the role of microbial biosurfactants in marine bioremediation are presented—marine algal and cyanobacterial surface-active compounds. In the book's final part, the authors discuss marine biosurfactants' application in agriculture, cosmetics, oil remediation, and environmental bioremediation—extensive studies on the fermentative production of biosurfactants from extremophilic marine microbes.

I am fully aware that I cannot completely satisfy the interest of various scientists working in this area. Still, I hope this book will bring forward new avenues in this ever-growing field. In the future, I intend to fully address any inadvertent inadequacies and welcome all suggestions that can be included in coming editions. Finally, I am thankful to Dr. Venkatesan J, Yenepoya (Deemed to be University), India, who has extended his helping hand to complete this book.

Prof. Se-Kwon Kim,
South Korea

Acknowledgments

Dr. Jae-Chul Kim is the chairman, president, and founder of the Dongwon group. The company was established in 1969 to explore and utilize ocean and marine resources. Chairman Kim is a pioneer in deep-sea fishing boats in South Korea. He started tuna fishing in 1958 as the first mate of Korea's first deep-sea fishing vessel.

After graduating from the National Fisheries University of Busan, he jumped into tuna fishing, and the university was renamed Pukyong National University. Further, he became the captain of a fleet of deep-sea fishing boats using a new tuna fishing method. He succeeded in catching large amounts of tuna in the South Pacific and Indian Oceans.

He founded Dongwon Industries in 1969 at 35 and was the first president. After that, he built a tuna processing plant and started to produce canned food. He immersed himself in business management in earnest based on his experiences at sea in his youth.

Dongwon Group has successfully expanded its business from the fishery industry as the primary industry to the manufacturing industry as the secondary industry and financial services as the tertiary industry. Currently, Mr. Kim runs 30 affiliated companies.

To contribute to social welfare, he also established the Dongwon Educational Foundation. He devoted himself, his heart, and his soul to nurturing competent people who are the backbone of our society. He has provided scholarships to numerous college students and grants R&D funds to researchers. He recognized the great value and potential of underutilized marine resources from his youth. He dedicated himself to publishing technical books emphasizing the scientific importance of marine life and related research.

With his help, this book has been published, providing readers with knowledge on how the scientific values of marine life can enhance human health and wellbeing.

I want to thank him sincerely for his support in publishing this book.

Prof. Se-Kwon Kim
Distinguished Professor
Dept. of Marine Science & Convergence Engineering
Hanyang University ERICA, South Korea

Editor Biographies

Se-Kwon Kim, Ph.D., is presently working as a distinguished professor in Hanyang University and was previously Korea Maritime and Ocean University and Research Advisor of Kolmar Korea Company. He worked as Professor at the Department of Marine Bio Convergence Science and Technology and Director of the Marine Bioprocess Research Center (MBPRC) at Pukyong National University, Busan, South Korea.

He received his M.Sc. and Ph.D. degrees from Pukyong National University and conducted his postdoctoral studies at the Laboratory of Biochemical Engineering, University of Illinois, Urbana-Champaign, Illinois, USA. Later, he became a visiting scientist at the Memorial University of Newfoundland and University of British Columbia in Canada.

Dr. Kim served as President of the Korean Society of Chitin and Chitosan in 1986–1990 and the Korean Society of Marine Biotechnology in 2006–2007. He won the best paper award from the American Oil Chemists' Society in 2002. Dr. Kim was also the chairman of the 7th Asia-Pacific Chitin and Chitosan Symposium, which was held in South Korea in 2006. He was the Chief Editor of the Korean Society of Fisheries and Aquatic Science during 2008–2009. In addition, he is a board member of the International Society of Marine Biotechnology Associations (IMBA) and International Society of Nutraceuticals and Functional Food (ISNFF).

His major research interests are the investigation and development of bioactive substances from marine resources. His immense experience in marine bioprocessing and mass-production technologies for marine bio-industry is a key asset of majorly funded marine bio projects in Korea. Furthermore, he expended his research fields to the development of bioactive materials from marine organisms for their applications in oriental medicine, cosmeceuticals, and nutraceuticals. To date, he has authored around 750 research papers, 70 books, and 120 patents.

Kyung-Hoon Shin, Ph.D., is currently Professor at the Department of Marine Sciences and Convergent Technology and Director of the Institute of Ocean and Atmospheric Sciences, Hanyang University. He received his B.Sc. and M.Sc. degrees from Hanyang University, South Korea, and got his M.Sc. from Nagoya University and Ph.D. from Hokkaido University, Japan. He worked as a research scientist in the Frontier Observational Research System of Global Change/International Arctic Research Center, University of Alaska Fairbanks (dispatched from Japan Marine Science and Technology Center, JAMSTEC), and obtained the Outstanding Research

Accomplishment Award of Frontier Observational Research System for Global Change/JAMSTEC in 2002.

He served as the Editor-in-Chief of the *Korean Journal of Ecology and Environment* in 2011–2018, is the Associated Editor of *Geochemical Journal*, *Marine Drugs*, and *Frontiers in Marine Science*, and also served the Guest Editor of *Polar Science* and *Marine Drugs*. Based on his research performance (around 210 research papers, six patents), he has received numerous academic awards during the last 10 years. He won the best paper award of the Korean Society of Oceanography in 2011, the Outstanding Research Academic Award of the Korean Society of Limnology, the Best Paper Award of the Korean Federation of Science and Technology Societies in 2012, the Best Academic Award of Hanyang University, the Yeochon Academic Award for Ecology in 2015, and the Best Cited Paper Award of Hanyang University in 2018.

His broad scientific interests include stable isotope ecology, marine biogeochemistry, the organic carbon cycle, environmental safety and food forensics, organic geochemistry, pollution tracing, environmental chemistry and contaminant bioaccumulation, and the production and fate of bio-toxins.

Contributors

Arturo Abreu
Universidad Autónoma
Metropolitana-Cuajimalp, Mexico

Edidiong Okokon Atakpa
Zhejiang University, China

Khouloud M. Barakat
National Institute of Oceanography
and Fisheries (NIOF), Egypt

Tirth Bhatt
Marwadi University, Rajkot, India

Camelia Bhattacharyya
Amity University, Kolkata, India

Avani Bhimani
Marwadi University, Rajkot, India

L. Blandón
Marine and Coastal Research
Institute, Colombia

Gabriela Coelho Breda
Federal University of Rio de Janeiro,
Rio de Janeiro, Brazil

Siti Khodijah Chaerun
Institut Teknologi Bandung, Indonesia

Evans M.N. Chirwa
University of Pretoria, Pretoria

Osama M. Darwesh
National Research Centre, Egypt

Sumitra Datta
Amity University, Kolkata, India

Ria Desai
Navsari Agricultural University, India

Asmita Detroja
Marwadi University, Rajkot, India

Deviany Deviany
Institut Teknologi Sumatera, Indonesia

Swasti Dhagat
NIT Raipur, India

Dulce Celeste López Díaz
Universidad Autónoma de Querétaro,
México

Leticia Dobler
St. Anne's University Hospital, Brno,
Czech Republic

Everaldo Silvino dos Santos
Federal University of Rio Grande do
Norte, Natal, Brazil

Dhruv Gevariya
Marwadi University, Rajkot, India

Brian Gidudu
National Institute of Pharmaceutical
Education and Research
(NIPER-R), India

J. Gómez-León
Marine and Coastal Research
Institute, Colombia

Jorge Gracida
Universidad Autónoma de Querétaro,
México

Q. Helmy
Institute of Technology Bandung,
  Indonesia

Sapna Jaiswar
NIMS University (Rajasthan), India

Lijia Jiang
Zhejiang University, China

Satya Eswari Jujjavarapu
NIT Raipur, India

E. Kardena
Institute of Technology Bandung,
  Indonesia

J.J. Mehjabin
Hokkaido University, Japan

Chiamaka Linda Mgbechidinma
Zhejiang University, China

Mithrambigai
Hindustan College of Arts and
  Science, India

Nabya Nehal
Banasthali Vidyapith, India

T. Okino
Hokkaido University, Japan

Ricardo R. Oliveira
St. Anne's University Hospital, Brno,
  Czech Republic

E.L. Otero-Tejada
Marine and Coastal Research
  Institute, Colombia

Evelyn Zamudio Pérez
Universidad Autónoma de Querétaro,
  México

M. Quintero
Marine and Coastal Research
  Institute, Colombia

C. Elizabeth Rani
Hindustan College of Arts and
  Science, India

Shashank Reddy
National Institute of Pharmaceutical
  Education and Research
  (NIPER-R), India

Patrícia Maria Rocha
Federal University of Rio Grande do
  Norte, Natal, Brazil

Gaurav Sanghvi
Marwadi University, Rajkot, India

Priyanka Singh
NIMS University (Rajasthan), India

Sonal
NIMS University (Rajasthan), India

Nidhi Srivastava
National Institute of Pharmaceutical
  Education and Research
  (NIPER-R), India

Sukandar
Institute of Technology Bandung,
  Indonesia

Vartika Verma
Banasthali Vidyapith, India

Wilza Kímilly Vital de Paiva
Federal University of Rio Grande do
  Norte, Natal, Brazil

Trupti K Vyas
Navsari Agricultural University, India

Guiling Wang
Guilin University of Technology,
    China

Chunfang Zhang
Zhejiang University, China

Xiaoyan Zhang
Guilin University of Technology,
    China

A. Zuleta-Correa Marine and Coastal
    Research Institute, Colombia

1 Surfactin and Surfactin-Like Production, Purification and Application in Marine Environments

Ricardo R. Oliveira, Gabriela Coelho Breda, Patrícia Maria Rocha, Wilza Kímilly Vital de Paiva, Everaldo Silvino dos Santos, and Leticia Dobler

CONTENTS

DOI: 10.1201/9781003307464-1

INTRODUCTION

Biosurfactants are considered a promising alternative to non-biologically syn-thesized surfactants due to their environmentally friendly characteristics and remarkable physicochemical properties, especially the ability to lower surface tension to very small values. This class of molecules has advantages in terms of biodegradability, toxicity, structural diversity, surfactant activity and stability under extreme conditions of temperature, pH and ionic strength (Dobler et al. 2016). In this context, surfactin and their derivatives stand out because of their high surface activity.

The chemical composition, physicochemical properties and crystalline struc-ture of surfactin were obtained in 1968 (Arima, Kakinuma, and Tamura 1968), produced by *Bacillus subtilis*. The crystal morphology is composed of white nee-dles revealed through micrograph. Also, the authors demonstrate that surfactin is a peptidelipid with a molecular weight of 1050 Da. Moreover, the primary struc-ture of surfactin was revealed in 1969 by a combination of different techniques such as elemental analysis, infrared (IR), nuclear magnetic resonance (NMR) and mass spectra (Kakinuma et al. 1969).

Although the structure was elucidated more than 50 years ago, several works on molecular geometry of different congeners have been performed applying experimental and theoretical protocols. For example, two-dimensional proton nuclear magnetic resonance (1H-NMR) analysis and molecular dynamics simu-lations were performed in order to reveal the three-dimensional structure of pro-tonated (Leu71)-surfactin (leucine in position number 7) (Bonmatin et al. 1994). Also, a model of surfactin was determined in sodium dodecyl sulfate (SDS) micellar solution, which is a more biomimetic environment than simple solutions of organic compounds (Tsan et al. 2007).

It is worth mentioning that there are several natural congeners, but an abun-dant one is [Glu1, Asp5]-iso-C15-surfactin (C15-surfcatin), that is, glutamic and aspartic acids in positions 1 and 5, respectively. Also, the lateral chain is com-posed of a β-hydroxyl fatty acid moiety containing 15 carbon atoms (Kowall et al. 1998; Shao et al. 2015; Gang, Liu, and Mu 2011; Gang, Liu, and Mu 2015; Gang et al. 2020; Peypoux, Bonmatin, and Wallach 1999; Seydlová and Svobodová 2008). A simplified 2D structure is present in Figure 1.1.

Surfactin is naturally synthesized by non-ribosomal peptide synthases (NRPSs) from microorganisms such *Bacillus subtilis* and other bacteria (Marahiel, Stachelhaus, and Mootz 1997). The biosynthesis of surfactin is the reason there is a high molecular biodiversity of molecules. However, aspartic acid is always present in position 5 and D-Leucine in positions 3 and 6 in all surfactin molecules (Théatre et al. 2021). Furthermore, the side (aliphatic) chain length can vary from 12 to 17 carbon atoms. Consequently, the different isomers can be linear, branched, iso or anteiso (Théatre et al. 2021). Several natural con-geners are shown in Figure 1.2.

Beyond natural diversity, chemical modifications of surfactin can produce several derivatives with particular properties due to different carboxylic sites.

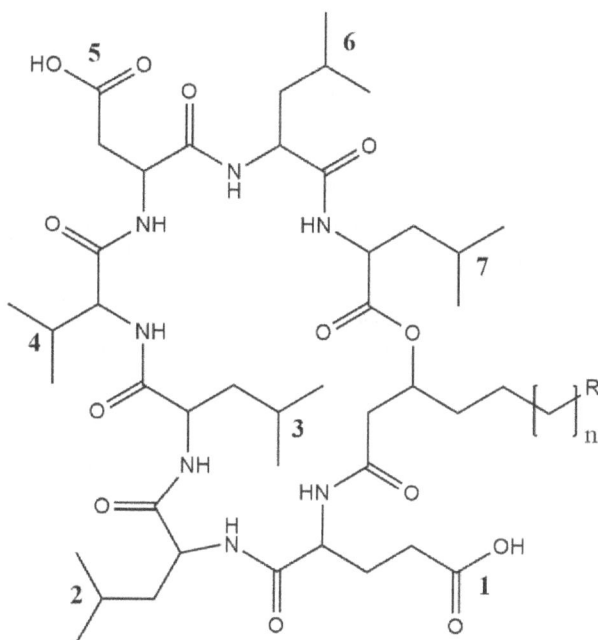

FIGURE 1.1 Basic structure of surfactin.

In particular, esterification reactions lead to high-yield products being a promising strategy for the synthesis of new macromolecules (Shao et al. 2015). These reactions can be catalyzed by HCl forming mono- and di-ester. Also, the product ratio can be controlled by the selection of different HCl concentrations (Shao et al. 2015), and hydrolysis reactions for ring opening were also reported, but surface activity was reduced when compared to the cyclic forms (Dufour et al. 2005; Eeman et al. 2006). Other chemical modifications can be done by alkaline treatment, esterification, amidation or through genetic engineering (Théatre et al. 2021; Hu, Liu, and Li 2019).

The lipopeptide family was named "surfactin" due to the unusually high surface activeness, in some cases even higher than that of common synthetic surfactants (Arima, Kakinuma, and Tamura 1968). When surfactin is added to aqueous solutions, a remarkable decrease in surface tension of water from 72 to 27 mN/m occurs at a concentration of 10 μM (Maget-Dana and Ptak 1992; Seydlová and Svobodová 2008; Heerklotz and Seelig 2001), a value much smaller than the critical micellar concentration (CMC), 3×10^{-4} M (Heerklotz and Seelig 2001). For the sake of comparison, most biosurfactants barely reach 30 mN/m (Cooper et al. 1981). Onazi (2018) investigated the adsorption mechanism of surfactin at the air/water interface and concluded that, at the initial stage of adsorption processes, a pure diffusion mechanism dominates. Further, there is a shift to a mixed diffusion-barrier mechanism (Onaizi 2018).

The surfactin structure at the air/water interface was revealed by the neutron reflectometry technique at pH = 7.5 combining different isotopic compositions (Shen et al. 2009). At the surface, surfactin adopts a ball-like structure, forming a compact surface layer with a thickness of approximately 14 Å (Shen et al. 2009). The structure at the surface has a small influence on pH, but the aggregates in the bulk are strongly influenced (Shen et al. 2011). At high pH values and above the CMC, surfactin forms spherical micelles, but at low values, rod-like or lamellar structures dominate (Shen et al. 2011). Other works also reported micellar structures and dynamics emphasizing rod-like (Maget-Dana and Ptak 1992; Seydlová and Svobodová 2008) and spherical (She, Gang, and Mu 2012; Shen et al. 2009) aggregate formation. Globular micelles can exhibit diameters of 10–20 nm (Gang

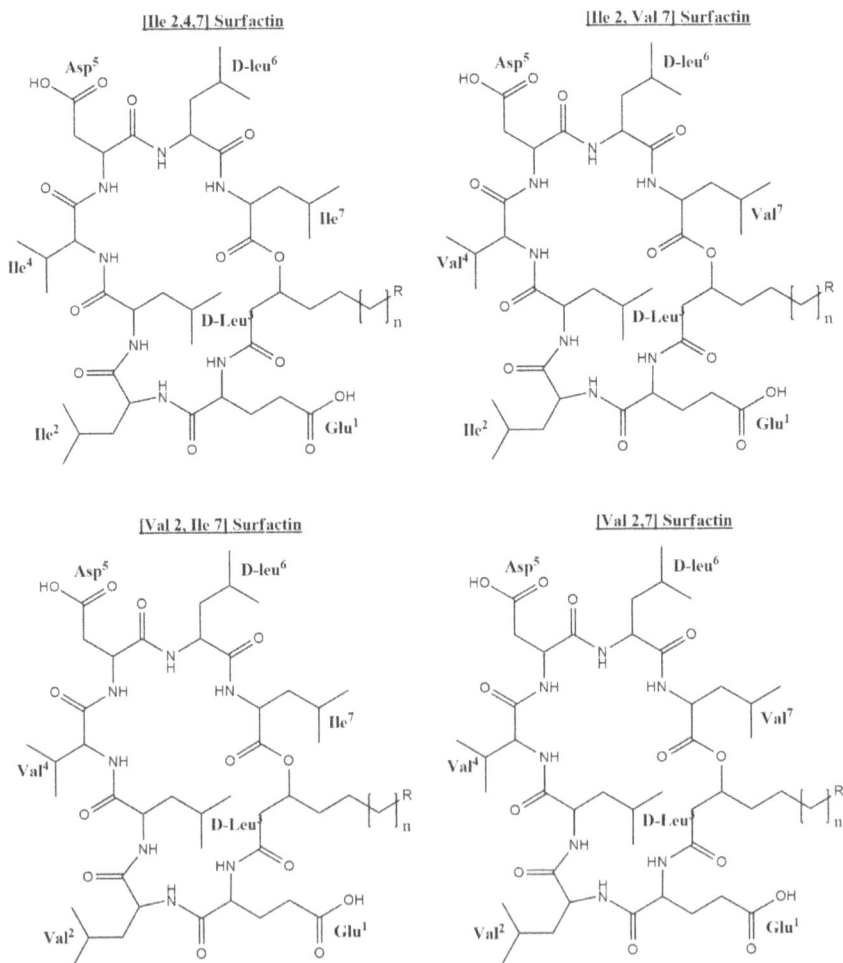

FIGURE 1.2 Ten most common natural surfactin structures.

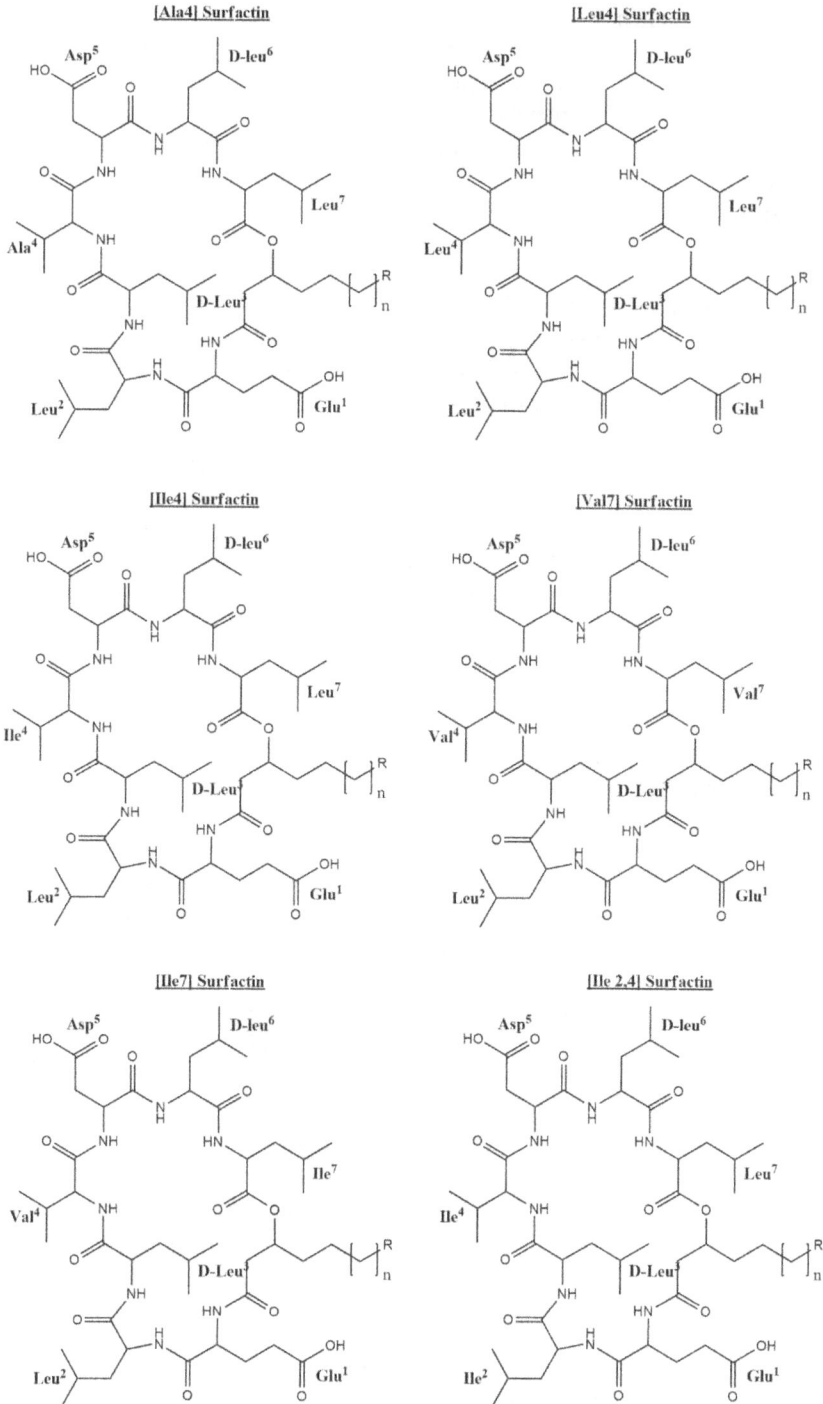

FIGURE 1.2 (Continued)

et al. 2020; Knoblich et al. 1995), but beyond the pH, aggregation formation is dependent on the temperature, environment and surfactin structure (Gang et al. 2020; Knoblich et al. 1995; She, Gang, and Mu 2012).

Due to structural complexity, several theoretical works applying molecular dynamics methods emerged (Gang, Liu, and Mu 2011; Gang, Liu, and Mu 2015; Gang et al. 2020). An example of a 3D structure of C4-surfactin obtained with CREST software (Pracht, Bohle, and Grimme 2020) using the extended tight binding method (xtb) (Bannwarth, Ehlert, and Grimme 2019) (in gas phase) is shown in Figure 1.3. The "horse saddle" topology proposed by Bonmatin and co-workers (1994) can be seen in Figure 1.3 (left structure). This structure was also confirmed by several molecular dynamics works by Gang and co-workers (Gang, Liu, and Mu 2011; Gang, Liu, and Mu 2015; Gang et al. 2020; Gang, Liu, and Mu 2010a; Gang, Liu, and Mu 2010b) in air/water and decane/water interfaces for C15-surfactin.

In the decane/water interface, the molecular area can vary from 110 to 133 Å, and the surfactin motion is limited by the strong interaction between the polar groups and water molecules (Gang, Liu, and Mu 2010a; Gang, Liu, and Mu 2010b). Also, the structural behavior of surfactin depends on surface concentration (Gang, Liu, and Mu 2015) and pH (Maget-Dana and Ptak 1992). It is worth mentioning that the pKa of surfactin is about 6.0 at the air/water interface and ~4.0 in bulk solution (Maget-Dana and Ptak 1992).

She and co-workers found from molecular dynamics simulations that micelle diameter is around 4.4 nm (2012). Furthermore, discrepancies on the number of molecules are reported; rod-like micelles can have aggregation numbers of 170 (Seydlová and Svobodová 2008; Heerklotz and Seelig 2001). In contrast, spherical/globular micelles can have up to 70 (Gang et al. 2020). This difference illustrates how complex the aggregation geometry and formation of surfactin systems are.

In addition to how surfactin's structure contributes to its amphiphilicity, stability, surface activity and density, these properties also give this family of lipopeptides high anti-inflammatory, viral and bacterial activity. Moreover,

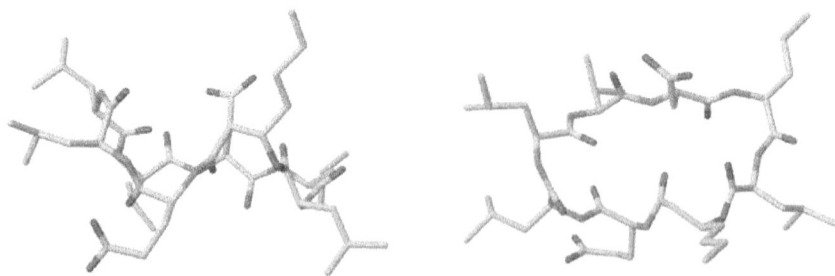

FIGURE 1.3 The structure of C4-surfactin obtained with CREST software using the GFN2-xTB method. Lateral (left) and front (right) views are shown. Light gray are carbon atoms, black are oxygen atoms and dark gray are nitrogen atoms. Hydrogen atoms are omitted.

environmental applications such as bioremediation and biodegradation of oils have also been reported (Chen, Juang, and Wei 2015).

PRODUCTION BY *BACILLUS SUBTILIS*

Bacillus subtilis is a spore-forming gram-positive bacterium found in soil and the gastrointestinal tracts of many animals. Although the genus *Bacillus* generally has the ability to produce surfactin, among other cyclic lipopeptide biosurfactants, such as iturin and fengycin, the species *B. subtilis* is reported in the vast majority of cases. In this topic, we will address the main aspects of surfactin production by *B. subtilis* with the different strategies explored to reach this bioprocess to a large scale, as well as the main challenges encountered.

METABOLIC PATHWAYS

The synthesis of surfactin in *Bacillus* spp. occurs nonribosomally, such as the biosynthesis of other bacterial peptide secondary metabolites. Nonribosomal peptide synthetases are the enzymatic machinery responsible for this assembly line that catalyzes stepwise peptide condensation in a multiple carrier thiotemplate mechanism (Kluge et al. 1988). Since it is not a ribosomal synthesis, the components integrated are not restricted to the 20 common amino acids, and several post-synthetic modifications, such as glycosylation and oxidative crosslinking, are now known. The incorporation of nonproteinogenic amino acids, such as D-isomers, carboxy acids, N-methylated residues, heterocyclic rings and fatty acids, is typical (Shaligram and Singhal 2010).

The inducible operon named *srfA* (25 kb), which is also responsible for sporulation and competence development, codes the surfactin synthetase complex composed of four enzymatic subunits: SrfA (E1A, 402 kDa), SrfB (E1B, 401 kDa), SrfC (E2, 144 kDa) and SrfD (E3, 40 kDa); the last plays an important role to initiate surfactin synthesis. The enzymes SrfA, SrfB and SrfC form seven modules that comprise 24 catalytic domains, and each module incorporates and modifies one specific amino acid into the peptide product (Seydlová and Svobodová 2008).

Every module comprises at least three domains. The first is the adenylation domain that catalyzes the formation of an aminoacyl adenosine and pyrophosphate from the cognate amino acid and adenosine triphosphate (ATP), and this step is also known as amino acid activation. The second domain covalently binds the activated amino acid to the 4'-phosphopantetheinyl prosthetic group that is present on the peptidyl carrier protein (PCP) via a thioester linkage. The last domain is the condensation domain that catalyzes the direct condensation of thioesterified intermediates in a growing chain that is chemically different for each module (Shaligram and Singhal 2010).

Besides the *srfA* operon, another essential gene to produce surfactin is *sfp* (4 kb), which encodes the enzyme Sfp (224 amino acids—superfamily of 4'-phosphopantetheinases), which presents the 4'-phosphopantetheinyl prosthetic group with a previously mentioned important function for peptide synthesis. Furthermore, phosphopantetheinyl transferase (Sfp) converts inactive apo-forms of PCP domains to active cofactor containing holo-forms. The gene *YerP* is believed to provide resistance to surfactin (Tsuge et al. 2001).

In wild-type *Bacillus* spp., surfactin synthesis only occurs in the late exponential phase, when SrfA operon expression is induced by the global regulatory mechanisms ComP-ComA (competence development) and Spo0A-AbrB (sporulation) (Hamoen et al. 2003). Both systems are sensitive to nutritional stress, modulating a variety of genes, including the induction of the *srfA* operon. The metabolic pathways and regulatory aspects related to surfactin synthesis are important to consider when determining the parameters of the bioprocess.

MAIN STRAINS

More than 30 natural strains of *B. subtilis* are reported for surfactin production, many of them isolated from soil (Modabber et al. 2020). Considering studies with a focus on high production strategies, the ATCC 21332 strain from the American Type Culture Collection (ATCC) is the most frequently reported, as it presents great natural surfactin production of 0.3–0.4 mg/mL in shake flasks without optimization (Wei and Chu 2002), but the strain DSM 10T also stands out (Willenbacher et al. 2014). Table 1.1 summarizes the main reports of high surfactin production yields using *B. subtilis* wild-type strains with different cultivation modes and conditions that will be discussed further.

CULTIVATION MODES

The fermentation technologies to produce metabolites of interest can be divided in two different classes: submerged fermentation (SMF) and solid-state fermentation (SSF). For the SMF processes, the microorganisms are suspended in a liquid medium, so a uniform system is available. This allows greater transfer of mass and monitoring and control of important parameters (temperature, pH and dissolved oxygen, among others), which affords thorough automation and reduces labor costs. Furthermore, the engineering aspects of SMF are well known and described, presenting different conduction modes (batch, fed-batch and continuous) and machinery (stirred tank, bench top, airlift and others).

On the other hand, SSF presents a solid matrix that acts as a substrate for the growth of microorganisms in the absence of free water. The solid matrix provides carbon, nitrogen, nutrients, minerals and salts as a proper solid medium. The heterogeneity of the medium in SSF processes can be more challenging, as it implies difficulties in controlling the fermentation parameters, such as temperature. However, SSF has been studied and explored in the past decades, as it can be done with simpler equipment, such as tray-type bioreactors; lower energy and space requirements; and, most importantly, using agro-industrial wastes as matrices, which could result in lower costs for biomolecule production (Kumar et al. 2021; Mitchell et al. 2019).

Industrial bioprocesses generally use SMF, and the majority of surfactin production studies also use this cultivation mode, which can be seen from the summarized data in Table 1.1. However, some studies have demonstrated SSF as a
potential alternative, reaching a surfactin production of 4 g/kg using larger fragments of milled oats as the solid matrix (Szmigiel et al. 2021). Some other studies even compared both cultivation modes (Ohno et al. 1992; Ohno et al. 1995).
For cultivations performed by the *B. subtilis* RB14 strain (Ohno et al. 1992),
the results were comparable with surfactin production using SMF with synthetic
media (350 mg/L) and SSF with soybean curd residue (250 mg/kg wet biomass).
In this case, SSF could be considered superior because of the low cost of the feedstock. It is worth noting that, although SSF is well known for using alternative
feedstock materials, SMF can also be done using liquid industrial by-products,
as will be discussed further.

MOST COMMON CONDITIONS IN SUBMERGED CULTIVATION

In SMF, several parameters influence the growth and production yields, such as
inoculum age and size, temperature, pH, media composition (carbon and nitrogen sources, concentration and ratio; mineral salt presence and concentration)
and dissolved oxygen when in aerobic culture. The inoculum or seed culture
is an essential component of the fermentation process. The variability of this
input in age and size influences the duration of the lag phase, specific growth
rate, biomass yield, sporulation and production yield. Furthermore, inoculum
preparation through various stages generally presents a significant impact, and
a two-stage inoculum is reported as a standard to produce surfactin (Sen and
Swaminathan 2004).

Sen and Swaminathan (2004) optimized surfactin production by evaluating the inoculum age and size of *B. subtilis* using response surface modeling.
The results obtained proved the interconnection of these twin parameters, with
a strong interaction between the variables primary inoculum size and secondary inoculum age. The optimal age and size of the primary inoculum were 56 h
(late stationary growth phase) and 5.5% (v/v), and those of secondary inoculum
were 4.5 h (early exponential growth phase) and 9.5% (v/v), respectively. Another
study showed that surfactin production could be increased upon increasing the
inoculum size up to 6% (v/v), although a negative effect was observed in bacterial
growth (Abdel-Mawgroud et al. 2008). This effect is consistent, as the production
of this biosurfactant is not associated with growth.

B. subtilis is reported to produce surfactin at temperatures from 25 to 37°C
and pH values of 6.0 to 9.0 (Chen et al. 2015). pH control is important in several
bioprocesses and can be easily performed by buffered growth media or adding an
alkaline/acid solution in the case of cultures in bioreactors. For the production of
surfactin, the maintenance of a constant pH value was also favorable (Wei and Chu
1998). Sen and Swaminathan (1997) studied the effect of several parameters such
as pH and temperature for the *B. subtilis* DSM 3256 strain. The results showed
that 37.4°C and a pH value of 6.75 maximized surfactin production. However,

the most frequently reported conditions are 30°C, with pH 6.8. It is important to emphasize that the optimal conditions of temperature and pH to produce surfactin can vary according to other cultivation parameters and also the strain used.

The most common media reported are synthetic, frequently minimal salt media (MSM), such as the one defined, after evaluation of metal cation addition, by Cooper and co-workers (1981) (composition: 4% glucose, 8.0×10^{-4} M $MgSO_4$, 7.0×10^{-6} M $CaCl_2$, 4.0×10^{-6} M $FeSO_4$, 4.0×10^{-6} M Na2 EDTA, 1×10^{-6} M $MnSO_4$, 0.05 M NH_4NO_3, 0.03 M KH_2PO_4 and 0.04 M Na_2HPO_4), also known as Cooper media. Semi-synthetic media are also used for surfactin production with *B. subtilis* and generally present, in addition to chemically formulated substances, complex compounds rich in carbon and nitrogen, such as yeast extract and peptone (Jokari et al. 2013; Huang et al. 2015; Zhao et al. 2021). A decrease in surfactin concentration is observed after depletion of glucose, as the cells utilize surfactin as the carbon source for their growth (Yeh et al. 2005). In a bioprocess, the degradation of the bioproduct by the microorganism is a critical feature, and the determination of an appropriate time to terminate the culture is essential to avoid losses.

The nature of the growth-limiting nutrient showed relevance to enhance the production of many secondary metabolites (Davis et al. 1999), such as surfactin. As mentioned before, the *srfA* operon is induced with nutritional stress at the late stationary growth phase. Davis et al. (1999) used a synthetic medium to produce surfactin and studied carbon and nitrogen limitations in conjunction with another important and correlated condition, the presence and deficiency of oxygen. Although *B. subtilis* is widely reported for surfactin production in aerobic conditions (Table 1.1), it can also grow anaerobically in the presence of nitrate as a final electron acceptor. In the absence of external electron acceptors, it grows by fermentation (Nakano et al. 1997).

TABLE 1.1
Surfactin High-Producing Natural *B. subtilis* Strains with the Product Yield Achieved, with the Main Parameters Used in the Reported Bioprocess

B. subtilis Strain	Surfactin Production	Cultivation mode	Media	Reactor Type	Reference
#309	2.8 g/L	Aerobic SMF[1]	MSM[3]	Flasks	Janek et al. 2021
87Y	4 g/kg	Aerobic SSF[2]	Larger fragments of milled oats	Flasks	Szmigiel et al. 2021
ATCC 21332	6.45 g/L	Aerobic SMF[1]	MSI[4]	5 L reactor	Yeh et al. 2006
ATCC 21332	9.4 g/L	Aerobic SMF[1]	MSM[3]	Flasks	Mei et al. 2021
ATCC 6051	1.1 g/L	Aerobic SMF[1]	Semi-synthetic with brewery waste	5 L reactor	Nazareth et al. 2021a

B. subtilis Strain	Surfactin Production	Cultivation mode	Media	Reactor Type	Reference
ATCC 6051	2.1 g/L	Aerobic SMF[1]	Semi-synthetic with brewery waste	5 L reactor	Nazareth et al. 2021b
BS5	1.12 g/L	Aerobic SMF[1]	MSM[3]	Flasks	Abdel-Mawgoud, Aboulwafa, and Hassouna 2008
CWS1	3.8 g/L	Aerobic SMF[1]	Synthetic with sucrose	Flasks	Kan et al. 2017
DSM 10[T]	4 g/L	Aerobic SMF[1]	Cooper medium	2.5 L reactor	Willenbacher et al. 2014
DSM 10[T]	0.87 g/L	Anaerobic SMF	Cooper medium	2.5 L reactor	Willenbacher et al. 2015
Inaquosorum 61	6 g/L	Aerobic SMF[1]	MSM[3]	5 L reactor	Modabber et al. 2020
MTCC 2415	1.75 g/L	Aerobic SMF[1]	MSM[3]	Flasks	Ganesan and Rangarajan 2021
MTCC 2423	4 g/kg	Aerobic SMF[1]	Rice mill processing residue	Flasks	Gurjar and Sengupta 2015

[1] Submerged fermentation
[2] Solid state fermentation
[3] Minimal salt medium
[4] Iron-enriched minimal salt media

The results obtained by Davis et al. (1999) finally showed the highest surfactin production in a defined medium with ammonium nitrate as nitrogen source in the absence of oxygen. During aerobic growth, ammonium was utilized preferentially over nitrate. Under anaerobic growth conditions, even though ammonium was present in the culture, nitrate utilization occurred, as it is used as the terminal electron acceptor. The study also indicated that by providing ammonium during the growth phase and implementing nitrate in the feed-batch approach, nitrate utilization can be prolonged and surfactin production increased.

Aerobic cultures generally present higher growth rates but consume more energy to promote continuous high-speed agitation and air sparging, which can provide high amounts of foam when producing biosurfactants (Yeh et al. 2006). In anaerobic cultivation, on the other hand, the growth is highly limited by the absence of oxygen, but the use of this kind of cultivation can be strategic to produce some biomolecules (Willenbacher et al. 2015). Surfactin production by B. subtilis anaerobically has shown potential, as it represents a process that completely avoids foam formation (Davis et al. 1999; Willenbacher et al. 2015).

Excessive foaming represents a challenge in a bioprocess, as it can decrease the oxygen transfer and thus surfactin yield (Yeh et al. 2005) and collapse the bioreactor. However, it is important to note that several foam containment and recovery techniques (such as foam fractionating—Cooper et al. 1981; Willenbacher et al. 2014; Alonso and Martin 2016) are already used in the production of surfactin and may even represent an advantage to the downstream process. Furthermore, it was reported that foaming stimulated surfactin production, resulting in higher productivities and higher biomass yields in comparison to fermentation performances achieved by using cells not subjected to foaming (Alonso and Martin 2016).

Another important method reported to enhance surfactin production is the addition of solid carriers (Yeh et al. 2005). In this method, a proper number of solids is added into the fermentation broth, promoting numerous micro surfaces to facilitate bacterial growth and surfactant production. Activated carbon was the best solid carrier tested for surfactin production in *B. subtilis* (Yeah et al. 2005), and bioreactor cultures achieved 6.45 g/L of surfactin, a very high yield compared to other studies, as shown in Table 1.1 (Yeh et al. 2006).

Several reports highlight the importance of some mineral salt presence and concentration in the medium to produce and achieve high yields of surfactin in *B. subtilis* cultivations. Response surface and Taguchi methodology were employed to evaluate the significance of mineral salts and optimize them in the medium (Sen et al. 1997; Wei et al. 2007). Surfactin production increased from 0.33 to 2.6 g/L simply by adding 0.01 mM of Mn^{2+} to a defined glucose medium (Wei and Chu 2002).

Iron enrichment in excessive concentrations of 2–4 mM of Fe^{2+} resulted in the production of 3.5 g/L of surfactin, which represented a seven-fold increase. Iron addition could be performed either initially or during batch culture, and pH maintenance above 5.0 (NaOH addition) proved essential in this case (Wei and Chu 1998). The excessive iron requirement might be due to a poor iron transport system and/or sequestering agent production by *B. subtilis*, which makes iron unavailable (Cooper et al. 1981). With the mentioned reports, it was possible to recognize the role of iron (Fe^{2+}) as a growth stimulator and that of manganese (Mn^{2+}) as a spore inducer and production-enhancing factor. Iron and manganese active transport systems have been observed in *B. subtilis*, and both ions act as major cofactors for enzymes in the surfactin biosynthetic pathway (Shaligram and Singhal 2010).

The optimization of a trace element solution for surfactin production using the Taguchi method found that the most significant elements, besides Mn^{2+} and Fe^{2+}, previously mentioned, are potassium (K^+) and magnesium (Mg^{2+}). Calcium (Ca^{2+}) presented no significance for cell growth and production (Wei et al. 2007). Optimized concentrations for the four metal ions were 2.4 mM of Mg^{2+}, 10 mM of K^+, 0.008 mM of Mn^{2+} and 7 mM of Fe^{2+}, with surfactin production of 3.34 g/L. The absence of Mn^{2+} or Fe^{2+} slightly inhibited cell growth but did not affect surfactin production. The trace element levels were determined individually. However, when optimal concentrations of the ions were used together in the medium, lower surfactin production was observed, indicating significant interactions between the four metal ions. Furthermore, the statistical data proved that Mg^{2+} and K^+ ions were more critical than the other two metal ions. Kinsinger and co-workers (2003) proposed that Mg^{2+} is a cofactor of Sfp, an essential protein in the synthesis of surfactin for the activation of the PCP domains, as mentioned before. K^+ was reported in the stimulation of surfactin secretion (Kinsinger et al. 2005).

Despite all the advances obtained by the aforementioned studies in the production of surfactin, its large-scale commercialization is still not competitive due to the high cost of production (Zanotto et al. 2019). For example, the production cost price of a petrochemical-derived surfactant was estimated at US$1/lb (Makkar et al. 2011), while the surfactin standard price marketed by Sigma Chemical Company is approximately US$17.6/mg, as stated on the Sigma-Aldrich website in 2022. Since the culture medium usually represents 30–50% of the total cost of production in a bioprocess, an important alternative to reduce costs is the use of agro-industrial by-products and residues, adding value to them and, at the same time, minimizing impacts from their inappropriate disposal.

Several studies have already proven the possibility to produce surfactin in submerged fermentation using liquid agro-industrial by-products and residues, such as sugarcane molasses (0.199 g/L—Rocha et al. 2020), crude glycerol from the biodiesel industry (0.23 g/L—de Faria et al. 2011) and treated palm oil mill effluent, also known as POME (0.03–0.035 g/L—Abas et al. 2013). As shown in Table 1.1, even higher productions were achieved by using a semi-synthetic medium with brewery waste (2.1 g/L—Nazareth et al. 2021b) and rice mill processing residue (4 g/kg—Gurjar et al. 2015), which demonstrates the potential of these alternative feedstocks.

PRODUCTION BY OTHER ORGANISMS

Mains aspects and results for surfactin production recently reported in other species of the genus *Bacillus*, as well as in genetically modified strains of *Bacillus* spp., will be addressed in this section. Table 1.2 summarizes the main reports with high surfactin production yields that will be discussed further.

Despite having a much smaller number of reports in the literature compared to *B. subtilis*, the species *Bacillus amyloliquefaciens* and *Bacillus velezensis* stood out for the high production values obtained without any genetic modification, which can be observed in Table 1.2. Moreover, two of the studies used alternative feedstock to produce surfactin. Zhu and co-workers (2013) explored the aforementioned solid-state fermentation using a *B. amyloliquefaciens* strain, with advantages in several aspects. Pan et al. (2021) used an innovative strategy from a consortium formed by a *B. amyloliquefaciens* strain together with the lipase producer *Aspergillus nidulan*. The lipidic feedstock used, kitchen waste, could be hydrolyzed by *A. nidulan* and assimilated by *B. amyloliquefaciens*, resulting in simultaneous enzymolysis and surfactin fermentation.

Furthermore, the endophytic *Bacillus mojavensis* species presented potential cyclic lipopeptide production, mainly fengycin, with very powerful biosurfactant activity (Snook et al. 2009; Hmidet et al. 2017). Another interesting

TABLE 1.2
Surfactin High-Producing *Bacillus* spp. Strains with the Product Yield Achieved, Showing the Main Parameters Used in the Reported Bioprocess

Bacillus sp. Strain	Strain Details	Surfactin Production	Cultivation Mode	Media	Reactor Type	Reference
B. subtilis BS-37	723 strain mutated with ARTP[5]	7.8 g/L	SMF[1]	Synthetic with glycerol	5 L reactor	Yi et al. 2017
B. subtilis JABs32	3NA srf+ strain	23.7 g/L	SMF[1]	MSM[3] modified	30 L reactor	Klausmann et al. 2021
B. subtilis BBG131	168 strain with constitutive expression of *srfA*	6.5 g/L	SMF[1]	Synthetic optimized by RSM[4]	Bubble-less membrane microbioreactors	Dos Santos et al. 2016
B. subtilis 168S35	168 strain with the double mutations of *comQXPAPsrfA* and *codY*	12.8 g/L	SMF[1]	MSM[3] modified	Flasks	Wu et al. 2019
B. amyloliquefaciens XZ-173	Native strain	6.25 g/kg of dry solid	SSF[2]	Soybean flour and rice straw with glycerol and maltose	Flasks	Zhu et al. 2013
B. amyloliquefaciens MT45	Native strain	¨7.15 g/L	SMF[1]	MSM[3] modified	7 L reactor	Yang et al. 2020
B. amyloliquefaciens HM618	Native strain	75.7 mg/g of dry KW6	SMF[1]	Semi-synthetic with KW[6]	Flasks	Pan et al. 2021
B. velezensis NRC-1	Native strain	8.92 g/l	SMF[1]	MSM[3]	7 L reactor	Atwa et al. 2013

[1] Submerged fermentation
[2] Solid-state fermentation
[3] Minimal salt media
[4] Response surface methodology
[5] Atmospheric and room-temperature plasma
[6] Kitchen waste

species recently reported is *Bacillus halotolerans*, with a promising ability to withstand harsh environments. For example, the strain KKD1 was isolated from Qinghai–Tibet Plateau, also known as the "third pole of the world" (Wu et al. 2021). However, more studies focused on surfactin production are needed to better assess the potential of these *Bacillus* spp.

As can be seen in Table 1.2, the production of surfactin by genetically modified *B. subtilis* strains reached the highest reported production values. Strain genetic engineering is indeed an important advance that has gained a lot of attention in almost all sectors that employ or have the potential to employ bioprocesses using microorganisms. Optimized strains with extraordinary production yields will be able to pay the costs of bioprocesses, and there is a consensus in the literature that metabolic engineering in conjunction with synthetic biology will bring us those strains. (Contesini et al. 2020).

According to Hu et al. (2019), when it comes to strain modification to improve the production of surfactin, there are three main strategies. The first one is the promotor engineering, as in the substitution of the inductive native promoters of the *srfA* operon for a constitutive one (Dos Santos et al. 2016). The second one is the enhancement of surfactin efflux by overexpression of assistant proteins and surfactin transporters. The last one is to modify the transcriptional regulatory genes of the *srfA* operon.

Klausmann and co-workers (2021) invested on a strategy widely explored in other bioprocesses, high cell density. The 3NA non-surfactin-producing strain was genetically modified with the *srfA* operon addition, producing the *sfp+* variant of this strain, named JABs32. The impact of achieving high cell density, 41.3 g/L of biomass, resulted in the highest surfactin production reported, 23.7 g/L. It is possible to conclude that joining genetic engineering techniques with the previously mentioned approaches of medium optimization, cultivation methods and alternative feedstock use is the best strategy to achieve a commercially viable surfactin production bioprocess.

RECOVERY AND PURIFICATION

For many biotechnological products, the downstream processing stage accounts for up to 60% of the total cost of production, which implies barriers to its commercial application (Satpute et al. 2010). A major obstacle to the commercialization of surfactin is its extraction and purification from complex fermentation broths (Isa et al. 2007).

The raw broth resulting from the fermentation of *B. subtilis* consists of macromolecules (peptides, proteins, polysaccharides and surfactin micelles), medium molecules and micromolecules (alcohols, mineral salts of the culture medium, phthalic acid, amino acids, etc.) (Chen and Juang 2008b; Dhanarajan et al. 2015).

Depending on the intended application, the concentration and purification of surfactin from fermentation broths may be necessary. For example, some applications in the petrochemical, environmental and textile industries allow the use of a raw product, that is, they do not require high-purity biosurfactants. On the other hand, biosurfactants with a higher purity are required for pharmaceutical, cosmetic and food purposes (Chen et al. 2015; Biniarz et al. 2016; Dhanarajan et al. 2015). To obtain high-purity surfactin, a single step downstream is often not enough. In such cases, a sequence of concentration and purification steps is required (Chen et al. 2008).

The recovery of biosurfactants depends mainly on their ionic charge, solubility (water/organic solvents) and location (intracellular, extracellular or cell-bound) (Satpute et al. 2010).

Conventional and unconventional methods for surfactin recovery have been reported in recent years (Chen et al. 2008). Among them are acid precipitation, solvent extraction, foam fractionation, liquid membrane extraction, membrane-based techniques and adsorption (Chen et al. 2008; Biniarz et al. 2016; Vicente et al. 2021), which will be described in the following.

ACID PRECIPITATION

Biosurfactants that have electronic charge, such as surfactin produced by *B. subtilis*, are isolated by acid precipitation in order to reach the isoelectric point of the biosurfactant under study (Cooper et al. 1981).

Acid precipitation is performed by adding concentrated hydrochloric acid to the cell-free supernatant. Under strongly acidic conditions (pH 2–4), the carboxyl group is protonated, and the surfactin becomes less soluble in water, allowing its recovery (Long et al. 2017). To promote better recovery, the precipitate should be stored overnight at 4°C, a process that minimizes the co-aggregation of other macromolecules (Rangarajan and Clarke 2016). Then centrifugation is performed, and the precipitate is solubilized in alkaline solution (Chen and Juang 2008b).

This method is one of the most traditional and easy to operate (Satpute et al. 2010). This technique can achieve a high recovery factor (>97%), but it presents a relatively low level of purity of biosurfactant (< 60%) as a drawback, which makes additional purification techniques necessary (Chen et al. 2007; Chen and Juang 2008a).

SOLVENT EXTRACTION

Solvent extraction is often used as an intermediate recovery operation in downstream processing of bioproducts, facilitating their concentration and purification (Rangarajan and Clarke 2016). For example, the first patent to describe the surfactin purification process was filed in 1972, where acid precipitation was used, proceeding from solvent extraction and finally column purification (Sephadex) (Arima 1972).

Surfactin recovery through this method depends on the polarity of both the biomolecule and the solvent employed. Because it is an amphiphilic molecule, its polar and nonpolar groups will interact according to the type of solvent (Yang et al. 2015; Otzen 2017). Surfactin is soluble primarily in polar solvents and partially soluble in some nonpolar solvents (Arima et al. 1972).

In practice, a wide variety of organic solvents, including methanol, ethanol, ethyl acetate, butanol, diethyl ether, n-pentane, n-hexane, acetone, acetic acid, chloroform and dichloromethane, can be employed in this technique, alone or in combination (Satpute et al. 2010). Among them, the most effective solvent would be the mixture of chloroform and methanol, in different proportions, which facilitates the adjustment of the polarity of the extraction agent for the target compound (Santos et al. 2016).

Surfactin recovery by solvent extraction has been performed by several strategies, standing out among them: liquid-liquid extraction (LLE) and solid-liquid extraction (SLE). In LLE, the surfactin of aqueous mixture (solubilized precipitate or cell-free broth) is purified by extraction with organic solvent. Generally, insoluble organic solvent in water is used to allow easy separation of organic and aqueous phases (Juang et al. 2012; Hu et al. 2022).

In SLE there is a mixture of a surfactin impure in solid form (precipitate) with an organic solvent for dissolution of the biosurfactant (Chen and Juang 2008b). Unlike LLE, emulsification does not occur in this technique, which is an undesirable process that often happens during solvent extractions (Chen and Juang 2008b; Hu et al. 2022). Table 1.3 shows studies conducted with LLE and SLE, in which several solvents were evaluated.

The extraction of surfactin of culture *B. subtilis* ATCC 21332 showed a higher percentage of recovery with ethyl acetate (99%) than with n-hexane (21%) in the LLE system, because n-hexane is a low-polarity solvent. However, under these same conditions, the polarities of the solvents used (ethyl acetate and n-hexane) had no significant effect on the purity of surfactin after LLE, which presented similar values (58–60%) (Chen and Juang 2008b).

This study proves that the polarity of solvents has a significant effect on surfactin recovery; however, this factor did not significantly influence the percentages

---

TABLE 1.3

Studies on Surfactin Recovery by Liquid–Liquid Extraction and Solid–Liquid Extraction

Solvent Extraction Type	Organic Solvent	Recovery (%)	Purity (%)	Reference
Liquid–liquid extraction	Ethyl acetate	99	58–60	Chen and Juang 2008b
Liquid–liquid extraction	n-hexane	21	58–60	Chen and Juang 2008b
Solid–liquid extraction	Ethyl acetate	78	84	Chen and Juang 2008b
Solid–liquid extraction	n-hexane	62	60	Chen and Juang 2008b
Liquid–liquid extraction	Chloroform-methanol (2:1, v/v)	99.6	N/A	Geissler et al. 2017

of purity obtained. This means that the high-polarity solvent (ethyl acetate) promoted greater surfactin recovery and large amounts of impurities, which can be attributed to the fact that surfactin, and at least other protein macromolecules present in broths, have highly polar hydrophobic groups (Chen and Juang 2008b).

Using the SLE system, the same authors recovered about 78% and 62% of surfactin using ethyl acetate and n-hexane, respectively, with purity around 84% (ethyl acetate) and 60% (n-hexane) (Chen and Juang 2008b). Such phenomena result from the differences between the miscibility of these solvents and the hydrophilic groups of surfactin. As an example, ethyl acetate has acceptable solubility in water (9.7% v/v), while n-hexane is almost insoluble in water (Chen and Juang 2008b). Therefore, more surfactin molecules are dissolved in ethyl acetate than in n-hexane.

Among the studies presented, the LLE of surfactin from *B. methylotrophicus* DSM 23117 cultures, using chloroform-methanol (2:1 v/v) as a solvent, showed a high percentage of recovery (99.6%) (Geissler et al. 2017). However, chloroform is a high-toxicity compound and is considered harmful to the environment and human health (Chen and Juang 2008b). Thus, less-toxic and low-cost solvents are more suitable in the extraction of biosurfactants for industrial applications. Ethyl acetate can be a good substitute for chloroform, and n-heptane may be safer than n-hexane (Kosaric and Vardar-Sukan 2021).

The solvent extraction technique has advantages because it is effective in the partial recovery and purification of surfactin, besides being relatively less complex compared to adsorption and ultrafiltration. However, the large volume of solvents needed in this process is harmful to the environment, in addition to increasing operating costs (Kosaric and Vardar-Sukan 2021; Biniarz et al. 2016).

Moreover, due to the similarities in the physical-chemical properties between the families of lipopeptides, it is difficult to separate surfactin from other coexisting families. For this reason, there is a need for additional processes to improve recovery and purity (Rangarajan and Clarke 2016).

LIQUID MEMBRANE EXTRACTION

Liquid membrane extraction, also known as pertraction, is based on solvent extraction and operates in three-phase systems. It consists of two aqueous phases separated by an organic phase representing the liquid membrane M, formed by an insoluble organic solvent in both aqueous phases. The aqueous phases are called feed solution F (fermentation broth containing biosurfactant) and a receiving solution R (Dimitrov et al. 2008; Chtioui et al. 2010). The mechanism involves transferring the target compound from the feed solution to the receiving solution, through the organic liquid membrane (Dimitrov et al. 2008).

As shown in Figure 1.4, the lower part of the contactor consists of four compartments: two for feed solution F and two for the receiving solution R. In each compartment, a disc coated with hydrophilic material rotates, and its rotation induces the formation and continuous renewal of films of the aqueous solutions F and R on the surfaces of the discs, leading to the agitation of all three phases.

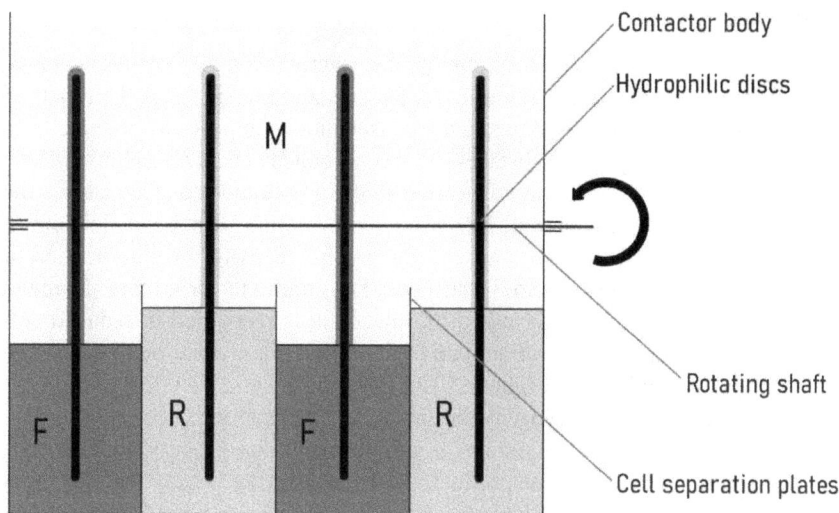

FIGURE 1.4 Experimental device used in liquid membrane extraction, in which the dark grey and light gray areas correspond to the feed solution F and receiving solution R, respectively. At the top, M, is the liquid membrane (Image adapted from Dimitrov et al. 2008).

The organic liquid membrane M occupies the common upper part of the rotation shaft covering both aqueous solutions.

The continuous movement of the three liquids provides an intense transfer of solute. First, interaction between the surfactin molecules of the feeding solution and the organic solvent at the F/M interface occurs, making possible the partial extraction of surfactin for the organic phase. Due to the concentration gradient, the surfactin extracted from the organic membrane phase is transferred to the second interface M/R (Dimitrov et al. 2008).

Surfactin recovery by pertraction was tested using several organic solvents (n-heptane, n-octane and 1-octanol), with n-heptane being the most appropriate solvent. In this technique, surfactin recovery using n-heptane in acidic conditions (pH = 5.6) was almost complete (97%). This result was higher than those found in conventional LLE and SLE when a non-polar solvent was used. With this, it is possible to conclude that liquid membrane extraction improved surfactin recovery. Another advantage of pertraction over classical LLE is the use of smaller amounts of organic solvent, due to the possibility of continuous regeneration of organic solvent (Dimitrov et al. 2008).

The use of liquid membranes presents itself as a promising alternative for the recovery of high-quality bioproducts at reduced costs, making it possible to use less potent, less toxic and cheaper organic solvents (Dimitrov et al. 2008). Despite these advantages, there are few studies in the literature on the recovery of biosurfactants using this technique.

FOAM FRACTIONATION

Foam fractionation is an efficient strategy for surfactin recovery from fermentative medium or diluted solutions (Winterburn, Russell, and Martin 2011). This method consists of the introduction of gas in aqueous solutions in order to generate bubbles, where the adsorption of surface-active compounds occurs (Burghoff 2012).

In this technique, surfactin adsorption in the gas-liquid interface causes the decrease of surface tension and increases the formation of bubbles, leading to the formation of an elastic film around them. This leads to the stabilization of the bubbles, so that they rise to the surface of the liquid to form foam (foamate) (Burghoff 2012). Thus, the adsorbed biosurfactant is recovered in reduced volume and at a higher concentration than in the initial liquid solution (Winterburn Russell, and Martin 2011; Burghoff 2012).

Due to the high surface activity, biosurfactant production is always accompanied by continuous foam formation. In any fermentation process, the presence of foam in the broth affects the productivity by interfering largely with the mass and heat transfer processes. However, foam formation with regard to biosurfactant production is beneficial, as it aids in the continuous removal of product and thereby both processes; that is, production and recovery can be accomplished in a single stage (Rangarajan and Sen 2013). This process is called *in situ* recovery because it occurs when the fractionation foam column is integrated into the production stage, allowing surfactin recovery during production. In *ex situ* recovery, surfactin is recovered after production (Chen et al. 2006).

Integrated fermentation with *in-situ* foam fractionation is capable of achieving high surfactin recoveries (≥90%) (Chen et al. 2006; Noah et al. 2002, Silva et al. 2015).

In research using *in situ* foam fractionation for surfactin recovery of *B. subtilis* BBK006 cultures, the concentration of surfactin in foamate was approximately 50 times higher than in culture broth. An increase in surfactin concentration was also observed in the integrated reactor, reaching 136 mg/L compared to shake flasks (92 mg/L) (Chen et al. 2006).

In another *in situ* study, the authors used a foam fractionation column to recover the surfactin produced by *Bacillus* sp. ITP-001, and the bioreactor was integrated in order to save in the stages of obtaining the product. Under these conditions, the average surfactin enrichment reached the value of 28 times, and recovery was 94% (Silva et al. 2015).

Davis et al. (2001) investigated the *ex situ* recovery of surfactin from *B. subtilis* ATCC 21332 cultures by foam fractionation, having obtained high recoveries (up to 95%). Foam fractionation was also analyzed for both cell-free broths and broths containing cells; the presence of cells improved the foam-forming capacity of the solution, which led to an increase in surfactin recovery. The maximum surfactin enrichment was 8.4 and 51.6 for cell-free and cell-containing broths, respectively.

In more recent research, the combination of foam fractionation and acid precipitation was used. For this, the culture broth of *B. subtilis* MTCC 2423 (cell-free) was submitted to several foam fractionation steps, and the biosurfactant was recovered from the foam by acid precipitation. Under these conditions, the biosurfactant recovered in foamate by this technique accounted for 69% of the total yield (Gurjar and Sengupta 2015). It is worth noting that, for application on an industrial scale, this method becomes inadequate, due to the use of highly acidic conditions and the generation of highly acidic residues.

The main advantages of foam fractionation include high efficacy, low cost, fast operation, the possibility of obtaining enriched solutions and *in situ* recovery (Sarachat et al. 2010; Alonso and Martin 2016). This technique has been regarded as a potential "green" alternative to solvent extraction, as it uses only air or inert gases (Burghoff 2012). However, this technique is not enough when one wishes to obtain surfactin with a high degree of purity.

MEMBRANE-BASED TECHNIQUES

Ultrafiltration (UF) is a process of separation of biosurfactants by membranes, driven by pressure for dissolved and suspended species, based on their size. Its selectivity is related to the dimensions of the molecule or particle and to the size of the pore, as well as to the diffusivity of the solute in the matrix and the associated electrical charges (Jauregi et al. 2013; Matsuura and Ismail 2021).

Separation of biosurfactants by membrane-based techniques requires knowledge of the geometry of the species to be separated, because the most important parameters that affect the separation are the size and shape of the molecules (Jauregi et al. 2013). UF is generally characterized by its molecular weight cut-off (MWCO), which is defined as the lowest molecular weight at which the membrane can reject more than 90% of a solute (Moo-Young 2011).

The ability of biosurfactants to aggregate into supramolecular structures is used to retain them in membranes of high molecular weight; that is, the UF system can be used to concentrate biosurfactants without losses, due to the formation of mycelial structures at concentrations above critical micelle concentration (Hamley et al. 2013; Jauregi et al. 2013).

When the concentrations are above CMC, surfactin molecules tend to associate, forming supramolecular structures, such as micelles or vesicles that have diameters ranging from 5 to 105 nm (two to three times larger than that of their monomers) (Chen et al. 2007; Jauregi et al. 2013). This allows the retention of surfactin micelles by UF membranes with higher MWCO, allowing them to be separated from smaller molecules such as mineral salts, amino acids, glucose, alcohols, organic acids and other by-products derived from metabolism (de Andrade et al. 2016; Jauregi et al. 2013).

In surfactin recovery by UF, different types of membrane materials can be used; among them, the most studied are polyethersulfone (PES) and regenerated cellulose membranes (RC and YM), with several MWCOs (Vicente et al. 2021; Isa et al. 2008). Table 1.4 presents performances of UF processes using different types of membranes to obtain surfactin.

Due to the electrostatic interactions between the membrane and surfactin, PES membranes are more suitable for high recovery and selectivity of these

TABLE 1.4

Summary of Studies Found in the Literature Using Ultrafiltration (Adapted from Vicente et al. 2021)

Membrane Material	Filtration Process	Membrane Rejection	Solvent	Recovery (%)	Purity (%)	Reference
PES 100 kDa	First ultrafiltration step	0.95	N/A	95	92	Jauregi et al. 2013
PES 100 kDa	Second ultrafiltration step	0.07	Ethanol 75% (v/v)	92	94	Jauregi et al. 2013
PES 300 kDa	First ultrafiltration step	0.22	N/A	93	89	Jauregi et al. 2013
PES 10 kDa	First ultrafiltration step	0.99	N/A	93	88	Isa et al. 2008
PES 10 kDa	Second ultrafiltration step	0.23	Methanol 50% (v/v)	94	96	Isa et al. 2008
RC 10 kDa	First ultrafiltration step	0.98	N/A	95	90	Isa et al. 2008
RC 10 kDa	Second ultrafiltration step	0.34	Methanol 50% (v/v)	92	96	Isa et al. 2008
PES 100 kDa	First ultrafiltration step	0.87	N/A	80	67	de Andrade et al. 2016
PES 50 kDa	Second ultrafiltration step	0.02	Ethanol 75% (v/v)	78	80	de Andrade et al. 2016

molecules, especially when included in the second stage of UF. In a study conducted by Isa et al. (2008), in the second stage of UF, the PES membrane showed lower surfactin rejection (0.23) and higher flow in relation to the cellulosic membrane RC (0.34) (Table 1.4).

De Andrade et al. (2016) investigated several purification strategies by UF, having obtained better results using pre-purified biosurfactant solution through acid precipitation and extraction proceeding from UF with PES membranes. In the first stage, the micelles were efficiently recovered as a product retained in the PES membrane with 100 kDa MWC. Then, 75% ethanol (v/v) was added to the retained substances in order to disrupt the surfactin micelles to monomers. In the second stage of UF, membranes with MWCO of 50 kD were used for separation of surfactin monomers, which were obtained in the permeate. At the end of the process, total recovery and purity of 78% and 80%, respectively, were achieved. However, these values were lower than those presented in other studies (Table 1.4) in which synthetic culture media were used.

The authors attribute this result to impurities related to surfactin production by the use of cassava wastewater as a culture medium. It was reported that the excess of proteins presented in the culture medium hindered the purification process by UF and suggested application of additional purification steps or pre-treatment of cassava wastewater to increase the degree of purity (De Andrade et al. 2016).

The characteristics of UF that make it excellent for the recovery and purification of surfactin include reduced physical damage and minimal denaturation of biomolecules, high recovery values and high yield. Moreover, this technique does not require the use of large volumes of organic solvents, which allows its application on a large scale (Coutte et al. 2017; Vicente et al. 2021).

ADSORPTION

Adsorption is a physical-chemical phenomenon in which a component in a gaseous or liquid phase is transferred to the surface of a solid phase. In many cases, adsorption of compounds occurs as a result of the degree of hydrophobicity, porosity, molecular structure of the compounds and volume variation of adsorbent material. A great advantage of adsorption is that, when compared to other separation processes, it presents high selectivity at the molecular level, allowing the separation of several components (Najafpour 2008).

Separation of biosurfactants by adsorption presents itself as an important and economically viable alternative in many cases (Santos et al. 2016; Vicente et al. 2021). For this purpose, the adsorption techniques used in the recovery and purification of surfactin make use of chromatographic columns of resins or activated carbon.

Chen et al. (2008) were able to increase the surfactin purity of previously treated fermentation broth by means of a two-step UF process. For this, columns of charged ion-exchange resin (AG1-X4) and neutral adsorption resin (macroporous XAD-7) were used. Under the tested conditions, impurities present in the broth, such as macromolecules (peptides, polysaccharides, proteins) or smaller molecules (alanine, serine, glycine, threonine), were adsorbed or exchanged in the 5 h interval of operation. The authors found that surfactin recovery exceeded 95% and purity increased from 76% (the feed) to 88% using XAD-7 resin.

Another approach to surfactin recovery present in aqueous solutions is adsorption in activated carbon. Liu et al. (2007) described activated carbon as an effective adsorbent for surfactin purification, evaluating parameters such as agitation rate, particle size, temperature, pH, adsorbent amount, initial adsorbate concentration and ionic strength of the solution. Adsorption was possible in an optimal pH range between 6.5 and 8.5, indicating that fermented broth can be used directly for surfactin recovery. The strong impact of temperature in this process, 30°C being the optimum adsorption temperature, was also observed. The results also showed that the smaller the size of the adsorbent particles used, the faster the adsorption rate. Through these parameters, it was demonstrated that activated carbon acted as an appropriate adsorbent for surfactin recovery.

In more recent research, HP-20 resin column chromatography was used for recovery and purification of three lipopeptide families (surfactin, iturin and fengycin) from cell-free culture broth. For this, the simultaneous variation of the solvent composition and pH of the mobile phase were performed, and surfactin with purity of up to 91% was obtained (Dhanarajan et al. 2015).

The advantages that make this technique promising involve its ability to operate in a continuous mode to recover biosurfactants with a high level of purity
while allowing the rapid recovery and reuse of extraction materials (Mukherjee
et al. 2006; Biniarz et al. 2016). Table 1.5 presents a summary of the main advantages and disadvantages of the techniques presented here.

TABLE 1.5
Advantages and Disadvantages of Techniques Commonly Used for Recovery and Purification of Surfactin

Procedure	Advantage	Disadvantage	References
Acid precipitation	Low cost, easy and efficient method for the recovery of crude biosurfactant for large-scale application	Low purity	Chen and Juang 2008a; Biniarz 2016
Solvent extration	Efficient in the recovery of crude biosurfactant for large-scale application, possibility of reuse of organic solvents, effective in recovery and partial purification	High cost, use of solvents that are toxic to health and the environment	Chen and Juang 2008b; Satpute et al. 2010; Biniarz et al. 2016
Liquid membrane extraction	Possibility of regeneration, low cost, posibility of using less toxic and cheaper organic solvents, high-quality surfactin recovery	Use of solvents	Dimitrov et al. 2008; Chtioui et al. 2010
Foam fractionation	Low cost, recovery from continuous culture, fast recovery, recovery from continuous culture, high percentage of recovery	—	Sarachat et al. 2010; Alonso and Martin 2016; Gurjar and Sengupta 2015
Ultrafiltration	Scale-widening possibility, fast recovery, high percentage of recovery and high level of purity	High cost, can be operated under high pressure	Chen et al. 2007; Andrade et al. 2016; Vicente et al. 2021
Adsorption on polystyrene resins	Possibility of reuse, fast operation, recovery from continuous culture, high percentage of recovery and high level of purity	Needs organic solvents for desorption	Mukherjee et al. 2006; Chen et al. 2008
Adsorption on wood-activated carbon	Possibility of reuse, fast operation, recovery from continuous culture, high percentage of recovery and high level of purity	Needs organic solvents for desorption	Mukherjee et al. 2006; Montastruc et al. 2008

PROSPECTS OF SURFACTIN USE IN MARINE ENVIRONMENTS

Surfactins have been reported as great potential substitutes in a wide range of applications, including environmental applications such as bioremediation, microbial-enhanced oil recovery (MEOR) and anti-pesticide treatment against plant pathogens (Hubbard et al. 2014; Santos, Silveira, and Pereira 2019). The main reasons for that are the great capability to decrease superficial or interfacial tension, their high degradability rate, the fact they are produced from renewable sources and their low toxicity to human tissues and animals.

Santos, Silveira and Pereira (2018) present a great review on the toxicological implications of surfactins and their ecotoxicological risk assessment. In summary, a lot still needs to be studied to address surfactant use in health and the environment.

For instance, in both in vitro and in vivo systems of mice, lipopeptide does not seem to exert genotoxic potential. Further, alterations were not detected in the relative maternal organ weights, including liver, brain, spleen, kidneys, heart and placenta (Santos, Silveira, and Pereira 2018 apud Hwang et al. 2008). On the other hand, a second study showed that oral administration in rats can generate a body weight loss and increased serum activity levels of alanine transaminase (ALT), alkaline phosphatase (ALP) and aspartate transaminase (AST), indicating necrosis of rat hepatocytes (Santos, Silveira, and Pereira 2018 apud Hwang et al. 2009).

In ecotoxicological assessments, Deravel et al. (2014) reported a low toxicity of surfactin, indicating an EC_{50}–48 hr value higher than 100 mg/L under the model microcrustacean *Daphnia magna* (Santos, Silveira, and Pereira 2018 apud Deravel et al. 2014). Also using *D. magna*, in addition to *Selenastrum capricornutum*, de Oliveira et al. reported no acute effects at concentrations 11-fold greater than its CMC (de Oliveira et al. 2016).

Despite many works exploring surfactins as anticancer drugs, their physicochemical characteristics draw attention to use in marine environments. Also, as few marine microorganisms are known to produce surfactin (Ramalingam et al. 2019; Jakinala et al. 2019; Kalinovskaya et al. 1995), it is expected that this surfactant will have activity under tough marine conditions such as high salinity and high pressure depending on the location of the water column and temperature.

BIODEGRADABILITY AND DECONTAMINATION

Surfactins have been studied at the laboratory scale, aiming at the recovery of impacted environments (water and soil) with atrazine, phenol, cooper and mainly hydrocarbons, petroleum and other oils.

Jakinala et al. (2019) studied the effect of surfactin produced by the marine strain *B. velezensis* MHNK1 on atrazine biodegradation. *B. velezenis* is reported to profoundly inhabit marine water, and atrazine is a widely used toxic herbicide and considered a serious environmental contaminant worldwide due to its long-term use in crop production. An 87.10% atrazine biodegradation was reported within 5 days and 100% degradation within 4 days using a combination of *B. velezensis* MHNK1 (2%) and surfactin. The authors also described the metabolic pathway of atrazine degradation, which can prompt new efforts in gene manipulation using recombinant DNA technology, aiming at the construction of strains with higher degradation capability or to degrade similar compounds (Jakinala et al. 2019).

The decontamination of copper and phenol (hydroquinone) from artificial samples was attained by treatment with residues from surfactin production by Silveira et al. (2021). The authors used cell leftovers (biomass) from surfactin production to perform environmental assays, while surfactins on supernatants were intended for health science applications. Copper was removed by 97% using the cell lysate and 46–51% using the water that was used to wash those cells prior to lysis. Hydroquinone, in other hand, was removed up to 51% using biomass. Results indicate that there is insoluble surfactin adsorbed to biomass. In a separate experiment, the group added 40% and 80% more surfactin to wash water, increasing copper removal by 9% and 12%, respectively (Silveira et al. 2021).

The use of surfactin to remediate oil-contaminated soil was analyzed and displayed up to $43.6 \pm 0.08\%$ and $46.7 \pm 0.01\%$ remediation of heavy engine oil–contaminated soil at 10 and 40 mg/L concentrations, respectively. The values were slightly higher from those obtained using SDS, a non-environmentally friendly surfactant (Phulpoto et al. 2020).

In a second work, samples of soils were collected from a heavy oil-polluted site. By using 0.2 mass% of rhamnolipids, surfactin, the total petroleum hydrocarbon removal for the soil contaminated with ca. 3000 mg TPH/kg dry soil was 23%, 14%, respectively, while removal efficiency increased to 63%, 62%, respectively, for the soil contaminated with ca. 9000 mg TPH/kg dry soil. Removal efficiency also increased with an increase in biosurfactant concentration (from 0 to 0.2 mass%), but it did not changed significantly by the time (1 and 7 days) (Lai et al. 2009).

Liu et al. (2015) studied the oil-washing efficiency of surfactin on quartz sand. At concentrations of 300 and 30 mg/L, surfactin could remove more than 95% and 88.5%, respectively, of oil from the sand. In a second-stage recovery, a sand-pack test was conducted. After injecting a volume of 300 mg/L biosurfactant solution, flooding could recover up to 14.21% extra crude oil. Also, studying the removal of crude oil from sand in Erlenmeyer flasks, up to 22.1% of light Arabian oil was recovered (Pereira et al. 2013).

MICROBIAL-ENHANCED OIL RECOVERY

As discussed, surfactin can highly decrease surface and/or interfacial tension, a property very important in the oil recovery process. The interfacial tension reduction permits the pressure required to release the oil trapped in the rock pores by capillary forces to drop, which displaces oil from the pores into the mobile liquid phase (Liu et al. 2015). This process is called enhanced oil recovery (EOR), or microbial-enhanced oil recovery, when the whole biosurfactant-producer microorganism or its biosurfactant product is used. When MEOR is processed *in situ*, it employs indigenous microorganisms; *ex situ*, bioproducts are produced outside of oil wells and then injected into it (Al-Wahaibi et al. 2014).

In core-flood studies, crude biosurfactant preparation enhanced light oil recovery by 17–26% and heavy oil recovery by 31% (Al-Wahaibi et al. 2014). A sand pack column study yielded up to 60% oil recovery (Varadavenkatesan and Murty 2013).

To predict surfactin actuation on contact with oil, several works report its potency as an emulsifier. It is a useful tool to predict its behavior in different complex environments. It has been shown that it displays great emulsifying potential for kerosene (68%), engine oil (85.21%), castor oil (79.03%), toluene (75.10%) and crude oil (68%) (Phulpoto et al. 2020; Jakinala et al. 2019; Haddad, Wang, and Mu 2009) and stability under environmental factors such as salinity, pH and temperature variations (Phulpoto et al. 2020).

Nevertheless, besides several laboratory trials that have been carried out in the past years, the real potential of biosurfactants in MEOR applications can only be fully assessed in field-scale applications. This scale assay had not been reported yet but is expected to be seen in the next few years (Al-Wahaibi et al. 2014).

Anti-Pesticide Treatment against Algae Cultivation Pathogens

It is found that *Bacillus* spp. colonizes plant roots and is known that surfactin production by this microorganism is crucial for biocontrol activity and systemic resistance in plants (including algae). It is expected that the dispersion of surfactin in the aerial parts of plants will also play a pesticidal role. For instance, the biocontrol activity of *Bacillus atrophaeus* 176s was studied under lettuce, sugar beet and tomato infected with *Rhizoctonia solani*. Untreated plants were susceptible to fungal infections, and the majority of the plants died due to damping off within 7 days after infection. Plants inoculated with *B. amyloliquefaciens* FZB42 and *B. atrophaeus* 176s or both had a higher resistance to fungal infection, reduced symptoms and increased plant recovery (Aleti et al. 2016). Is expected that soon surfactins can be used on closed cultivars of algae to human consumption and molecular bioproduct production such biofuels, additives for animal nutrition and the production of carotenes.

REFERENCES

Abas, M. R., A. J. A. Kader, M. S. Khalil, A. A. Hamid, and M. H. M. Isa. 2013. Production of Surfactin from *Bacillus subtilis* ATCC 21332 by Using Treated Palm Oil Mill Effluent (POME) as Fermentation Media. *International Conference on Food and Agricultural Sciences* 55 (17): 87–93.

Abdel-Mawgoud, A. Mohammad, M. Mabrouk Aboulwafa, and Nadia Abdel-Haleem Hassouna. 2008. "Optimization of Surfactin Production by *Bacillus subtilis* Isolate BS5." *Applied Biochemistry and Biotechnology* 150 (3): 305–25. https://doi.org/10.1007/s12010-008-8155-x.

Aleti, Gajender, Sylvia Lehner, Markus Bacher, Stéphane Compant, Branislav Nikolic, Maja Plesko, Rainer Schuhmacher, Angela Sessitsch, and Günter Brader. 2016. "Surfactin Variants Mediate Species-Specific Biofilm Formation and Root Colonization in Bacillus." *Environmental Microbiology* 18 (8): 2634–45. https://doi.org/10.1111/1462-2920.13405.

Alonso, S., and P. J. Martin. 2016. Impact of Foaming on Surfactin Production by *Bacillus subtilis*: Implications on the Development of Integrated In Situ Foam Fractionation Removal Systems. *Biochemical Engineering Journal* 110: 125–33. https://doi.org/10.1016/j.bej.2016.02.006.

Al-Wahaibi, Yahya, Sanket Joshi, Saif Al-Bahry, Abdulkadir Elshafie, Ali Al-Bemani, and Biji Shibulal. 2014. "Biosurfactant Production by *Bacillus subtilis* B30 and Its Application in Enhancing Oil Recovery." *Colloids and Surfaces B: Biointerfaces* 114 (February): 324–33. https://doi.org/10.1016/j.colsurfb.2013.09.022.

Arima, K., A. Kakinuma, and G. Tamura. 1968. "Surfactin, a Crystalline Peptidelipid Surfactant Produced by *Bacillus subtilis*: Isolation, Characterization and Its Inhibition of Fibrin Clot Formation." *Biochemical and Biophysical Research Communications* 31 (3): 488–94. https://doi.org/10.1016/0006-291x(68)90503-2.

Arima, K., G. Tamura, and A. Kakinuma. 1972. Surfactin. U.S. Patent No. 3, 687, 926. https://patents.google.com/patent/US3687926A/en

Atwa, N. A., E. El-Shatoury, A. Elazzazy, M. A. Abouzeid, and A. El-Diwany. 2013. Enhancement of Surfactin Production by *Bacillus velezensis* NRC-1 Strain Using a Modified Bench-Top Bioreactor. *Journal of Food, Agriculture and Environment* 11 (2): 169–74.

Bannwarth, Christoph, Sebastian Ehlert, and Stefan Grimme. 2019. "GFN2-XTB—An Accurate and Broadly Parametrized Self-Consistent Tight-Binding Quantum Chemical Method with Multipole Electrostatics and Density-Dependent Dispersion Contributions." *Journal of Chemical Theory and Computation* 15 (3): 1652–71. https://doi.org/10.1021/acs.jctc.8b01176.

Biniarz, P., M. Łukaszewicz, and T. Janek. 2016. Screening Concepts, Characterization and Structural Analysis of Microbial-Derived Bioactive Lipopeptides: A Review. *Critical Reviews in Biotechnology* 37 (3): 393–410. http://doi.org/10.3109/07388551.2016.1163324.

Bonmatin, Jean-Marc-M., Monique Genest, Henri Labbé, and Marius Ptak. 1994. "Solution Three-Dimensional Structure of Surfactin: A Cyclic Lipopeptide Studied by 1H-nmr, Distance Geometry, and Molecular Dynamics." *Biopolymers* 34 (7): 975–86. https://doi.org/10.1002/bip.360340716.

Burghoff, B. 2012. Foam Fractionation Applications. *Journal of Biotechnology* 161 (2): 126–37. http://doi.org/10.1016/j.jbiotec.2012.03.008.

Chen, H., Y. Chen, and R. Juang. 2007. Separation of Surfactin from Fermentation Broths by Acid Precipitation and Two-Stage Dead-End Ultrafiltration Processes. *Journal of Membrane Science* 299 (1–2): 114–21. http://doi.org/10.1016/j.memsci.2007.04.031.

Chen, H., and R. Juang. 2008a. Extraction of Surfactin from Fermentation Broth with n-Hexane in Microporous PVDF Hollow Fibers: Significance of Membrane Adsorption. *Journal of Membrane Science* 325 (2): 599–604. http://doi.org/10.1016/j.memsci.2008.08.017.

Chen, H., and R. Juang. 2008b. Recovery and Separation of Surfactin from Pretreated Fermentation Broths by Physical and Chemical Extraction. *Biochemical Engineering Journal* 38 (1): 39–46. http://doi.org/10.1016/j.bej.2007.06.003.

Chen, H., Y. Lee, Y. Wei, and R. Juang. 2008. Purification of Surfactin in Pretreated Fermentation Broths by Adsorptive Removal of Impurities. *Biochemical Engineering Journal* 40 (3): 452–59. http://doi.org/10.1016/j.bej.2008.01.020.

Chen, W. C., S. C. Baker, and R. C. Darton. 2006. Batch Production of Biosurfactant with Foam Fractionation. *Journal of Chemical Technology & Biotechnology* 81 (12): 1923–31. http://doi.org/10.1002/jctb.1625.

Chen, W. C., R. S. Juang, and Y. H. Wei. 2015. Applications of a Lipopeptide Biosurfactant, Surfactin, Produced by Microorganisms. *Biochemical Engineering Journal* 103: 158–69. https://doi.org/10.1016/j.bej.2015.07.009.

Chtioui, O., K. Dimitrov, F. Gancel, and I. Nikov. 2010. Biosurfactants Production by Immobilized Cells of *Bacillus subtilis* ATCC 21332 and Their Recovery by Pertraction. *Process Biochemistry* 45 (11): 1795–99. http://doi.org/10.1016/j. procbio.2010.05.012.

Contesini, F. J., M. G. Davanço, G. P. Borin, K. G. Vanegas, J. P. G. Cirino, R. R. de Melo, U. H. Mortensen, K. Hildén, D. R. Campos, and Patricia de Oliveira Carvalho. 2020. Advances in Recombinant Lipases: Application in the Pharmaceutical Industry. *Catalysts* 10: 1032.

Cooper, D. G., C. R. Macdonald, S. J. B. Duff, and N. Kosaric. 1981. Enhanced Production of Surfactin from *Bacillus subtilis* by Continuous Product Removal and Metal Cation Additions. *Applied and Environmental Microbiology* 42 (3): 408–12. https:// doi.org/10.1128/aem.42.3.408-412.1981.

Coutte, F., D. Lecouturier, K. Dimitrov, J. Guez, F. Delvigne, P. Dhulster, and P. Jacques. 2017. Microbial Lipopeptide Production and Purification Bioprocesses, Current Progress and Future Challenges. *Biotechnology Journal* 12 (7): 1600566–76. http:// doi.org/10.1002/biot.201600566.

Davis, D. A., H. C. Lynch, and J. Varley. 1999. The Production of Surfactin in Batch Culture by *Bacillus subtilis* ATCC 21332 Is Strongly Influenced by the Conditions of Nitrogen Metabolism. *Enzyme and Microbial Technology* 25 (3–5): 322–29. http://doi.org/10.1016/S0141-0229(99)00048-4.

Davis, D. A., H. C. Lynch, and J. Varley. 2001. The Application of Foaming for the Recovery of Surfactin from *B. subtilis* ATCC 21332 Cultures. *Enzyme and Microbial Technology* 28 (4–5): 346–54. http://doi.org/10.1016/s0141-0229(00)00327-6.

de Andrade, C. J., F. F. C. Barros, L. M. de Andrade, S. A. Rocco, M. L. Sforça, G. M. Pastore, and P. Jauregi. 2016. Ultrafiltration Based Purification Strategies for Surfactin Produced by *Bacillus subtilis* LB5A Using Cassava Wastewater as Substrate. *Journal of Chemical Technology & Biotechnology* 91 (12): 3018–27. http://doi.org/10.1002/jctb.4928.

de Faria, A. F., D. S. Teodoro-Martinez, G. N. O. Barbosa, B. G. Vaz, I. S. Silva, J. S. Garcia, M. R. Tótola, et al. 2011. Production and Structural Characterization of Surfactin (C 14/Leu7) Produced by *Bacillus subtilis* Isolate LSFM-05 Grown on Raw Glycerol from the Biodiesel Industry. *Process Biochemistry* 46 (10): 1951–57. http://doi.org/10.1016/j.procbio.2011.07.001.

de Oliveira, Darlane W. F., Alejandro B. Cara, Manuela Lechuga-Villena, Miguel García-Román, Vania M. M. Melo, Luciana R. B. Gonçalves, and Deisi A. Vaz. 2016. Aquatic Toxicity and Biodegradability of a Surfactant Produced by *Bacillus subtilis* ICA56. *Journal of Environmental Science and Health, Part A* 52 (2): 174–81. https://doi.org/10.1080/10934529.2016.1240491.

Deravel, J., S. Lemiere, F. Coutte, F. Krier, N. Van Hese, M. Béchet, N. Sordeau, M. Hofte, A. Leprêtre, and P. Jacques. 2014. Mycosubtilin and Surfactin Are Efficient,

Low Ecotoxicity Molecules for the Biocontrol of Lettuce Downy Mildew. *Applied Microbiology and Biotechnology* 98: 6255–64. https://doi.org/10.1007/s00253-014-5663-1.

Dhanarajan, G., V. Rangarajan, and R. Sen. 2015. Dual Gradient Macroporous Resin Column Chromatography for Concurrent Separation and Purification of Three Families of Marine Bacterial Lipopeptides from Cell Free Broth. *Separation and Purification Technology* 143: 72–79. http://doi.org/10.1016/j.seppur.2015.01.025.

Dimitrov, K., F. Gancel, L. Montastruc, and I. Nikov. 2008. Liquid Membrane Extraction of Bio-Active Amphiphilic Substances: Recovery of Surfactin. *Biochemical Engineering Journal* 42 (3): 248–53. http://doi.org/10.1016/j.bej.2008.07.005.

Dobler, Leticia, Leonardo F. Vilela, Rodrigo V. Almeida, and Bianca C. Neves. 2016. Rhamnolipids in Perspective: Gene Regulatory Pathways, Metabolic Engineering, Production and Technological Forecasting. *New Biotechnology* 33 (1): 123–35. https://doi.org/10.1016/j.nbt.2015.09.005.

Dos Santos, L. F. M., F. Coutte, R. Ravallec, P. Dhulster, P. Jacques, and L. Tournier-Couturier. 2016. An Improvement of Surfactin Production by *B. subtilis* BBG131 Using Design of Experiments in Microbioreactors and Continuous Process in Bubbleless Membrane Bioreactor. *Bioresource Technology* 218: 944–52. http://doi.org/10.1016/j.biortech.2016.07.053.

Dufour, Samuel, Magali Deleu, Katherine Nott, Bernard Wathelet, Philippe Thonart, and Michel Paquot. 2005. Hemolytic Activity of New Linear Surfactin Analogs in Relation to Their Physico-Chemical Properties. *Biochimica et Biophysica Acta—General Subjects* 1726 (1): 87–95. https://doi.org/10.1016/j.bbagen.2005.06.015.

Eeman, M., A. Berquand, Y. F. Dufrêne, M. Paquot, S. Dufour, and M. Deleu. 2006. "Penetration of Surfactin into Phospholipid Monolayers: Nanoscale Interfacial Organization." *Langmuir* 22 (26): 11337–45. https://doi.org/10.1021/la061969p.

Ganesan, N. G., and V. Rangarajan. 2021. A Kinetics Study on Surfactin Production from *Bacillus subtilis* MTCC 2415 for Application in Green Cosmetics. *Biocatalysis and Agricultural Biotechnology* 33: 102001. http://doi.org/10.1016/j.bcab.2021.102001.

Gang, Hong Ze, Hao He, Zhou Yu, Zhenyu Wang, Jinfeng Liu, Xiujuan He, Xinning Bao, Yingcheng Li, and Bo Zhong Mu. 2020. "A Coarse-Grained Model for Microbial Lipopeptide Surfactin and Its Application in Self-Assembly." *Journal of Physical Chemistry B* 124 (9): 1839–46. https://doi.org/10.1021/acs.jpcb.9b11381.

Gang, Hong-Ze, Jin-Feng Liu, and Bo-Zhong Mu. 2010a. Molecular Dynamics Simulation of Surfactin Derivatives at the Decane/Water Interface at Low Surface Coverage. *The Journal of Physical Chemistry B* 114 (8): 2728–37. https://doi.org/10.1021/jp909202u.

Gang, Hong-Ze, Jin-Feng Liu, and Bo-Zhong Mu. 2010b. Interfacial Behavior of Surfactin at the Decane/Water Interface: A Molecular Dynamics Simulation. *The Journal of Physical Chemistry B* 114 (46): 14947–54. https://doi.org/10.1021/jp1057379.

Gang, Hong Ze, Jin-Feng Liu, and Bo-Zhong Mu. 2011. Molecular Dynamics Study of Surfactin Monolayer at the Air/Water Interface. *Journal of Physical Chemistry B* 115 (44): 12770–77. https://doi.org/10.1021/jp206350j.

Gang, Hong Ze, Jin-Feng Liu, and Bo-Zhong Mu. 2015. Binding Structure and Kinetics of Surfactin Monolayer Formed at the Air/Water Interface to Counterions: A Molecular Dynamics Simulation Study. *Biochimica et Biophysica Acta—Biomembranes* 1848 (10): 1955–62. https://doi.org/10.1016/j.bbamem.2015.05.016.

Geissler, M., C. Oellig, K. Moss, W. Schwack, M. Henkel, and R. Hausmann. 2017. High-Performance Thin-Layer Chromatography (HPTLC) for the Simultaneous

Quantification of the Cyclic Lipopeptides Surfactin, Iturin A and Fengycin in Culture Samples of Bacillus Species. *Journal of Chromatography B* 1044–1045: 214–24. http://doi.org/10.1016/j.jchromb.2016.11.013.

Gurjar, Jigar, and Bina Sengupta. 2015. Production of Surfactin from Rice Mill Polishing Residue by Submerged Fermentation Using *Bacillus subtilis* MTCC 2423. *Bioresource Technology* 189: 243–49. http://doi.org/10.1016/j.biortech.2015.04.013.

Haddad, Namir, Ji Wang, and Bozhong Mu. 2009. Identification of a Biosurfactant Producing Strain: *Bacillus subtilis* HOB2. *Protein & Peptide Letters* 16 (1): 7–13. https://doi.org/10.2174/092986609787049358.

Hamley, I. W., A. Dehsorkhi, P. Jauregi, Jani Seitsonen, J. Ruokolainen, F. Coutte, G. Chataigné, and P. Jacques. 2013. Self-Assembly of Three Bacterially-Derived Bioactive Lipopeptides. *Soft Matter* 9 (40): 9572. http://doi.org/10.1039/c3sm51514a.

Hamoen, L. W., G. Venema, and O. P. Kuipers. 2003. Controlling Competence in *Bacillus subtilis*: Shared Use of Regulators. *Microbiology* 149: 9–17. http://doi.org/10.1099/mic.0.26003-0.

Heerklotz, Heiko, and Joachim Seelig. 2001. Detergent-Like Action of the Antibiotic Peptide Surfactin on Lipid Membranes. *Biophysical Journal* 81 (3): 1547–54. https://doi.org/10.1016/S0006-3495(01)75808-0.

Hmidet, N., H. B. Ayed, P. Jacques, and M. Nasri. 2017. Enhancement of Surfactin and Fengycin Production by *Bacillus mojavensis* A21: Application for Diesel Biodegradation. *BioMed Research International* 2017: ID 5893123. http://doi.org/10.1155/2017/5893123.

Huang, X., J. Liu, Y. Wang, J. Liu, and L. Lu. 2015. The Positive Effects of Mn^{2+} on Nitrogen Use and Surfactin Production by *Bacillus subtilis* ATCC 21332. *Biotechnology and Biotechnological Equipment* 29 (2): 381–89. http://doi.org/10.1080/13102818.2015.1006905.

Hubbard, Michelle, Russell K. Hynes, Martin Erlandson, and Karen L. Bailey. 2014. "The Biochemistry behind Biopesticide Efficacy." *Sustainable Chemical Processes* 2 (1). https://doi.org/10.1186/s40508-014-0018-x.

Hu, Fangxiang, Yuyue Liu, and Shuang Li. 2019. Rational Strain Improvement for Surfactin Production: Enhancing the Yield and Generating Novel Structures. *Microbial Cell Factories* 18 (1). https://doi.org/10.1186/s12934-019-1089-x.

Hu, M., J. Yu, H. Zhang, and Q. Xu. 2022. An Efficient Method for the Recovery and Separation of Surfactin from Fermentation Broth by Extraction-Back Extraction. *Process Biochemistry* 114: 59–65. http://doi.org/10.1016/j.procbio.2022.01.014.

Hwang, Youn-Hwan, Byung-Kwon Park, Jong-Hwan Lim, Myoung-Seok Kim, In-Bae Song, Seung-Chun Park, and Hyo-In Yun. 2008. "Evaluation of Genetic and Developmental Toxicity of Surfactin C from *Bacillus subtilis* BC1212." *Journal of Health Science* 54 (1): 101–6. https://doi.org/10.1248/jhs.54.101.

Hwang, Youn-Hwan, Myoung-Seok Kim, In-Bae Song, Byung-Kwon Park, Jong-Hwan Lim, Seung-Chun Park, and Hyo-In Yun. 2009. "Subacute (28 Day) Toxicity of Surfactin C, a Lipopeptide Produced by *Bacillus subtilis*, in Rats." *Journal of Health Science* 55 (3): 351–55. https://doi.org/10.1248/jhs.55.351.

Isa, M. H., D. Coraglia, R. Frazier, and P. Jauregi. 2007. Recovery and Purification of Surfactin from Fermentation Broth by a Two-Step Ultrafiltration Process. *Journal of Membrane Science* 296 (1–2): 51–57. http://doi.org/10.1016/j.memsci.2007.03.023.

Isa, M. H., R. A. Frazier, and P. Jauregi. 2008. A Further Study of the Recovery and Purification of Surfactin from Fermentation Broth by Membrane Filtration.

Separation and Purification Technology 64 (2): 176–82. http://doi.org/10.1016/j. seppur.2008.09.008.

Jakinala, Parameshwar, Nageshwar Lingampally, Archana Kyama, and Bee Hameeda. 2019. Enhancement of Atrazine Biodegradation by Marine Isolate *Bacillus velezensis* MHNK1 in Presence of Surfactin Lipopeptide. *Ecotoxicology and Environmental Safety* 182 (October): 109372. https://doi.org/10.1016/j.ecoenv.2019.109372.

Janek, T., E. J. Gudiña, X. Polomska, P. Biniarz, D. Jama, L. R. Rodrigues, W. Rymowicz, and Z. Lazar. 2021. Sustainable Surfactin Production by *Bacillus subtilis* Using Crude Glycerol from Different Wastes. *Molecules* 26: 3488. http://doi.org/10.3390/molecules26123488.

Jauregi, P., F. Coutte, L. Catiau, D. Lecouturier, and P. Jacques. 2013. Micelle Size Characterization of Lipopeptides Produced by *B. subtilis* and Their Recovery by the Two-Step Ultrafiltration Process. *Separation and Purification Technology* 104: 175–82. http://doi.org/10.1016/j.seppur.2012.11.017.

Jokari, S., H. Rashedi, G. Amoabediny, S. N. Dilmaghani, and M. M. Assadi. 2013. Optimization of Surfactin Production by *Bacillus subtilis* ATCC 6633 in a Miniaturized Bioreactor. *International Journal of Environmental Research* 7 (4): 851–58.

Juang, R., H. Chen, and S. Tsao. 2012. Recovery and Separation of Surfactin from Pretreated *Bacillus subtilis* Broth by Reverse Micellar Extraction. *Biochemical Engineering Journal* 61: 78–83. http://doi.org/10.1016/j.bej.2011.12.008.

Kakinuma, Atsushi, Hiromu Sugino, Masao Isono, Gakuzo Tamura, and Kei Arima. 1969. "Determination of Fatty Acid in Surfactin and Elucidation of the Total Structure of Surfactin." *Agricultural and Biological Chemistry* 33 (6): 973–76. https://doi.org/1 0.1080/00021369.1969.10859409.

Kalinovskaya, N. I., T. A. Kuznetsova, Ya. V. Rashkes, Yu. M. Mil'grom, E. G. Mil'grom, R. H. Willis, A. I. Wood, et al. 1995. Surfactin-Like Structures of Five Cyclic Depsipeptides from the Marine Isolate of *Bacillus pumilus*. *Russian Chemical Bulletin* 44 (5): 951–55. https://doi.org/10.1007/bf00696935.

Kan, S. C., C. C. Lee, Y. C. Hsu, Y. H. Peng, C. C. Chen, J. J. Huang, J. W. Huang, C. J. Shieh, T. Y. Juang, and Y. C. Liu. 2017. Enhanced Surfactin Production via the Addition of Layered Double Hydroxides. *Journal of the Taiwan Institute of Chemical Engineers* 80: 10–15. http://doi.org/10.1016/j.jtice.2017.06.017.

Kinsinger, R. F., D. B. Kearns, M. Hale, and R. Fall. 2005. Genetic Requirements for Potassium Ion-Dependent Colony Spreading in *Bacillus subtilis*. *Journal of Bacteriology* 187 (24): 8462–69. http://doi.org/10.1128/JB.187.24.8462-8469.2005.

Kinsinger, R. F., M. C. Shirk, and R. Fall. 2003. Rapid Surface Motility in *Bacillus subtilis* Is Dependent on Extracellular Surfactin and Potassium Ion. *Journal of Bacteriology* 185 (18): 5627–31. http://doi.org/10.1128/JB.185.18.5627-5631.2003.

Klausmann, P., K. Hennemann, M. Hoffmann, C. Treinen, M. Aschern, L. Lilge, K. M. Heravi, M. Henkel, and R. Hausmann. 2021. *Bacillus subtilis* High Cell Density Fermentation Using a Sporulation-Deficient Strain for the Production of Surfactin. *Applied Microbiology and Biotechnology* 105 (10): 4141–51. http://doi.org/10.1007/s00253-021-11330-x.

Kluge, B., J. Vater, J. Salnikow, and K. Eckart. 1988. Studies on the Biosynthesis of Surfactin, a Lipopeptide Antibiotic from *Bacillus subtilis* ATCC 21332. *FEBS Letters* 231 (1): 107–10. http://doi.org/10.1016/0014-5793(88)80712-9.

Knoblich, A., M. Matsumoto, R. Ishiguro, K. Murata, Y. Fujiyoshi, Y. Ishigami, and M. Osman. 1995. "Electron Cryo-Microscopic Studies on Micellar Shape and Size of Surfactin, an Anionic Lipopeptide." *Colloids and Surfaces B: Biointerfaces* 5 (1–2): 43–48. https://doi.org/10.1016/0927-7765(95)01207-Y.

Kosaric, N., and F. Vardar-Sukan. 2021. *Biosurfactants: Production and Utilization— Processes, Technologies, and Economics.* Boca Raton: CRC Press.

Kowall, Martin, Joachim Vater, Britta Kluge, Torsten Stein, Peter Franke, and Dieter Ziessow. 1998. "Separation and Characterization of Surfactin Isoforms Produced by *Bacillus subtilis* OKB 105." *Journal of Colloid and Interface Science* 204 (1): 1–8. https://doi.org/10.1006/jcis.1998.5558.

Kumar, V., V. Ahluwalia, S. Saran, J. Kumar, A. K. Patel, and R. R. Singhania. 2021. Recent Developments on Solid-State Fermentation for Production of Microbial Secondary Metabolites: Challenges and Solutions. *Bioresource Technology* 323: 124566. http://doi.org/10.1016/j.biortech.2020.124566.

Lai, Chin-Chi, Yi-Chien Huang, Yu-Hong Wei, and Jo-Shu Chang. 2009. Biosurfactant-Enhanced Removal of Total Petroleum Hydrocarbons from Contaminated Soil. *Journal of Hazardous Materials* 167 (1–3): 609–14. https://doi.org/10.1016/j. jhazmat.2009.01.017.

Liu, Qiang, Junzhang Lin, Weidong Wang, He Huang, and Shuang Li. 2015. "Production of Surfactin Isoforms by *Bacillus subtilis* BS-37 and Its Applicability to Enhanced Oil Recovery under Laboratory Conditions." *Biochemical Engineering Journal* 93 (January): 31–37. https://doi.org/10.1016/j.bej.2014.08.023.

Liu, T., L. Montastruc, F. Gancel, L. Zhao, and I. Nikov. 2007. Integrated Process for Production of Surfactin: Part 1: Adsorption Rate of Pure Surfactin onto Activated Carbon. *Biochemical Engineering Journal* 35 (3): 333–40. http://doi.org/10.1016/j. bej.2007.01.025.

Long, X., N. He, Y. He, J. Jiang, and T. Wu. 2017. Biosurfactant Surfactin with Ph-Regulated Emulsification Activity for Efficient Oil Separation When Used as Emulsifier. *Bioresource Technology* 241: 200–206. http://doi.org/10.1016/j. biortech.2017.05.120.

Maget-Dana, Régine, and Marius Ptak. 1992. "Interfacial Properties of Surfactin." *Journal of Colloid and Interface Science* 153 (1): 285–91. https://doi. org/10.1016/0021-9797(92)90319-H.

Makkar, R. S., S. S. Cameotra, and I. M. Banat. 2011. Advances in Utilization of Renewable Substrates for Biosurfactant Production. *AMB Express* 1 (5): 1–19. http://doi.org/10.1186/2191-0855-1-5.

Marahiel, Mohamed A., Torsten Stachelhaus, and Henning D. Mootz. 1997. "Modular Peptide Synthetases Involved in Nonribosomal Peptide Synthesis." *Chemical Reviews* 97 (7): 2651–74. https://doi.org/10.1021/cr960029e.

Matsuura, T., and A. F. Ismail. 2021. *Membrane Separation Processes: Theories, Problems, and Solutions.* Amsterdam: Elsevier.

Mei, Y., Z. Yang, Z. Kang, F. Yu, and X. Long. 2021. Enhanced Surfactin Fermentation via Advanced Repeated Fed-Batch Fermentation with Increased Cell Density Stimulated by EDTA—Fe (II). *Food and Bioproducts Processing* 127: 288–94. http://doi.org/10.1016/j.fbp.2021.03.012.

Mitchell, D. A., M. H. Sugai-Guérios, and N. Krieger. 2019. *Solid-State Fermentation. Reference Module in Chemistry, Molecular Sciences and Chemical Engineering.* Waltham, MA: Elsevier Inc.

Modabber, G., A. A. Sepahi, F. Yazdian, and H. Rashedi. 2020. Surfactin Production in the Bioreactor: Emphasis on Magnetic Nanoparticles Application. *Engineering in Life Sciences* 20 (11): 466–75. http://doi.org/10.1002/elsc.201900163.

Montastruc, L., T. Liu, F. Gancel, L. Zhao, and I. Nikov. 2008. Integrated Process for Production of Surfactin: Part 2. Equilibrium and Kinetic Study of Surfactin Adsorption onto Activated Carbon. *Biochemical Engineering Journal* 38 (3): 349–54. http://doi.org/10.1016/j.bej.2007.07.023.

Moo-Young, M. 2011. *Comprehensive Biotechnology* (Second Edition). Amsterdam: Elsevier Science.

Mukherjee, S., P. Das, and R. Sen. 2006. Towards Commercial Production of Microbial Surfactants. *Trends in Biotechnology* 24 (11): 509–15. http://doi.org/10.1016/j.tibtech.2006.09.005.

Najafpour, G. D. 2008. *Biochemical Engineering and Biotechnology*. Amsterdam: Elsevier.

Nakano, M. M., Y. P. Dailly, P. Zuber, and D. P. Clark. 1997. Characterization of Anaerobic Fermentative Growth of *Bacillus subtilis*: Identification of Fermentation End Products and Genes Required for Growth. *Journal of Bacteriology* 179 (21): 6749–55. http://doi.org/10.1128/jb.179.21.6749-6755.1997.

Nazareth, T. C., C. P. Zanutto, D. Maass, A. A. U. Souza, and S. M. A. G. U. Souza. 2021a. Bioconversion of Low-Cost Brewery Waste to Biosurfactant: An Improvement of Surfactin Production by Culture Medium Optimization. *Biochemical Engineering Journal* 172: 108058. http://doi.org/10.1016/j.bej.2021.108058.

Nazareth, T. C., C. P. Zanutto, D. Maass, A. A. U. Souza, and S. M. A. G. U. Souza. 2021b. Impact of Oxygen Supply on Surfactin Biosynthesis Using Brewery Waste as Substrate. *Journal of Environmental Chemical Engineering* 9: 105372. http://doi.org/10.1016/j.jece.2021.105372.

Noah, K. S., S. L. Fox, D. F. Bruhn, D. N. Thompson, and G. A. Bala. 2002. Development of Continuous Surfactin Production from Potato Process Effluent by *Bacillus subtilis* in an Airlift Reactor. *Biotechnology for Fuels and Chemicals*, 803–13. http://doi.org/10.1007/978-1-4612-0119-9_65.

Ohno, A., T. Ano, and M. Shoda. 1992. Production of a Lipopeptide Antibiotic Surfactin with Recombinant *Bacillus subtilis*. *Biotechnology Letters* 14 (12): 1165–68. http://doi.org/10.1007/BF01027022.

Ohno, Akihiro, Takashi Ano, and Makoto Shoda. 1995. "Production of a Lipopeptide Antibiotic, Surfactin, by Recombinant *Bacillus subtilis* in Solid State Fermentation." *Biotechnology and Bioengineering* 47 (2): 209–14. https://doi.org/10.1002/bit.260470212.

Onaizi, Sagheer A. 2018. "Dynamic Surface Tension and Adsorption Mechanism of Surfactin Biosurfactant at the Air–Water Interface." *European Biophysics Journal* 47 (6): 631–40. https://doi.org/10.1007/s00249-018-1289-z.

Onaizi, Sagheer A., Lizhong He, and Anton P. J. Middelberg. 2009. "Rapid Screening of Surfactant and Biosurfactant Surface Cleaning Performance." *Colloids and Surfaces B: Biointerfaces* 72 (1): 68–74. https://doi.org/10.1016/j.colsurfb.2009.03.015.

Otzen, Daniel E. 2017. Biosurfactants and Surfactants Interacting with Membranes and Proteins: Same But Different? *Biochimica Et Biophysica Acta (BBA)— Biomembranes* 1859 (4): 639–49. http://doi.org/10.1016/j.bbamem.2016.09.024.

Pan, F. D., S. Liu, Q. M. Xu, X. Y. Chen, and J. S. Cheng. 2021. Bioconversion of Kitchen Waste to Surfactin via Simultaneous Enzymolysis and Fermentation Using Mixed-Culture of Enzyme-Producing Fungi and *Bacillus amyloliquefaciens* HM618. *Biochemical Engineering Journal* 172: 108036. http://doi.org/10.1016/j.bej.2021.108036.

Pereira, Jorge F. B., Eduardo J. Gudiña, Rita Costa, Rui Vitorino, José A. Teixeira, João A. P. Coutinho, and Lígia R. Rodrigues. 2013. Optimization and Characterization of

Biosurfactant Production by *Bacillus subtilis* Isolates towards Microbial Enhanced Oil Recovery Applications. *Fuel* 111 (September): 259–68. https://doi.org/10.1016/j. fuel.2013.04.040.

Peypoux, F., J. M. Bonmatin, and J. Wallach. 1999. Recent Trends in the Biochemistry of Surfactin. *Applied Microbiology and Biotechnology* 51 (5): 553–63. https://doi. org/10.1007/s002530051432.

Phulpoto, Irfan Ali, Zhisheng Yu, Bowen Hu, Yanfen Wang, Fabrice Ndayisenga, Jinmei Li, Hongxia Liang, and Muneer Ahmed Qazi. 2020. Production and Characterization of Surfactin-Like Biosurfactant Produced by Novel Strain *Bacillus nealsonii* S2MT and Its Potential for Oil Contaminated Soil Remediation. *Microbial Cell Factories* 19 (1). https://doi.org/10.1186/s12934-020-01402-4.

Pracht, Philipp, Fabian Bohle, and Stefan Grimme. 2020. Automated Exploration of the Low-Energy Chemical Space with Fast Quantum Chemical Methods. *Physical Chemistry Chemical Physics* 22 (14): 7169–92. https://doi.org/10.1039/c9cp06869d.

Ramalingam, Vaikundamoorthy, Krishnamoorthy Varunkumar, Vilwanathan Ravikumar, and Rajendran Rajaram. 2019. Production and Structure Elucidation of Anticancer Potential Surfactin from Marine Actinomycete *Micromonospora marina*. *Process Biochemistry* 78 (March): 169–77. https://doi.org/10.1016/j. procbio.2019.01.002.

Rangarajan, V., and K. G. Clarke. 2015. Process Development and Intensification for Enhanced Production of Bacillus Lipopeptides. *Biotechnology and Genetic Engineering Reviews* 31 (1–2): 46–68. http://doi.org/10.1080/02648725.2016.1166335.

Rangarajan, V., and R. Sen. 2013. An Inexpensive Strategy for Facilitated Recovery of Metals and Fermentation Products by Foam Fractionation Process. *Colloids and Surfaces B: Biointerfaces* 104: 99–106. http://doi.org/10.1016/j.colsurfb.2012.12.007.

Rocha, P. M., A. C. S. Mendes, S. D. O. Júnior, C. E. A. Padilha, A. L. O. S. Leitão, C. C. Nogueira, G. R. Macedo, and E. S. Santos. 2021. Kinetic Study and Characterization of Surfactin Production by *Bacillus subtilis* UFPEDA 438 Using Sugarcane Molasses as Carbon Source. *Preparative Biochemistry and Biotechnology* 51 (3): 300–08. http://doi.org/10.1080/10826068.2020.1815055.

Santos, D., R. Rufino, J. Luna, V. Santos, and L. Sarubbo. 2016. Biosurfactants: Multifunctional Biomolecules of the 21st Century. *International Journal of Molecular Sciences* 17 (3): 401. http://doi.org/10.3390/ijms17030401.

Santos, Vanessa Santana Vieira, Edgar Silveira, and Boscolli Barbosa Pereira. 2018. Toxicity and Applications of Surfactin for Health and Environmental Biotechnology. *Journal of Toxicology and Environmental Health, Part B* 21 (6–8): 382–99. https:// doi.org/10.1080/10937404.2018.1564712.

Santos, Vanessa Santana Vieira, Edgar Silveira, and Boscolli Barbosa Pereira. 2019. Toxicity and Applications of Surfactin for Health and Environmental Biotechnology. *Journal of Toxicology and Environmental Health, Part B*, 1–18. https://doi.org/10. 1080/10937404.2018.1564712. https://pubmed.ncbi.nlm.nih.gov/30614421/

Sarachat, T., O. Pornsunthorntawee, S. Chavadej, and R. Rujiravanit. 2010. Purification and Concentration of a Rhamnolipid Biosurfactant Produced by *Pseudomonas aeruginosa* SP4 Using Foam Fractionation. *Bioresource Technology* 101 (1): 324–30. http://doi.org/10.1016/j.biortech.2009.08.012.

Satpute, S. K., A. G. Banpurkar, P. K. Dhakephalkar, I. M. Banat, and B. A. Chopade. 2010. Methods for Investigating Biosurfactants and Bioemulsifiers: A Review. *Critical Reviews in Biotechnology* 30 (2): 127–44. http://doi.org/10.3109/07388550903427280.

Sen, R., and T. Swaminathan. 1997. Application of Response-Surface Methodology to Evaluate the Optimum Environmental Conditions for the Enhanced Production of Surfactin. *Applied Microbiology and Biotechnology* 47: 358–63. http://doi.org/10.1007/s002530050940.

Sen, R., and T. Swaminathan. 2004. Response Surface Modeling and Optimization to Elucidate and Analyze the Effects of Inoculum Age and Size on Surfactin Production. *Biochemical Engineering Journal* 21: 141–48. https://doi.org/10.1016/j.bej.2004.06.006.

Seydlová, Gabriela, and Jaroslava Svobodová. 2008. "Review of Surfactin Chemical Properties and the Potential Biomedical Applications." *Central European Journal of Medicine* 3 (2): 123–33. https://doi.org/10.2478/s11536-008-0002-5.

Shaligram, N. S., and R. S. Singhal. 2010. Surfactin—a Review on Biosynthesis, Fermentation, Purification and Applications. *Food Technology and Biotechnology* 48 (2): 119–34.

Shao, Chuanshi, Lin Liu, Hongze Gang, Shizhong Yang, and Bozhong Mu. 2015. Structural Diversity of the Microbial Surfactin Derivatives from Selective Esterification Approach. *International Journal of Molecular Sciences* 16 (1): 1855–72. https://doi.org/10.3390/ijms16011855.

She, An Qi, Hong Ze Gang, and Bo Zhong Mu. 2012. Temperature Influence on the Structure and Interfacial Properties of Surfactin Micelle: A Molecular Dynamics Simulation Study. *Journal of Physical Chemistry B* 116 (42): 12735–43. https://doi.org/10.1021/jp302413c.

Shen, Hsin-Hui, Tsung-Wu Lin, Robert K. Thomas, Diana J. F. Taylor, and Jeffrey Penfold. 2011. Surfactin Structures at Interfaces and in Solution: The Effect of PH and Cations. *The Journal of Physical Chemistry B* 115 (15): 4427–35. https://doi.org/10.1021/jp109360h.

Shen, Hsin-Hui, Robert K. Thomas, Chien-Yen Chen, Richard C. Darton, Simon C. Baker, and Jeffrey Penfold. 2009. Aggregation of the Naturally Occurring Lipopeptide, Surfactin, at Interfaces and in Solution: An Unusual Type of Surfactant? *Langmuir* 25 (7): 4211–18. https://doi.org/10.1021/la802913x.

Silva, M. T. S., C. M. Soares, A. S. Lima, and C. C. Santana. 2015. Integral Production and Concentration of Surfactin from *Bacillus* sp. ITP-001 by Semi-Batch Foam Fractionation. *Biochemical Engineering Journal* 104: 91–97. http://doi.org/10.1016/j.bej.2015.04.010.

Silveira, Thais de Carvalho, Wyllerson Evaristo Gomes, Giovana Chinaglia Tonon, Thainá Godoy Beatto, Nicolas Spogis, Luiz Henrique Dallan Cunha, Bruno Pera Lattaro, et al. 2021. Residual Biomass from Surfactin Production Is a Source of Arginase and Adsorbed Surfactin That Is Useful for Environmental Remediation. *World Journal of Microbiology and Biotechnology* 37 (7). https://doi.org/10.1007/s11274-021-03094-3.

Snook, M. E., T. Mitchell, D. M. Hinton, and C. W. Bacon. 2009. Isolation and Characterization of Leu 7-Surfactin from the Endophytic Bacterium *Bacillus mojavensis* RRC 101, a Biocontrol Agent for *Fusarium Verticillioides*. *Journal of Agricultural and Food Chemistry* 57: 4287–92. http://doi.org/10.1021/jf900164h.

Szmigiel, I., D. Kwiatkowska, M. Lukaszewicz, and A. Krasowska. 2021. Xylan Decomposition in Plant Cell Walls as an Inducer of Surfactin Synthesis by *Bacillus subtilis*. *Biomolecules* 11: 239. http://doi.org/10.3390/biom11020239.

Théatre, Ariane, Carolina Cano-Prieto, Marco Bartolini, Yoann Laurin, Magali Deleu, Joachim Niehren, Tarik Fida, et al. 2021. The Surfactin-like Lipopeptides from

Bacillus spp.: Natural Biodiversity and Synthetic Biology for a Broader Application Range. *Frontiers in Bioengineering and Biotechnology* 9 (March). https://doi.org/10.3389/fbioe.2021.623701.

Tsan, Pascale, Laurent Volpon, Françoise Besson, and Jean Marc Lancelin. 2007. Structure and Dynamics of Surfactin Studied by NMR in Micellar Media. *Journal of the American Chemical Society* 129 (7): 1968–77. https://doi.org/10.1021/ja066117q.

Tsuge, K., Y. Ohata, and M. Shoda. 2001. Gene YerP, Involved in Surfactin Self-Resistance in *Bacillus subtilis*. *Antimicrobial Agents and Chemotherapy* 45 (12): 3566–73. http://doi.org/10.1128/AAC.45.12.3566-3573.2001.

Ullrich, C., B. Kluge, Z. Palacz, and J. Vater. 1991. Cell-Free Biosynthesis of Surfactin, a Cyclic Lipopeptide Produced by *Bacillus subtilis*. *Biochemistry* 30 (26): 6503–08. http://doi.org/10.1021/bi00240a022.

Varadavenkatesan, Thivaharan, and Vytla Ramachandra Murty. 2013. Production of a Lipopeptide Biosurfactant by a Novel *Bacillus* sp. and Its Applicability to Enhanced Oil Recovery. *ISRN Microbiology* 2013: 1–8. https://doi.org/10.1155/2013/621519.

Vicente, R., C. J. de Andrade, D. de Oliveira, and A. Ambrosi. 2021. A Prospection on Membrane-Based Strategies for Downstream Processing of Surfactin. *Chemical Engineering Journal* 415 (1): 129067. http://doi.org/10.1016/j.cej.2021.129067.

Wei, Y. H., and I. M. Chu. 1998. Enhancement of Surfactin Production in Iron-Enriched Media by *Bacillus subtilis* ATCC 21332. *Enzyme and Microbial Technology* 22: 724–28. http://doi.org/10.1016/S0141-0229(98)00016-7.

Wei, Y. H., and I. M. Chu. 2002. Mn2+ Improves Surfactin Production by *Bacillus subtilis*. *Biotechnology Letters* 24: 479–82. http://doi.org/10.1023/A:1014534021276.

Wei, Y. Hong, C. C. Lai, and J. S. Chang. 2007. Using Taguchi Experimental Design Methods to Optimize Trace Element Composition for Enhanced Surfactin Production by *Bacillus subtilis* ATCC 21332. *Process Biochemistry* 42: 40–45. http://doi.org/10.1016/j.procbio.2006.07.025.

Willenbacher, J., J. T. Rau, J. Rogalla, C. Syldatk, and R. Hausmann. 2015. Foam-Free Production of Surfactin via Anaerobic Fermentation of *Bacillus subtilis* DSM 10T. *AMB Express* 5 (21). http://doi.org/10.1186/s13568-015-0107-6.

Willenbacher, J., M. Zwick, T. Mohr, F. Schmid, C. Syldatk, and R. Hausmann. 2014. Evaluation of Different Bacillus Strains in Respect of Their Ability to Produce Surfactin in a Model Fermentation Process with Integrated Foam Fractionation. *Applied Microbiology and Biotechnology* 98: 9623–32. http://doi.org/10.1007/s00253-014-6010-2.

Winterburn, J. B., A. B. Russell, and P. J. Martin. 2011. Integrated Recirculating Foam Fractionation for the Continuous Recovery of Biosurfactant from Fermenters. *Biochemical Engineering Journal* 54 (2): 132–39. https://doi.org/10.1016/j.bej.2011.02.011.

Wu, Q., Y. Zhi, and Y. Xu. 2019. Systematically Engineering the Biosynthesis of a Green Biosurfactant Surfactin by *Bacillus subtilis* 168. *Metabolic Engineering* 52: 87–97. http://doi.org/10.1016/j.ymben.2018.11.004.

Wu, X., H. Wu, R. Wang, Z. Wang, Y. Zhang, Q. Gu, A. Farzand, et al. 2021. Genomic Features and Molecular Function of a Novel Stress-Tolerant *Bacillus halotolerans* Strain Isolated from an Extreme Environment. *Biology* 10: 1030. http://doi.org/10.3390/biology10101030.

Yang, H., X. Li, X. Li, H. Yu, and Z. Shen. 2015. Identification of Lipopeptide Isoforms by MALDI-TOF-MS/MS Based on the Simultaneous Purification of Iturin, Fengycin, and Surfactin by RP-HPLC. *Analytical and Bioanalytical Chemistry* 407 (9): 2529–42. http://doi.org/10.1007/s00216-015-8486-8.

Yang, N., Q. Wu, and Y. Xu. 2020. Fe Nanoparticles Enhanced Surfactin Production in *Bacillus amyloliquefaciens*. *ACS Omega* 5: 6321–29. http://doi.org/10.1021/acsomega.9b03648.

Yeh, M. S., Y. H. Wei, and J. S. Chang. 2005. Enhanced Production of Surfactin from *Bacillus subtilis* by Addition of Solid Carriers. *Biotechnology Progress* 21: 1329–34. http://doi.org/10.1021/bp050040c.

Yeh, Mao-Sung, Yu-Hong Wei, and Jo-Shu Chang. 2006. Bioreactor Design for Enhanced Carrier-Assisted Surfactin Production with *Bacillus subtilis*. *Process Biochemistry* 41 (8): 1799–1805. https://doi.org/10.1016/j.procbio.2006.03.027.

Yi, G., Q. Liu, J. Lin, W. Wang, H. Huang, and S. Li. 2017. Repeated Batch Fermentation for Surfactin Production with Immobilized *Bacillus subtilis* BS-37: Two-Stage PH Control and Foam Fractionation. *Journal of Chemical Technology and Biotechnology* 92: 530–35. http://doi.org/10.1002/jctb.5028.

Zanotto, A. W., A. Valério, C. J. Andrade, and G. M. Pastore. 2019. New Sustainable Alternatives to Reduce the Production Costs for Surfactin 50 Years after the Discovery. Applied Microbiology and Biotechnology 103. *Applied Microbiology and Biotechnology* 103: 8647–56. http://doi.org/10.1007/s00253-019-10123-7.

Zhao, F., H. Zhu, Q. Cui, B. Wang, H. Su, and Y. Zhang. 2021. Anaerobic Production of Surfactin by a New *Bacillus subtilis* Isolate and the *in situ* Emulsification and Viscosity Reduction Effect towards Enhanced Oil Recovery Applications. *Journal of Petroleum Science and Engineering* 201: 108508. http://doi.org/10.1016/j.petrol.2021.108508.

Zhu, Z., F. Zhang, Z. Wei, W. Ran, and Q. Shen. 2013. The Usage of Rice Straw as a Major Substrate for the Production of Surfactin by *Bacillus amyloliquefaciens* XZ-173 in Solid-State Fermentation. *Journal of Environmental Management* 127: 96–102. http://doi.org/10.1016/j.jenvman.2013.04.017.

2 Marine Microbe Surfactants
Future Implementations

Khouloud M. Barakat[1] and Osama M. Darwesh[2]
1 National Institute of Oceanography
and Fisheries (NIOF), Egypt
2 Environmental Biotechnology and
Nanotechnology, Agricultural Microbiology
Dept., National Research Centre, Cairo, Egypt
Tel. 002/01023347533
E. mail: kh2m2@yahoo.com

CONTENTS

DOI: 10.1201/9781003307464-2

INTRODUCTION

In 1950 the term surfactant was coined by Antara products and covered all traded products with wetting agents, surface activity, foaming agents, detergents, dispersants and emulsifiers (Xia et al., 2014). One of the major characteristics of surfactants is their ability to reduce surface tension. Surfactants are in huge demand worldwide. The global market for surfactants was estimated to be USD 30.64 billion in 2016 and expected to rise up to 39.86 billion USD by 2021 (Singh et al., 2018). Moreover, when petrochemical stocks were scarce, a drive towards identification of new renewable bioresources was created for efficient surfactant production (Foley et al., 2012). In recent years, scientists have discussed a continued environmental search for natural-source replacements for synthetic surfactants (Kalogerakis et al., 2015).

Biosurfactants are in demand by the worldwide market as natural materials that can be added to commercial products or used in environmental applications. These biomolecules decrease the surface tension between liquid phases and show superior stability to their chemical equivalents under different physico-chemical conditions. Biotechnological production of biosurfactants is still emerging (da Silva et al., 2021).

Biosurfactants (BSs) are amphiphilic surface-active molecules produced mainly by microbes' hydrocarbon-degrading bacteria. In addition to bioremediation properties, BSs can also have many biotechnological applications (Patiño et al., 2021). Biosurfactant market revenue generation was over $1.8 billion in 2016 and is expected to reach USD 2.6 billion by 2023 (540 kilotons by 2024) with the rhamnolipid market set to witness a gain of over 8%. Other market research projected the global biosurfactant market at over $5.52 billion by 2022, at a compound annual growth rate of 5.6% from 2017 to 2022 (Singh et al., 2018). Therefore, there is a need for new naturally occurring microbial sources of biosurfactant production.

BSs produced by microorganisms offer an ideal sustainable substitute for petrochemical-based surfactants. Biologically produced BSs have numerous potential applications in a wide variety of sectors: environmental, food, agriculture, biomedical and nanotechnology (Saha and Rao, 2017; Gaur and Manickam, 2021). BSs are synthesized from waste and renewable substrates such as hydrocarbon wastes, crude oil and vegetable oils in addition to being bio-compatible, non-toxic and biodegradable (Banat et al., 2010). They are usually classified based on their molecular weight; those surfactant molecules with low molecular weights are known as biosurfactants, while those with high molecular weights are known as bioemulsifiers (BEs). BSs are further categorized based upon their molecular structure, for example, glycolipids, lipopeptides, phospholipids, lipoprotein, fatty acids and polymeric BSs. They also have a range of different properties such as surface tension reduction, emulsification, foaming and wetting (Banat et al., 2014; Marchant and Banat, 2012).

During the catastrophic oil spills in the oceans between 1970 and 2016 (ITOPF, 2017), BSs naturally played a major role in bioremediation and acted as efficient dispersing agents, facilitating biodegradation by other microbial communities (De-Almeida et al., 2016). Despite BSs' versatile properties, a higher production cost compared to chemical surfactants and the pathogenicity of some BS-producing strains remain major obstacles for their large-scale production (Irorere et al., 2018), and the search for nonpathogenic strains remain an important research area (Elshikh et al., 2017). Therefore, an important question to achieve these applications is, "How can we evolve an ecofriendly and cost-effective process for BS production?"

BIOSURFACTANTS

In the last decade, the chemical surfactants have been part of several commercial products. These compounds structurally consist of both a hydrophilic and hydrophobic moiety, with variations in structure depending on the synthetic process (Cowan-Ellsberry et al., 2014). They play a vital role in various industrial market sections, including products currently needed due to the COVID-19 pandemic (Celik et al., 2020). Many products contain effective surfactants in their composition, like soaps, toothpastes, fabric softeners, detergents and so on. Also, petrochemical products are major sources of mostly chemical synthesized surfactants, so they are ecologically undesirable additives. However, biotechnological studies and chemical companies are continuously searching for safer environmentally friendly industrial bioprocesses using ecological biomolecules with superior structural and functional properties. Biosurfactants are naturally synthesized by biological systems such as plants and microorganisms; they represent a sustainable alternative to these chemical counterparts, offering lower toxicity and higher biodegradability (Geetha et al., 2018).

Biosurfactants have advantages over their synthetic counterparts:

a. *Low toxicity:* biosurfactants are mainly used in cleaning food and cosmetic products and in bioremediation, determining that they have low or no toxicity. Recent reports have demonstrated the absence of BS toxic effects against microorganisms and microcrustaceans or in the germination of seeds. Tests have been done for the toxic application of biosurfactants in detergents to verify acute oral toxicity (LD_{50} and LC_{50}), acute dermal irritation, washing efficiency, surface activity and finally compatibility tests with purified hard water (Sobrinho et al., 2013; Fei et al., 2019).

b. *High biodegradability:* biosurfactants are easily degradable in water or soil, which permits them to be used in the bioremediation process to release contaminants from soil like pesticide formulations, which is considered a biological control (Rodriguez-Lopez et al., 2020). Biosurfactants have received increased attention in scientific research for their commercial application regarding their tolerance to pH variation, salinity and temperature; thus, novel biosurfactants are able to perform efficiently under extreme temperatures, pH and salinity (Roy, 2017).

c. *Use of renewable substrates:* the use of economically cheaper substrates provides a cost-effective industrial biosurfactant production process (Banat et al., 2014).

Various types of biosurfactants have potential application in numerous areas, due to their antimicrobial, antiadhesive, emulsifying, antitumor and anticorrosion activities. These activities are of use to the textile, food and biomedical industries. Because of their structural diversity, higher biodegradability, environmental compatibility and lower critical micelle concentration (CMC), marine biosurfactants have drawn much attention and led to strengthened oil recovery pesticide and herbicide formulations, health care, detergents, pulp and paper, coal, textiles, uranium ore processing, ceramic processing and mechanical dewatering of peat (Shoeb et al., 2013). It has been reported marine biosurfactants are used in shampoos, detergents, toothpaste, oil additives and a number of other consumer and industrial products (Kaya et al., 2014).

Microbial compounds that exhibit marked surface activity are known as biosurfactants, agents capable of reducing surface tension at the air–water interface and between immiscible liquids or at the solid–liquid interface (Sarubbo et al., 2006). Biosurfactants are amphiphilic molecules consisting of hydrophobic and hydrophilic moieties and can reduce the surface and interfacial tension of solutions and increase solubility, mobility and bioavailability of hydrophobic or insoluble organic compounds (Xia et al., 2014). Biosurfactants may be intracellular or secreted outside (extracellular) (Antoniou et al., 2015).

Bioremediation is a biological treatment technique that uses microorganisms to degrade hydrophobic pollutants and has two major advantages: low-cost environmental technology and being an alternative technique in solving problems of hydrocarbon pollution (Silva et al., 2014). Diverse microorganisms are known to produce a number of surface-active agents primarily in order to adapt and grow on a variety of substrates, among other natural functions (Banat et al., 2010). These biosurfactants are produced under various growth and environmental conditions and are reported to be mainly involved in increasing the solubility and availability of various water-immiscible substrates (Shekhar et al., 2015; Luna et al., 2016; Varjani and Upasani, 2017).

MARINE BIOSURFACTANT CLASSES

Depending on the nature of the head group, surfactants are commonly classified into four categories, anionic (negatively charged), cationic (positively charged), nonionic (polymerization products) and amphoteric (both negatively and positively charged) (Figure 2.1). The anionic or cationic hydrophilic head is binding

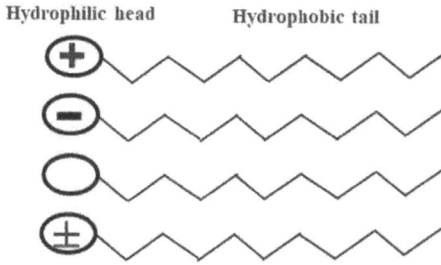

FIGURE 2.1 Classification of surfactants based on the charge of their head and tail group: cationic, anionic, nonionic and amphoteric.

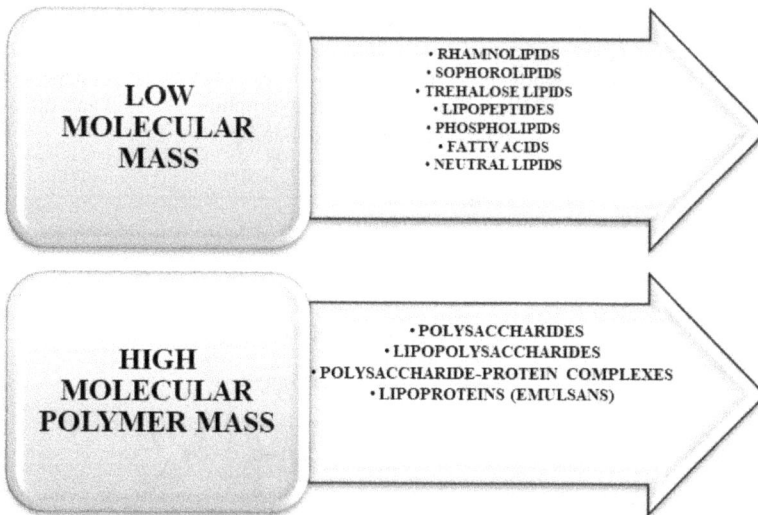

FIGURE 2.2 Classification of surfactants based on their molecular mass.

with water due to the presence of a negative or positive charge on it. Also, surfactants are grouped into two categories: low molecular mass molecules with lower surface and interfacial tensions (Mulligan, 2005) and high molecular mass, which binds tightly to surfaces (Rosenberg and Ron, 1999) (Figure 2.2).

Generally, the most known BSs among microbial communities are glycolipid molecules that consist of mono-, di-, tri- and tetra-saccharides including glucose, glucuronic acid, galactose, rhamnose, mannose and galactose sulfate in combination with long-chain aliphatic acids or hydroxyaliphatic acids (Plaza et al., 2015). Simple glycolipids (GLs) are amphiphilic molecules, as they comprise both hydrophilic glycosyl and lipophilic lipid residues (Figure 2.3). This amphiphilic nature confers surfactant activity to most GLs (Abdel-Mawgoud and Stephanopoulos, 2018). These glycolipids named rhamnolipids, trehalolipids

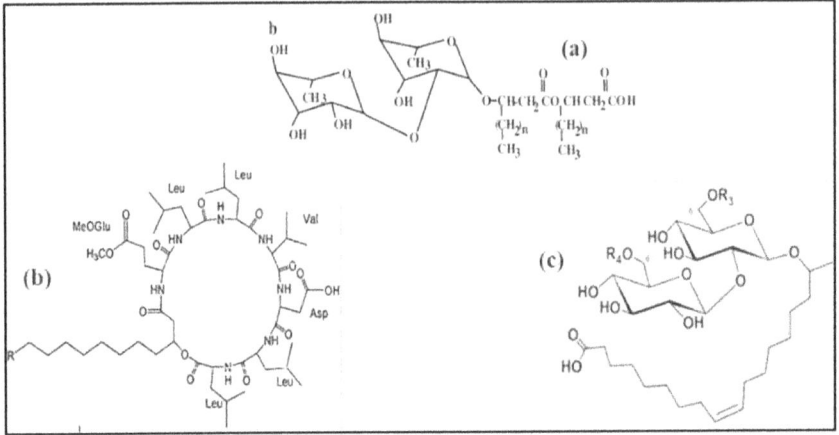

FIGURE 2.3 Three predominant simple glycolipid biosurfactant structures consisting of amphiphilic molecules, as they comprise both the hydrophilic glycosyl and lipophilic lipid residues rhamnolipid (a), trehalolipid (b) and sophorolipid (c).

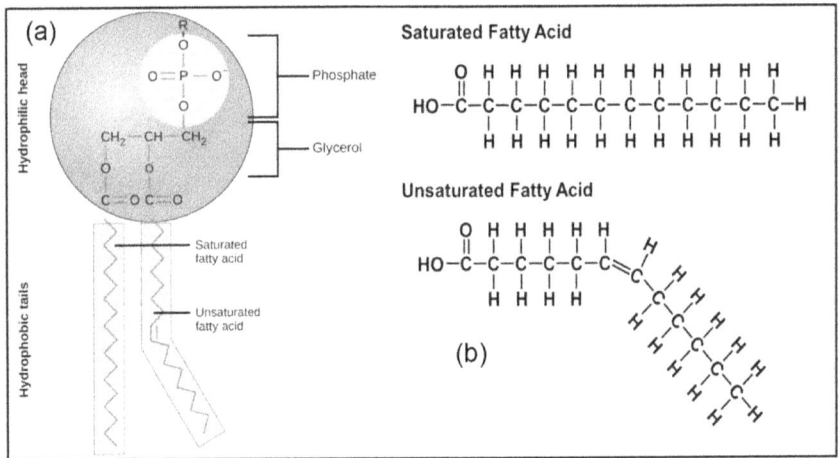

FIGURE 2.4 Structure of phospholipids (a) and fatty acid (b).

and sophorolipids that are the most reviewed disaccharides from *Pseudomonas* sp. (Rahman et al., 2007), *Arthrobacter* sp., *Corynebacterium, Mycobacterium, Nocardia, Rhodococcus erythropolis* (dos Reis et al., 2018) and finally *Candida* spp. (Mulligan, 2005).

Phospholipid, fatty acid and neutral lipid surfactants (Figure 2.4) represent the major microbial membrane components that are largely produced by hydrocarbon-degrading microorganisms such as *Acinetobacter* sp. *Capnocytophaga* sp., *Corynebacterium* sp., *Rhodococcus erythropolis, Penicillium spiculisporum, Thiobacillus thiooxidans* and yeasts (Rosenberg and Ron, 1999).

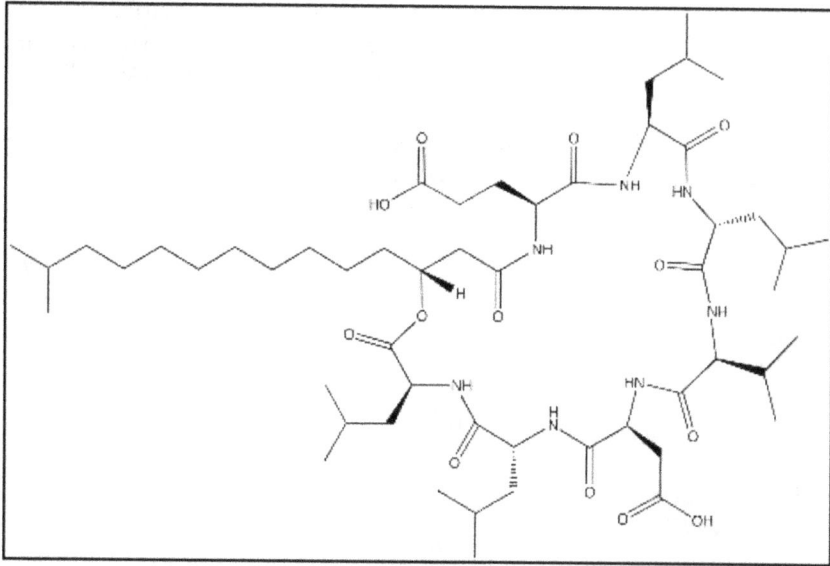

FIGURE 2.5 Structure of lipopeptide surfactins.

Lipopeptide biosurfactants are cyclic compounds, and they are mostly isolated from *Bacillus* and *Pseudomonas*. Lipopeptides mainly consist of hydrophilic peptides; generally they are seven and ten amino acids long, linked to a hydrophobic fatty acid structure. *Bacillus* cyclic lipopeptides consist of three major groups known as the surfactin, iturin and fengycin families. Surfactin (Figure 2.5) is the most commonly studied, and it contains seven amino acid cyclic sequences connected to a C13–C16 fatty acid (Ongena and Jacques, 2008).

Polymeric biosurfactants were reviewed by Uzoigwe et al. (2015) and classified as the following. Emulsan is considered a powerful emulsifying hydrocarbon agent in water, even as low as 0.001 to 0.01%. It is a poly-anionic hetero-polysaccharide bioemulsifier produced from *Candida lipolytica* (liposan), *Acinetobacter calcoaceticus* (biodispersan) and *Saccharomyces cerevisiae*. Mannan is a lipid-protein BS, also produced by *Candida tropicalis*. Another carbohydrate-lipid-protein BS is produced by *Pseudomonas fluorescens*. Alasan is the best polymeric biosurfactant protein. *Acinetobacter calcoaceticus* RAG-1 strain makes an integration of an extracellular strong polyanionic amphipathic hetero-polysaccharide bioemulsifier.

Regarding the unique chemical structure of BS, microbial surfactants share two fundamental characteristics: their ability to adsorb at the interface and to aggregate in solution (Wang, 2011). The major role of microbial surfactants in hydrocarbon uptake is the regulation of cell attachment to hydrophobic and hydrophilic surfaces by exposing allover cell surface to BSs, resulting in cell-surface hydrophobicity alteration (Franzetti et al., 2008). Das and Mukherjee (2007) confirmed the importance of biosurfactant-producing strains in bioremediation of sites highly polluted with crude petroleum-oil hydrocarbons. Marine microbial BSs from different marine sources (hot water springs, corals, sponges, sea and sediments) were studied by their production of several lipopeptide antibiotics with potent surface-active properties, including thermotolerant and halotolerant *Bacillus licheniformis*Bas50 (Yakimov et al., 1995), *Brevibacillus laterosporus* (Desjardine et al., 2007), *Nocardiopsis alba* MSA10 (Gandhimathi et al., 2009), *Brevibacterium aureum* MSA13 (Kiran et al., 2010a), *Providencia rettgeri*, *Psychrobacter* sp., *Bacillus flexus*, *Bacillus anthracis* and *Bacillus pumilus* (Padmavathi and Pandian, 2014), as well as *Bacillus subtilis* MB-7, *Bacillus amyloliquefaciens* MB-101, *Halomonas* sp. MB-30 and *Alcaligenes* sp. MB-I9 (Dhasayan et al., 2015).

In addition, a broad spectrum of BS bacteria isolates from various marine matrices (*Annelida*, sea pen Pteroeides, fish gut and Arctic and Antarctic contaminated sediments) able to produce glycolipids have been widely investigated (Figure 2.6);

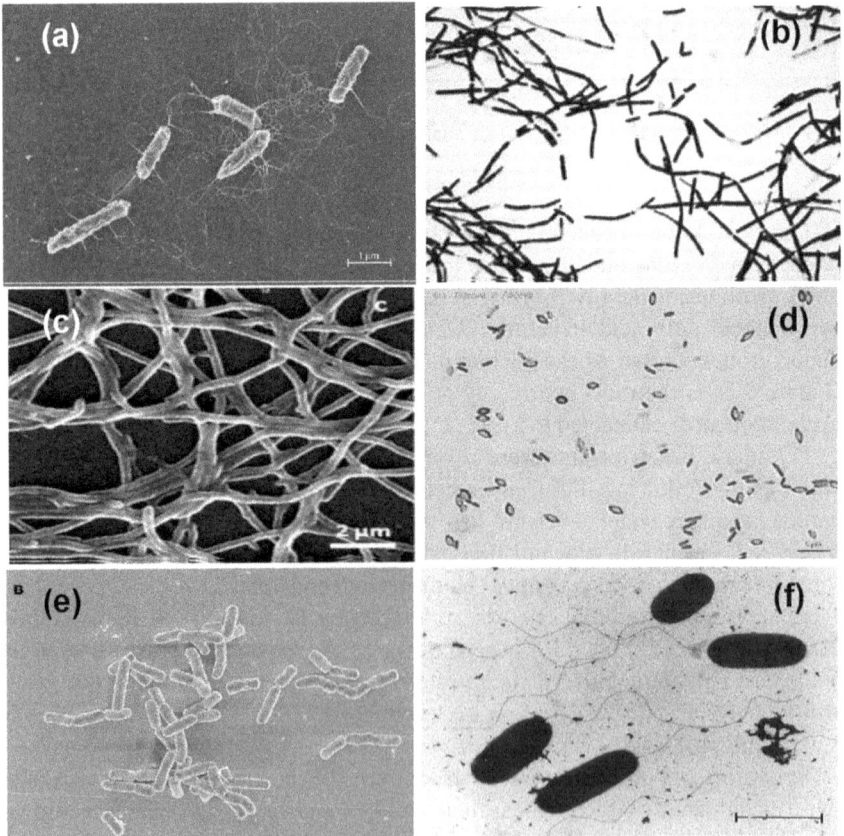

FIGURE 2.6 Electron microscope showing different marine biosurfactant-producing strains: *Halomonas* (a), *Bacillus licheniformis* (b), *Nocardiopsis* sp. (c), *Brevibacillus laterosporus* (d), *Bacillus amyloliquefaciens* (e) and *Maribacter* (f).

Pseudoalteromonas sp. TG12 (Gutierrez et al., 2008); *Cellulophaga, Cobetia, Cohaesibacter, Idiomarina, Pseudovibrio* and *Thalassospira* (Rizzo et al., 2013); *Citricoccus, Cellulophaga, Tenacibaculum, Maribacter, Psychrobacter, Vibrio* and *Pseudoalteromonas* (Rizzo et al., 2014); *Brachybacterium paraconglomeratum* (Kiran et al., 2014); *Bacillus* sp. E34 (Mabrouk et al., 2014); *Rhodococcus, Pseudomonas, Pseudoalteromonas* and *Idiomarina* spp. (Malavenda et al., 2015); *Brevibacterium* and *Vibrio* spp. (Graziano et al., 2016); and finally *Pseudomonas, Acinetobacter, Sphingomonas* and *Aeromonas* (Floris et al., 2018).

Heterogeneous high molecular weight exopolysaccharides (EPSs) with no strictly defined biosurfactants but furnished with interfacial properties originate from prokaryotic marine cyanobacteria in extreme environments (Kumar et al., 2007).

BIOSURFACTANT-PRODUCING MARINE BACTERIA

Biosurfactant-producing microbes are ubiquitous, inhabiting water (sea, fresh water and ground water) as well as environments characterized by extreme conditions as higher or lower pH and temperature or salinity (e.g., hypersaline sites and oil reservoirs) (Satpute et al., 2010; Ibacache-Quiroga et al., 2013). Due to these distinctive environmental conditions, marine habitats are a good source for new biosurfactant-producing microbe discovery. However, it has to be considered that the majority of the marine microbial assortment remains unexplored, maybe due to the failure to grow unculturable marine microorganisms under laboratory conditions (Gudiña et al., 2016). Marine bacteria are unique in the marine environment as well as extreme environments. These bacteria are known by metabolically adapting to survive under extreme pressure, temperature, pH and salinity conditions (Thavasi et al., 2014; Brasileiro et al., 2015). Large molecules known as exopolysaccharides are secreted by marine bacteria; these molecules may be proteins, lipids, nucleic acids polysaccharides and uronic acids. These EPSs strengthen the survival of bacterial cells through enhancing substrate adhesion, biofilm formation, protection against limited nutrient availability, detoxification of metals and the presence of antibiotics (Harimawan and Ting 2016). Some bacteria produce amphiphilic EPS, known as BSs that help in increasing the bioavailability of hydrophobic substrates such as hydrocarbons. These BSs encourage the growth of other bacteria capable of degrading aromatic and aliphatic hydrocarbons. Also, marine bacteria producing BS can ease hydrocarbon dispersion, emulsification, degradation and bioavailability (Mapelli et al., 2017).

BSs produced by psychrophilic marine bacteria can work efficiently at cold and freezing temperatures and are therefore adequate for laundry detergent formulations where washing conditions at low temperatures become a priority for energy conservation (Perfumo et al., 2018). The potential uses of BS are classified by their low toxicity, applicable for large-scale industrial production, followed by environmental disposal where they can easily biodegrade (Irorere et al., 2017). Hence, marine bacteria offer an excellent opportunity for the discovery of new BS molecules with distinctive properties. Although highly attractive, the biosynthesis of BSs from marine organisms has largely been overlooked. The mechanism of their regulation during synthesis is also not fully understood, adding further difficulties to the process for their production. Several approaches are required before the widespread application of marine-derived BS producing bacteria can be achieved: (i) isolation and identification of novel, non-pathogenic marine strains; (ii) optimization of culture conditions to achieve sufficient yields of BS; and (iii) characterization of genes involved in BS production from marine bacteria. These will allow the use of marine strains in largescale BS production processes while improving yield and cost-efficiency of BS production (Tripathi et al., 2018). Table 2.1 shows different marine bacterial strains producing different BSs.

Marine *Pseudomonas aeruginosa* isolated from seawater-polluted oil are able to break down octadecane, hexadecane, heptadecane and nonadecane after 28 days of incubation. The degradation ability of this bacterium has resulted in

TABLE 2.1
Biosurfactants Producing Marine Bacteria Strains

Bacterial Strain	Biosurfactants	References
Alcanivorax dieselolei	proline lipid	Qiao et al., 2010
Brevibacterium aureum	gly-gly-leu-pro	Kiran et al., 2010a
Serratia marcescens	glucosyl ester lipid	Dusane et al., 2011
Paenibacillus polymyxa	polymyxin B	Quinn et al., 2012
Tistrella mobilis	didemnin B	Xu et al., 2012
Cobetia sp. strain MM1IDA2H-1	3-hydroxy fatty acids	Ibacache-Quiroga et al., 2013
Rhodococcus sp. strain PML026	trehalose lipid	White et al., 2013
Brachybacterium paraconglomeratum	glycolipid	Kiran et al., 2014
Bacillus sp.	glycolipopeptide	Mabrouk et al., 2014
Providencia rettgeri, Psychrobacter sp., *Bacillus flexus, Bacillus anthracis, Bacillus pumilus*	lipopeptide/new BS	Padmavathi et al., 2014
Bacillus licheniformis NIOT-AMKV06	lipopeptide (unknown)	Lawrance et al., 2014
Streptomyces sp. MAB36	glycolipid (unknown)	Manivasagan et al., 2014
Streptomyces sp. IA49E	di-rhamnolipid	Yan et al., 2014
Rhodococcus sp. BS-15	tri-glucose lipid tetraester,	Konishi et al., 2014
Bacillus megaterium	Iturin	Dey et al., 2015
Bacillus licheniformis NIOT-06	surfactin	Anburajan et al., 2015
Paracoccus marcusii, Alcanivorax borkumensis	lipopeptide	Antoniou et al., 2015
Pseudomonas aeruginosa, Aeromonas hydrophila	ND	Shoeb et al., 2015

Bacterial Strain	Biosurfactants	References
Rhodococcus, Pseudomonas, Pseudoalteromonas, Idiomarina spp.	glycolipid	Malavenda et al., 2015
Bacillus sp., Halomonas sp., Alcaligenes sp.	lipopeptide	Dhasayan et al., 2015
Bacillus sp. KCB14S006	iturins	Son et al., 2016
Brevibacterium, Vibrio spp.	glycolipid	Graziano et al., 2016
Nocardiopsis alba	phenyl alanine dipeptide	Selvin et al., 2016
Achromobacter sp. HZ01	gly-gly-leu-met-leu-leu	Deng et al., 2016
Pontibacter korlensis	pontifactin(miao) ser-asp-val-ser-ser	Balan et al., 2016
Buttiauxella sp.	glucosyl ester lipid	Marzban et al., 2016
Buttiauxella sp.	glucosyl ester lipid	Marzban et al., 2016
Pseudomonas sp. BTN-1	rhamnolipid	Tedesco et al., 2016
Pseudomonas aeruginosa	rhamnolipid	Chakraborty et al., 2016
Pseudomonas aeruginosa	rhamnolipid	Cheng et al., 2017
Halobacterium salinarum	lipopeptide	Sumaiya et al., 2017
Aneurinibacillus aneurinilyticus	aneurinifactin	Balan et al., 2017
Bacillus amyloliquefaciens	didemnin B	Barakat et al., 2017
Staphylococcus lentus	threose diester	Hamza et al., 2017
Pseudomonas, Acinetobacter, Sphingomonas, Aeromonas	glycolipid/rhamnolipid	Floris et al., 2018
Brevibacterium luteolum	proline lipid	Unás et al., 2018
Bacillus pumilus	pumilacidin	Saggese et al., 2018
Bacillus siamensis	surfactin(miao)bacillomycin F	Xu et al., 2018
Bacillus amyloliquefaciens SH-B74	plipastatin A1	Ma et al., 2018
Pseudomonas sp. MCTG214(3b1)	rhamnolipid	Twigg et al., 2018
Bacillus stratophericus	surfactin(miao) pumilacidin	Hentati et al., 2019
Bacillus sp. CS30	surfactin	Wu et al., 2019
Cyberlindnera saturnus	cybersan (galactose lipid)	Balan et al., 2019
Actinoalloteichus hymeniacidonis	doktolipids(miao) (rhamnose lipids)	Choi et al., 2019
Pseudomonas aeruginosa	rhamnolipid	Du et al., 2019
Paracoccus sp. MJ9	rhamnolipid	Xu et al., 2020
Halomonas sp. INV PRT124, Halomonas sp. INVPRT125, Bacillus sp. INV FIR48, Pseudomonas sp. INV PRT82 and Streptomyces sp. INV ACT15	isoform surfactin	Patiño et al., 2021

ND: Not determined

biosurfactant production. It was also demonstrated that *P. aeruginosa* effectively degraded other hydrocarbons like 2-methylnaphthalene, tetradecane and pristine (Karlapudi et al., 2018).

Biosurfactants produced by marine bacteria are detected in biofilm formation, which interacts with an interface and changes the surface properties such as wettability. A biosurfactant-producing *Pseudomonas stutzeri* (SSASM1) strain was isolated from a sediment sample collected from the Pondicherry harbor region on the southeast coast of India. This strain showed emulsification activity of 77.6%, surface tension reduction of 33.5 mN/m and the produced biosurfactant characterized as rhamnolipid in nature (Shekhar et al., 2019). Two marine bacterial strains, *Marinobacter* sp. and *Pseudomonas mendocina*, have been screened for their ability to synthesize biosurfactant rhamnolipid detected by high-performance liquid chromatography—mass spectrometry and nuclear magnetic resonance (NMR) (Twigg et al., 2019). Fifty pure cultures of interest were obtained from seawater and sediment samples collected from six locations of the major industrial area Elefsina Bay, Attica, Aegean Sea, Greece, then screened for BS production by the drop collapse test. The isolated strains' phylogenetic identity, strains E8Y, E4D, E4F (*Alcanivorax borkumensis* SK2) and ESP-A (*Paracoccus marcusii*) achieved the highest records (Antoniou et al., 2015). Barakat et al. (2017) studied the isolation of BS-producing extremophilic marine *Bacillus amyloliquefaciens* SH20 and *Bacillus thuringiensis* SH24 collected from Shalateen, Red Sea, Egypt. The strains showed emulsification indexes of 57% and 56%, respectively.

A novel marine bacterium *Bacillus simplex* with promising biosurfactant production was isolated from petroleum hydrocarbon–contaminated coastal sea sediment samples of Nagapattinam fishing harbor, Tamil Nadu, India. The biosurfactant was identified as lipopeptide using thin layer chromatography (TLC), biochemical estimation methods, Fourier transform infrared, NMR, and matrix-assisted laser desorption ionization time-of-flight mass spectrometry analysis (Mani et al., 2016a). Among the isolated strains from an oil-spill area of the Arabian Sea, strain 2, identified as *Bacillus* sp., showed the highest biosurfactant activity. Various biosurfactant activity assay tests were performed to isolate the potent bacterial strain. The isolated culture filtrate was found to be highly effective in microbial-enhanced oil recovery (MEOR) (Dhail, 2017).

Surface-active agent-producing, oil-degrading marine bacteria were isolated using a modified Bushnell-Haas medium with high-speed diesel as a carbon source from three oil-polluted sites of Mumbai Harbor. These strains were screened and biochemically characterized, and nucleic acid sequencing methods identified them as *Acinetobacter, Alcanivorax, Bacillus, Comamonas, Chryseomicrobium, Halomonas, Marinobacter, Nesterenkonia, Pseudomonas* and *Serratia* (Mohanram et al., 2016). A potential biosurfactant producer, *Nocardiopsis lucentensis* MSA04, was isolated from the marine sponge *Dendrilla nigra*. Among the substrates screened, wheat bran increased the production significantly (E_{24} 25%). Enhanced biosurfactant production was achieved under solid-state cultivation conditions using kerosene as carbon source, beef extract as nitrogen source and wheat bran as substrate. The surface-active compound produced by MSA04 was characterized as glycolipid with a hydrophobic non-polar hydrocarbon chain (nonanoic acid methyl ester) and hydrophilic sugar, 3-acetyl 2,5 dimethyl furan (Kiran et al., 2010b).

BIOSURFACTANT-PRODUCING MARINE FUNGI

Fungi are the most active biosurfactant producers, with unique chemical structures, such as cellobiose lipids, xylolipids, sophorolipids, polyol lipids, mannosylerythritol lipids and hydrophobins (HFBs). Fungal biosurfactants, which represent only 19% out of the total microbes, are able to produce the widest chemical structural variant of biosurfactants; some of these BSs are exclusively produced by fungi (Sunde et al., 2017; Garay et al., 2018; Sanches et al., 2021). In general, the chemical structure versatility of fungal biosurfactants allows a wide range of applications such as in the personal care sector (Bae et al., 2018), food (Chieregato et al., 2019), agriculture (Shah and Daverey, 2021), pharmaceutical (Chuo et al., 2019), biomedicine (Guerfali et al., 2019), materials engineering (Ranjana et al., 2019), bioenergy (Menon et al., 2010) and environmental remediation (Ye et al., 2016).

Hydrophobins are high molecular weight fungal BSs and self-assembling proteins and are designated as the most powerful surface-active proteins. HFBs are small (about 100 amino acids) amphiphilic proteins that play vital roles in fungal biology by lowering the surface tension of the liquid medium in their soluble form and coating aerial structures such as hyphae, fruiting bodies and spores for easy growth and dispersal in the air, then fungal adhesion to surfaces and host-pathogen interactions (Lo et al., 2019). Protocols to isolate the most surface-active HFB known proteins were set up from fungal culture broth (Cicatiello et al., 2016). Generally, high molecular weight BS proteins from fungi can be better defined as BSs to stabilize emulsions more than to reduce the surface tension of water (Mujumdar et al., 2019).

Few studies on marine fungal producing BS strains have been carried out (Table 2.2). Marine endosymbiotic fungi *Aspergillus ustus* MSF3, which produces a high yield of biosurfactant, was isolated from the marine sponge *Fasciospongia cavernosa* collected from the peninsular coast of India. Maximum production of biosurfactant was obtained in Sabouraud dextrose broth. The optimized bioprocess conditions for the maximum production were pH 7.0, temperature 20°C, salt concentration 3%, glucose and yeast extract as carbon source and nitrogen sources, respectively. The biosurfactant produced by MSF3 was partially characterized as glycolipoprotein based on the estimation of macromolecules and TLC analysis. The partially purified biosurfactant showed a broad spectrum of antimicrobial activity (Kiran et al., 2009).

Two fungal strains, *Trichoderma harzianum* MUT 290 and *Aspergillus terreus* MUT 271, isolated from chronically pervaded oil spills in Mediterranean marine site, are able to use crude oil as a sole carbon source as biosurfactant producers. Both fungi secreted low molecular weight proteins identified as

TABLE 2.2
Biosurfactant-Producing Marine Fungal Strains

Fungal Strains	Biosurfactant	References
Aspergillus ustus	carbohydrate-lipid-protein complexes	Kiran et al., 2009
Aureobasidium pullulans	massoia lactone	Luepongpattana et al., 2017
Aspergillus terreus MUT 271 and *Trichoderma harzianum* MUT 290	cerato-platanins	Pitocchi et al., 2020
Aspergillus terreus MUT 271 and *Trichoderma harzianum* MUT 290	cerato-platanins	Bovio et al., 2017
Penicillium chrysogenum,		Cicatiello et al., 2019
Cyberlindnera saturnus SBPN-27	Cybersan	Balan et al., 2019
Fusarium sambucinum and *Trichoderma camerunense*	ND	Martinho et al., 2019
Meyerozyma guilliermondii L21, *Cryptococcus victoriae* L92 and *Leucosporidium scotti* L120	ND	Correa et al., 2020

ND: not determined

cerato-platanins (CPs), which are small, conserved, hydrophobic surface-active proteins. Both proteins were able to stabilize emulsions compared to other bio-surfactant proteins and commercially available chemical surfactants. Also, these proteins had the ability to work both as biosurfactant and bioemulsifier (Bovio et al., 2017). Indeed, the surface tension value tested in ThCP (*T. harzianum)* was much lower than that of AtCP (*A. terreus*). The marine strain of *Penicillium chrysogenum* showed another BE protein that was recently purified and measured in comparable conditions (Cicatiello et al., 2019).

Many reports on cerato-platanins as effective biosurfactants from different fungal strains have been deeply studied, including: *Trichodermaatroviride* (Frischmann et al., 2013), *Ceratocystis fimbriata* f. sp. platani (Pazzagli et al., 1999), *Ceratocystis platani* fungus (De Oliveira et al., 2011) and *Trichoderma virens* (Gaderer et al., 2014).

Among the 136 marine yeast strains isolated from three different stations in Tamil Nadu, India, *Cyberlindnera saturnus* SBPN-27 exhibited promising features for biosurfactant production. This marine yeast was purified and structurally characterized as cybersan based on different spectral analyses. Further, cybersan revealed surface tension of 28 mN m^{-1} at a critical micelle concentration of 30 mg L^{-1} and stability over broad pH and temperature conditions. Cybersan showed appreciable growth inhibition towards clinical bacterial pathogens and revealed no considerable cytotoxicity against mammalian 3T3 fibroblast cells, suggesting its biocompatible nature (Balan et al., 2019).

Thirty endophytic marine fungi were isolated from mangrove forest sampling in Cananeia, SP, Brazil. These microorganisms were analyzed for their production of bioactive secondary metabolites like biosurfactants and/or bioemulsifiers. The fungal isolates named *Fusarium sambucinum* and *Trichoderma camerunense* showed biosurfactant ability demonstrated by superficial tension decreasing to 38 mN/m. In addition, 15 fungi exhibited bioemulsifier activity, with E_{24} values up to 62.8% (Martinho et al., 2019). Recently, other yeasts isolated from Antarctic marine environments, *Meyerozyma guilliermondii* L21, *Cryptococcus victoriae* L92 and *Leucosporidium scotti* L120, have been reported as biosurfactant producers (Correa et al., 2020).

BIOSURFACTANT-PRODUCING MICROALGAE

Microalga cultivation has become a new source of food, biofuel and other products for industrial and pharmaceutical purposes because it grows fast and depletes CO_2 from the environment. Also, the benefits of biosurfactant production have recently been studied (Akubude and Mbab, 2021). Biosurfactants are an effective biological agent for the control of prodigious algae grazers that attack microalgae and drastically reduce their productivity.

A major advantage of cultivating microalgae for biosurfactant production is that many of these microorganisms fall into the Generally Recognized As Safe (GRAS) category. Such certified organisms have no risk of toxicity or pathogenicity and can be used for applications in the food and pharmaceutical industries (Soccol et al., 2013). Cyclic lipopeptides, apratoxins, are extracted from marine cyanobacteria exhibiting cytotoxic activity against many cancer cells (Nunnery et al., 2010). Also, the cyanobacterium *Moorea producens* produced two apratoxin analogues (apratoxin A sulfoxide and apratoxin H) showing great cytotoxicity on human NCI-H460 lung cancer cells (Thornburg et al., 2013). Wrasidlo and coworkers (2008) extracted the somocystinamide A lipopeptide from the cyanobacteria *Lyngbya majuscula*. Commercial biosurfactant production is limited due to the high costs involved, particularly with respect to culture media. The use of cheaper substrates, such as molasses and glucose, may reduce the cost factor and make production economically viable. Mixotrophic microalgae culture can significantly enhance the growth of microalgae, resulting in cell densities three to ten times higher than those obtained in autotrophic culture (Bhatnagar et al., 2011). The cyanobacteria strains *Arthrospira* sp. LEB 18 and *Synechococcus nidulans* LEB 25 and the chlorophyte strains *Chlorella minutissima* LEB 108, *Chlorella vulgaris* LEB 106 and *Chlorella homosphaera* were investigated for their biosurfactant production in autotrophic and mixotrophic cultivation (Radmann et al., 2015).

METHODS FOR MICROBIAL BS SCREENING

Marine biosurfactant-producing microbes can be screened using different assays depending on the performance of selected standard tests. Due to BS chemical diversity and different properties, the screening procedure has to examine all the multifaceted activities, from the interfacial to the emulsifying, from chelating to foaming stabilization functions. Therefore, screening methods could be divided based on direct (accurate) and indirect measurements.

DIRECT METHODS

Direct measurements of surface tension involve the evaluation of the force required to detach a ring or loop of wire (Du-Noüy method) or a platinum plate (Wilhelmy plate method) (Figure 2.7) from an interface or surface (Tadros, 2005). These methods secure the advantages of accuracy and easiness, despite the specialized equipment and the impossibility of performing measurements on different samples simultaneously.

INDIRECT METHODS

The main advantages of indirect methods are the possibility to screen more samples quickly, although they have low sensitivity and a strong dependence on BS concentration. The methods are based on distortion visual effects caused by the BS and are generally performed on supernatants and suggested coloring of BSs as visual effects (Walter et al., 2010).

Cetyltrimethylammonium Bromide Agar Plate Measuring

Cetyltrimethylammonium bromide (CTAB) agar plate measuring is an indirect method for screening surface/interfacial tension: it assays BS production by colorimetric measuring and studies the ability of BSs to form clear halos in a methylene blue/CTAB plate (Siegmund and Wagner, 1991). It is a semi-quantitative assay to detect extracellular glycolipids or other anionic surfactants. The CTAB agar assay is specific to anionic biosurfactants (Figure 2.8). The disadvantage of this method is that CTAB is toxic to microbial growth. In the process, microbes producing BS grow on the plate, producing anionic surfactants, which form a dark blue, insoluble ion that pairs with cethyltrimethylammonium bromide and methylene blue (basic dye).

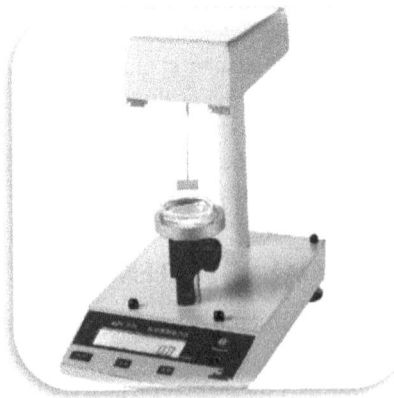

FIGURE 2.7   Automatic surface tensiometer interfacial tension meter platinum plate.

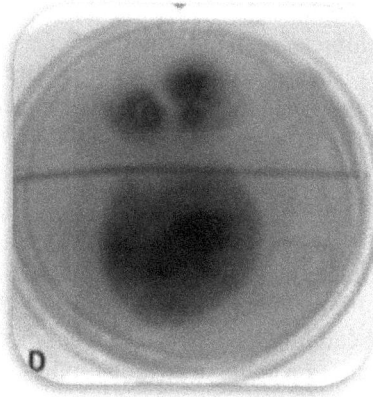

FIGURE 2.8   A dark blue halo around bacteria on a CTAB plate, indicating the interact of anionic biosurfactant with the cationic bromide salt. The complex is revealed by methylene blue in the agar (Darwesh et al., 2021).

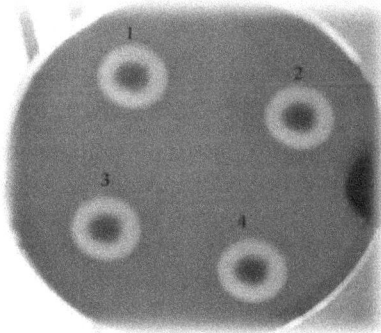

FIGURE 2.9   Surfactin production assayed on blood agar plates showing the hemolytic zone around the colony (Shannaq and Isa, 2013).

Blood Agar Hemolysis

This is a preliminary screening test to detect the ability of microbes to produce biosurfactants. A blood agar plate (5% sheep's blood) is used for the hemolytic activity test. Positive strains result in blood cell lysis and exhibit a colorless, transparent ring around the colonies (Fiebig et al., 1997). This assay is based on the hemolytic actions of biosurfactants—α, β and γ hemolysis—on solid medium containing defibrinated blood as greenish or clarification halos around the bacterial colonies (Figure 2.9). This preliminary screening test should be supported by other techniques based on surface activity measurements.

Thin-Layer Chromatography

The direct thin-layer chromatography technique is a rapid characterization of biosurfactant-producing bacterial colonies. A colony surrounded by an emulsified halo on an L-agar plate coated with oil was determined to be a biosurfactant producer strain, as described by Morikawa et al. (1992).

Drop-Collapsing

A drop-collapsing test is a rapid, sensitive method with an easy and fast procedure to assess microbial biosurfactant production. A small volume of microbial supernatant is tested. Drops of oil are placed on the slide; then 10 µl of the microbial sample is added by careful piercing using a micropipette without disturbing the dome of the oil. If oil drop collapses within 1 min, it is considered positive. Also, oil displacement and development of a clearing zone indicate the presence of a biosurfactant in the supernatant (Figure 2.10). The diameter of this clearing zone on the oil surface correlates to surfactant activity. This is also known as the oil spreading technique (Vandana and Peter, 2014).

Emulsification

Emulsification power (E-24) is measured by rigorously mixing an equal volume of the culture supernatant and kerosene for 1 min. The percentage of emulsion formation after 24 h is recorded, according to Haba et al. (2000). The evaluation of emulsifying activity depends on the volume of culture/supernatant, the tested hydrocarbon and the vortexing time. The emulsifying activity test is the most used BS test regarding a quick observation of emulsion occurrence that is stable over time and E24 index detection (Christova et al., 2004) (Figure 2.11). There is no correlation between surface tension activity and emulsification capacity (Plaza et al., 2006). Indeed, it was observed that different studies by Rizzo et al. (2013) and Malavenda et al. (2015) showed some BSs might stabilize (emulsifiers)

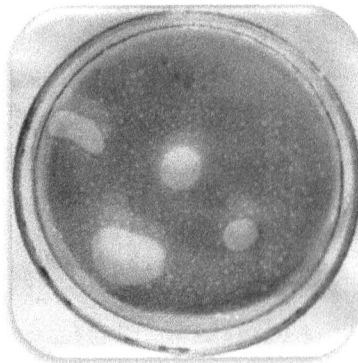

FIGURE 2.10   Different biosurfactants showing oil displacement and clearing zone formation.

FIGURE 2.11 Different supernatants of marine isolates after incubation mixed with mineral oil and the emulsification index being measured.

or destabilize (de-emulsifiers) the emulsion, so that using the emulsification test only is not accurate to identify active surfactant compounds. However, when surface activity indicates BS production, the detection of a stable emulsion index correlates with surfactant concentration.

According to Rizzo et al. (2014), surface tension measurement and emulsification activity assays could be complementary to each other and represent fundamental tests for screening procedures, resulting in both low molecular mass BSs with efficiency in surface and interfacial tension reduction and high molecular mass BSs more effective as emulsion stabilizers (Dhasayan et al., 2015; Dang et al., 2016; Graziano et al., 2016; Sumaiya et al., 2017; Rizzo et al., 2018).

Emulsion Formation Mechanism

The formation of stable emulsion was better in saline concentrations below 0.5%, pH values in the range of 6 to 9 and temperatures in the range of 35 to 40°C (two different interaction types commonly detected in the oil/hydrocarbon biodegradation processes; Figure 2.12). Pseudo-solubilization or oil adhesion and hydrocarbon degradation to form small oil droplets are two sequential steps found in one of the mechanisms. Microbial cells adhere to the drops of hydrocarbons of small size compared to the cells; then the uptake of substrate occurs either by active transport or by diffusion at the interference point between microbial cells and hydrocarbon molecules (Palecek et al., 2015).

a. Monolayer formation Surface tension is discontinuity of the bulk liquid at the interface. In this bulk liquid, molecules are encircled by other molecules on all sides, and the same forces are then initiated from different directions. However, at the interface, molecules are forcibly attracted by neighboring molecules in the bulk on one side of the interface only and thus form a net force, which creates motion of molecules from the bulk to the surface or generates the required surface energy. Hence, surface tension (γ) is identified as the amount of energy required to create a surface per unit area. The addition of surfactants or surface-active agents affects the surface tension. A main characteristic feature

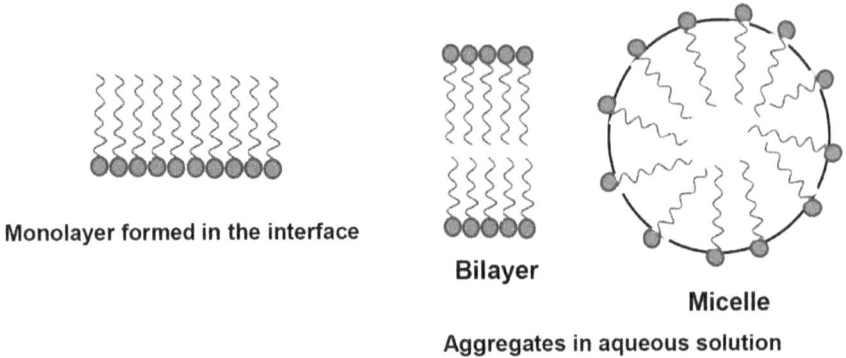

Monolayer formed in the interface

Bilayer

Micelle

Aggregates in aqueous solution

FIGURE 2.12 Two fundamental characteristics of surfactants: adsorption at the air–water interface and formation of aggregation in solution.

of surfactants is tending to replace solvent molecules and self-assemble at the interface as dipolar surfactants, which causes a substantial reduction of surface tension (Wang, 2011).

b. Micelle Formation Surfactants self-assemble at the interface; when it becomes saturated with these surfactants and there is no more chance for other surfactants to occupy space, then surfactant has its way inside the solution, forming micelles with its hydrophobic moiety that point toward the inside of the micelle, where its hydrophilic moiety comes into contact with a polar solvent such as water and leads to lower internal energy. The critical micelle concentration is the concentration at which micelles start to form. Micelle size differs according micelle radii, which range from 10 to 1000 Å. Micelle formation has a major role in BS processes like detergency and solubilization. Different micelle shapes can be formed in aqueous solution: spherical, disk-like lamellar, vesicle and elongated cylindrical with hemispherical ends. Many factors affect the shapes of micelles, including pH and temperature, ionic strength, solvent properties and surfactant concentration (Wang, 2011).

FACTORS AFFECTING PRODUCTION OF BSS FROM MARINE BACTERIA

The type of microbial strains and their growth conditions extensively affect the quality and productivity of BS. Optimized conditions are a major factor for maximum BS production. The type and productivity of BSs are also affected by the type and concentration of carbon and nitrogen sources, temperature, pH, salinity and agitation rate, among other factors.

Halophiles are marine microbes that require salts for their growth (Margesin and Schinner, 2001). Halophilic *Halomonas* are known to produce different types of glycoproteins and glycolipids with higher emulsifying activity than commonly used emulsifiers. These halophiles play a crucial role in oil contaminated saline environments by producing different types of surfactants and emulsifiers. Salt concentration is a critical factor for bacterial growth and BS production. The higher the salt concentration of halophilic bacteria, the lower the contamination risks, which can significantly reduce upstream fermentation costs of the produced BS (Tan et al., 2011). For large-scale industrial BS production, high salt concentration growth media would be regarded as undesirable due to the corrosive effects, so it must be considered in the production plant infrastructure.

EFFECT OF CARBON SOURCE

Maximum BS production depends on the different carbon sources used in the growth medium. Many types of carbon sources have been studied for BS production, including crude oil, diesel, glycerol, glucose, sucrose and other hydrocarbons. Previous studies show that marine bacteria has the ability to degrade these hydrocarbons to produce effective BS and could be utilized in the bioremediation of hydrocarbon-polluted environments. *Halomonas* sp. strain C2SS100 degraded polluted hydrocarbons and produced BS with high efficiency (Mnif et al., 2009), while *Brevibacterium luteolum* synthesized BS using mineral oil as a carbon source (Vilela et al., 2014).

Fifty-five biosurfactant producers belonging to the genera of *Alcanivorax, Exiguobacterium, Bacillus, Rhodococcus, Acinetobacter, Halomonas, Pseudomonas* and *Streptomyces* were isolated from water and sediment samples of North Atlantic Canada coastal and offshore contaminated sites either by n-hexadecane or diesel as the sole carbon source (Cai et al., 2014). Recently, Moshtagh et al. (2021) studied a marine *Acinetobacter calcoaceticus* P1–1A biosurfactant-producing strain isolated from the North Atlantic Ocean, using cooking oil refined waste as the sole carbon source.

For large-scale BS commercial production, the growth medium's cost may be a smaller contribution in terms of the total production costs and when the fermentation requires energy for a prolonged period, since many renewable plant oils are relatively cheap, readily available and of consistent composition (Tripathi et al., 2018). Therefore, shifting to industrial wastes as a carbon source for biosurfactant production can reduce the cost and also provide a sustainable management approach to wastes.

Various renewable sources for biosurfactant production have been studied and proven effective, such as agro-industrial wastes, frying oils, oil refinery wastes, molasses, starch-rich wastes, cassava wastewater, potato waste and distilled grape marc (Fleurackers, 2006; Zhu et al., 2007; Sobrinho et al., 2008). Crude glycerol has been used widely as a carbon source for biosurfactant production (Sousa et al., 2012). Waste cooking oil has been utilized as a carbon source for biosurfactant production (Luo et al., 2013) Various polar compounds and polymers are produced during the frying process. Adsorption has been used for improving the quality of used frying oil (Asri and Sari, 2015).

EFFECT OF NITROGEN SOURCE

Microbial growth and optimization of BS production depend on the type and concentration of nitrogen sources (Davis et al., 1999). Many types of nitrogen sources could be used for the production of BS: peptone, yeast extract, urea, ammonium nitrate ammonium sulfate and sodium nitrate. The best nitrogen source for the production of the BS is yeast extract using marine *Streptomyces* species B3 (Khopade et al., 2012a).

Yeast extract and sodium nitrate were preferred as nitrogen sources for BS-producing marine *B. subtilis* N3–4P (Zhu et al., 2016), while the maximum production of BS was studied using phenyl alanine as the nitrogen source for marine *Nocardiopsis* B4 (Khopade et al., 2012b). Lately, many reports have found that inorganic tested nitrogen sources: ammonium salts, nitrate, urea and L-amino acids (glutamic acid, aspartic acid, asparagine and glycine) were preferred for maximum biosurfactant production (MacElwee et al., 1990; Abu-Ruwaida et al., 1991; Peypoux et al., 1994; Yakimov et al., 1996).

EFFECT OF TEMPERATURE

The next important parameter is temperature, which forcibly influences BS production. Thermophilic microbes that produce BS are preferred in industrial sections regarding their thermostability (above 40°C); also, mesophiles producing BS have high levels of thermo-stability, followed by psychrophilic marine bacteria capable of producing BSs with potential applications for bioremediation in cold environments. *Rhodococcus* sp. hydrocarbon-degrading marine bacterium, isolated from the Norwegian coastline, produced BS at 20°C with n-hexadecane, kerosene or rapeseed oil as a carbon source (Dang et al., 2016).

EFFECT OF pH

A relatively narrow range of pH and salinity was found in marine environments, while areas of volcanic vents experience extreme conditions. These microorganisms are known to be metabolically and physiologically adapted to live under extreme pH and salinity conditions (Thavasi et al., 2014). The emulsification capacity of the biosurfactant for hexane demonstrated an increasing E24 value over the pH range of 5–12 and an appreciable decrease at pH < 5. The reduction in emulsification activity at low pH scales (<5) is due to the occurrence of precipitation, caused by the consequent insolubility of the biosurfactant produced at these pH values (Rocha et al., 2014).

Increase in emulsion stability with increasing pH, due to NaOH, could result in better stability of fatty acid surfactant micelles and further increase of pH, resulting in the precipitation of secondary metabolites (Khopade et al., 2012b). Extreme pH values could transform weak surface-active species into more active emulsifiers by increasing ionization (Abouseoud et al., 2008). Several studies also reported stability of biosurfactants in the range of pH values of 6 to 12, while for pH values lower than 6, surface tension starts increasing and the sample's turbidity increases due to partial precipitation of the biosurfactant (Al-Bahry et al., 2013; Joshi et al., 2016; Ali et al., 2021, Darwesh et al., 2021).

Based on surfactant nature, three general roles of biosurfactants were listed (Franzetti et al., 2010):

1. The BS increases the surface area of the substrate with hydrophobic water-immiscible growth.
2. The BS solubilizes and promotes the uptake of hydrophobic substrates.
3. The BS regulates the detachment and attachment of microorganisms to and from surfaces.

Therefore, biosurfactants are "green" replacements for synthetic chemical surfactants. However, biosurfactants have not yet been used extensively in industry for many technical and economic reasons.

MARINE BIOSURFACTANT APPLICATIONS

BIOREMEDIATION AND INDUSTRIAL APPLICATIONS

Bioremediation

A crude oil-polluted sea caused by tankers is considered an urgent and serious environmental subject worldwide. Also, working ships produce wastes that are collected in the lowest part of the hull, called the bilge area. This oil-containing bilge waste must be controlled properly to prevent environmental pollution (Olivera et al., 2009).

By using synthetic detergents to clean up this leakage, resulting in a more destructive and polluted environment that can only manage 40–45% of the discharged oil, new technologies are required for enhanced oil recovery (EOR) (Dastgheib et al., 2008). It is important that all substances discharged into the environment be degradable, so we must first evaluate their potential for environmental damage and second put rules in place to safeguard against the possibility of future harm.

Biosurfactants and bioemulsifiers, as mentioned, are new natural molecules considered the most versatile and effective byproduct of modern microbial technology (Perfumo et al., 2010). Many marine proteobacteria secrete cells with surface amphiphilic substances (BSs or BEs) that permit the solubilization of aromatic hydrocarbons. During the growth of oil pollution, microbial cells attach to oil droplets by secreting BS to enhance the bioavailability of hydrocarbons. These bacteria produce BS to increase dispersion of oil hydrocarbons to facilitate their degradation by other microbes, called non-BS microbes (McGenity et al., 2012). Marine *Bacillus subtilis*, *Torulopsis bombicola* and *Pseudomonas aeruginosa* are able to utilize crude oil as a sole carbon source and are used for oil spill clean-ups (Das and Mukherjee, 2007).

Marine *Marinobacter, Halomonas* and *Myroides* and tropical marine yeast *Yarrowia lipolytica* are able to remove hydrocarbon (or oil spill) compounds from contaminated sites by the production of BEs. *Halomonas* sp. also participates in the dispersion of spilled oil by synthesizing surface-active glycolipids and emulsifiers on their cell surface to enhance the solubility of hydrocarbons and increase their digestibility for biodegradation (Dhasayan et al., 2014). A report by Raddadi et al. (2017) showed that *Marinobacter* species produced a phospholipopeptide class of BS capable of emulsifying with low ecotoxicity, then the bioremediation of crude oil in artificial marine water was carried out by dispersion. *Marinobacter hydrocarbonoclasticus* strain SdK644 producing glycolipid BSs showed twofold dissolution of crude oil compared with Tween 80, showing potential bioremediation in marine environments (Zenati et al., 2018). Yansan is a fungal emulsifier produced by *Yarrowia lipolytica*, an aerobic yeast, that shows high emulsification activity and stability for potential applications in the formulation of perfluoro-carbons (PFCs) (Amaral et al., 2006). *Candida lipolytica* yeast-produced biosurfactants were used to formulate a commercial product used in oil bioremediation (Santos et al., 2017).

Marine *Rhodococcus* spp. was capable of reducing the surface tension of oily substrates, where the extracted BSs proved an optimal enhancer of tetradecane and n-hexadecane biodegradation at 13°C (Dang et al., 2016; Malavenda et al., 2015). The study described by Graziano et al. (2016) tested the potential of marine *Brevibacterium* and *Vibrio* spp. strains in the field of bioremediation for being able to utilize diesel oil as a better carbon source and evaluated BS production in the presence of these hydrocarbonic substrates. The bioremediation of hydrocarbon was also deeply investigated by marine *Alcanivorax* sp. A53, *Joostella* sp. A8 and *Pseudomonas* sp. A6 for BS production from diesel oil as carbon source in both pure culture and co-culture conditions, where the biodegradation rates and efficiency were 99.4 and 99.2%, respectively (Rizzo et al., 2018). Furthermore, Mabrouk et al. (2014) reported coral-associated *Bacillus* was able to produce a glycolipidic biosurfactant with a 45% removing capacity as an optimal candidate for oil removal.

The bioremediation potential in terms of chelating activity toward heavy metals using bacterial BSs has scarcely been studied. In fact, many bacteria, such as *Joostella* sp. A8 and *Alcanivorax* sp. A53, have been reported to be able to produce BSs in the presence of heavy metals such as Cu, Cd and Zn (Rizzo et al., 2014, 2018).

Biorefinery Processes

Despite the drawback of environmental pollution, petroleum-based products and crude oil still play an important role in our modern community. However, only 10–40% of the content of oil reservoirs is capable of being recovered using common oil extraction procedures, which leads to new ideas to enhance oil recovery (Patel et al., 2015). Microbially enhanced oil recovery is one concept that utilizes the natural proficiency of certain microorganisms to disperse oil using biosurfactants. This has been carried out by: (i) provoking the reservoir by growing indigenous hydrocarbonoclastic bacteria and supplying additional nitrogen sources; (ii) inoculation of a selected consortia of targeted bacteria into the reservoir; or (iii) addition of ex *situ* products, that is, biosurfactants, to reservoirs (Sen, 2008). Properly, many microbial biosurfactants have been demonstrated to enhance oil recovery (Khire et al., 2010).

Ice formations with molecules of gas trapped inside are called gas hydrates, considered another depository to store energy and carbon (Chong et al., 2016); therefore, current research efforts aim to develop technologies for effective and safe storage for transportation of gas hydrates. It was found that biosurfactants are capable of promoting methane hydrate re-formation and thus improving storage (Arora et al., 2014).

Food Processing

Due to their ability to decrease surface tension, biosurfactants play a vital role in food processing as anti-adhesive agents to promote and improve food formation and stabilization. Also, they are able to control clumps of fat globules, stabilize aerated systems, improve food texture, enhance the shelf life of starch-containing products, modify the viscoelasticity properties of wheat dough and ameliorate the consistency and texture of fat-based products (Krishnaswamy et al., 2008).

Food applications using marine-derived BSs have been studied so far. Emulsan produced from *Acinetobacter calcoaceticus* is a commercialized microbial emulsifier (Nerurkar et al., 2009). This emulsifier can be used as a stabilizing agent in food processing; it can improve texture, consistency and solubilization of fat globules and aroma. Microbial biosurfactants incorporated by food may improve dough rheology, increasing the emulsification and volume of fat to find usefulness in meat processing and the bakery industries (Tripathi et al., 2018). For instance, a marine *Antarctobacter* sp. TG22–produced glycoprotein BS stabilizer can make stable oil-in-water emulsions with commercial food-grade oils (Gutierrez et al., 2007a).

Two marine *Halomonas* species, TG39 and TG67, likely produced glycoprotein emulsifiers showing higher emulsifying properties than their commercial counterparts, with stable activity even in acidic conditions or high temperatures (Gutierrez et al., 2007b). The marine *Nesterenkonia* species produced BS lipopeptide MSA31 with an effective emulsifier and good antioxidant activity and, when added to muffins, it improved softness and kept food quality (Kiran et al., 2017). A bioemulsifier produced by marine bacterium *Enterobacter cloaca* was reported to develop the viscosity of acidic food products (Iyer et al., 2006). Recently, there has been a wide range of marine-derived BSs as stabilizing agents, emulsifiers, antiadhesives and antimicrobial and antibiofilm agents (Patiño et al., 2021).

Cosmetic Industry

Chemically synthesized surfactants cause skin allergies and irritations. There is great interest in using natural cosmetic products among consumers. BSs are used as an alternative to chemically synthesized surfactants in cosmetic products to reduce such harmful effects. BSs are used in foaming agents, emulsifiers, solubilizers, cleansers, wetting agents, antimicrobial agents, bath products, mediators of enzyme action, acne pads, anti-dandruff products, insect repellents, baby products, toothpaste, mascara, dentine cleansers, lipsticks and contact lens solutions (Gharaei-Fathabad, 2011).

The type of BS compound to be incorporated in formulations can be selected based on emulsifying ability and/or surface activity such as hydrophilic–lipophilic balance (HLB) and critical micelle concentration, respectively. A high HLB value indicates that a BS is highly hydrophilic, while a low HLB value shows a highly lipophilic character. Based on HLB values, a BS will be an emulsifier, antifoaming agent and wetting agent, which are desirable properties in cosmetic products. CMC is the minimum concentration of BS required to lower the surface tension of water. At the CMC, surfactant molecules form micelles to reduce surface and interfacial tension. The surface-active properties of a BS are determined by its side chain length, unsaturated bonds and size of hydrophilic groups. With increasing hydrophobicity, the CMC of BS molecules tends to decrease. That means a lower concentration of BS is required for micelle formation (Tripathi et al., 2018). It is reported that the biosurfactant obtained from marine *Nocardiopsis* VITSISB was used in the cosmetic formulation of toothpaste, replacing sodium lauryl sulfate, which is normally used in commercial toothpaste as a surfactant. The results indicate that this biosurfactant is a more efficient and less toxic surfactant compared to the chemical one and could be used in other cosmetic formulations like shampoo, face wash and so on. (Das et al., 2013). Considering the foaming and emulsifying properties, marine BSs can be used for different applications in health care products, including cleansers, moisturizers, toothpaste and personal care products (Vecino et al., 2017; Patiño et al., 2021).

Consumer Products

Regarding their intrinsic surface-active and emulsifying properties, there are applications for biosurfactants as a nature-derived alternative compound to chemical detergents (Van Renterghem et al., 2018). Moreover, BS biodegradability makes them less harmful during environmental discharge. Also, biosurfactants used in such applications are frequently observed by their activities over a broad range of pH, temperature and salinity (Mukherjee and Das, 2010). Cold-adapted microbial species in marine habitats are able to produce biosurfactants that depend on particular interests requiring less energy-demanding and low-temperature applications (Dinamarca et al., 2013; Collins and Margesin, 2019).

Despite all the advantages, the establishment of biosurfactants as a sustainable alternative to inexpensive conventional surfactants is still impeded by the fact that the bulk detergent market is strongly driven by cost-effectiveness. Although biosurfactants currently appear too expensive to compete, a few cleaning products containing glycolipids have been commercialized as niche products and successfully marketed emphasizing sustainability and biodegradability (Kubicki et al., 2019).

Industrial Processes

Industrial processes using biosurfactants are extensively discussed, and one of their uses is cooling. BSs for cooling, cold storage and air conditioning systems may depend on ice slurry, a homogenous mixture of water and small ice particles. However, the particle size is deleterious to flow and equipment. Di-acetylated mannosylerythritol lipids (di-acetylated MEL) BSs were shown to prevent clumping in ice-water slurries by balancing small ice particles and thus inhibiting the bigger crystal formation. At low temperatures, MEL additives were shown to enhance the flow properties of biodiesel and improve its performance (Madihalli et al., 2016).

BIOMEDICAL APPLICATIONS

Antimicrobial Agents

Several biosurfactants exhibit antimicrobial activity against viruses, bacteria, fungi and algae (Krishnaswamy et al., 2008). Following increased pathogenic microorganism resistance to existing drugs, the high demand for new antimicrobial agents has spotlighted biosurfactants (Coates et al., 2011). Lipopeptide and glycolipid bacterial surfactants are fabricated by formation of biofilm or motility processes through colonization on surfaces. The antimicrobial activity of BSs has been examined *in vitro* and *in vivo* showing broad-spectrum activity against Gram-positive and Gram-negative bacteria, fungi, viruses and algae (Vatsa et al., 2010). These biosurfactants have been broadly studied from microorganisms isolated from terrestrial samples or hydrocarbon-polluted areas; however, marine biosurfactant microorganisms have been less investigated (Graziano et al., 2016). BSs of marine origin have been proved active against several bacterial pathogens considering antimicrobial activity including rhamnolipids, flocculosin sophorolipids, bacillomycin, surfactin, iturin, fengycin, pumilacidin, lichenysin, cyclic lipopeptide (i.e., daptomycin and viscosin), mannosyl erythritol, polymyxin B and mycosubtilins (Tripathi et al., 2018). A marine *B. circulans*–produced lipopeptide biosurfactant was active against *Alcaligens faecalis*, *Proteus vulgaris*, methicillin-resistant *Staphylococcus aureus* (MRSA) and other multidrug-resistant pathogenic strains (Das et al., 2008). Indeed, the lipopeptide produced by marine *Nocardiopsis alba* exhibited antibacterial activity against *B. subtilis*, *E. faecalis* and *C. albicans* (Gandhimathi et al., 2009). Different fengycin isoforms in the crude biosurfactant including C15-, C16- and C17-fengycin produced by marine *B. circulans* DMS-2 was responsible for antimicrobial activity (Sivapathasekaran et al., 2009). Sponge-associated *Brachybacterium paraconglomeratum* MSA21 and *Brevibacterium aureum* MSA13 were reported as BS producers with broad antibacterial activity toward pathogenic bacteria and fungi such as *B. subtilis, E. coli, E. faecalis, K. pneumonia, M. luteus, P. aeruginosa, P. mirabilis, Streptococcus* sp., *S. aureus, S. epidermidis* and *C. albicans*, (Kiran et al., 2010a, 2014). *Halobacterium salinarum* also produces BSs that have antibacterial and antifungal activities (Sumaiya et al., 2017).

The mechanism by which BS producers act is quite interesting. Biosurfactants exhibit antimicrobial activity against different pathogenic microbes; additionally, these active materials display anti-biofilm and anti-adhesive activities to reduce the colonization of pathogenic microorganisms on the surface and remove pre-formed biofilms (Rodrigues, 2011). In this regard, Ron and Rosenberg (2001) reported the mechanism of bioemulsifier-producing microorganisms that managed biofilm formation. Interestingly, the biofilm of *P. aeruginosa* ATCC10145 formation was inhibited by coral mucus-associated microbes that seem to be BS producer strains (Padmavathi et al., 2014).

Different marine isolate phylogeny indicates that they are isolated from sponges and other invertebrates considered sources of novel bioactive compounds, including anti-adhesive, antimicrobial and anti-biofilm agents. These active compounds play a vital role in defense against predators and biofilm-forming microorganisms. These BSs are synthesized by symbiotic relations between macro-organisms and their associated microbes (Gudiña et al., 2016). *Nocardiopsis dassonvillei* MAD08 produced biosurfactants more active against *E. coli* and *Staphylococcus epidermidis* compared with chloramphenicol (Selvin et al., 2009). The *Brevibacterium casei* MSA19–associated marine *Dendrilla nigra* sponge produced active antimicrobial glycolipids against *Escherichia coli, Proteus mirabilis, Pseudomonas aeruginosa, Klebsiella pneumoniae, Vibrio parahaemolyticus, Vibrio vulnificus* hemolytic and *Streptococcus* and acted as an anti-biofilm agent against mixed and individual cultures of *E. coli, P. aeruginosa* and *Vibrio* spp., where it removed preformed biofilms at 30 μg mL^{-1} (Kiran et al., 2010c). Marine *Serratia* marcescens biosurfactant exhibited higher inhibitory action against *C. albicans* and *Pseudomonas aeruginosa* compared to the traditional antimicrobials fluconazole and streptomycin, respectively (Dusane et al., 2011). Glycolipids produced by marine *Streptomyces* sp. B3 had activity against *C. albicans, E. coli, P. aeruginosa* and *Staphylococcus aureus* (Khopade et al., 2012a) Similarly, *Streptomyces* sp. MAB36 from marine sediment possessed inhibitory activity against *Aspergillus niger* and *C. albicans* compared with the conventional antifungal nystatin (Manivasagan et al., 2014). The first rhamnolipid-(Rha-Rha-C10-C10) biosurfactant, produced by marine *Streptomyces* sp. ISP2–49E, possessed a broad spectrum of antimicrobial and anti-adhesive activities (Yan et al., 2014). Marine sponge *Acanthella* sp.–associated *Bacillus licheniformis* NIOT-AMKV06 had lipopeptide BSs against *K. pneumoniae, E. faecalis, P. mirabilis, M. luteus, Shigella flexineri, Salmonella typhi, Vibrio cholera* and *S. aureus* (Lawrance et al., 2014). Among the three different lipopeptides produced by marine-derived *B. mojavensis*, the antifungal activity of fengycins (C16 and C17) was stronger than that of mojavensin A (C15) (Ma and Hu, 2014; Ma et al., 2012).

B. circulans biosurfactants of marine origin had forceful antimicrobial action against Gram-positive, Gram-negative and semi-pathogenic microbial strains (MDR strains) (Kügler et al., 2015). These biosurfactants the antagonistic activities against human pathogens were used as an alternative to traditional antibiotics. Unfortunately, despite their great potential, none of these compounds are being used as human infection treatment (Gudiña et al., 2016).

Yuliani et al. (2018) isolated marine *Bacillus subtilis* that produces surfactin lipopeptide biosurfactants that have great potential antimicrobial activity against human pathogenic microbes *Salmonella enterica* typhi, *Escherichia coli, Pseudomonas aeruginosa, Listeria monocytogenes, Staphylococcus aureus* and *Candida albicans*. Another glycolipid BS produced by marine *Staphylococcus saprophyticus* SBPS 15 showed antimicrobial activity against different human clinical isolates (Mani et al., 2016b). Balan et al. (2016) studied the production of a new lipopeptide pontifactin produced by a marine *Pontibacter korlensis* strain SBK-47 that showed different biological activities.

Similarly, a new lipopeptide aneurinifactin produced by marine *Aneurinibacillus aneurinilyticus* SBP-11 showed antimicrobial activities against pathogenic microbes (Balan et al., 2017). Recently, a research work exhibited the most relevant marine-derived microorganisms with BS antimicrobial activities over the last decade and talked about their potential as new agents against catheter-associated urinary tract infections (CAUTIs), providing a prospective proposal for researchers (Zhang et al., 2021).

Minimum inhibitory concentration (MIC) and minimum bactericidal concentration (MBC) are common parameters used to compare antimicrobial activity shown by different naturally producing compounds. Purified lipopeptide fractions produced by *B. circulans* exhibited lower MICs and MBCs of 10–60 μg mL^{-1} when compared with penicillin and streptomycin at 40 and 900 μg mL^{-1} (Das et al., 2008). The same authors exhibited the anti-adhesive activity of *B. circulans* producing partially purified biosurfactant at concentrations between 0.1 and 10 mg mL^{-1}, where microbial adhesion was inhibited with efficiencies between 59% and 94% at the highest BS concentration (Das et al., 2009).

Before establishing these BSs for industrial purposes, it is important to find the toxicity of these surface-active complexes (SACs) on cells or animal models. For instance, glycolipid BS (BS-SLSZ2) produced by marine epizootic bacterium *Staphylococcus lentus* was examined for its toxicity towards eukaryotic model organisms. This glycolipid showed effective inhibition towards the biofilms formed by *Vibrio harveyi* and *P. aeruginosa*. Also, in *in vivo* experiments, BS-SLSZ2 was shown to be non-toxic towards the biomarker *Artemia salina* (Hamza et al., 2017).

In spite of their potential applications, the use of biosurfactants in marketing is still limited due to their low productivities. To increase their production, the optimization of culture conditions is the key role of marine microbial BS yields (Gandhimathi et al., 2009). Furthermore, the components of the culture medium can change the activity and structure of the produced biosurfactant. Different isoforms are produced depending on different culture media. Higher antimicrobial activity was obtained when marine *B. circulans* produced the biosurfactant in culture media containing glycerol, starch or sucrose rather than using a medium containing glucose (Das et al., 2009). However, it is difficult to cultivate marine microorganisms in the lab or in industrial fermenters for massive production of biosurfactants. Therefore, an alternative method for biosurfactant production is using heterologous hosts. In this case, *Bacillus licheniformis* NIOT-AMKV06– produced lipopeptide biosurfactant brought three genes required for the biosynthesis of biosurfactants: sfp, sfpO and srfA, that were cloned and expressed in *E. coli*, so the productivity increased from 3 up to 11.7 g L^{-1} (Lawrance et al., 2014).

Anti-Cancer Agents

Cancer is an enormous important health risk that affects millions of people. Traditional medical chemotherapy uses highly cytotoxic drugs that non-specifically target any dividing cells, resulting in a moderate improvement in patient health; however, most patients' prognosis remains dismal using this non-selective, non-specific and toxic treatment (Siegel et al., 2015).

In this regard, many anti-cancer drugs used for clinical purpose are natural products or derivatives (Bolhassani, 2015). So regular exploration of natural sources, like marine microbiota, will guide researchers to new compounds with fascinating anti-cancer activity (Janakiram et al., 2015). Biosurfactants, in particular glycolipids and lipopeptides, have been featured for their potential for being anti-cancer agents that interfere with processes of cancer progression (Gudiña et al., 2013).

Biosurfactants as anti-cancer agents have been exposed to many intercellular molecular recognition steps: cell differentiation, signal transduction and cell immune response. They also exhibit low toxicity, high efficacy and easy biodegradability. Biosurfactant mechanisms have been suggested: (i) inhibition of crucial signaling pathways; (ii); delay of cell cycle progression (iii) activation of natural killer T (NKT) cells; (iv) reduction of angiogenesis; (v) induction of apoptosis through death receptors in cancer cells; (vi) disruption of cell membranes, leading to lysis, increased membrane permeability and leakage of metabolite (Rodrigues et al., 2006).

Marine lipopeptide surfactins have been widely studied as anti-cancer agents (Sivapathasekaran et al., 2010). Surfactin anti-cancer activity has been related to the hydrophobic moiety that forcibly interacts with the polar heads of cancer cell membrane lipids, then penetrates more efficiently inside the cell membrane (Liu et al., 2010). Different mechanisms have been proposed for surfactin anti-cancer activity (Kim et al., 2007; Cao et al., 2009; Park et al., 2013; Das et al., 2015).

Marine *Bacillus circulans* DMS-2 is able to produce different lipopeptides, fengycin and surfactin isoforms, which display a selective and significant anti-proliferative activity against the colon cancer cell lines HT-29 (IC_{50} 120 gmL^{-1}) and HCT-15 (IC_{50} 80 gmL^{-1}) (Sivapathasekaran et al., 2010). Different marine bacilli strains produce different iturins (iturin A/C bacillomycins and mycosub-tilins, mixirins, hallobacillin), a cyclic peptide amphiphilic molecule. Iturin A, produced by a marine *Bacillus megaterium* strain, which significantly damaged cell proliferation and ceased the Akt signaling network conducting apoptosis in breast cancer cells (MCF-7 and MDA-MB-231), also inhibited epidermal growth factor-induced Akt phosphorylation and its downstream targets FoxO3 and GSK3. In addition, iturin A was able to inhibit tumor growth in a breast cancer xenograft model (Dey et al., 2015). Hallobacillin, produced by marine *Bacillus* sp., exhibited cytotoxic effects against colon cancer cell line HCT-116 (IC_{50} 0.98 gmL^{-1}) (Zhang et al., 2004). Mixirins (A, B and C), produced by a marine *Bacillus* sp. strain, are cyclic octapeptides that have a mixture of L- and D-amino acids and an unusual alkanoic amino acid, considering colon cancer cell cytotoxic activity (Zhang et al., 2004).

Among marine compounds, some new biosurfactant structures have been widely studied: somocystinamide A (Suyama et al., 2008), fellutamides (Lee et al., 2010, 2011), rakicidin (Poulsen, 2011) and apratoxin (Robertson et al., 2012). The e-poly-L-lysine (e-PL) surfactant produced by marine-derived *B. subtilis* SDNS showed anti-cancer activity against the HeLa S3 cell line in humans (El-Sersy et al., 2012).

The significant cytotoxicity of marine biosurfactants was studied against leukemia, lung, breast and prostate cancer cells by IC_{50} values ranging from 1.3 µM to 970 nM depending on the cancer type (Gudiña et al., 2016). Fellutamides C and F produced by an *Aspergillus versicolor*–associated sponge-derived fungus displayed cytotoxicity against XF498 CNC cancer, SK-MEL-2 skin cancer, A549 lung cancer, HCT-15 colon cancer and SK-OV-3 ovarian cancer cell lines, with IC_{50} values ranging from 3.1 to 33.1 µM and between 0.2 and 3.1 µM for fellutamide C and F, respectively. (Lee et al., 2010, 2011).

Marine *Micromonospora* exhibited selective cytotoxicity using rakicidin A as an anti-cancer lipopeptide agent and against many cancer cell lines, such as PANC-1 and HCT-8 (Takeuchi and Rakicidin, 2011). It was reported that rakicidin B from marine *Micromonospora chalcea* FW523–3 was active against lung cancer cells (A549 and 95D), gastric cancer cells (SGC7901), esophageal squamous carcinoma cells (EC109), hepatocellular carcinoma cells (HepG2) (HeLa) and uterine cervix cancer cells (Xie et al., 2011). Short lipid chains of rakicidin derivatives C and D were non-cytotoxic; however, derivative D could alter the invasiveness of aggressive breast cancer cells (Poulsen et al., 2011). Another cyclic lipopeptide BS, "apratoxins," (derivatives A to G) isolated from marine cyanobacteria exhibited significant cytotoxicity against a number of cancer cells. Apratoxin A induced apoptosis through caspase activation and inhibited the IL-6 signaling pathway in human bone osteosarcoma U2OS cells (Nunnery et al., 2010). Robertson et al. (2012) also reported strong cytotoxicity against this cell line using apratoxin derivatives (D, F and G), while apratoxins E, F and G were active against HCT-116 colon cancer cells in a mouse model (Tidgewell et al., 2010).

Glycolipids extracted from marine sponge *Myrmekioderma dendyi*–associated microbial strains have been studied. Myrmekioside (E-1, E-2 and E-3) showed anti-cancer activity toward two human non-small-cell lung cancer cells (A549 and NSCLC-N6). Also, the glycolipid trikentroside, produced by *Trikentrion* sponge-associated strains, inhibited human non-small-cell lung cancer A549 cell proliferation (Farokhi et al., 2012). A lipopeptide BS extracted from the Arctic strain of *P. fluorescens* BD5, pseudofactin II (PFII), was compared with normal human dermal fibroblast (NHDF). PFII prompted melanoma skin cancer cell apoptosis, while NHDF had less effect under the same conditions. BS micelles may increase plasma membrane permeability, resulting in melanoma cell death, where PFII activity was higher above CMC (130–140 lM) (Janek et al., 2013). Many marine biosurfactant antimicrobial and anti-cancer activities need to be discovered to open up opportunities for more developments that will benefit humankind (Patiño et al., 2021).

Anti-Phytopathogenic Agents
In fact, the ongoing increase of the human population means there is a need to assess how to increase agricultural productivity. Marine biosurfactants can furnish this by (i) preventing phytopathogens (biocides and insecticides), (ii) improving beneficial microbe–plant interactions, (iii) upgrading soil and (iv) inducing effective foliar fertilizer uptake. However, there are only rare reports of marine biosurfactants being studied (Vatsa et al., 2010; Mulligan et al., 2014; Liu et al., 2016).

Biosurfactants from marine habitats, in particular, may be of interest due to cold and saline environments, which open up extended application fields (Perfumo et al., 2018). Also, reports on novel biosurfactant-producing isolates are scarce because of temperature or salinity. Therefore, the physicochemical characters of those BSs cannot be necessarily concluded from their source and have to be tested *in vitro*. The development of appropriate production strategies and processes for sufficient amounts of novel marine BS compounds highlights present-day difficulties to supply industries with their requirements. New structural elucidation of the physicochemical properties of such marine novel biosurfactants will lead to more novel applications (Gomes et al., 2016).

CONCLUSION

Recently, there has been great awareness of biosurfactant production from marine microbes, but they are less explored than their terrestrial counterparts. These biosurfactants show some specific properties, and, undoubtedly, their chemical variety is much greater than described to date, and many biosurfactants' structures still remain undisclosed. Marine biosurfactants represent a promising source of novel bioactive compounds for application in multidisciplinary fields, including several industrial and therapeutic areas. Furthermore, marine life is present in heterogenous habitats that should accommodate many marine microbes producing biosurfactants in promising communities, either accompanying filter-feeding invertebrates or sites affected by toxic materials. However, there is a difficulty in isolating and growing these marine microbes, resulting in unexplored marine microbial strains producing BSs. Using more advanced metagenomics establishes a prospective study on the genetic resources of unreachable marine microbes, allowing the development of synthetic biology-derived products to build up efficient strains with recombinant production. All these integrated research efforts will lead to the identification, production and application of novel marine biosurfactants.

REFERENCES

Abdel-Mawgoud, A.M., G. Stephanopoulos. 2018. Simple glycolipids of microbes: Chemistry, biological activity and metabolic engineering. *Synthetic and Systems Biotechnology* 3:3–19.

Abouseoud, M., A. Yataghene, A. Amrane, R. Maachi. 2008. Biosurfactant production by free and alginate entrapped cells of *Pseudomonas fluorescens*. *Journal of Industrial Microbiology and Biotechnology* 35(11):1303–1308.

Abu-Ruwaida, A.S., I.M. Banat, S. Haditirto, A. Salem, M. Kadri. 1991. Isolation of biosurfactant-producing bacteria, product characterization, and evaluation. *Engineering in Life Sciences* 11(4):315–324.

Akubude, V.C. and B.A. Mbab. 2021. Application of biosurfactants in algae cultivation systems. In Chapter 4 — *Green Sustainable Process for Chemical and Environmental Engineering and Science Microbially-Derived Biosurfactants for Improving Sustainability in Industry*. Elsevier: Amsterdam, pp. 97–108. ISBN: 978-0-12-823380-1.

Al-Bahry, S., Y.M. Al-Wahaibi, A. Elshafie, A. Al-Bemani, S. Joshi, H.S. Al-Makhmari, H. Al-Sulaimani. 2013. Biosurfactant production by *Bacillus subtilis* B20 using date molasses and its application in enhanced oil recovery. *International Biodeterioration & Biodegradation* 81:141–146.

Ali, F., S. Das, T.J. Hossain, S.I. Chowdhury, S.A. Zedny, T. Das, M.N.A. Chowdhury, M.S. Uddin. 2021. Production optimization, stability and oil emulsifying potential of biosurfactants from selected bacteria isolated from oil-contaminated sites. *Royal Society Open Science* 8(10):1–14.

Amaral, P.F.F., J.M. da Silva, M. Lehocky, A.M.V. Barros-Timmons, M.A.Z. Coelho, I.M. Marrucho, J.A.P. Coutinho. 2006. Production and characterization of a bioemulsifier from *Yarrowia lipolytica*. *Process BioChemistry* 41:1894–1898.

Anburajan, L., B. Meena, R.V. Raghavan, D. Shridhar, T.C. Joseph, N.V. Vinithkumar, G. Dharani, P.S. Dheenan, R. Kirubagaran. 2015. Heterologous expression, purification, and phylogenetic analysis of oil-degrading biosurfactant biosynthesis genes from the marine sponge-associated *Bacillus licheniformis* NIOT-06. *Bioprocess and Biosystems Engineering* 38:1009–1018.

Antoniou, E., S. Fodelianakis, E. Korkakaki, N. Kalogerakis. 2015. Biosurfactant production from marine hydrocarbon-degrading consortia and pure bacterial strains using crude oil as carbon source. *Frontiers in Microbiology* 6:274.

Arora, A., S.S. Cameotra, R. Kumar, P. Kumar, C. Balomajumder. 2014. Effects of biosurfactants on gas hydrates. *Journal of Petroleum & Environmental Biotechnology* 5:1–7.

Asri, N.P. and D.A.P. Sari. 2015. Pre-treatment of waste frying oils for biodiesel production. *Modern Applied Science* 9(7):99–106.

Bae, I., E.S. Lee, J.W. Yoo, S.H. Lee, J.Y. Ko, Y.J. Kim, T.R. Lee, D.-Y. Kim, C.S. Lee. 2018. Mannosylerythritol lipids inhibit melanogenesis via suppressing ERK-CREB-MiTF-tyrosinase signaling in normal human melanocytes and a three-dimensional human skin equivalent. *Experimental Dermatology*. 28:738–741.

Balan, S.S., C.G. Kumar, S. Jayalakshmi. 2016. Pontifactin, a new lipopeptide biosurfactant produced by a marine *Pontibacter korlensis* strain SBK-47: Purification, characterization and its biological evaluation. *Process Biochemistry* 51:2198–2207.

Balan, S.S., C.G. Kumar, S. Jayalakshmi. 2017. Aneurinifactin, a new lipopeptide biosurfactant produced by a marine *Aneurinibacillus aneurinilyticus* SBP-11 isolated from Gulf of Mannar: Purification, characterization and its biological evaluation. *Microbioligical Research* 194:1–9.

Balan, S.S., G.C. Kumar, S. Jayalakshmi. 2019. Physicochemical, structural and biological evaluation of Cybersan (trigalactomargarate), a new glycolipid biosurfactant produced by marine yeast, *Cyberlindnera saturnus* strain SBPN-27. *Process Biochemistry* 80:171–180.

Banat, I.M., A. Franzetti, I. Gandolfi, G. Bestetti, M.G. Martinotti, L. Fracchia, T.J. Smyth, R. Marchant. 2010. Microbial biosurfactants production, applications and future potential. *Applied Microbiology and Biotechnology* 87:427–444.

Banat, I.M., S.K. Satpute, S.S. Cameotra, R. Patil, N.V. Nyayanit. 2014. Cost effective technologies and renewable substrates for biosurfactants' production. *Frontiers in Microbiology* 5:1–18

Barakat, K.M., S.W.M. Hassan, O.M. Darwesh. 2017. Biosurfactant production by halo-alkaliphilic *Bacillus strains* isolated from Red Sea, Egypt. *Egyptian Journal of Aquatic Research* 43:205–211.

Bhatnagar, A., S. Chinnasamy, M. Singh, K.C. Das. 2011. Renewable biomass production by mixotrophic algae in the presence of various carbon sources and wastewaters. *Applied Energy* 88:3425–3431.

Bolhassani, A. 2015. Cancer chemoprevention by natural carotenoids as an efficient strategy. *Anti-Cancer Agents in Medicinal Chemistry* 15:1026–1031.

Bovio, E., G. Gnavi, V. Prigione, F. Spina, R. Denaro, M. Yakimov, R. Calogero, F. Crisa, G. Cristina. 2017. The culturable mycobiota of a Mediterranean marine site after an oil spill: Isolation, identification and potential application in bioremediation. *Science of the Total Environment* 576:310–318.

Brasileiro, P.P.F., D.G. Almeida, J.M. Luna, R.D. Rufino, V.A. Santos, L.A. Sarubbo. 2015. Optimization of biosurfactant production from *Candida guiliermondii* using a rotate central composed design. *Chemical Engineering Transactions* 43:1411–1416.

Cai, Q., B. Zhang, B. Chen, Z. Zhu, W. Lin, T. Cao. 2014. Screening of biosurfactant producers from petroleum hydrocarbon contaminated sources in cold marine environments. *Marine Pollution Bulletin* 86(1–2):402–410.

Cao, X.H., A.H. Wang, R.Z. Jiao, C.L. Wang, D.Z. Mao, L. Yan, B. Zeng. 2009. Surfactin induces apoptosis and G (2)/M arrest in human breast cancer MCF-7 cells through cell cycle factor regulation. *Cell Biochemistry and Biophysics* 55:163–171.

Celik, P.A., E.B. Manga, A. Cabuk, I.M. Banat. 2020. Biosurfactants' potential role in combating COVID-19 and similar future microbial threats. *Applied Sciences* 11:334.

Chakraborty, J. and S. Das. 2016. Characterization of the metabolic pathway and catabolic gene expression in biphenyl degrading marine bacterium *Pseudomonas aeruginosa* JP-11. *Chemosphere* 144:1706–1714.

Cheng, T., J. Liang, J. He, X. Hu, Z. Ge, J. Liu. 2017. A novel rhamnolipid-producing *Pseudomonas aeruginosa* ZS1 isolate derived from petroleum sludge suitable for bioremediation. *AMB Express* 7:1–14.

Chieregato, M.B., D.A. Laroque, L.M. de Andrade, B.A.M. Carciofi, J.A.S. Tenório, C.J. de Andrade. 2019. Production of active cassava starch films; effect of adding a biosurfactant or synthetic surfactant. *Reactive & Functional Polymers* 144:104368.

Choi, B.-K., H.-S. Lee, J.S. Kang, H.J. Shin, B-K. Choi, H.-S. Lee, J.S. Kang, H.J. Shin. 2019. Dokdolipids A-C, hydroxylated rhamnolipids from the marine-derived Actinomycete *Actinoalloteichus hymeniacidonis*. *Marine Drugs* 17:237.

Chong, Z.R., S.H.B. Yang, P. Babu, P. Linga, X.-S. Li. 2016. Review of natural gas hydrates as an energy resource: Prospects and challenges. *Applied Energy* 162:1633–1652.

Christova, N., B. Tuleva, Z. Lalchev, A. Jordanovac, B. Jordanov. 2004. Rhamnolipid biosurfactants produced by *Renibacterium salmoninarum* 27BN during growth on n-hexadecane. *Zeitschrift für Naturforschung. Section C* 59(1–2):70–74.

Chuo, S.C., A. Ahmad, S.H. Mohd-Setapar, S.F. Mohamed, M. Rafatullah. 2019. Utilization of green sophorolipids biosurfactant in reverse micelle extraction of antibiotics: Kinetic and mass transfer studies. *Journal of Molecular Liquids* 276:225–232.

Cicatiello, P., A.M. Gravagnuolo, G. Gnavi, G.C. Varese, P. Giardina. 2016. Marine fungi as source of new hydrophobins. *International Journal of Biological Macromolecules* 92:1229–1233.

Cicatiello, P., I. Stanzione, P. Dardano, L. De Stefano, L. Birolo, A. De Chiaro, D.M. Monti, G. Petruk, G. D'Errico, P. Giardina. 2019. Characterization of a surface-active protein extracted from a marine strain of *Penicillium chrysogenum*. *International Journal of Biological Macromolecules* 20:3242.

Coates, A.R.M., G. Halls, Y. Hu. 2011. Novel classes of antibiotics or more of the same? *British Journal of Pharmacology* 163:184–194.

Collins, T., R. Margesin. 2019. Psychrophilic lifestyles: Mechanisms of adaptation and biotechnological tools. *Applied Microbiology and Biotechnology* 103: 2857–2871.

Correa, H.T., W.F. Vieira, T.M.A. Pinheiro, V.L. Cardoso, E. Silveira, L.D. Sette, A.P. Junior, U.C. Filho. 2020. L-asparaginase and biosurfactants produced by extremophile yeasts from Antarctic environments. *Industrial Biotechnology* 16:107–116.

Cowan-Ellsberry, C., S. Belanger, P. Dorn, S. Dyer, D. McAvoy, H. Sanderson, D. Versteeg, D. Ferrer, K. Stanton. 2014. Environmental safety of the use of major surfactant classes in North America. *Critical Reviews in Environmental Science and Technology* 44:1893–1993.

Dang, N.P., B. Landfald, N.P. Willassen. 2016. Biological surface-active compounds from marine bacteria. *Environmental Technology* 37(9):1151–1158.

Darwesh, O.M., M.S. Mahmoud, K.M. Barakat, A. Abuellil, M.S. Ahmad. 2021. Improving the bioremediation technology of contaminated wastewater using biosurfactants produced by novel bacillus isolates. *Heliyon* 7(12): e08616.

Das, I., S. Roy, S. Chandni, L. Karthik, G. Kumar, B. Rao. 2013. Biosurfactant from marine actinobacteria and its application in cosmetic formulation of toothpaste. *Der Pharmacia Lettre* 5(5):1–6.

Das, K., A.K. Mukherjee. 2007. Crude petroleum-oil biodegradation efficiency of *Bacillus subtilis* and *Pseudomonas aeruginosa* strains Isolated from a petroleum-oil contaminated soil from North-East India. *Bioresource Technology* 98(7):1339–45.

Das, P., S. Mukherjee, R. Sen. 2008. Antimicrobial potential of a lipopetide biosurfactant derived from a marine *Bacillus circulans*. *Journal of Applied Microbiology* 104:1675–1684.

Das, P., S. Mukherjee, R. Sen. 2009. Antiadhesive action of a marine microbial surfactant. *Colloids and Surfaces B: Biointerfaces* 71:183–186.

Das, P., S. Sarkar, M. Mandal, R. Sen. 2015. Green surfactant of marine origin exerting a cytotoxic effect on cancer cell lines. *RSC Advances* 5:53086–53094.

da Silva, A.F., I.M. Banat, A.J. Giachini, D. Robl. 2021. Fungal biosurfactants, from nature to biotechnological product: Bioprospection, production and potential applications. *Bioprocess and Biosystems Engineering* 44:2003–2034.

Dastgheib, S.M.M., M.A. Amoozegar, E. Elahi, S. Asad, I.M. Banat. 2008. Bioemulsifier production by a halothermophilic *Bacillus* strain with potential applications in microbially enhanced oil recovery. *Biotechnology Letters* 30:263–270.

Davis, D.A., H.C. Lynch, J. Varley. 1999. The production of surfactin in batch culture by *Bacillus subtilis* ATCC 21332 is strongly influenced by the conditions of nitrogen metabolism. *Enzyme and Microbial Technology* 25:322–329.

De-Almeida, D.G., R. d. C. F. Soares Da Silva, J.M. Luna, R.D. Rufino, V.A. Santos, I.M. Banat, L.A. Sarubbo. 2016. Biosurfactants: Promising molecules for petroleum biotechnology advances. *Frontiers in Microbiology* 7:1–14.

Deng, M.C., J. Li, Y.H. Hong, X.M. Xu, W.X. Chen, J.P. Yuan, J. Peng, M. Yi, J.H. Wang. 2016. Characterization of a novel biosurfactant produced by marine hydrocarbon-degrading bacterium *Achromobacter* sp. HZ01. *Journal of Applied Microbiology* 120:889–899.

De Oliveira, A.L., M. Gallo, L. Pazzagli, C.E. Benedetti, G. Cappugi, A. Scala, B. Pantera, A. Spisni, T.A. Pertinhez, D.O. Cicero. 2011. The structure of the elicitor cerato-platanin (CP), the first member of the CP fungal protein family, reveals a double ψβ-barrel fold and carbohydrate binding. *Journal of Biological Chemistry* 286:17560–17568.

Desjardine, K., A. Pereira, H. Wright, T. Matainaho, M. Kelly, R.J. Andersen. 2007. Tauramamide, a lipopeptide antibiotic produced in culture by *Brevibacillus laterosporus* isolated from a marine habitat. *Journal of Natural Products* 70: 1850–1853.

Dey, G., R. Bharti, G. Dhanarajan, S. Das, K.K. Dey, B.N.P. Kumar, R. Sen, M. Mandal. 2015. Marine lipopeptideiturin A inhibits Akt mediated GSK3 and FoxO3a signaling and triggers apoptosis in breast cancer. *Scientific Report* 5:10316.

Dhail, S. 2017. Characterization of biosurfactant produced by *Bacillus* sp. isolated from oil spilled water samples of Arabian sea. *International Journal of Petrochemical Science & Engineering* 2(3):95–97.

Dhasayan, A., G.S. Kiran, J. Selvin. 2014. Production and characterisation of glycolipid biosurfactant by *Halomonas* sp. MB-30 for potential application in enhanced oil recovery. *Applied Biochemistry and Biotechnology* 174(7):2571–2584.

Dhasayan, A., J. Selvin, S. Kiran. 2015. Biosurfactant production from marine bacteria associated with sponge *Callyspongia diffusa*. *Biotechnology* 5:443–454.

Dinamarca, M.A., C. Ibacache-Quiroga, J. Ojeda, J. Troncoso. 2013. Marine microbial biosurfactants: Biological functions and physical properties as the basis for innovations to prevent and treat infectious diseases in aquaculture. In *Microbial Pathogens and Strategies for Combating Them: Science, Technology and Education*, edited by A. Méndez-Vilas. Formatex Research Center: Badajoz, Spain, vol. 2, pp. 1135–1144.

dos Reis, C.B.L.D., L.M.B. Morandini, C.B. Bevilacqua, F. Bublitz, G. Ugalde, M.A. Mazutti, R.J.S. Jacques. 2018. First report of the production of a potent biosurfactant with ß-trehalose by *Fusarium fujikuroi* under optimized conditions of submerged fermentation. *Brazilian Journal of Microbiology* 49(17):185–192.

Du, J., A. Zhang, X. Zhang, X. Si, J. Cao. 2019. Comparative analysis of rhamnolipid congener synthesis in neotype *Pseudomonas aeruginosa* ATCC 10145 and two marine isolates. *Bioresource Technology* 286:121380.

Dusane, D.H., V.S. Pawar, Y.V. Nancharaiah, V.P. Venugopalan, A.R. Kumar, S.S. Zinjarde. 2011. Anti-biofilm potential of a glycolipid surfactant produced by a tropical marine strain of *Serratia marcescens*. *Biofouling* 27:645–654.

El-Sersy, N.A. 2012. Plackett-Burman design to optimize biosurfactant production by marine *Bacillus subtilis* N10. *Romanian Biotechnology Letters* 17(2):7049–7064.

Elshikh, M., S. Funston, A. Chebbi, S. Ahmed, R. Marchant, I.M. Banat. 2017. Rhamnolipids from non-pathogenic *Burkholderia thailandensis* E264: Physicochemical characterization, antimicrobial and antibiofilm efficacy against oral hygiene related pathogens. *New Biotechnology* 36:26–36.

Farokhi, F., G. Wielgosz-Collin, A. Robic, C. Debitus, M. Malleter, C. Roussakis, J.-M. Kornprobst, G. Barnathan. 2012. Antiproliferative activity against human non-small cell lung cancer of two O-alkyl-diglycosylglycerols from the marine sponges *Myrmekioderma dendyi* and *Trikentrion laeve*. *European Journal of Medicinal Chemistry* 49:406–410.

Fei, D., G.-W. Zhou, Z.-Q. Yu, H.-Z. Gang, J.-F. Liu, S.-Z. Yang, R.-Q. Ye, B.-Z. Mu. 2019. Low-toxic and nonirritant biosurfactant surfactin and its performances in detergent formulation. *Journal of Surfactants and Detergents* 23:109–118.

Fiebig, R., D. Schulze, J.-C. Chung, S.-T. Lee. 1997. Biodegradation of polychlorinated biphenyls (PCBs) in the presence of a bioemulsifier produced on sunflower oil. *Biodegradation* 8:67–75.

Fleurackers, S.J.J. 2006. On the use of waste frying oil in the synthesis of sophorolipids. *European Journal of Lipid Science and Technology* 108:5–12

Floris, R., G. Scanu, N. Fois, C. Rizzo, R. Malavenda, N. Spano, L.A. Giudice. 2018. Intestinal bacterial flora of Mediterranean gilthead seabream (*Sparus aurata*, L.) as a novel source of natural surface-active compounds. *Aquaculture Research* 49:1262–1273.

Foley, P., A. Kermanshahi, E.S. Beach, J.B. Zimmerman. 2012. Derivation and synthesis of renewable surfactants. *Chemical Society Reviews* 41:1499–1518.

Franzetti, A., G. Bestetti, P. Caredda, P. La Colla, E. Tamburini. 2008. Surface-active compounds and their role in the access to hydrocarbons in *Gordonia* strains. *FEMS Microbiology Ecology* 63:238–248.

Franzetti, A., E. Tamburini, I.M. Banat. 2010. Application of biological surface active compounds in remediation technologies. In: Sen R (ed) Biosurfactants. *Advances in Experimental Medicine and Biology* 672:121–134.

Frischmann, A., S. Neudl, R. Gaderer, K. Bonazza, S. Zach, S. Gruber, O. Spadiut, G. Friedbacher, H. Grothe, V. Seidl-Seiboth. 2013. Self-assembly at air/water interfaces and carbohydrate binding properties of the small secreted protein EPL1 from the fungus *Trichoderma atroviride*. *Journal of Biological Chemistry* 288:4278–4287.

Gaderer, R. K. Bonazza, V. Seidl-Seiboth. 2014. Cerato-platanins: A fungal protein family with intriguing properties and application potential. *Applied Microbiology and Biotechnology* 98:4795–4803.

Gandhimathi, R., S.G. Kiran, T.A. Hema, J. Selvin, T.R. Raviji, S. Shanmughapriya. 2009. Production and characterization of lipopeptide biosurfactant by a sponge associated marine actinomycetes *Nocardiopsis alba* MSA10. *Bioprocess and Biosystems Engineering* 32:825–835.

Garay, L.A., I.R. Sitepu, T. Cajka, J. Xu, H.E. Teh, J.B. German, Z. Pan, S.R. Dungan, D.E. Block, K.L. Boundy-Mills. 2018. Extracellular fungal polyol lipids: A new class of potential high value lipids. *Biotechnology Advances* 36:397–414.

Gaur, V.K. and N. Manickam. 2021. Microbial biosurfactants: Production and applications in circular bioeconomy. *In Biomass, Biofuels, Biochemicals—Circular
Bioeconomy—Current Status and Future Outlook.* Elsevier: Amsterdam,
pp. 353–378.

Geetha, S.J., I.M. Banat, S.J. Joshi. 2018. Biosurfactants: Production and potential applications in microbial enhanced oil recovery (MEOR). *Biocatalysis and Agricultural
Biotechnology* 14:23–32.

Gharaei-Fathabad, E. 2011. Biosurfactants in pharmaceutical industry: A mini-review.
American Journal of Drug Discovery and Development 1:58–69.

Gomes, N.G.M., R. Dasari, S. Chandra, R. Kiss, A. Kornienko. 2016. Marine invertebrate
metabolites with anticancer activities: Solutions to the supply problem. *Marine
Drugs* 14:98.

Graziano, M., C. Rizzo, L. Michaud, E.M.D. Porporato, E. De Domenico, N. Spano, L.A.
Giudice. 2016. Biosurfactant production by hydrocarbon-degrading *Brevibacterium*
and *Vibrio* isolates from the sea pen *Pteroeides spinosum* (Ellis, 1764). *Journal of
Basic Microbiology* 56:963–974.

Gudiña, E.J., V. Rangarajan, R. Sen, L.R. Rodrigues. 2013. Potential therapeutic applications of biosurfactants. *Trends Pharmacological Science* 34(12):667–75.

Gudiña, E.J., J.A. Teixeira, L.R. Rodrigues. 2016. Biosurfactants produced by marine
microorganisms with therapeutic applications. *Marine Drugs* 14:38.

Guerfali, M., I. Ayadi, N. Mohamed, W. Ayadi, H. Belghith, M.R. Bronze, M.H.L.
Ribeiro, A. Gargouri. 2019. Triacylglycerols accumulation and glycolipids secretion by the oleaginous yeast *Rhodotorula babjevae* Y-SL7 Structural identification
and biotechnological applications. *Bioresource Technology* 273:326–334.

Gutierrez, T., B. Mulloy, C. Bavington, K. Black, D.H. Green. 2007a. Partial purification and chemical characterization of a glycoprotein (putative hydrocolloid) emulsifier produced by a marine bacterium *Antarctobacter. Applied Microbiology
Biotechnology* 76:1017–1026.

Gutierrez, T., B. Mulloy, K. Black, D.H. Green. 2007b. Glycoprotein emulsifiers from
two marine *Halomonas* species: Chemical and physical characterization. *Journal
of Applied Microbiology* 103:1716–1727.

Gutierrez, T., T. Shimmied, C. Haidon, K. Black, D.H. Green. 2008. Emulsifying and metal
ion binding activity of a glycoprotein exopolymer produced by *Pseudoalteromonas*
sp. Strain TG12. *Applied and Environmental Microbiology* 74:4867–4876.

Haba, E., M.J. Espuny, M. Busquets, A. Manresa. 2000. Screening and production of
rhamnolipids by *Pseudomonas aeruginosa* 47T2 NCIB 40044 from waste frying
oils. *Journal of Applied Microbiology* 88:379–387.

Hamza, F., S. Satpute, A. Banpurkar, A.R. Kumar, S. Zinjarde. 2017. Biosurfactant from
a marine bacterium disrupts biofilms of pathogenic bacteria in a tropical aquaculture system. *FEMS Microbiology Ecology* 93(11). http://doi.org/10.1093/femsec/
fix140.

Harimawan, A. and Y.-P. Ting. 2016. Investigation of extracellular polymeric substances
(EPS) properties of *P. aeruginosa* and *B. subtilis* and their role in bacterial adhesion. *Colloids and Surfaces B* 146:459–467.

Hentati, D., A. Chebbi, F. Hadrich, I. Frikha, F. Rabanal, S. Sayadi, A. Manresa, M.
Chamkha. 2019. Production, characterization and biotechnological potential of
lipopeptide biosurfactants from a novel marine *Bacillus stratosphericus* strain
FLU5. *Ecotoxicology and Environmental* 167:441–449.

Ibacache-Quiroga, C., J. Ojeda, G. Espinoza-Vergara, P. Olivero, M. Cuellar, M.A.
Dinamarca. 2013. The hydrocarbon-degrading marine bacterium *Cobetia* sp. strain

MM1IDA2H-1 produces a biosurfactant that interferes with quorum sensing of fish pathogens by signal hijacking. *Microbial Biotechnology* 6:394–405.

Irorere, V.U., T.J. Smyth, D. Cobice, S. McClean, R. Marchant, I.M. Banat. 2018. Fatty acid synthesis pathway provides lipid precursors for rhamnolipid biosynthesis in *Burkholderia thailandensis* E264. *Applied Microbiology and Biotechnology* 102(14):6163–6174.

Irorere, V.U., L. Tripathi, R. Marchant, S. McClean, I.M. Banat. 2017. Microbial rhamnolipid production: A critical re-evaluation of published data and suggested future publication criteria. *Applied Microbiology and Biotechnology* 101:3941–3951.

ITOPF-International Tanker Owners Pollution Federation. 2017. www.itopf.com. Accessed April 2017.

Iyer, A., K. Mody, B. Jha. 2006. Emulsifying properties of a marine bacterial exopolysaccharide. *Enzyme and Microbial Technology* 38:220–222.

Janakiram, N.B., A. Mohammed, C.V. Rao. 2015. Sea cucumbers metabolites as potent anti-cancer agents. *Marine Drugs* 13:2909–2923.

Janek, T., A. Krasowska, A. Radwanska, M. Lukaszewicz. 2013. Lipopeptide biosurfactant pseudofactin II induced apoptosis of melanoma A375 cells by specific interaction with the plasma membrane. *PLoS ONE* 8:e57991.

Joshi, S.J., Y.M. Al-Wahaibi, S.N. Al-Bahry, A.E. Elshafie, A.S. Al-Bemani, A. Al-Bahri, M.S. Al-Mandhari. 2016. Production, characterization, and application of *Bacillus licheniformis* W16 biosurfactant in enhancing oil recovery. *Frontiers in Microbiology* 7(1324).

Kalogerakis, N., J. Arff, I.M. Banat, et al. 2015. The role of environmental biotechnology in exploring, exploiting, monitoring, preserving, protecting and decontaminating the marine environment. *New Biotechnology* 32:157–167.

Karlapudi, A.P., T.C. Venkateswarulu, J. Tammineedi, L. Kanumuri, B.K. Ravuru, V. Dirisala, V.P. Kodali. 2018. Role of biosurfactants in bioremediation of oil pollution-a review. *Petroleum* 4(3):241–249.

Kaya, T., B. Aslım, E. Kariptaş. 2014. Production of biosurfactant by *Pseudomonas* spp. isolated from industrial waste in Turkey. *Turkish Journal of Biology* 38:307–317.

Khire, J.M. 2010. Bacterial biosurfactants, and their role in microbial enhanced oil recovery (MEOR). In *Biosurfactants: Advances in Experimental Medicine and Biology*, edited by R. Sen. Springer: New York, NY, p. 672.

Khopade, A., R. Biao, X. Liu, K.R. Mahadik, L. Zhang, C. Kokare. 2012b. Production and stability studies of the biosurfactant isolated from marine *Nocardiopsis* sp. *B4 Desalination* 285(11).

Khopade, A., B. Ren, X.-Y. Liu, K. Mahadik, L. Zhang, C. Kokare. 2012a. Production and characterization of biosurfactant from marine *Streptomyces* species B3. *Journal of Colloid and Interface Science* 367:311–318.

Kim, S.Y., J.Y. Kim, S.H. Kim, H.J. Bae, H. Yi, S.H. Yoon, B.S. Koo, M. Kwon, J.Y. Cho, C.E. Lee, S. Hong. 2007. Surfactin from *Bacillus subtilis* displays anti-proliferative effect *via* apoptosis induction, cell cycle arrest and survival signaling suppression. *FEBS Letters* 581:865–871.

Kiran, G.S., T.T. Anto, J. Selvin, B. Sabarathnam. 2010a. Optimization and characterization of a new lipopeptide biosurfactant produced by marine *Brevibacterium aureum* MSA13 in solid state culture. *Bioresource Technology* 101:2389–2396.

Kiran, G.S., T.A. Hema, R. Gandhimathi, J. Selvin, T.A. Thomas, T. Rajeetha Ravji, K. Natarajaseenivasan. 2009. Optimization and production of a biosurfactant from the sponge-associated marine fungus *Aspergillus ustus* MSF3. *Colloids and Surfaces B: Biointerfaces* 73:250–256.

Kiran, G.S., S. Priyadharsini, A. Sajayan, G.B. Priyadharsini, N. Poulose, J. Selvin. 2017. Production of lipopeptide biosurfactant by a Marine *Nesterenkonia* sp. and its application in food industry. *Frontiers in Microbiology* 28(8):1138.

Kiran, G.S., B. Sabarathnam, J. Selvin. 2010c. Biofilm disruption potential of a glycolipid biosurfactant from marine *Brevibacterium casei. FEMS Immunology and Medical Microbiology* 59:432–438.

Kiran, G.S., B. Sabarathnam, N. Thajuddin, J. Selvin. 2014. Production of glycolipid biosurfactant from sponge associated marine actinobacterium *Brachybacterium paraconglomeratum*MSA21. *Journal of Surfactants and Detergents* 17:531–542.

Kiran, G.S., T.A. Thomas, J. Selvin. 2010b. Production of a new glycolipid biosurfactant from marine *Nocardiopsis lucentensis* MSA04 in solid-state cultivation. *Colloids Surface B Biointerfaces* 78(1):8–16.

Konishi, M., S. Nishi, T. Fukuoka, D. Kitamoto, T. Watsuji, Y. Nagano, A. Yabuki, S. Nakagawa, Y. Hatada, J. Horiuchi. 2014. Deep-sea *Rhodococcus* sp. BS-15, lacking the phytopathogenic fas genes, produces a novel glucotriose lipid biosurfactant. *Marine Biotechnology* 16:484–493.

Krishnaswamy, M., G. Subbuchettiar, T.K. Ravi, S. Panchaksharam. 2008. Biosurfactants properties, commercial production and application. *Current Science* 94:736–747.

Kubicki, S., A. Bollinger, N. Katzke, K.-E. Jaeger, A. Loeschcke, S. Thies. 2019. Marine biosurfactants: Biosynthesis, structural diversity and biotechnological applications. *Marine Drugs* 17(7):408.

Kügler, J.H., M. Le Roes-Hill, C. Syldatk, R. Hausmann. 2015. Surfactants tailored by the class actinobacteria. *Frontiers in Microbiology* 6(212):1–23.

Kumar, A.K., K. Mody, B. Jha. 2007. Bacterial exopolysaccharides—A perception. *Journal of Basic Microbiology* 47:103–117.

Lawrance, A., M. Balakrishnan, T.C. Joseph, D.P. Sukumaran, V.N. Valsalan, D. Gopal, K. Ramalingam. 2014. Functional and molecular characterization of a lipopeptide surfactant from the marine sponge-associated eubacteria *Bacillus licheniformis* NIOT-AMKV06 of Andaman and Nicobar Islands, India. *Marine Pollution Bulletin* 82:76–85.

Lee, Y.M., H.T. Dang, J. Hong, C.O. Lee, K.S. Bae, D.K. Kim, J.H. Jung. 2010. A cytotoxic lipopeptide from the sponge-derived fungus *Aspergillus versicolor. Bulletin of the Korean Chemical Society* 31:205–208.

Lee, Y.M., H.T. Dang, J. Li, P. Zhang, J. Hong, C.O. Lee, J.H. Jung. 2011. A cytotoxic fellutamide analogue from the sponge-derived fungus *Aspergillus versicolor. Bulletin of the Korean Chemical Society* 32:3817–3820.

Liu, H., B. Shao, X. Long, Y. Yao, Q. Meng. 2016. Foliar penetration enhanced by biosurfactant rhamnolipid. *Colloids and Surfaces B: Biointerfaces* 145:548–554.

Liu, X., X. Tao, A. Zou, S. Yang, L. Zhang, B. Mu. 2010. Effect of the microbial lipopeptide on tumor cell lines: Apoptosis induced by disturbing the fatty acid composition of cell membrane. *Protein Cell* 1:584–594.

Lo, V., J.I. Lai, M. Sunde. 2019. Fungal hydrophobins and their self-assembly into functional nanomaterials. *Advances in Experimental Medicine and Biology* 1174:161–185.

Luepongpattana, S., J. Thaniyavarn, M. Morikawa. 2017. Production of massoia lactone by *Aureobasidium pullulans* YTP6–14 isolated from the Gulf of Thailand and its fragrant biosurfactant properties. *Journal of Applied Microbiology* 123:1488–1497.

Luna, J.M., R.D. Rufino, L. Sarubbo. 2016. Biosurfactant from *Candida sphaerica* UCP0995 exhibiting heavy metal remediation properties. *Process Safety and Environmental Protection* 102:558–566.

Luo, Z., X.-Z. Yuan, H. Zhong, G.-M. Zeng, Z.-F. Liu, X.-L. Ma, Y.-Y. Zhu. 2013. Optimizing rhamnolipid production by *Pseudomonas aeruginosa* ATCC 9027 grown on waste frying oil using response surface method and batch-fed fermentation. *Journal of Central South University* 20:1015–1021.

Ma, Z. and J. Hu. 2014. Production and characterization of iturinic lipopeptides as antifungal agents and biosurfactants produced by a marine *Pinctada martensii*-derived *Bacillus mojavensis* B0621A. *Applied Biochemistry and Biotechnology* 173:705–715.

Ma, Z. and J. Hu. 2018. Plipastatin A1 produced by a marine sediment-derived *Bacillus amyloliquefaciens* SH-B74 contributes to the control of gray mold disease in tomato. *3 Biotech* 8:125.

Ma, Z., N. Wang, J. Hu, S. Wang. 2012. Isolation and characterization of a new iturinic lipopeptide, mojavensin a produced by a marine-derived bacterium *Bacillus mojavensis* B0621A. *Journal of Antibiotics* 65:317–322.

Mabrouk, M.E.M., E.M. Youssif, S.A. Sabry. 2014. Biosurfactant production by a newly isolated soft coral-associated marine *Bacillus* sp. E34: Statistical optimization and characterization. *Life Science Journal* 11(10):756–768.

MacElwee, C.G., H. Lee, J.T. Trevors. 1990. Production of extracellular emulsifying agent by *Pseudomonas aeruginosa* UG 1. *Journal of Industrial Microbiology & Biotechnology* 5:25–32.

Madihalli, C., H. Sudhakar, M. Doble. 2016. Mannosylerythritol lipid-A as a pour point depressant for enhancing the low-temperature fluidity of biodiesel and hydrocarbon fuels. *Energy Fuels* 30:4118–4125.

Malavenda, R., C. Rizzo, L. Michaud, B. Gerçe, V. Bruni, C. Syldatk, R. Hausmann, L.A. Giudice. 2015. Biosurfactant production by Arctic and Antarctic bacteria growing on hydrocarbons. *Polar Biology* 38:1565–1574.

Mani, P., G. Dineshkumar, T. Jayaseelan, K. Deepalakshmi, G.C. Kumar, S.S. Balan. 2016b. Antimicrobial activities of a promising glycolipid biosurfactant from a novel marine *Staphylococcus saprophyticus* SBPS15. *3 Biotech* 6(163):1–9.

Mani, P., P. Sivakumar, S.S. Balan. 2016a. Economic production and oil recovery efficiency of a lipopeptide biosurfactant from a novel marine bacterium *Bacillus simplex*. *Achievements in the Life Sciences* 10(1):102–110.

Manivasagan, P., P. Sivasankar, J. Venkatesan, K. Sivakumar, S.K. Kim. 2014. Optimization, production and characterization of glycolipid biosurfactant from the marine actinobacterium, *Streptomyces* sp. MAB36. *Bioprocess and Biosystems Engineering* 37:783–797.

Mapelli, F., A. Scoma, G. Michoud, F. Aulenta, N. Boon, S. Borin, N. Kalogerakis, D. Daffonchio. 2017. Biotechnologies for marine oil spill cleanup: Indissoluble ties with microorganisms. *Trends Biotechnology* 35:860–887.

Marchant, R. and I.M. Banat. 2012. Biosurfactants: A sustainable replacement for chemical surfactants? *Biotechnology Letters* 34:1597–1605.

Margesin, R. and F. Schinne. 2001. Biodegradation and bioremediation of hydrocarbons in extreme environments. *Applied Microbiology and Biotechnology* 56(5–6):650–663.

Martinho, V., L.M.D-S. Lima, C.A. Barros, V.B. Ferrari, M.R.Z. Passarini, L.A. Santos, F.L. D-S. Sebastianes, P.T. Lacava, S.P. de Vasconcellos. 2019. Enzymatic potential and biosurfactant production by endophytic fungi from mangrove forest in Southeastern Brazil. *AMB Express* 9(1):130.

Marzban, A., G. Ebrahimipour, A. Danesh. 2016. Bioactivity of a novel glycolipid produced by a halophilic *Buttiauxella* sp. and improving submerged fermentation using a response surface method. *Molecules* 21:1256.

McGenity, T.J., B.D. Folwell, B.A. McKew, G.O. Sanni. 2012. Marine crude-oil biodegradation: A central role for interspecies interactions. *Aquatic Biosystems* 8 (10):1–19.

Menon, V., G. Prakash, A. Prabhune, M. Rao. 2010. Biocatalytic approach for the utilization of hemicellulose for ethanol production from agricultural residue using thermostable xylanase and thermotolerant yeast. *Bioresource Technology* 101:5366–5373.

Mnif, S., M. Chamkha, S. Sayadi. 2009. Isolation and characterization of *Halomonas* sp. strain C2SS100, a hydrocarbon-degrading bacterium under hypersaline conditions. *Journal of Applied Microbiology* 107(3):785–794.

Mohanram, R., C. Jagtap, P. Kumar. 2016. Isolation, screening, and characterization of surface-active agent-producing, oil-degrading marine bacteria of Mumbai Harbor. *Marine Pollution Bulletin* 105(1):131–138.

Morikawa, M., M. Ito, T. Imanaka. 1992. Isolation of a new surfactin producer *Bacillus pumilus* A-1, and cloning and nucleotide sequence of the regulator gene, psf-1. *Journal of Fermentation and Bioengineering* 74:255–261.

Moshtagh, B., K. Hawboldt, B. Zhang. 2021. Biosurfactant production by native marine bacteria (*Acinetobacter calcoaceticus* P1–1A) using waste carbon sources: Impact of process conditions Canadian Society for Chemical Engineering. *Canadian Journal of Chemical Engineering* 99:2386–239.

Mujumdar, S., P. Joshi, N. Karve. 2019. Production, characterization, and applications of bioemulsifiers (BE) and biosurfactants (BS) produced by *Acinetobacter* spp.: A review. *Journal of Basic Microbiology* 59:277–287.

Mukherjee, A.K., K. Das. 2010. Microbial surfactants and their potential applications: An overview. In *Biosurfactants. Advances in Experimental Medicine and Biology*, edited by R. Sen. Springer: New York, NY, Vol 672, pp. 54–64.

Mulligan, C.N. 2005. Environmental applications for biosurfactants. *Environmental Pollution* 133:183–198.

Mulligan, C.N., S.K. Sharma, A. Mudhoo, K. Makhijan. 2014. Green chemistry and biosurfactants. In *Biosurfactants: Research Trends and Applications*. 1st ed., edited by C.N. Mulligan, S.K. Sharma, A. Mudhoo. CRC Press: Boca Raton, FL, pp. 1–30.

Nerurkar, A., K.S.H. Ingurao, H. Suthar. 2009. Bioemulsifiers from marine microorganisms *Journal of Scientific and Industrial Research* 68(4):273–277.

Nunnery, J.K., E. Meyers, W.H. Gerwick. 2010. Biologically active secondary metabolites from marine cyanobacteria. *Current Opinion in Biotechnology* 21:787–793.

Olivera, N.L., M.L. Nievas, M. Lozada, G. del Prado, H.M. Dionisi, F. Sineriz, 2009. Isolation and characterization of biosurfactant-producing *Alcanivorax* strains: Hydrocarbon accession strategies and alkane hydroxylase gene analysis. *Research in Microbiology* 160:19–26.

Ongena, M., P. Jacques. 2008. *Bacillus* lipopeptides: Versatile weapons for plant disease biocontrol. *Trends in Microbiology* 16:115–125.

Padmavathi, A.R., S.K. Pandian. 2014. Antibiofilm activity of biosurfactant producing coral associated bacteria isolated from Gulf of Mannar. *Indian Journal of Microbiology* 54:376–382.

Palecek, E., J. Tkac, M. Bartosik, T. Bertok, V. Ostatna, J. Palecek. 2015. Electrochemistry of non-conjugated proteins and glycoproteins; toward sensors for biomedicine and glycomics. *Chemical Reviews* 115(5):2045–2108.

Park, S.Y., J.H. Kim, Y.J. Lee, S.J. Lee, Y. Kim. 2013. Surfactin suppresses TPA-induced breast cancer cell invasion through the inhibition of MMP-9 expression. *International Journal of Oncology* 42:287–296.

Patel, J., S. Borgohain, M. Kumar, V. Rangarajan, P. Somasundaran, R. Sen. 2015. Recent developments in microbial enhanced oil recovery. *Renewable and Sustainable Energy Reviews* 52:1539–1558.

Patiño, A.D., M. Montoya-Giraldo, M. Quintero, L.L. López-Parra, L.M. Blandón, J. Gómez-León. 2021. Dereplication of antimicrobial biosurfactants from marine bacteria using molecular networking. *Scientific Reports* 11(16286):1–12.

Pazzagli, L., G. Cappugi, G. Manao, G. Camici, A. Santini, A. Scala. 1999. Purification, characterization, and amino acid sequence of cerato-platanin, a new phytotoxic protein from *Ceratocystis fimbriata* f. sp. platani. *Journal of Biological Chemistry* 274:24959–24964.

Perfumo, A., I.M. Banat, R. Marchant. 2018. Going green and cold: Biosurfactants from low-temperature environments to biotechnology applications. *Trends Biotechnology* 36:277–289.

Perfumo, A., T.J. Smyth, R. Marchant, I.M. Banat. 2010. Production and roles of biosurfactants and bioemulsifiers in accessing hydrophobic substrates. In: *Handbook of Hydrocarbon and Lipid Microbiology.* Springer: Berlin, pp. 1501–1512.

Peypoux, F., J.M. Bonmatin, H. Labbe, I. Grangemard, B.C. Das, M. Ptak, J. Wallach, G. Michel. 1994. [Ala4] surfactin, a novel isoform from *Bacillus subtilis* studied by mass and NMR spectroscopies. *European Journal of Biochemistry* 224:89–99.

Pitocchi, R., P. Cicatiello, L. Birolo, A. Piscitelli, E. Bovio, G.C. Varese, P. Giardina. 2020. Cerato-platanins from marine fungi as effective protein biosurfactants and bioemulsifiers. *International Journal of Molecular Sciences* 21(8):2913.

Plaza, G., J. Chojniak, K. Rudnicka, K. Paraszkiewicz, P. Bernat. 2015. Detection of biosurfactants in *Bacillus* species: genes and products identification. *Journal of Applied Microbiology* 119:1023–1034.

Plaza, G., I. Zjawiony, I. Banat. 2006. Use of different methods for detection of thermophilic biosurfactant producing bacteria from hydrocarbon contaminated bioremediated soils. *Journal of Petroleum Science and Engineering* 50(1):71–77.

Poulsen, T.B. 2011. A concise route to the macrocyclic core of the rakicidins. *Chemical Communications (Camb.)* 47:12837–12839.

Qiao, N. and Z. Shao. 2010. Isolation and characterization of a novel biosurfactant produced by hydrocarbon-degrading bacterium *Alcanivorax dieselolei* B-5. *Journal of Applied Microbiology* 108:1207–1216.

Quinn, G.A., A.P. Maloy, S. McClean, B. Carney, J.W. Slater. 2012. Lipopeptide biosurfactants from *Paenibacillus polymyxa* inhibit single and mixed species biofilms. *Biofouling* 28:1151–1166.

Raddadi, N., L. Giacomucci, G. Totaro, F. Fava. 2017. *Marinobacter* sp. from marine sediments produces highly stable surface-active agents for combatting marine oil spills. *Microbial Cell Factories* 16:186.

Radmann, E.M., E.G. de Morais, C.F. de Oliveira, K. Zanfonato, J.A.V. Costa. 2015. Microalgae cultivation for biosurfactant production. *African Journal of Microbiology Research.* 9(47):2283–2289.

Rahman, K.S.M., T.J. Rahman, I.M. Banat, R. Lord, G. Street. 2007. Bioremediation of petroleum sludge using bacterial consortium with biosurfactant. In *Environmental Bioremediation Technologies*, edited by S.N. Singh, R.D. Tripathi. Springer: Berlin, pp. 391–408.

Ranjana, M., V.V.E. Ramesh, S.T.G. Babu, D.V.R. Kumar. 2019. Sophorolipid induced hydrothermal synthesis of Cu nanowires and its modulating effect on Cu nanostructure. *Nano-Structures & Nano-Objects* 18:100285.

Rizzo, C., L. Michaud, B. Hörmann, B. Gerçe, C. Syldatk, R. Hausmann, E. De Domenico, A. Lo Giudice. 2013. Bacteria associated with Sabellids (Polychaeta: *Annelida*) as a novel source of surface-active compounds. *Marine Pollution Bulletin* 70:125–133.

Rizzo, C., L. Michaud, C. Syldatk, R. Hausmann, E. De Domenico, A. Lo Giudice. 2014. Influence of salinity and temperature on the activity of biosurfactants by polychaete associated isolates. *Environmental Science and Pollution Research* 21:2988–3004.

Rizzo, C., C. Syldatk, R. Hausmann, B. Gerçe, C. Longo, M. Papale, A. Conte, E. De Domenico, L. Michaud, A. Lo Giudice. 2018. The demosponge *Halichondria* (*Halichondria*) *panacea* (Pallas, 1766) as a novel source of biosurfactant-producing bacteria. *Journal of Basic Microbiology*:1–11.

Robertson, B.D., S.E. Wengryniuk, D.M. Coltart. 2012. Asymmetric total synthesis of apratoxin D. *Organic Letters* 14:5192–5195.

Rocha, M.V.P., J.S. Mendes, M.E.A. Giro, V.M.M. Melo, L.R.B. Gonçalves. 2014. Biosurfactant production by *Pseudomonas aeruginosa* MSIC 02 in cashew apple juice using a 24 full factorial experimental design *Chemical Industry & Chemical Engineering Quarterly* 20:49–58.

Rodrigues, L.R. 2011. Inhibition of bacterial adhesion on medical devices. In *Bacterial Adhesion: Biology, Chemistry, and Physics, Series: Advances in Experimental Medicine and Biology*, edited by D. Linke, A. Goldman. Springer: Berlin, Germany, 715, pp. 351–367.

Rodrigues, L.R., I.M. Banat, J.A. Teixeira, R. Oliveira. 2006. Biosurfactants: Potential applications in medicine. *Journal of Antimicrobial Chemotherapy* 57:609–618.

Rodriguez-Lopez, L., M. Rincon-Fontan, X. Vecino, A.B. Moldes, J.M. Cruz. 2020. Biodegradability study of the biosurfactant contained in a crude extract from corn steep water. *Journal of Surfactants and Detergents* 23:79–90.

Ron, E.Z., E. Rosenberg. 2001. Natural roles of biosurfactants. *Environmental Microbiology* 3:229–236.

Rosenberg, E., E.Z. Ron. 1999. High and low molecular mass microbial surfactants. *Applied Microbiology and Biotechnology* 52:154–162.

Roy, A. 2017. A review on the biosurfactants: Properties, types and its applications. *Journal of Fundamentals of Renewable Energy and Applications* 8(1):1–5.

Saggese, A., R. Culurciello, A. Casillo, M. Corsaro, E. Ricca, L. Baccigalupi. 2018. A marine isolate of *Bacillus pumilus* secretes a pumilacidin active against *Staphylococcus aureus*. *Marine Drugs* 16:180.

Saha, P. and B. Rao. 2017. Biosurfactants—a current perspective on production and applications. *Bioremediation* 16(1):181–188.

Sanches, M.A., I.G. Luzeiro, A.C.A. Cortez, É.S. de Souza, P.M. Albuquerque, H.K. Chopra, J.V.B. de Souza. 2021. Production of biosurfactants by ascomycetes. *International Journal of Microbiology* 2021:6669263.

Santos, D.K.F., A.H.M. Resende, D.G. de Almeida, R. de C.F.S. da Silva, R.D. Rufino, J.M. Luna, I.M. Banat, L.A. Sarubbo. 2017. *Candida lipolytica* UCP0988 biosurfactant: Potential as a bioremediation agent and in formulating a commercial related product. *Frontiers Microbiology* 8:1–11.

Sarubbo, L.A., J.M. Luna, G.M. Campos-Takaki. 2006. Production and stability studies of thebioemulsifier obtained from a new strain of *Candida glabrata* UCP 1002. *Eletronic Journal of Biotechnology* 9:400–406.

Satpute, S.K., I.M. Banat, P.K. Dhakephalkar, A.G. Banpurkar, B.A. Chopade. 2010. Biosurfactants, bioemulsifiers and exopolysaccharides from marine microorganisms. *Biotechnoloy Advances* 28:436–450.

Selvin, J., G. Sathiyanarayanan, A.N. Lipton, N.A. Al-Dhabi, M. Valan Arasu, G.S. Kiran. 2016. Ketide synthase (KS) domain prediction and analysis of iterative type II PKS gene in marine sponge-associated actinobacteria producing biosurfactants and antimicrobial agents. *Frontiers in Microbiology* 7:63.

Selvin, J., S. Shanmughapriya, R. Gandhimathi, G.S. Kiran, T.R. Ravji. 2009. Optimization and production of novel antimicrobial agents from sponge associated marine actinomycetes *Nocardiopsis dassonvillei* MAD08. *Applied Microbiology and Biotechnology* 83:435–445.

Sen, R. 2008. Biotechnology in petroleum recovery: The microbial EOR. *Progress in Energy and Combustion Science* 34:714–724.

Shah, V., and A. Daverey. 2021. Effects of sophorolipids augmentation on the plant growth and phytoremediation of heavy metal contaminated soil. *Journal of Cleaner Production* 280:124406.

Shannaq, M.A.-H. and M.H.M. Isa. 2013. Isolation and molecular identification of surfactin-producing. *Bacillus subtilis.* www.researchgate.net/publication/258517200_2013-421.

Shekhar, S., A. Sundaramanickam, T. Balasubramanian. 2015. Biosurfactant producing microbes and their potential applications: A review. *Critical Reviews in Environmental Science and Technology* 45:1522–1554.

Shekhar, S., A. Sundaramanickam, K. Saranya, M. Meena, S. Kumaresan, T. Balasubramanian. 2019. Production and characterization of biosurfactant by marine bacterium *Pseudomonas stutzeri* (SSASM1). *International Journal of Environmental Science and Technology* 16:4697–4706.

Shoeb, E., N. Ahmed, J. Akhter, U. Badar, K. Siddiqui, F.A. Ansari, M. Waqar, S. Imtiaz, N. Akhtar, Q.A. Shaikh, R. Baig, S. Butt, S. Khan, S. Khan, S. Hussain, B. Ahmed, M.A. Ansari. 2015. Screening and characterization of biosurfactant producing bacteria isolated from the Arabian Sea coast of Karachi. *Turkish Journal of Biology* 39:210–216.

Shoeb, E., F. Akhlaq, U. Badar, J. Akhter, S. Imtiaz. 2013. Classification and industrial applications of biosurfactants. *Academic Research International* 4:243–252.

Siegel, R.L., K.D. Miller, A. Jemal. 2015. Cancer statistics. *Cancer Journal for Clinicians* 65:5–29.

Siegmund, I. and F. Wagner. 1991. New method for detecting rhamnolipids excreted by *Pseudomonas* species during growth on mineral agar. *Biotechnology Techniques* 5:265–268.

Silva, E.J., N.M.P. Rocha e Silva, R.D. Rufino, J.M. Luna, R.O. Silva, L.A. Sarubbo. 2014. Characterization of a biosurfactant produced by *Pseudomonas cepacia* CCT6659 in the presence of industrial wastes and its application in the biodegradation of hydrophobic compounds in soil. *Colloids and Surfaces B* 117:36–41.

Singh, P., Y. Patil, V. Rale. 2018. Biosurfactant production: Emerging trends and promising strategies. *Journal of Applied Microbiology* 126(1):2–13.

Sivapathasekaran, C., P. Das, S. Mukherjee, J. Saravanakumar, M. Mandal, R. Sen. 2010. Marine bacterium derived lipopeptides: Characterization and cytotoxic activity against cancer cell lines. *International Journal for Peptide Research & Therapeutics* 16:215–222.

Sivapathasekaran, C., S. Mukherjee, R. Samanta, R. Sen. 2009. High-performance liquid chromatography purification of biosurfactant isoforms produced by a marine bacterium. *Analytical and Bioanalytical Chemistry* 395:845–854.

Sobrinho, H.B.S., J.M. Luna, R.D. Rufino, A.L.F. Porto, L.A. Sarubbo. 2013. Assessment of toxicity of a biosurfactant from *Candida sphaerica* UCP 0995 cultivated with industrial residues in a bioreactor. *Electronic Journal of Biotechnology* 16(4):1–12.

Sobrinho, H.B.S., R.D. Rufino, J.M. Luna, A.A. Salgueiro, G.M. Campos-Takaki, L.F.C. Leite, L.A. Sarubbo. 2008. Utilization of two agroindustrial by-products for the production of a surfactant by *Candida sphaerica* UCP0995. *Process Biochemistry* 43(9):912–917.

Soccol, C.R., A. Pandey, C. Larroche. 2013. *Fermentation Processes Engineering in the Food Industry.* Taylor and Francis group, Broken Sound Parkway Northwest, CRC Press: Boca Raton, FL, p. 510.

Son, S., S.-K. Ko, M. Jang, J. Kim, G. Kim, J. Lee, E. Jeon, Y. Futamura, I.-J. Ryoo, J.-S. Lee, et al. 2016. New cyclic lipopeptides of the iturin class produced by saltern-derived *Bacillus* sp. KCB14S006. *Marine Drugs* 14(4):72.

Sousa, M., V.M.M. Melo, S. Rodrigues, H.B. Sant'ana, L.R.B. Gonçalves. 2012. Screening of biosurfactant-producing *Bacillus* strains using glycerol from the biodiesel synthesis as main carbon source. *Bioprocess and Biosystems Engineering* 35:897–906.

Sumaiya, M., A.C. Devi, K. Leela. 2017. A study on biosurfactant production from marine bacteria. *International Journal of Scientific and Research Publications* 7(12):139–145.

Sunde, M., C.L.L. Pham, A.H. Kwan. 2017. Molecular characteristics and biological functions of surface-active and surfactant proteins. *Annual Review of Biochemistry* 86:585–608.

Suyama, T.L. and W.H. Gerwick. 2008. Stereospecific total synthesis of somocystin-amide A. *Organic Letters* 10:4449–4452.

Tadros, T. 2005. Adsorption of surfactants at the air/liquid and liquid/liquid interfaces. In *Applied Surfactants: Principles and Applications,* edited by T.F. Tadros. Wiley-VCH Verlag GmbH & Co. KGaA: Weinheim, pp. 73–84. ISBN: 3-527-30629-373.

Takeuchi, M.I. 2011. Rakicidin A effectively induces apoptosis in hypoxia adapted Bcr-Abl positive leukemic cells. *Cancer Science* 102:591–596.

Tan, D., Y.S. Xue, G. Aibaidula, G.Q. Chen. 2011. Unsterile and continuous production of polyhydroxybutyrate by *Halomonas* TD01. *Bioresource Technology* 102(17):8130–8136.

Tedesco, P., I. Maida, P.F. Esposito, E. Tortorella, K. Subko, C. Ezeofor, Y. Zhang, J. Tabudravu, M. Jaspars, R. Fani, D. de Pascale. 2016. Antimicrobial activity of monoramnholipids produced by bacterial strains isolated from the Ross Sea (Antarctica). *Marine Drugs* 14(5):83.

Thavasi, R., S. Jayalakshmi, I.M. Banat. 2014. Biosurfactants and bioemulsifiers from marine sources. In: *Biosurfactants,* edited by C.N. Mulligan, S.K. Sharma, A. Mudhoo. Hardback—CRC Press: Boca Raton, pp. 125–146.

Thornburg, C.C., E.S. Cowley, J. Sikorska, L.A. Shaala, J.E. Ishmael, D.T. Youssef, K.L. McPhail. 2013. Apratoxin H and apratoxin A sulfoxide from the Red Sea cyanobacterium *Moorea producens. Journal of Natural Products* 76(9):1781–1788.

Tidgewell, K., N. Engene, T. Byrum, J. Media, T. Doi, F.A. Valeriote, W.H. Gerwick. 2010. Evolved diversification of a modular natural product pathway: Apratoxins F and G, two cytotoxic cyclic depsipeptides from a Palmyra collection of *Lyngbya bouillonii. ChemBioChem* 11:1458–1466.

Tripathi, L., V.U. Irorere, R. Marchant, I.M. Banat. 2018. Marine derived biosurfactants: A vast potential future resource. *Biotechnology Letters* 40:1441–1457.

Twigg, M.S., L. Tripathi, A. Zompra, K. Salek, V.U. Irorere, T. Gutierrez, G.A. Spyroulias, R. Marchant, I.M. Banat. 2018. Identification and characterisation of short chain rhamnolipid production in a previously uninvestigated, non-pathogenic marine pseudomonad. *Applied Microbiology and Biotechnology* 102:8537–8549.

Twigg, M.S., L. Tripathi, A. Zompra, K. Salek, V.U. Irorere, T. Gutierrez, G. Spyroulias, R. Marchant, I. Banat. 2019. Surfactants from the sea: Rhamnolipid production by marine bacteria. *Microbiology Society* 1(1A). https://doi.org/10.1099/acmi.ac2019. po0066

Unás, J.H., D.D.-A. Santos, E.B. Azevedo, M. Nitschke. 2018. *Brevibacterium luteolum* biosurfactant: Production and structural characterization. *Biocatalysis and Agricultural Biotechnology* 13:160–167.

Uzoigwe, C., J.G. Burgess, C.J. Ennis, P.K.S.M. Rahman. 2015. Bioemulsifiers are not biosurfactants and require different screening approaches. *Frontiers in Microbiology* 6:1–6.

Vandana, P. and J.K. Peter. 2014. Peter production, partial purification and characterization of biosurfactant produced by *Pseudomonas fluorescens*. *International Journal of Advanced Technology in Engineering and Science* 2(7):258–264.

Van Renterghem, L., S.L.K.W. Roelants, N. Baccile, K. Uyttersprot, M.C. Taelman, B. Everaert, S. Mincke, S. Ledegen, S. Debrouwer, K. Scholtens, C. Stevens, W. Soetaert. 2018. From lab to market: An integrated bioprocess design approach for new-to-nature biosurfactants produced by *Starmerella bombicola*. *Biotechnology and Bioengineering* 115:1195–1206.

Varjani, S.J. and V.N. Upasani. 2017. Critical review on biosurfactant analysis, purification and characterization using rhamnolipid as a model biosurfactant. *Bioresource Technology* 232:389–397.

Vatsa, P., L. Sanchez, C. Clement, F. Baillieul, S. Dorey. 2010. Rhamnolipid biosurfactants as new players in animal and plant defense against microbes. *International Journal of Molecular Sciences* 11(12):5095–5108.

Vecino, X., J.M. Cruz, A.B. Moldes, L.R. Rodrigues. 2017. Biosurfactants in cosmetic formulations: Trends and challenges. *Critical Reviews in Biotechnology* 37:911–923.

Vilela, W.F.D., S.G. Fonseca, F. Fantinatti-Garboggini, M. Nitschke. 2014. Production and properties of a surface-active lipopeptide produced by a new marine *Brevibacterium luteolum* strain. *Applied Biochemistry and Biotechnology* 174(6):2245–2256.

Walter, V., C. Syldatk, R. Hausmann. 2010. Screening concepts for the isolation of biosurfactant producing microorganisms. *Advances in Experimental Medicine and Biology* 672:1–13.

Wang, H. 2011. *Solution and Interfacial Characterization of Biosurfactant Rhamnolipid, presentation College of Science and the College of Medicine Tucson*, Ph.D. in Chemistry and Biochemistry, The University of Arizona. https://repository.arizona.edu/handle/10150/193778

White, D.A., L.C. Hird, S.T. Ali. 2013. Production and characterization of a trehalolipid biosurfactant produced by the novel marine bacterium *Rhodococcus* sp. strain PML026. *Journal of Applied Microbiology* 115:744–755.

Wrasidlo, W., A. Mielgo, V.A. Torres, S. Barbero, K. Stoletov, T.L. Suyama, R.L. Klemke, W.H. Gerwick, D.A. Carson, D.G. Stupack. 2008. The marine lipopeptide somocystinamide a triggers apoptosis *via* caspase 8. *Proceedings of the National Academy of Sciences* 105:2313–2318.

Wu, S., G. Liu, S. Zhou, Z. Sha, C. Sun, S. Wu, G. Liu, S. Zhou, Z. Sha, C. Sun. 2019. Characterization of antifungal lipopeptide biosurfactants produced by marine bacterium *Bacillus* sp. CS30. *Marine Drugs* 17(4):199.

Xia, W.J., Z.F. Du, Q.F. Cui, H. Dong, F.Y. Wang, P.Q. He, Y.-C. Tang. 2014. Biosurfactant produced by novel *Pseudomonas* sp. WJ6 with biodegradation of n-alkanes and polycyclic aromatic hydrocarbons. *Journal of Hazardous Materials* 276:489–498.

Xie, J.J., F. Zhou, E.M. Li, H. Jiang, Z.P. Du, R. Lin, D.S. Fang, L.Y. Xu. 2011. FW523–3, a novel lipopeptide compound, induces apoptosis in cancer cells. *Molecular Medicine Reports* 4:759–763.

Xu, B.-H., Z.-W. Ye, Q.-W. Zheng, T. Wei, J.-F. Lin, L.-Q. Guo. 2018. Isolation and characterization of cyclic lipopeptides with broad-spectrum antimicrobial activity from *Bacillus siamensis* JFL15. *3 Biotech* 8:444.

Xu, M., X. Fu, Y. Gao, L. Duan, C. Xu, W. Sun, Y. Li, X. Meng, X. Xiao. 2020. Characterization of biosurfactant-producing bacteria isolated from Marine environment: Surface activity, chemical characterization and biodegradation. *Journal of Environmental Chemical Engineering* 8(5):104277.

Xu, Y., R.D. Kersten, S-J. Nam, L. Lu, A.M. Al-Suwailem, H. Zheng, W. Fenical, P.C. Dorrestein, B.S. Moore, P.-Y. Qian. 2012. Bacterial biosynthesis and maturation of the didemn in anti-cancer agents. *Journal of the American Chemical Society* 134:8625–8632.

Yakimov, M.M., H.L. Fredrickson, K.N. Timmis. 1996. Effect of hydrophobic moieties on surface activity of Lichenysin A, a lipopeptide biosurfactant from *Bacillus licheniformis* BAS50. *Applied Biochemistry and Biotechnology* 23:13–18.

Yakimov, M.M., K.N. Timmis, V. Wray, H.L. Fredrickson. 1995. Characterization of a new lipopeptide surfactant produced by thermotolerant and halotolerant subsurface *Bacillus licheniformis* Bas50. *Applied and Environmental Microbiology* 61:1706–1713.

Yan, X., J. Sims, B. Wang, M.T. Hamann. 2014. Marine actinomycete *Streptomyces* sp. ISP2–49E, a new source of rhamnolipid. *Biochemical Systematics and Ecology* 55:292–295.

Ye, M., M. Sun, J. Wan, Y. Feng, Y. Zhao, D. Tian, F. Hu, X. Jiang. 2016. Feasibility of lettuce cultivation in sophoroliplid-enhanced washed soil originally polluted with Cd, antibiotics, and antibiotic-resistant genes. *Ecotoxicology and Environmental Safety* 124:344–350.

Yuliani, H., M.S. Perdani, I. Savitri, M. Manurung, M. Sahlan, A. Wijanarko, H. Hermansyah. 2018. Antimicrobial activity of biosurfactant derived from *Bacillus subtilis* C19. *Energy Procedia* 153:274–278.

Zenati, B., A. Chebbi, A. Badis, K. Eddouaouda, H. Boutoumi, M. El Hattab, D. Hentati, M. Chelbi, S. Sayadi, M. Chamkha, A. Franzetti. 2018. A non-toxic microbial surfactant from *Marinobacter hydrocarbono-clasticus* SdK644 for crude oil solubilization enhancement. *Ecotoxicology and Environmental Safety* 154:100–107.

Zhang, H.L., H.M. Hua, Y.H. Pei, X.S. Yao. 2004. Three new cytotoxic cyclic acylpeptides from marine *Bacillus* sp. *Chemical and Pharmaceutical Bulletin* 52:1029–1030.

Zhang, S., X. Liang, G.M. Gadd, Q. Zhao. 2021. Marine microbial-derived antibiotics and biosurfactants as potential new agents against catheter-associated urinary tract infections. *Marine Drugs* 19:1–21.

Zhu, Y., G. Jun-Jiang, G.L. Zhang, J. Bin, W.-J.M. Qin. 2007. Reuse of waste frying oil for production of rhamnolipids using *Pseudomonas aeruginosa* zju.u1M. *Journal of Zhejiang University-SCIENCE A* 8:1514–1520.

Zhu, Z., B. Zhang, B. Chen, Q. Cai, W. Lin. 2016. Biosurfactant production by marine-originated bacteria *Bacillus subtilis* and its application for crude oil removal. *Water Air and Soil Pollution* 227(9):328.

3 Marine Bacteria Surfactants

Bioremediation and Production Aspects

L. Blandón, A. Zuleta-Correa, M. Quintero,
E.L. Otero-Tejada and J. Gómez-León

CONTENTS

DOI: 10.1201/9781003307464-3

INTRODUCTION

Uncontrolled and increased discharge of recalcitrant toxic compounds into the environment poses severe threats to ecosystem conservation and human health (Samanta, Singh, and Jain 2002; Häder et al. 2020). Anthropogenic activities such as mining, agriculture, and hydrocarbon extraction, among others, generate large amounts of xenobiotic compounds or release naturally occurring elements in their more mobile forms, causing known or suspected adverse effects in receiving environments. Greener production practices to reduce pollution generation and detoxification of such compounds have consequently been considered top priorities aligned with international commitments to promote conservation and restoration of terrestrial and marine habitats. Among environmentally friendly approaches to remove and degrade a wide diversity of contaminants, bioremediation techniques have gathered considerable attention in the past years. Bioremediation consists of a wide arrange of biotechnological processes that use whole cells—or their components—to recover a polluted environment by degrading or transforming recalcitrant toxic compounds into less toxic or non-toxic forms (Shukla et al. 2014; Azubuike, Chikere, and Okpokwasili 2016). Changes in toxicity or conversion into inert substances are achieved through pollutant incorporation into microbial metabolic activities as sources of macronutrients (i.e., carbon, nitrogen, etc.) or micronutrients (i.e., copper, iron, etc.) (Gadd 2010). Moreover, microorganisms can develop tolerance to the contaminants, adapting their genomic machinery as part of their defense mechanism, which researchers may exploit for increased detoxification or degradation capacities (Shukla et al. 2014).

Today, there is an increased interest in studying microorganisms from underexplored locations like marine environments for their application in bioremediation. These ecosystems are frequently considered extreme due to critical conditions like high atmospheric pressure, high salinity, nutrient scarcity, and various temperature scales depending on the depth. The successful growth and survival of marine microorganisms in these conditions result from multiple physiological and biochemical adaptations that lead to the production of attractive bioactive and stable compounds (de Carvalho and Fernandes 2010). Among metabolites of industrial importance are biosurfactants: molecules of renewed interest due to their wide applications in different industries (i.e., food, pharmaceutical, cosmetic, etc.) and bioremediation processes (Bonugli-Santos et al. 2015; Dalmaso, Ferreira, and Vermelho 2015; Velásquez et al. 2018). Biosurfactants are molecules of amphiphilic nature produced by microorganisms to increase their chances of survival and proliferation. They are highly efficient in solubilizing hydrophobic organic compounds facilitating their uptake (bioavailability) by cells (Santos et al. 2016). Besides, they can also induce changes in the cellular membranes, increasing microbial adherence (biofilm formation)(Ławniczak, Marecik, and Chrzanowski 2013). Moreover, in marine ecosystems, biosurfactants are also involved in natural selection (antimicrobial activity) and cell-to-cell communication (quorum sensing) (Ibacache-Quiroga et al. 2013).

Despite the existence of practical low-cost surfactants of chemical origin, biosurfactants have more attractive features such as degradability, flexibility, and low toxicity (Xu et al. 2020b). This environmental friendliness of biosurfactants explains the current interest in exploring high-yielding microorganisms, like marine bacteria, and developing cost-effective biosurfactant production processes. This chapter offers an overview of the principal biosurfactant-producing marine bacteria, the structural characteristics of produced molecules, culture conditions implemented during biosurfactant production by marine strains, and biosurfactant applications on bioremediation of different pollutants.

BIOSURFACTANT-PRODUCING MARINE BACTERIA

In marine ecosystems, microbial communities and biosurfactant-producing strains can be found in diverse locations such as the water column, sediments, and forming symbiotic associations with macroorganisms (Ward and Bora 2006). Biosurfactant-producing indigenous bacteria are frequently found in sediments and seawater contaminated with hydrocarbons or other insoluble substrates (Santos-Gandelman et al. 2014; Blandón et al. 2022; Domingues et al. 2020; Ravindran et al. 2020; Lee et al. 2018). This is not surprising since they are expected to thrive in those polluted environments and one of the primary roles of biosurfactants is solubilizing hydrophobic organic compounds to increase their mobility and accessibility into the cell (Adetunji and Olaniran 2021). In a general context, biosurfactant-producing marine bacteria have been isolated from uncontaminated seawater and sediments (54%), followed by contaminated seawater and sediments (20%), and in lesser proportion from deep-sea sediments (14%), marine invertebrates (8%), and estuarine environments (4%) (Figure 3.1). The isolated bacteria from those environments belong to diverse genera, with *Bacillus* and *Pseudomonas* the most widely reported (Santos et al. 2016). Members of the phyla Firmicutes, Proteobacteria, and Actinobacteria are the main producers reported in the last decade (Figure 3.1).

Biosurfactants produced by *Bacillus* and *Pseudomonas* have been studied to a greater extent than those made by other bacterial genera, specifically some cyclic lipopeptides derived from the *Bacillus* genus known as surfactin, iturin, and fengycin (Ohadi et al. 2018) (see "Classification, Chemical Isolation, and

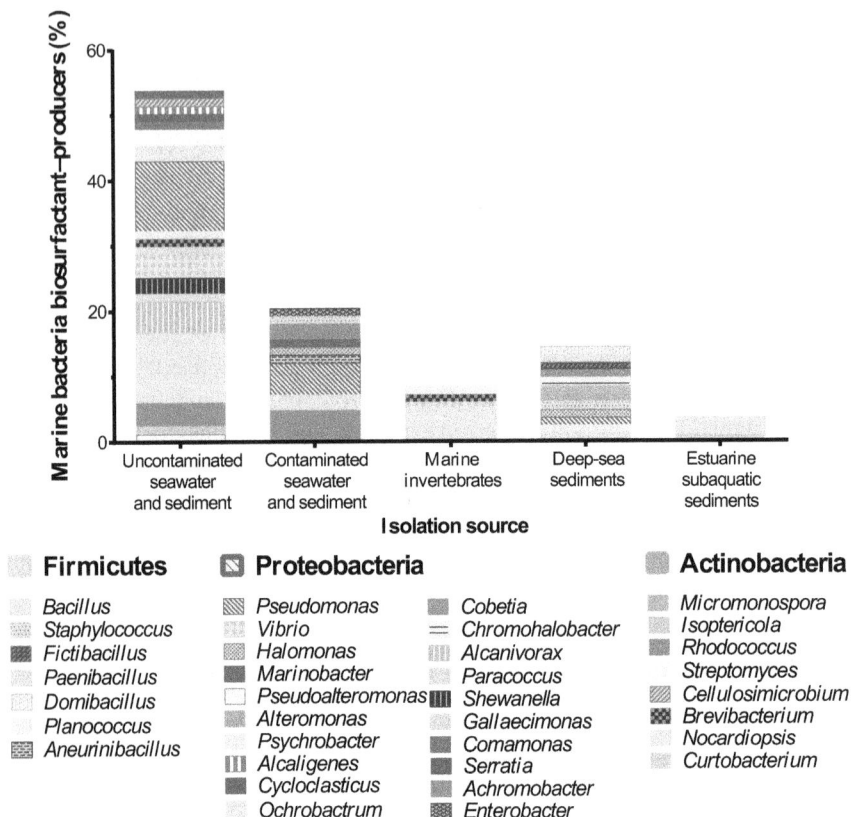

FIGURE 3.1 Percentages of biosurfactant-producing marine bacteria grouped according to their isolation source.

Principal Characterization Techniques of Biosurfactants"). Some members of *Bacillus* genus recently reported as lipopeptide producers are *B. methylotrophicus* UCP1616, *B. cereus* UCP 1615, *Bacillus* sp. MSI 54 (Chaprão et al. 2018; Ravindran et al. 2020; Durval et al. 2020), and *Bacillus velezensis* MHNK1 (Jakinala et al. 2019). On the other hand, rhamnolipids are the best-known biosurfactant representatives of the *Pseudomonas* genera (Kaczorek et al. 2018; Jakinala et al. 2019).

Another bacterium recently reported as a biosurfactant producer is *Marinobacter hydrocarbonoclasticus* SdK644, a hydrocarbonoclastic strain isolated from hydrocarbon-contaminated sediments (Zenati et al. 2018). These bacteria are characterized by using hydrocarbons exclusively as growth substrates (Cafaro et al. 2013). The marine sponge-associated *Planococcus* sp. MMD26 strain (Hema et al. 2019) and Actinobacteria strains like *Streptomyces* sp. B3 and *Streptomyces* sp. MAB36 isolated from marine sediments are also producers of glycolipids (see "Classification, Chemical Isolation, and Principal Characterization

Techniques of Biosurfactants") (Khopade et al. 2012; Manivasagan et al. 2014). Other reports have found rhamnolipid producers isolated from the water column belonging to the genera *Pseudoalteromonas, Alcanivorax, Rhodococcus, Isoptericola, Bacillus,* and *Nocardiopsis* (Antoniou et al. 2015; Lee et al. 2018; Roy et al. 2015; Shekhar et al. 2019; Chen et al. 2021). Additionally, *Rhodococcus* sp. PML026 and *Paracoccus marcusii* ESP-A have been reported as producers of trehalolipids and sophorolipids, respectively (White, Hird, and Ali 2013; Antoniou et al. 2015). Lipopeptide production has been also reported by the genera *Aneurinibacillus* (Wu et al. 2019), *Achromobacter* (Deng et al. 2016), *Rhodococcus, Alcaligenes, Cellulosimicrobium* (Yalaoui-Guella et al. 2020), *Enterobacter* (Muneeswari et al. 2022), *Halomonas* (Cheffi et al. 2020), and *Brevibacterium* (Vilela et al. 2014). Additionally, the production of lichenysin and iturine by members of the *Staphylococcus* genus was recently reported (Hentati et al. 2021).

Different reports evaluating biosurfactant production from deep-sea bacteria included members of the phyla Proteobacteria, Actinobacteria, and Firmicutes. A recent study carried out by Blandón et al. (2022) isolated 20 strains with the capacity to produce biosurfactants from the Colombian Caribbean Sea. Molecular analysis of those culturable strains revealed they belonged to the genera *Alteromonas* sp., *Pseudomonas* sp., *Chromohalobacter* sp., *Cobetia* sp., *Halomonas* sp., *Micromonospora* sp., *Bacillus* sp., *Paenibacillus* sp., *Domibacillus* sp., and *Fictibacillus* sp. On the other hand, Domingues et al. (2020) isolated six strains from deep-sea sediments from the Gulf of Cadiz (Western Mediterranean) that showed biosurfactant activity even under conditions of oxygen limitation, which belonged to the genera *Curtobacterium* sp. R33, *Ochrobactrum* sp. (R98 and R114), *Psychrobacter* sp. (DS104), and *Staphylococcus* sp. (DS140).

BIOSURFACTANT DETECTION METHODS

Exploring new sources for the isolation of marine bacteria is a current challenge due to the limited accessibility of some locations and difficulties in simulating oceanic nutritional culture requirements in a laboratory setting (Tripathi et al. 2018). Nevertheless, the strains that can be isolated and grown under laboratory conditions are invaluable sources of desired metabolites and represent novel metabolisms and genomes to explore.

The most frequently used culture media for screening for biosurfactant-producing strains include the artificial seawater mineral salt medium (ONR7) (Antoniou et al. 2015; Hassanshahian 2014), the mineral salt medium (MSM) (Lee et al. 2018; Muneeswari et al. 2022), the Bushnell Hass broth (BH) (Shekhar et al. 2019; Blandón et al. 2022; Bushnell and Haas 1941; El-Sheshtawy et al. 2014), nutrient broth (NB) (Hema et al. 2019; Cheffi et al. 2020), and artificial seawater (ASW) (Khopade et al. 2012; Vilela et al. 2014). These basal media are usually supplemented with carbohydrates, insoluble substrates, or different industrial wastes as carbon sources to promote rapid growth and biosurfactant production (Santos et al. 2016; Nurfarahin, Mohamed, and Phang 2018). Changes in culture media and physicochemical parameters can influence biosurfactant yield (Banat et al. 2014) (see "Culture Conditions for Biotechnological Production of Biosurfactants"). Therefore, screening of producing strains is generally based on the interfacial activity of the liquid culture supernatant under different growing conditions.

Once growth is achieved in the laboratory, different strategies are implemented to detect if the microorganisms produce biosurfactant molecules and compare the extent of that production under the studied growing conditions. The most common reported methodologies for detecting extracellular biosurfactant production include 1) oil-spreading assay, 2) emulsification index, 3) surface tension measurements, and 4) drop collapsing. Other detection methods reported in the literature include the hemolysis assay, cetyltrimethylammonium bromide-methylene blue agar, and the stalagmometric method. However, they are considered less reliable for detecting biosurfactant production due to disadvantages such as high frequency of false positive or false negatives, high variability in the results, and toxicity (Walter, Syldatk, and Hausmann 2013). The procedures of the most applied detection methods are briefly detailed in the following.

Oil-Spreading Assay

The oil-spreading assay is a rapid and easy method for biosurfactant detection developed by Morikawa, Hirata, and Imanaka (2000). For this assay, 20 µL of crude oil is added to 50 mL of distilled water in a Petri plate. Then 10 µL of liquid culture supernatant is added to the center of the oil layer until an oil displacement is observed. The clearing zone due to oil displacement indicates the presence of biosurfactants in the liquid culture supernatant (Satpute, Banat et al. 2010) (Figure 3.2).

Emulsification Index

Mixing an immiscible liquid into another liquid in the form of droplets is a property of emulsifiers; these can reduce the surface tension until the solubility of the insoluble substrate increases (Peele, Ch, and Kodali 2016). In the emulsification index assay, kerosene or oil/hydrocarbon is mixed with the liquid culture supernatant (1:2 v/v) and vortexed at high speed for 2 minutes (Satpute, Banpurkar et al. 2010). After 24 hours, the emulsification index (E_{24}) is calculated as relation

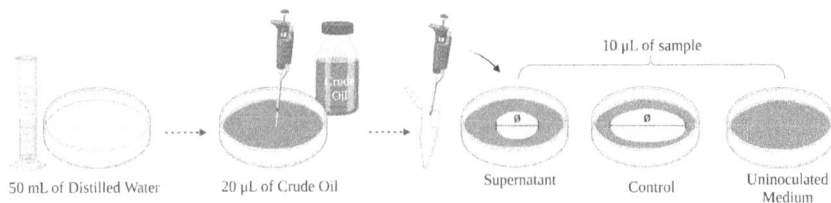

FIGURE 3.2   Oil-spreading assay description. Created with BioRender.com.

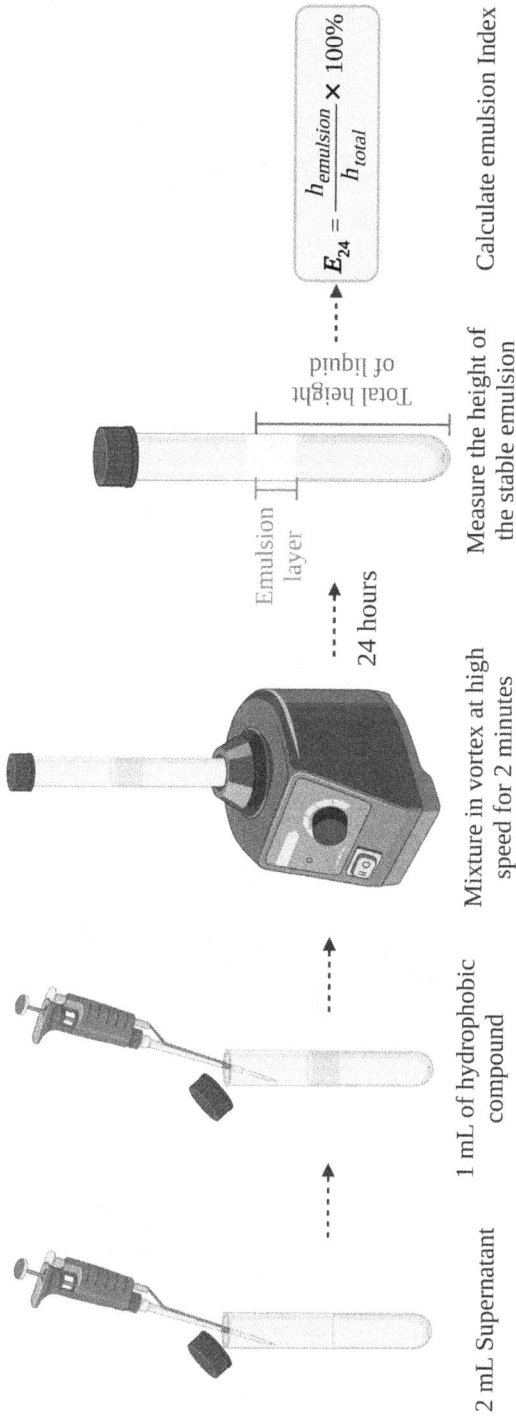

$$E_{24} = \frac{h_{emulsion}}{h_{total}} \times 100\%$$

Calculate emulsion Index

Total height of liquid

Emulsion layer

24 hours

Measure the height of the stable emulsion

Mixture in vortex at high speed for 2 minutes

1 mL of hydrophobic compound

2 mL Supernatant

FIGURE 3.3 Emulsification index assay description. Created with BioRender.com.

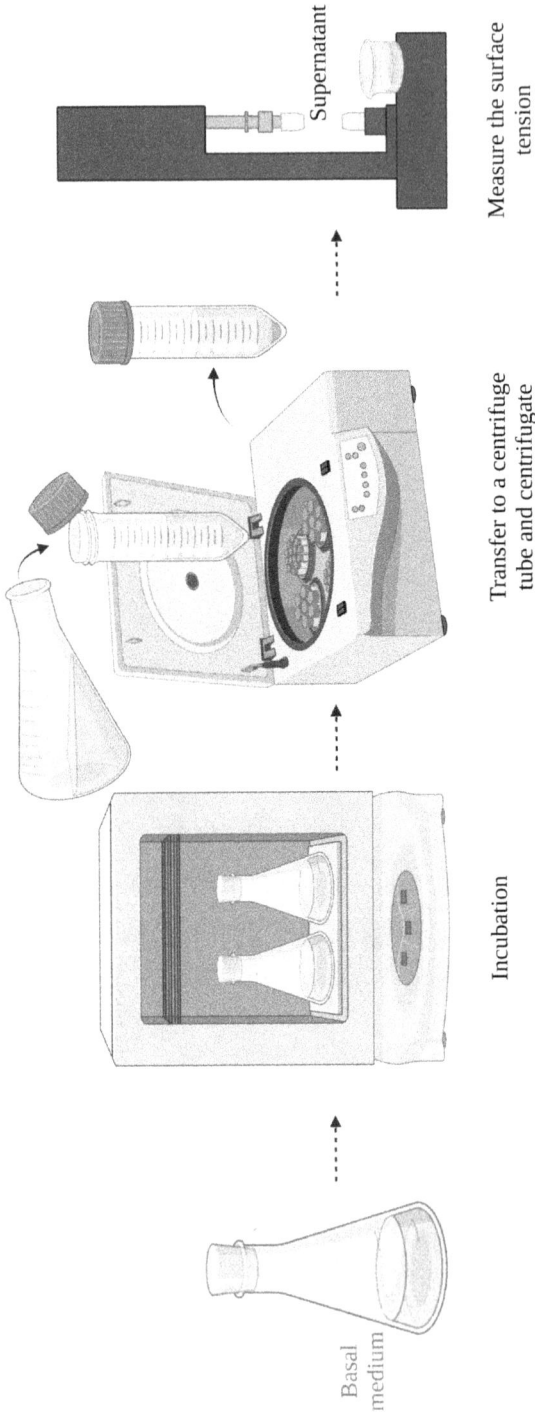

FIGURE 3.4 Surface tension measurements description. Created with BioRender.com.

between the height of the emulsion layer and the total height of the liquid (Walter, Syldatk, and Hausmann 2013) (Figure 3.3).

Surface Tension Measurements

For this screening assay, a piece of automated equipment known as tensiometer is required to measure the force to detach a ring or loop of wire from an interface or surface. Surface tension values of ≤40 mN/m indicate good biosurfactant activity (Walter, Syldatk, and Hausmann 2013). Measurements can be done using the cell-free liquid culture supernatant or serial dilutions of biosurfactant extracts or solutions. This measure is crucial to obtain the critical micelle concentration (CMC) (Esmaeili et al. 2021; Guo et al. 2022), which will be discussed in the next section. Factors such as pH, ionic strength, and the presence of oils or insoluble substrates can strongly affect the measurements; in the latter case, removing oils is recommended before any measures are taken (Walter, Syldatk, and Hausmann 2013) (Figure 3.4).

Drop Collapsing

Drop collapsing is a rapid method for biosurfactant-producing screening developed by Jain et al. (1991). For this assay, a water drop is placed on an oil-coated solid surface; then a drop of liquid culture supernatant or of cell suspension is added. If the water drop collapses, it is an indication of biosurfactant production by the cultured microorganism (Satpute, Banpurkar et al. 2010) (Figure 3.5).

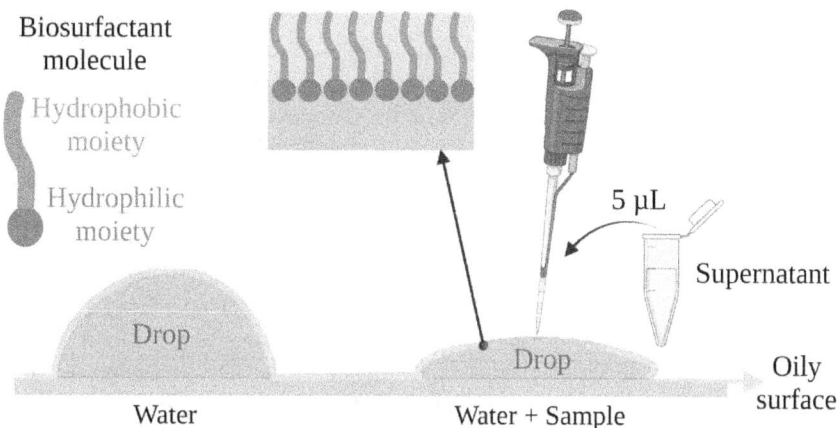

FIGURE 3.5 Drop-collapsing method description. Created with BioRender.com.

CLASSIFICATION, CHEMICAL ISOLATION, AND PRINCIPAL CHARACTERIZATION TECHNIQUES OF BIOSURFACTANTS

CLASSIFICATION OF BIOSURFACTANTS

Biosurfactants are mainly categorized by their mode of action, microbial origin, and molecular weight. Peptides, amino acids, monosaccharides, disaccharides, or polysaccharides are the main components of the hydrophilic head, whereas the hydrophobic tail is mainly composed of fatty acids (Drakontis and Amin 2020). Low molecular weight surfactants are free fatty acids, phospholipids, lipopeptides, and glycolipids; they are more effective in reducing interfacial tension at the oil–water interface as well as surface tension at the air–water interface. High molecular weight polymers are proteins, amphipathic polysaccharides, lipopolysaccharides, lipoproteins, or complex mixtures of these biopolymers, and they are more effective in the stabilization of oil in water emulsions; that is why they are also called bioemulsifiers (Satpute, Banpurkar et al. 2010; Drakontis and Amin 2020). On the other hand, the general classification of biosurfactants is based on the parent chemical structure, and their surface properties. The major classes of biosurfactants are: 1) glycolipids, 2) phospholipids and fatty acids, 3) lipopeptide or lipoproteins, and 4) polymeric surfactants (Palashpriya Das, Mukherjee, and Sen 2008a).

Glycolipids

Glycolipids consist of individual sugars (monosaccharides) or a short-chain sugars (oligosaccharides) linked to a lipid. Typical sugars forming the hydrophilic part of glycolipids are glucose, mannose, galactose, glucuronic acid, or rhamnose. On the other hand, the hydrophobic portion of glycolipids consists of saturated or unsaturated fatty acids, hydroxyl fatty acids, or fatty alcohols. The best-explored groups are sophorolipids, mannosylerythritol lipids, trehalose lipids, and rhamnolipids (Kubicki et al. 2019); the last two are widely reported as produced by bacteria.

Trehalose lipids are composed of the disaccharide trehalose combined with fatty acid groups. Their hydrophobic moieties are diverse and include aliphatic acids and hydroxylated branched-chain fatty acids (mycolic acids) of varying chain lengths. These glycolipids have been mainly derived from Actinobacteria like *Mycobacterium*, *Arthrobacter*, and *Rhodococcus* species (Kügler et al. 2015) (Figure 3.6).

Rhamnolipids are the most-investigated biosurfactants to date, particularly those produced by different strains of *Pseudomonas aeruginosa*; however, they have also been reported by bacteria from the genera *Burkholderia* (Abdel-Mawgoud, Lépine, and Déziel 2010). They comprise two forms, monorhamnolipid and di-rhamnolipid. Monorhamnolipids are formed by one rhamnose molecule and two hydroxyl fatty acid molecules. On the other hand, in di-rhamnolipid, the rhamnose molecules are linked to two molecules of hydroxyl fatty acids of varying chain lengths from 8 to 14, of which β-hydroxydecanoic acid is predominant (Nitschke, Costa, and Contiero 2005) (Figure 3.7).

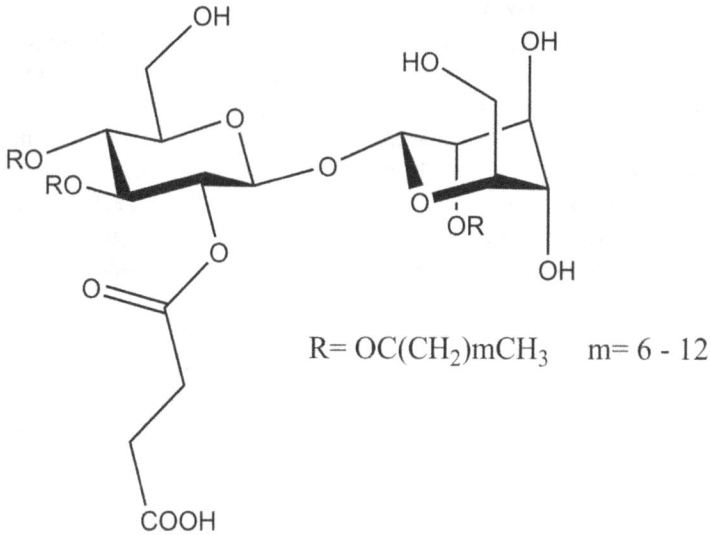

FIGURE 3.6 Structure of the anionic trehalose tetraester produced by *Arthrobacter* sp. (Adapted from: Franzetti et al. 2010).

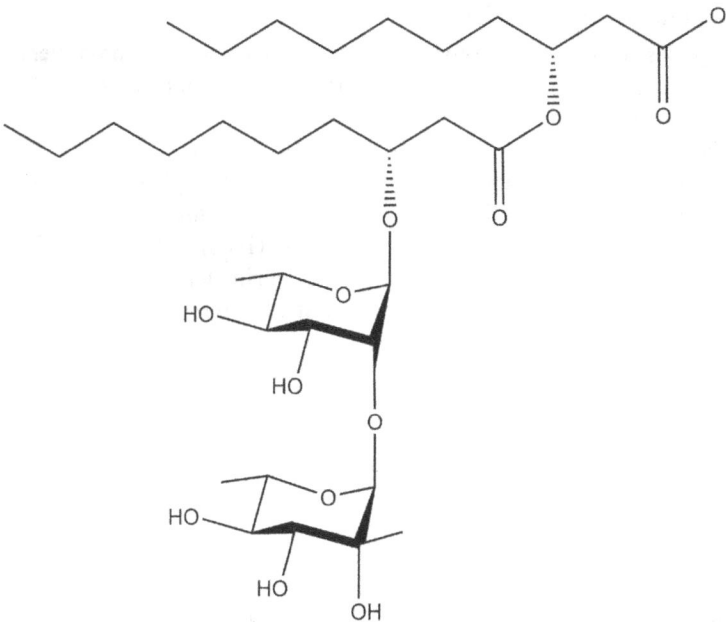

FIGURE 3.7 Structure of di-rhamnolipid from *Pseudomonas aeruginosa* (adapted from Kubicki et al. 2019).

Phospholipids and Fatty Acids

Fatty acids are hydrophobic compounds consisting of carboxylic acid with an aliphatic chain. Phospholipids are amphipathic molecules containing a hydrophilic polar phosphate head and a hydrophobic tail composed of two fatty acids. Phospholipids are known as major components of microbial membranes and play crucial roles as permeability barriers, affecting the transportation of nutrients and other molecules in and out of the cell. It has been reported that when certain hydrocarbon-degrading bacteria or yeast are grown on hydrocarbons, their level of phospholipid synthesis increases considerably (Konwar 2022). For instance, phosphatidylethanolamine produced by *Rhodococcus erythropolis* grown on n-alkane resulted in a decrease of interfacial tension between water and hexadecane to less than 1 m/Nm and critical micelle concentration (CMC) of 30 mg/L (Hatha, Edward, and Pattanathu Rahman 2007; Kretschmer, Bock, and Wagner 1982). Fatty acids produced from alkanes due to microbial oxidations have been considered surfactants; they are saturated fatty acids in the range of C_{12} to C_{14}. Complex fatty acids containing hydroxyl groups, alkyl branches (MacDonald and Chandler 1981; Konwar 2022), or fatty acids bound to proteinogenic or non-proteinogenic amino acids form lipoamino acid biosurfactants (Figure 3.8). Examples of such molecules are ornithine lipids, lysine lipids, and N-acyltyrosines (Kishimoto et al. 1993; Maneerat et al. 2006; Kubicki et al. 2019).

Lipopeptides or Lipoproteins

Lipopeptides consist of a lipid connected to a peptide; they are a prominent group of low molecular weight biosurfactants derived from amino acids. The lipopeptides known as surfactin, iturin, and fengycin are commonly produced by bacteria from the *Bacillus* genus (Ohadi et al. 2018). Surfactin is one of the most effective biosurfactants known so far; it is a cyclic lipopeptide and is composed of seven amino-acid rings joined to a fatty acid chain by means of lactone linkage (Figure 3.9). It was first reported by Arima et al. (1968) in *B. subtilis*. Surfactin was initially identified as a potent inhibitor of fibrin clotting and later found to lyse erythrocytes, protoplasts, and spheroplasts. It lowers the surface tension of

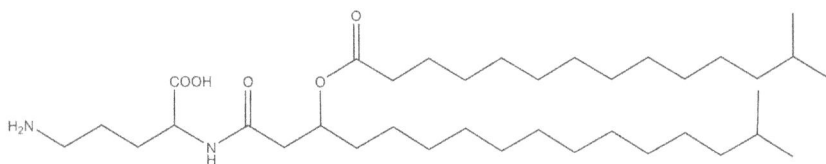

FIGURE 3.8 Structure of ornithine lipid from *Myroydes* sp. strain SM1 (adapted from Maneerat et al. 2006).

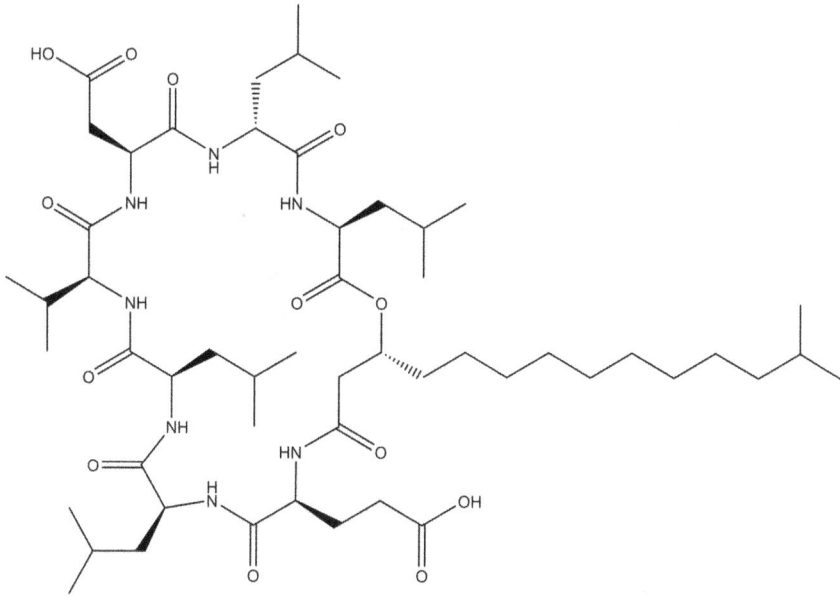

FIGURE 3.9 Structure of surfactin from *Bacillus subtilis* (adapted from Sharma et al. 2020).

water from 72 to 27 mN/m, making it a very powerful biosurfactant (Shaligram and Singhal 2010).

Polymeric

Polymeric surfactants are macromolecules with hydrophobic and hydrophilic groups in their structural makeup. They constitute polysaccharides, proteins, lipopolysaccharides, lipoproteins, or complex mixtures of these compounds referred to as lipoheteropolysaccharides (Kubicki et al. 2019). Emulsan is the most common polymeric biosurfactant produced by marine bacteria; its production has been reported by the marine bacteria *Acinetobacter calcoaceticus* RAG-1 isolated from the Mediterranean Sea (Sar and Rosenberg 1983). This consists of a heteropolysaccharide backbone with a repeating tri-saccharide unit (probably composed of N-acetyl-d-galactosamine, N-acetylgalactosamine uronic acid, and a di-amino-6-deoxy-d-glucose with fatty acids covalently linked) (Figure 3.10).

CHEMICAL ISOLATION

Researchers have employed different kinds of procedures for extracting biosurfactants. The most commonly used approaches include acid precipitation and

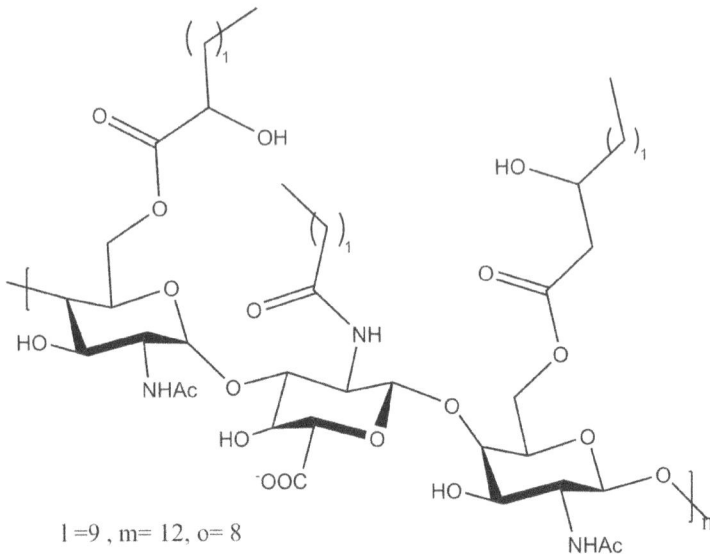

FIGURE 3.10   Structure of emulsan (adapted from Romsted 2014).

solvent extraction, methodologies briefly described in the following (Shekhar et al. 2019; Patiño et al. 2021).

Acid precipitation

a. After biosurfactant production, remove cells from culture broth by centrifuging.
b. Acidify collected supernatant using hydrochloric acid solution to bring the culture pH to 2.0, allowing it to settle for precipitation formation overnight.
c. Finally, centrifuge to obtain a precipitate.

Solvent extraction

a. Remove cells by centrifuging and acidify the collected supernatant using hydrochloric acid solution to bring the culture pH to 2.0, allowing it to settle for precipitation formation overnight.
b. Centrifuge to obtain a precipitate.
c. Wash the precipitate with HCl, and suspend it in distillate water and adjust the pH at 8.0.
d. Freeze-dry the suspension.
e. Finally, extract the biosurfactant with a mix of chloroform: methanol (2:1) and allow the two layers to separate in the separating funnel.
f. Add approximately 0.5 g of magnesium sulfate per 100 ml of chloroform: methanol (2:1) portion, filter, and evaporate under reduced pressure to obtain a solid.

TABLE 3.1
Biosurfactant Extraction and Characterization Methods Recently Reported

Extraction Method	Biosurfactant Type	Yield (g/L)	Characterization Method	Reference
Precipitation with chilled acetone + solvent extraction	Mixture of fatty acids	N/A	FT-IR, GC-MS, ^{1}H ^{13}C-NMR	(Ibacache-Quiroga et al. 2013)
Acid precipitation + solvent extraction (CHCl$_3$/MeOH 2:1)	Lipopeptide	10	FT-IR, ^{1}H ^{13}C NMR	(Chaprão et al. 2018)
Solvent extraction (acetate:/methanol 4:1)	phospholipid	N/A	TLC, FT-IR	(Zhou et al. 2021)
Solvent extraction (methyl tert-butyl ether)	Trehalose lipids	1,4	Orcinol method, TLC, FT-IR, LC-MS	(White, Hird, and Ali 2013)
Solvent extraction (CHCl$_3$/MeOH 1:1)	ornithine lipids	0,0124	FT-IR, MS, ^{1}H ^{13}C–NMR	(Maneerat et al. 2006)

BIOSURFACTANT CHARACTERIZATION TECHNIQUES

Once an extract or a fraction rich in biosurfactants is obtained or a purified bio-surfactant is separated, further analyses are carried out to establish structural features and quantify production. Techniques such as infrared spectroscopy (FTIR), nuclear magnetic resonance (NMR), and liquid chromatography with tandem mass spectrometry (LC-MS/MS) offer significant information about the produced compound, such as the presence of impurities, spatial conformation, and so on. Depending on the selected technique, these analyses can be completed with the intact compound or by breaking down the molecule into its components (i.e., fatty acid and carbohydrate moiety) (Soberón-Chávez 2010). Various extraction methods and technologies used to characterize biosurfactants are shown in Table 3.1.

CULTURE CONDITIONS FOR BIOTECHNOLOGICAL PRODUCTION OF BIOSURFACTANTS

Recent interest in biosurfactant production is a result of their biodegradability, effectiveness, and low-toxicity features that promote their use for bioremediation purposes and implementation in other industries (Xu et al. 2020; Alemán-Vega et al. 2020). Despite significant advances in biosurfactant-producing strain isolation and characterization, process optimization and downstream purification are still in their infancy, hindering a cost-effective large-scale production of these molecules (Ray, Kumar, and Banerjee 2022). Strategies to increase the competitiveness of biotechnological approaches include the implementation of sustainable, low-cost substrates (e.g., agro-industrial residues, oil wastes, molasses, etc.)

instead of expensive growth media components during the fermentation process (Vigneshwaran et al. 2021). Other reports have focused on assessing the effects of physicochemical parameters on bacterial biosurfactant yields using either one-factor-at-a-time methods (Prakash et al. 2021; Vigneshwaran et al. 2018; Ali et al. 2021) or a more efficient statistical experimental design that simultaneously analyzes multiple factors (Nalini et al. 2021). The latter includes response surface methodology approaches that allow finding the statistical significance of variables, the interaction among factors, and optimal conditions for an optimal response (e.g., production) on the evaluated range. Regardless of the selected methodology, common factors evaluated during optimization of biosurfactant production are pH, temperature, and medium components (mainly types and concentrations of carbon and nitrogen sources). Salinity, often referred to as NaCl concentration, has also been assessed during product maximization trials. The reasoning behind these experiments is that biosurfactants are growth-associated products; therefore, maximized growth will result in increased product yield. Some recent findings on the impact of culture conditions on biosurfactant production are discussed in the following.

EFFECTS OF NITROGEN AND CARBON SOURCE ON BIOSURFACTANT PRODUCTION

The selection of nitrogen and carbon source type and concentration plays an essential role in increasing the production of different kinds of biosurfactants by many bacteria (Vigneshwaran et al. 2018). Indeed, the carbon to nitrogen ratio (C/N) has been proposed by some scientists as a critical factor directing microbial metabolism, including biosurfactant synthesis (Khopade et al. 2012). However, others argue that selecting proper macronutrient sources, rather than carbon/nitrogen ratios, is more crucial for biosurfactant production (Unás et al. 2018). Fortunately, there is a general agreement that both macronutrients play key roles in cell development and as constituents of the biosurfactant molecule, resulting in multiple reports evaluating numerous carbon and nitrogen sources to find the optimal ones for their respective microorganism.

Recent reports of biosurfactant production with marine-derived bacteria use various substrates as carbon sources, with the most relevant medium component selected based on screening experiments or statistical design results. Pure sugars, mainly glucose, are widely implemented as carbon sources for marine bacteria. For instance, *Planococcus sp.* MMD26 showed improved biosurfactant production using glucose as a carbon source compared to other substrates such as paddy straw, olive oil, glycerol, and kerosene (Hema et al. 2019). Conversely, *Vibrio* sp. 3B-2 did not produce biosurfactant when growing on glucose or sucrose, but it efficiently consumed other sugars, namely lactose, maltose, and xylose (Hu, Wang, and Wang 2015). Marine *Streptomyces youssoufiensis* SNSAA03 could also metabolize different sugars as carbon sources; however, glucose at 15 g/L was the most suitable for maximizing product yield, with further concentration increases negatively impacting production (Nalini et al. 2021). Similarly,

Moshtagh, Hawboldt, and Zhang (2021) found a maximum carbon source concentration during the evaluation of seven different substrates for *Acinetobacter*
calcoaceticus P1–1A; their results showed crude glycerol as the poorest substrate
for biosurfactant generation, while refined waste cooking oil represented a cost-
effective alternative for increasing product yield. Other medium optimization
studies have found engine oil (Nayak et al. 2020), starch (Javee, Karuppan, and
Subramani 2020), karanja oil (Deepika et al. 2016), olive oil (Kim et al. 2019),
frying oil (Durval et al. 2019), vegetable oil (RamyaDevi et al. 2018), shrimp
shell waste (Kadam and Savan 2019), glycerol (Dhasayan, Selvin, and Kiran
2015), molasses (Mabrouk, Youssif, and Sabry 2014), and peanut oil cake + corn
oil (Ekramul and Saqib 2020) the most appropriate carbon sources for biosurfactant production by marine strains of *Bacillus licheniformis* LRK1, *Streptomyces*
sp. SNJASM6, *Pseudomonas aeruginosa* KVD-HR42, *Rhodococcus fascians*
SDRB-G7, *B. cereus* BCS0, *P. guguanensis*, *P. stutzeri*, *B. amyloliquefaciens*
MB-101, *Bacillus* sp.E34, and *P. aeruginosa* ENO14, respectively.

Nitrogen sources and concentration levels have also been described as essential factors in maximizing biosurfactant production. For instance, yeast extract
as a nitrogen source led to the highest biosurfactant production by *B. licheniformis* LRK1, whereas peptone showed the lowest productivity (Nayak et al. 2020).
Conversely, peptone increased biosurfactant yield during *B. cereus* BCS0 culture
on frying oil (Durval et al. 2019). Gharaei et al. (2022) found ammonium chloride
(NH_4Cl) as the primary nitrogen source compared to peptone and sodium nitrate
($NaNO_3$) in glycolipid production by the marine bacteria *Shewanella algae* B12
using crude oil as substrate. The authors argued that ammonium sources might
be readily adsorbed precursors helping synthesize biosurfactant molecules.
Others have also described a positive contribution of ammonium salts such as
ammonium sulfate (Moshtagh, Hawboldt, and Zhang 2021) and ammonium
nitrate (NH_4NO_3) in biosurfactant productivity (Hema et al. 2019; Moshtagh,
Hawboldt, and Zhang 2019; Mukherjee et al. 2008). Moshtagh, Hawboldt, and
Zhang (2021) even reported that increasing ammonium sulfate concentration up
to ≈6 g/L was beneficial to product generation. Not surprisingly, other authors
reported detrimental effects of ammonium salts on biosurfactant productivity
due to a rapid decrease of media pH (Ekramul and Saqib 2020; Khopade et al.
2012). Conflicting reports demonstrate that multiple interacting factors (e.g.,
selection of strain and culture conditions) contribute to optimizing biosurfactant
production by a specific bacterium, pointing to the need to evaluate the most
appropriate media components for a given strain.

Effects of pH on Biosurfactant Production

Several reports have demonstrated that the production of different biosurfactants
is markedly affected by initial culture pH. Indeed, some authors have identified
pH as the most significant factor during optimization experiments (Moshtagh,
Hawboldt, and Zhang 2019, 2021). The importance of such a factor arises from
its vast influence during bacterial growth by modulating the transportation of

molecules through the cell membrane and affecting the activity of numerous enzymes involved in the metabolism of substrates (Zhang, Zhang, and Cui 2015). Recently, the influence of pH effect on biosurfactant production by marine bacteria has been commonly assessed from 4 to 9. However, increased production of different types of biosurfactant by a variety of strains was generally found at neutral pH (Khopade et al. 2012; Hema et al. 2019; Javee, Karuppan, and Subramani 2020; Nayak et al. 2020; Nalini et al. 2021; Kadam and Savan 2019) or at values close to neutrality (≈6.4–6.5) (Moshtagh, Hawboldt, and Zhang 2019; Tripathi et al. 2019). Some authors have proposed that strong acidic conditions (pH < 5) may induce the precipitation of the molecules of interest, drastically reducing their effectivity (Nayak et al. 2020). More basic higher optimum pH values around 7.7–8.0 were also reported for marine-derived bacteria (RamyaDevi et al. 2018; Moshtagh, Hawboldt, and Zhang 2021; Deepika et al. 2016; Kim et al. 2019)

Effects of Temperature on Biosurfactant Production

The temperature of culture broth is another physical variable with significant effects on biosurfactant production by bacteria. Similar to other physical parameters like pH, the temperature impacts enzyme activity and transport processes that influence microbial development. It is widely observed that different strains have an optimal temperature range that promotes growth and product assembly; therefore, culture temperatures outside such limits reduce cell development and product release (Zhang, Zhang, and Cui 2015). Recently, biosurfactant production by marine bacteria has been evaluated in temperatures ranging from 5 to 45°C. Most studies have found mesophilic temperatures between 27 to 35°C the most appropriate for improving biosurfactant yield; however, 30°C is the most widely reported as the optimum point for production (Moshtagh, Hawboldt, and Zhang 2021; Nayak et al. 2020; Javee, Karuppan, and Subramani 2020; Hema et al. 2019; Moshtagh, Hawboldt, and Zhang 2019; Kadam and Savan 2019; Hu, Wang, and Wang 2015; Khopade et al. 2012).

Effects of Salinity and Other Components on Biosurfactant Production

In addition to carbon and nitrogen sources, other growing media components known as micronutrients are critical factors during biosurfactant production. Thus, supplementation of minerals is a common step during trials due to their impact on cell growth and enzymatic activity. For instance, some authors have evaluated the significance of ferrous sulfate concentrations (Mukherjee et al. 2008; Dhasayan, Selvin, and Kiran 2015), magnesium sulfate, and calcium chloride (Mabrouk, Youssif, and Sabry 2014). Evaluation of the concentration of sodium chloride (NaCl) is a common practice since it is an important factor affecting the growth and product yields of microorganisms from marine environments. Marine bacteria have developed adaptive mechanisms that help them avoid plasmolysis due to high salt concentration outside the cell. Such mechanisms include the production of specialized lipids that change membrane composition to maintain its integrity in hypertonic environments and the release of biosurfactants, which increase the bioavailability of substrates and their transportation inside the cell (De Carvalho and Fernandes 2010). That may explain why some marine bacteria have shown poor growth and negligible biosurfactant production in the absence of sodium chloride, as noted by Hu, Wang, and Wang (2015) during the culture of *Vibrio* sp. 3B-2. Different strains have shown variable optimal concentrations of NaCl. As noted with other medium components, an increase in its concentrations beyond a maximum point had an adverse effect on biosurfactant production. For instance, *Streptomyces youssoufiensis* SNSAA03 showed the highest biosurfactant yield at a NaCl concentration of 2.25 g/L (Nalini et al. 2021), whereas *Streptomyces sp.* SN JASM6 had its highest production at 30 g/L (Javee, Karuppan, and Subramani 2020). Salinities in the range of 27 to 35 g/L were reported as the optimal points for biosurfactant production by *A. calcoaceticus, P. stutzeri, Bacillus* sp. E34, *Nocardiopsis* sp. B4, and *Marinobacter* sp. MCTG107b (Moshtagh, Hawboldt, and Zhang 2021; Tripathi et al. 2019; Kadam and Savan 2019; Mabrouk, Youssif, and Sabry 2014; Khopade et al. 2012).

CULTURE CONDITIONS IN BIOREACTORS

Bioreactors, or fermenters, are vessels with reliable measurement and control tools that provide an optimal environment for growing bacterial cells and producing metabolites of interest. Bioreactors are frequently implemented for scaling up of fermentation processes since they allow monitoring and controlling of important variables during cell growth. As expected, previously discussed parameters such as temperature, pH, and concentrations of NaCl and carbon and nitrogen sources are critical factors during the production of biosurfactants in bioreactors (Fooladi et al. 2018). However, other parameters such as aeration, dissolved oxygen, and agitation speed gain importance when fermentations are carried out at higher working volumes. For instance, air injection and agitation are pivotal during large-scale processes because they allow assuming "perfect mixing" in the reactor, which means that homogenous conditions are predominant throughout the whole culture broth (i.e., same pH; temperature; and concentrations of cells, nutrients, products, and oxygen) (Doran 1995). On the other hand, dissolved oxygen is crucial in producing biosurfactants by aerobic microorganisms, which show limited growth at low gas concentrations in the liquid medium (Coutte et al. 2010). Desired oxygen levels in bioreactors have been achieved by injecting sterile air and controlling mechanical agitation. Both air injection and rapid mixing can lead to foam formation during the fermentation, which should be controlled to avoid contamination, overflow, and loss of media (Guez et al. 2021). However, some authors have argued the transfer of biosurfactants from the liquid medium to the foam formed in the reactor constitutes a method of pre-separation of the molecules of interest (Yao et al. 2015).

Few reports have recently assessed biosurfactant production by marine-derived bacteria in bioreactors. For instance, Silva et al. (2018) reported an improvement in biosurfactant production by *P. aeruginosa* UCP 0992 in a 1.2-L fermenter by optimizing aeration rate, agitation speed, inoculum size, and batch time. Others have compared bioreactor operation modes during biosurfactant production by marine bacteria (Fooladi et al. 2018; Sivapathasekaran and Sen 2013). In both cases, the authors argued that fed-batch processes would be more suitable than a batch approach since adding a concentrated medium at the end of the exponential phase restimulated growth and therefore increased product yield.

BIOREMEDIATION APPLICATIONS OF MARINE BACTERIA-DERIVED BIOSURFACTANTS

Many complex pollutants—namely halogenated compounds, petroleum hydro-carbons, heavy metals, and metalloids—are constantly introduced to the marine environments as a result of anthropogenic activities, posing a threat to marine biodiversity and its ecosystem services. There are different physicochemical techniques used to reduce the concentration of pollutants such as ion exchange, electro-dialysis, reverse osmosis, and ultrafiltration, as well as others based on thermal and electrochemical strategies. Nevertheless, the search more for eco-compatible technologies for decontaminating polluted environments is increasing, resulting in recognizing microorganisms and their active compounds (e.g., biosurfactants) as promising agents to reduce contaminant concentrations and toxicity (Dell' Anno et al. 2021).

Contamination by petroleum and its derivatives is one of the most severe environmental problems worldwide. In marine areas, this type of pollution is caused by anthropogenic activities as well as natural disasters like tsunamis and earthquakes (Primeia, Inoue, and Chien 2020). When an oil spill occurs, the immediate treatment is the application of highly stable chemically derived dispersants like surfactants or emulsifiers. Unfortunately, the use of these synthetic compounds is related to potential lasting toxicity to ecosystems. In this sense, bioremediation strategies using natural-based biosurfactants have emerged as promising solutions, since they have shown high efficiencies and are readily biodegradable. Biosurfactants help with the solubilization of hydrophobic organic compounds like hydrocarbons, facilitating cell uptake of pollutants and their utilization as carbon sources. Thus, it is unsurprising that numerous bioremediation studies using biosurfactant molecules or biosurfactant-producing marine bacteria have been highly reported. Biosurfactants derived from marine microorganisms are of particular interest since these strains, and their metabolites, are already adapted to thrive under generally considered harsh conditions of oceanic environments (Antoniou et al. 2015; Primeia, Inoue, and Chien 2020; Satpute, Banpurkar et al. 2010).

Heavy metals such as cadmium, copper, chromium, zinc, arsenic, nickel, and mercury represent another category of major recalcitrant pollutants. Their mobilization from natural deposits and accumulation in water bodies and soils represents a potential hazard for ecosystems, animals, and humans. The traditional remediation techniques of heavy metals involve using water, organic and inorganic acids, metal chelating agents, and chemical surfactants; unfortunately, these methods do not ensure the proper removal of metals from polluted sites (Palashpriya Das, Mukherjee, and Sen 2009). Biosurfactants can act as mediators facilitating the biological treatment of environments contaminated with these pollutants due to active chemical groups in their structure like hydroxyl, carbonyl, or amine that can form complexes with heavy metals (Ławniczak, Marecik, and Chrzanowski 2013).

IMPORTANT CONSIDERATIONS FOR SUCCESSFUL BIOREMEDIATION

A successful bioremediation process primarily depends on the bioavailability of pollutants, meaning the ease of movement of the toxic molecules from the environment into the microbial cells (Palashpriya Das, Mukherjee, and Sen 2008a). Another critical factor is the stimulation of the natural attenuation of contaminants, which is generally considered a slow process compared to pollutant generation (Ray et al. 2022). Acceleration normally implicates human interventions in the degradation process following bioaugmentation and biostimulation approaches (Dell' Anno et al. 2021). During biostimulation, nutritional compounds are provided to stimulate assemblages of autochthonous strains, while bioaugmentation is achieved by introducing specific strains (i.e., native or exogenous) with desired biodegradation/detoxification capacities (Dell' Anno et al. 2021).

Bioavailability can be significantly improved in the presence of biosurfactants through the formation of micelles. Micelles are formed as a consequence of biosurfactant self-assembly and aggregation in water and some polar solvents; additionally, fibrils and vesicles can also be created (Ohadi et al. 2018). Identifying the point at which micelles start to form requires the determination of essential parameters like the critical micelle concentration and the Krafft temperature (Ohadi et al. 2018; Cai et al. 2014).

Critical Micelle Concentration

Critical micelle concentration is the concentration at which there is no further decrease in surface tension (Heryani and Putra 2017). At the same time, CMC is the lowest concentration of surfactant molecules in a solution needed to form micelles; this means that the lower CMC value, the more efficient the biosurfactant (Primeia, Inoue, and Chien 2020). This parameter is obtained by graphing the surface tension at different concentrations (Guo et al. 2022); when a decrease in surface tension values is not further observed, and the curve becomes asymptotic from a certain concentration value, that concentration becomes the CMC value. Table 3.2 shows some CMC values of different biosurfactants obtained from marine bacteria. It is essential to highlight that biosurfactant CMC values are lower compared to a common chemical surfactant like sodium dodecyl sulfate (SDS). Lower CMC values are crucial features indicating that less concentration of biosurfactants is necessary to achieve micelle formation.

TABLE 3.2

CMC Values of Marine Bacteria Biosurfactants

Marine Bacteria	Sampling Site	Isolation Source	Biosurfactant Type	Critical Micelle Concentration (mg/L)	Reference
Sodium dodecyl sulfate (SDS)				*1000*	
Bacillus circulans	Andaman and Nicobar Islands, India	Marine water	Lipopeptide	40	(Raddadi et al. 2017) (Palashpriya Das, Mukherjee, and Sen 2008a; Palashpriya Das, Mukherjee, and Sen 2008b, 2009)
Streptomyces sp. B3	West coast of India	Marine sediments	Glycolipid	110	(Khopade et al. 2012b)
Rhodococcus sp. PML026	Western English Channel	Seawater	Trehalolipid	250	(White, Hird, and Ali 2013)
Streptomyces sp. MAB36	Tuticorin harbor, southeast coast of India	Sediments	Glycolipid	36	(Manivasagan et al. 2014)
Brevibacterium luteolum	North coast of São Paulo State, Brazil	Marine invertebrates	Lipopeptide	40	(Vilela et al. 2014)
Achromobacter sp. HZ01	Daya Bay, South China Sea	Crude oil–contaminated seawater	Lipopeptide	48	(Deng et al. 2016)
Bacillus subtilis N3–4P	the Northern Region Persistent Organic Pollution Control (NRPOP)—Canada	Oily contaminated seawater samples	Lipopeptide and Glycolipid	507	(Zhu et al. 2016)
Marinobacter hydrocarbonoclasticus SdK644	Bou-Ismail bay, Tipaza, Algeria	Hydrocarbon-contaminated sediments	Glycolipid	788	(Zenati et al. 2018)
Bacillus cereus BCS0	Port area, Atlantic Ocean of the state of Pernambuco, Brazil	Seawater contaminated with petroleum derivatives	N/A	500	(Durval et al. 2019)

B. cereus UCP 1615	Port area, Atlantic Ocean of the state of Pernambuco, Brazil	Water samples contaminated with petroleum derivatives spilled from ships	Lipopeptide	500	(Durval et al. 2020)
Bacillus sp. MSI 54	Coastal regions of Palk Bay, Rameswaram	Marine sponge Agelas clathrodes	Lipopeptide	10	(Ravindran et al. 2020)
Halomonas pacifica Cnaph3	Ataya fishing harbor, Kerkennah islands, Tunisia	Seawater samples contaminated with hydrocarbons	Lipopeptide	500	(Cheffi et al. 2020)
Pseudomonas aeruginosa	Coastal intertidal zone, East China Sea	Intertidal sediment	Rhamnolipid	43.73	(Chen et al. 2021)
Staphylococcus sp. CO100	Vizhinjam coast, southwest coast of India	Contaminated sediment	Lipopeptide	65–750	(Hentati et al. 2021)
Vibrio sp. LQ2	Marine cold-seep region	Sediments	Phospholipid	200	(Zhou et al. 2021)
Planococcus sp. XW-1	The Dalian Port cruise terminal, west coast of the north Yellow Sea	Surface seawater	Glycolipid	60	(Guo et al. 2022)

* Commercial surfactant

Krafft Point

Krafft temperature or Krafft point is the minimum temperature from which micelle formation occurs. As stated by Nakama 2017, Kraft temperature is defined as "the triple point of the CMC temperature curve, the surfactant monomer's solubility curve, and the transition line or hydrated solids to micelles." It is expected that a biosurfactant with a low Krafft temperature will be applied in a cold environment; on the contrary, in a hot environment it would remain crystalline and without its surface-active functions. In this sense, cold-adapted microorganisms could produce biosurfactants with desirable traits such as low Krafft temperatures, which have been the subject of recent studies. For instance, Cai et al. (2014) isolated 55 biosurfactant-producing bacteria from water and sediment samples from a petroleum-contaminated site in North Atlantic Canada. Two strains of *Rhodococcus* genera could reduce water surface tension as low as 28 m/Nm. Other researchers reported the isolation of the cold-adapted *Planococcus* sp. XW-1 from the Yellow Sea, a strain able to produce a glycolipid with a critical micelle concentration of 60 mg/L and the capacity to reach a crude oil degradation up to 54% by adding the bacterium at 4°C (Guo et al. 2022).

APPROACHES FOR BIOREMEDIATION USING BIOSURFACTANTS

Bioremediation of contaminated environments using biosurfactants can follow two main approaches: *ex-situ* or *in-situ*. In this chapter, *ex-situ* refers to the addition of biosurfactants directly on the affected site; in this sense, they are obtained outside the polluted environment and added externally after production. *In-situ* refers to the biosurfactant production *on-site*, so the biosurfactants are produced by adding producing microorganisms to the affected area (Ławniczak, Marecik, and Chrzanowski 2013; Solís-González and Loza-Tavera 2019). In both cases, microorganism mediation is required to complete the bioremediation process.

The microbial cells involved in the process can be free, immobilized, present as axenic cultures, or participants in a microbial consortium. In the case of microbial consortia, it has been reported that the nutrient consumption rates of implicated strains greatly influence biodegradation of some compounds. For instance, fast-growing consortia rapidly form biofilms after using hydrocarbons as carbon sources; however, the biofilm limits the contact of microorganisms with remaining hydrocarbons, which halts its degradation. On the contrary, slow-growing consortia continuously solubilize and uptake pollutants, which results in increased biodegradation (Ławniczak, Marecik, and Chrzanowski 2013).

Ex-Situ Methods

To date, most studies can be considered *ex-situ* approaches since they are conducted in laboratory settings under simulated conditions, or they involve the addition of biosurfactants to contaminated material to increase the bioavailability of pollutants to present microbiota. For instance, Palashpriya Das, Mukherjee, and Sen (2008a) used a lipopeptide produced by the marine bacterium *Bacillus circulans* to degrade anthracene—a model polyaromatic hydrocarbon (PAH). Their findings indicated that the biosurfactant trapped the anthracene molecules in its micellar structure, increasing its solubility and bioavailability. Interestingly, it was noted that the bacterium could not use anthracene as the sole carbon source, needing the presence of glycerol to metabolize and excrete the PAH in a non-toxic form. In a follow-up study aiming to elucidate the role of the produced biosurfactant in removing heavy metals, the authors found a near-complete removal of 100 ppm of cadmium with a concentration of the lipopeptide of 5 × CMC (5 × 40 mg/L) (Palashpriya Das, Mukherjee, and Sen 2009).

In-Situ Methods

In-situ methods expect an improvement of the remediation process after adding biosurfactant-producing strains to the contaminated environment. Degradation can be achieved by the added microorganisms or by combining efforts with other bacteria or consortia. Wu et al. (2019) improved diesel degradation by mixing diesel-degrading bacteria with a surfactant-producing bacterium from the genus *Aneurinibacillus* isolated from a marine environment. Other researchers immobilized cells of a marine bacterium identified as *Vibrio* sp. LQ2 to treat marine water contaminated with diesel oil (Zhou et al. 2021). Removal efficiency reached 94.7% at 7 days; however, the authors highlighted the need to recover the biochar-immobilized cells systems after their application; otherwise, it could cause secondary pollution due to adsorption of pollutants on the particles. In addition, the authors recommended applying containment booms as a follow-up strategy to the physicochemical treatments because the complexity of the marine environment could limit the efficiency of the bioremediation process under certain conditions (Zhou et al. 2021).

The use of immobilized cells in bioremediation approaches has been shown to diminish biomass loss, enabling higher cell density and reusability during degradation processes (Zhou et al. 2021). Especially in the bioremediation of seawater, immobilized cells are a promising solution. Immobilization can be achieved through cell entrapment or attachment into organic, inorganic, and composite materials known as carriers. Desired characteristics of such carriers are insolubility, non-toxicity, non-biodegradability by the cells, ease of handling and regeneration, high cell loading capacity, and allowing optimum nutrient diffusion (Bouabidi, El-Naas, and Zhang 2019). It has also been reported that biosurfactant presence can induce changes in the properties of cellular membranes, increasing microbial adherence to surfaces (Ławniczak, Marecik, and Chrzanowski 2013).

Application of consortia is another promising *in situ* bioremediation method proposed for multi-molecule pollutants (e.g., oils spills), mainly because it is difficult to reach a broad spectrum in degradation capabilities using single bacteria. There are two general ways to approach artificial microbial consortia design: top down (complex to simple) and bottom up (simple to complex). In the first method, the optimum consortium comprises the surviving strains from environmental samples, which are considered key players. The second approach may mix engineered microorganisms and isolated strains from other environments, which may exhibit different metabolic pathways (González et al. 2021). Primeia, Inoue, and Chien (2020), evaluated six bacterial consortia from oil-spilled beach areas with high potential in degrading heavy oil. One of the consortia showed superior petroleum hydrocarbon degradation ability. This microbial community had a strong presence of biosurfactant-producing bacteria, members of the Pseudomonadaceae family.

COMPLEX SURFACES FOR BIOREMEDIATION: SEDIMENTS AND POROUS SOLIDS

Sediments accumulate high water-insoluble pollutants due to the attachment of these contaminants to solid particles, which eventually become deposited in the bottom of water bodies (Dell'Anno et al. 2018). Numerous studies show that the bioremediation efficiency of sediments depends on the properties of pollutants and sediment particle size. For instance, it has been reported that adsorbed long-chain hydrocarbons are more challenging to biodegrade than readily metabolizable short-chain pollutants. On the other hand, sediment particle size affects mass transfer processes; particularly, it can limit the access of microorganisms to the contaminant molecule and restrict the diffusion of other nutrients and dissolved gases (Dell' Anno et al. 2021). Biosurfactant-producing bacteria are promising for sediment bioremediation since biosurfactants can assist in the desorption of pollutants from solid particles, increasing their susceptibility to microbial attack (Dell'Anno et al. 2018).

Another critical factor affecting sediment bioremediation *in situ* is oxygen concentration; hydrocarbon and other pollutant degradation in anoxic sediments (probably the deepest) has been considered a challenge because most biosurfactant-producing bacteria are aerobic organisms. Recently, Domingues et al. (2020) isolated biosurfactant-producing bacteria from deep sub-seafloor sediments and estuarine sub-surface sediments, finding 12 strains of the genera *Pseudomonas, Bacillus, Ochrobactrum, Brevundimonas, Psychrobacter, Staphylococcus*, and *Curtobacterium*. Anaerobic biosurfactant-producing strains of the first two genera were previously reported, but it was the first time that the other genera were described as anaerobic producers.

Some authors have tried other valuable methods, such as soil washing, to treat contaminated sediments with biosurfactants. As its name indicates, soil washing or flushing is an *ex-situ* technology commonly implemented for soil remediation. It consists of separating and promoting contaminant leaching from solid particles using liquid solutions (Mulligan 2021). For instance, Zhu et al. (2016) studied the washing process of sediments contaminated with crude oil using a lipopeptide and a glycolipid produced by the marine bacterium *Bacillus subtilis* N3–4P. The authors reported 58 and 65.2% removal rates at biosurfactant concentrations of 4 and 8 g/L, respectively.

Porous surfaces like marine rocks, coral reefs, and so on represent other challenging surfaces for bioremediation approaches. The conventional chemical and physical methods used to remediate contaminants' presence in these surfaces can cause damage to the corals and marine ecosystems. Chaprão et al. (2018) used the biosurfactant produced by the estuarine bacterium *Bacillus methylotrophicus* UCP 1616 to test its capacity in the removal of motor oil from contaminated soil and marine rocks. In the last case, 70% of the pollutant was removed from the surface after 5 minutes of contact with the biosurfactant.

CONCLUSIONS AND PERSPECTIVES

There is a general agreement that oceanic ecosystems embrace one of the wealthiest diversities of microorganisms with incredible adaptations to survive and thrive in these harsh environments. Adaptation to oceanic conditions results in tremendous genetic flexibility in diverse metabolic processes; consequently, marine microorganisms, especially bacteria, have become an attractive source of new natural products with potential biotechnological applications in industrial and environmental processes.

As discussed in this chapter, the surfactants produced by different genera of marine bacteria have been the subject of study, particularly for their potential use in the bioremediation of environments contaminated by hydrocarbons. Bacterial surfactants have received particular attention due to their high diversity of chemical structures, stability under extreme conditions, and their yet-untapped potential for large-scale production. The selection of new microbial-derived surfactants has generally consisted of three simple but crucial steps: environmental sampling, strain isolation, and production investigation (efficiency evaluation, identification of compounds of interest, increased production). However, the discovery of new biosurfactants is limited by the fact that the vast majority of marine bacteria are non-cultivable under traditional approaches, diminishing knowledge about bacterial diversity and also the ability to discover and harness their full biotechnological potential.

Reducing this knowledge gap requires multidisciplinary research efforts that advance the discovery and isolation of new types of microbial strains, especially those active under anaerobic conditions with rapid biomass and surfactant production. One way could be the use of genetic engineering of promising strains to make them more efficient, for example, for the *ex-situ* production of surfactants of interest. Another way could also be metagenomics approaches, which allow the identification of surfactants from genetic material, either from isolated strains or extracted directly from environmental samples.

Large-scale process development implementing marine-derived bacteria is currently limited, becoming another critical bottleneck for achieving practical application of their biosurfactants at real-scale pollution catastrophes. Valuable information gathered from bench-top experimental results reviewed in this chapter can be the base for advancing culture media formulation and selecting optimal physicochemical parameters at higher volumes. Implementation of low-cost substrates, particularly locally generated residues, stands out as the most promising line of research for developing real eco-friendly and cost-effective biosurfactant production processes with marine bacteria.

The prospects are excellent; the ocean and its microbial biodiversity offer a potential resource for obtaining biosurfactants. However, it is vital to establish interdisciplinary alliances among research groups that accelerate strain isolation, new promising metabolite identification, and large-scale biosurfactant production to achieve a reliable bioremediation process to protect and restore marine and terrestrial ecosystems.

ACKNOWLEDGMENTS

The authors acknowledge the "Ministerio de Ambiente y Desarrollo Sostenible de Colombia (Minambiente)," the Marine and Coastal Research institute "Jose' Benito Vives de Andréis"—INVEMAR and are also thankful to the evaluation and use of Marine and Coastal Resources Program-VAR (Contribution number CTRB-1333—from the Evaluation and Use of Marine and Coastal Resources Program—VAR, Marine Bioprospecting Line).

REFERENCES

Abdel-Mawgoud, Ahmad Mohammad, François Lépine, and Eric Déziel. 2010. "Rhamnolipids: Diversity of Structures, Microbial Origins and Roles." *Applied Microbiology and Biotechnology* 86 (5): 1323–36. doi:10.1007/s00253-010-2498-2.

Adetunji, Adegoke Isiaka, and Ademola Olufolahan Olaniran. 2021. "Production and Potential Biotechnological Applications of Microbial Surfactants: An Overview." *Saudi Journal of Biological Sciences* 28 (1): 669–79. doi:10.1016/j.sjbs.2020.10.058.

Alemán-Vega, Monserrat, Ilse Sánchez-Lozano, Claudia J. Hernández-Guerrero, Claire Hellio, and Erika T. Quintana. 2020. "Exploring Antifouling Activity of Biosurfactants Producing Marine Bacteria Isolated from Gulf of California." *International Journal of Molecular Sciences* 21 (17): 1–19. doi:10.3390/ijms21176068.

Ali, Ferdausi, Sharup Das, Tanim Jabid Hossain, Sumaiya Islam Chowdhury, Subrina Akter Zedny, Tuhin Das, Mohammad Nazmul Ahmed Chowdhury, and Mohammad Seraj Uddin. 2021. "Production Optimization, Stability and Oil Emulsifying Potential of Biosurfactants from Selected Bacteria Isolated from Oil-Contaminated Sites." *Royal Society Open Science* 8 (10). doi:10.1098/rsos.211003.

Antoniou, Eleftheria, Stilianos Fodelianakis, Emmanouela Korkakaki, and Nicolas Kalogerakis. 2015. "Biosurfactant Production from Marine Hydrocarbon-Degrading Consortia and Pure Bacterial Strains Using Crude Oil as Carbon Source." *Frontiers in Microbiology* 6 (April). doi:10.3389/fmicb.2015.00274.

Arima, Kei, Atsushi Kakinuma, and Gakuzo Tamura. 1968. "Surfactin, a Crystalline Peptidelipid Surfactant Produced by *Bacillus subtilis*: Isolation, Characterization and Its Inhibition of Fibrin Clot Formation." *Biochemical and Biophysical Research Communications* 31 (3). Elsevier: 488–94.

Azubuike, Christopher Chibueze, Chioma Blaise Chikere, and Gideon Chijioke Okpokwasili. 2016. "Bioremediation Techniques-Classification Based on Site of Application: Principles, Advantages, Limitations and Prospects." *World Journal of Microbiology & Biotechnology* 32 (11): 180. doi:10.1007/s11274-016-2137-x.

Banat, Ibrahim M., Surekha K. Satpute, Swaranjit S. Cameotra, Rajendra Patil, and Narendra V. Nyayanit. 2014. "Cost Effective Technologies and Renewable Substrates for Biosurfactants' Production." *Frontiers in Microbiology* 5 (December). doi:10.3389/fmicb.2014.00697.

Blandón, Lina Marcela, Mario Alejandro Marín, Marynes Quintero, Laura Marcela Jutinico-Shubach, Manuela Montoya-Giraldo, Marisol Santos-Acevedo, and Javier Gómez-León. 2022. "Diversity of Cultivable Bacteria from Deep-Sea Sediments of the Colombian Caribbean and Their Potential in Bioremediation." *Antonie Van Leeuwenhoek* 115 (January): 421–31. doi:10.1007/s10482-021-01706-4.

Bonugli-Santos, Rafaella C., Maria R. dos Santos Vasconcelos, Michel R.Z. Passarini, Gabriela A.L. Vieira, Viviane C.P. Lopes, Pedro H. Mainardi, Juliana A. dos Santos, et al. 2015. "Marine-Derived Fungi: Diversity of Enzymes and Biotechnological Applications." *Frontiers in Microbiology* 6 (April). doi:10.3389/fmicb.2015.00269.

Bouabidi, Zineb B., Muftah H. El-Naas, and Zhien Zhang. 2019. "Immobilization of Microbial Cells for the Biotreatment of Wastewater: A Review." *Environmental Chemistry Letters* 17 (1): 241–57. doi:10.1007/s10311-018-0795-7.

Bushnell, L.D., and H.F. Haas. 1941. "The Utilization of Certain Hydrocarbons by Microorganisms 1." *Journal of Bacteriology* 41 (5): 653–73. www.ncbi.nlm.nih.gov/pmc/articles/PMC374727/.

Cafaro, Valeria, Viviana Izzo, Eugenio Notomista, and Alberto Di Donato. 2013. "Marine Hydrocarbonoclastic Bacteria." In *Marine Enzymes for Biocatalysis*, edited by Antonio Trincone, 373–402. Woodhead Publishing Series in Biomedicine. Woodhead Publishing. doi:10.1533/9781908818355.3.373.

Cai, Qinhong, Baiyu Zhang, Bing Chen, Zhiwen Zhu, Weiyun Lin, and Tong Cao. 2014. "Screening of Biosurfactant Producers from Petroleum Hydrocarbon Contaminated Sources in Cold Marine Environments." *Marine Pollution Bulletin* 86 (1–2): 402–10. doi:10.1016/j.marpolbul.2014.06.039.

Carvalho, Carla C.C.R. de, and Pedro Fernandes. 2010. "Production of Metabolites as Bacterial Responses to the Marine Environment." *Marine Drugs* 8 (3): 705–27. doi:10.3390/md8030705.

Chaprão, Marco José, Rita de Cássia F. Soares da Silva, Raquel D. Rufino, Juliana M. Luna, Valdemir A. Santos, and Leonie A. Sarubbo. 2018. "Production of a Biosurfactant from *Bacillus methylotrophicus* UCP1616 for Use in the Bioremediation of Oil-Contaminated Environments." *Ecotoxicology (London, England)* 27 (10): 1310–22. doi:10.1007/s10646-018-1982-9.

Cheffi, Meriam, Dorra Hentati, Alif Chebbi, Najla Mhiri, Sami Sayadi, Ana Maria Marqués, and Mohamed Chamkha. 2020. "Isolation and Characterization of a Newly Naphthalene-Degrading *Halomonas pacifica*, Strain Cnaph3: Biodegradation and Biosurfactant Production Studies." *3 Biotech* 10 (3): 89. doi:10.1007/s13205-020-2085-x.

Chen, Qingguo, Yijing Li, Mei Liu, Baikang Zhu, Jun Mu, and Zhi Chen. 2021. "Removal of Pb and Hg from Marine Intertidal Sediment by Using Rhamnolipid Biosurfactant Produced by a *Pseudomonas aeruginosa* Strain." *Environmental Technology & Innovation* 22 (May): 101456. doi:10.1016/j.eti.2021.101456.

Coutte, François, Didier Lecouturier, Saliha Ait Yahia, Valérie Leclère, Max Béchet, Philippe Jacques, and Pascal Dhulster. 2010. "Production of Surfactin and Fengycin by *Bacillus subtilis* in a Bubbleless Membrane Bioreactor." *Applied Microbiology and Biotechnology* 87 (2): 499–507. doi:10.1007/s00253-010-2504-8.

Dalmaso, Gabriel Zamith Leal, Davis Ferreira, and Alane Beatriz Vermelho. 2015. "Marine Extremophiles: A Source of Hydrolases for Biotechnological Applications." *Marine Drugs* 13 (4): 1925–65. doi:10.3390/md13041925.

Das, Palashpriya, Soumen Mukherjee, and Ramkrishna Sen. 2008a. "Improved Bioavailability and Biodegradation of a Model Polyaromatic Hydrocarbon by a Biosurfactant Producing Bacterium of Marine Origin." *Chemosphere* 72 (9): 1229–34. doi:10.1016/j.chemosphere.2008.05.015.

Das, Palashpriya, Soumen Mukherjee, and Ramkrishna Sen. 2008b. "Antimicrobial Potential of a Lipopeptide Biosurfactant Derived from a Marine *Bacillus circulans*." *Journal of Applied Microbiology* 104 (6): 1675–84. doi:10.1111/j.1365-2672.2007.03701.x.

Das, Palashpriya, Soumen Mukherjee, and Ramkrishna Sen. 2009. "Biosurfactant of Marine Origin Exhibiting Heavy Metal Remediation Properties." *Bioresource Technology* 100 (20): 4887–90. doi:10.1016/j.biortech.2009.05.028.

Deepika, K.V., Sadaf Kalam, P. Ramu Sridhar, Appa Rao Podile, and P.V. Bramhachari. 2016. "Optimization of Rhamnolipid Biosurfactant Production by Mangrove Sediment Bacterium *Pseudomonas aeruginosa* KVD-HR42 Using Response Surface Methodology." *Biocatalysis and Agricultural Biotechnology* 5. Elsevier: 38–47. doi:10.1016/j.bcab.2015.11.006.

Dell' Anno, Filippo, Eugenio Rastelli, Clementina Sansone, Christophe Brunet, Adrianna Ianora, and Antonio Dell' Anno. 2021. "Bacteria, Fungi and Microalgae for the Bioremediation of Marine Sediments Contaminated by Petroleum Hydrocarbons in the Omics Era." *Microorganisms* 9 (8). Multidisciplinary Digital Publishing Institute: 1695. doi:10.3390/microorganisms9081695.

Dell'Anno, F., C. Sansone, A. Ianora, and A. Dell'Anno. 2018. "Biosurfactant-Induced Remediation of Contaminated Marine Sediments: Current Knowledge and Future Perspectives." *Marine Environmental Research* 137 (June): 196–205. doi:10.1016/j.marenvres.2018.03.010.

Deng, Jingdan, Bi He, Daohang He, and Zhifen Chen. 2016. "A Potential Biopreservative: Chemical Composition, Antibacterial and Hemolytic Activities of Leaves Essential Oil from *Alpinia guinanensis*." *Industrial Crops and Products* 94 (December): 281–87. doi:10.1016/j.indcrop.2016.09.004.

Dhasayan, Asha, Joseph Selvin, and Seghal Kiran. 2015. "Biosurfactant Production from Marine Bacteria Associated with Sponge *Callyspongia diffusa*." *3 Biotech* 5 (4). Springer Berlin Heidelberg: 443–54. doi:10.1007/s13205-014-0242-9.

Domingues, Patrícia M., Vanessa Oliveira, Luísa Seuanes Serafim, Newton C.M. Gomes, and Ângela Cunha. 2020. "Biosurfactant Production in Sub-Oxic Conditions Detected in Hydrocarbon-Degrading Isolates from Marine and Estuarine Sediments." *International Journal of Environmental Research and Public Health* 17 (5): E1746. doi:10.3390/ijerph17051746.

Doran, Pauline M. 1995. *Bioprocess Engineering Principles*. Elsevier Science and Technology Books. doi:10.1016/B978-012220855-3/50001-8.

Drakontis, Constantina Eleni, and Samiul Amin. 2020. "Biosurfactants: Formulations, Properties, and Applications." *Current Opinion in Colloid & Interface Science, Formulations and Cosmetics*, 48 (August): 77–90. doi:10.1016/j.cocis.2020.03.013.

Durval, Italo José B., Ana Helena R. Mendonça, Igor V. Rocha, Juliana M. Luna, Raquel D. Rufino, A. Converti, and L.A. Sarubbo. 2020. "Production, Characterization, Evaluation and Toxicity Assessment of a *Bacillus cereus* UCP 1615 Biosurfactant for Marine Oil Spills Bioremediation." *Marine Pollution Bulletin* 157 (August): 111357. doi:10.1016/j.marpolbul.2020.111357.

Durval, Italo José B., Ana Helena M. Resende, Mariana A. Figueiredo, Juliana M. Luna, Raquel D. Rufino, and Leonie A. Sarubbo. 2019. "Studies on Biosurfactants Produced Using *Bacillus cereus* Isolated from Seawater with Biotechnological

Potential for Marine Oil-Spill Bioremediation." *Journal of Surfactants and Detergents* 22 (2): 349–63. doi:10.1002/jsde.12218.

Ekramul, Haque, and Hassan Saqib. 2020. "Statistical Optimization of Rhamnolipid Biosurfactant for Cost-Effective Production Using Box-Behnken Design." *Research Journal of Biotechnology* 15 (12): 143–55.

El-Sheshtawy, H.S., N.M. Khalil, W. Ahmed, and R.I. Abdallah. 2014. "Monitoring of Oil Pollution at Gemsa Bay and Bioremediation Capacity of Bacterial Isolates with Biosurfactants and Nanoparticles." *Marine Pollution Bulletin* 87 (1–2): 191–200. doi:10.1016/j.marpolbul.2014.07.059.

Esmaeili, Hossein, Seyyed Mojtaba Mousavi, Seyyed Alireza Hashemi, Chin Wei Lai, Wei-Hung Chiang, and Sonia Bahrani. 2021. "Chapter 7—Application of Biosurfactants in the Removal of Oil from Emulsion." In *Green Sustainable Process for Chemical and Environmental Engineering and Science*, edited by Inamuddin and Charles Oluwaseun Adetunji, 107–27. Elsevier. doi:10.1016/B978-0-12-822696-4.00008-5.

Fooladi, Tayebeh, Peyman Abdeshahian, Nasrin Moazami, Mohammad Reza Soudi, Abudukeremu Kadier, Wan Mohtar Wan Yusoff, and Aidil Abdul Hamid. 2018. "Enhanced Biosurfactant Production by *Bacillus pumilus* 2IR in Fed-Batch Fermentation Using 5-L Bioreactor." *Iranian Journal of Science and Technology, Transaction A: Science* 42 (3). Springer International Publishing: 1111–23. doi:10.1007/s40995-018-0599-4.

Franzetti, Andrea, Isabella Gandolfi, Giuseppina Bestetti, Thomas J.P. Smyth, y Ibrahim M. Banat. 2010. "Production and Applications of Trehalose Lipid Biosurfactants." *European Journal of Lipid Science and Technology* 112 (6): 617–27. doi:10.1002/ejlt.200900162.

Gadd, Geoffrey Michael. 2010. "Metals, Minerals and Microbes: Geomicrobiology and Bioremediation." *Microbiology (Reading, England)* 156 (Pt 3): 609–43. doi:10.1099/mic.0.037143-0.

Gharaei, Sanaz, Mandana Ohadi, Mehdi Hassanshahian, Sara Porsheikhali, and Hamid Forootanfar. 2022. "Isolation, Optimization, and Structural Characterization of Glycolipid Biosurfactant Produced by Marine Isolate *Shewanella algae* B12 and Evaluation of Its Antimicrobial and Anti-Biofilm Activity." *Applied Biochemistry and Biotechnology*, no. 0123456789. doi:10.1007/s12010-021-03782-8.

González, Camila, Yajie Wu, Ana Zuleta-Correa, Glorimar Jaramillo, and Juliana Vasco-Correa. 2021. "Biomass to Value-Added Products Using Microbial Consortia with White-Rot Fungi." *Bioresource Technology Reports* 16 (December): 100831. doi:10.1016/j.biteb.2021.100831.

Guez, Jean Sébastien, Antoine Vassaux, Christian Larroche, Philippe Jacques, and François Coutte. 2021. "New Continuous Process for the Production of Lipopeptide Biosurfactants in Foam Overflowing Bioreactor." *Frontiers in Bioengineering and Biotechnology* 9 (May). doi:10.3389/fbioe.2021.678469.

Guo, Ping, Weiwei Xu, Shi Tang, Binxia Cao, Danna Wei, Manxia Zhang, Jianguo Lin, and Wei Li. 2022. "Isolation and Characterization of a Biosurfactant Producing Strain *Planococcus* Sp. XW-1 from the Cold Marine Environment." *International Journal of Environmental Research and Public Health* 19 (2). Multidisciplinary Digital Publishing Institute: 782. doi:10.3390/ijerph19020782.

Häder, Donat-P., Anastazia T. Banaszak, Virginia E. Villafañe, Maite A. Narvarte, Raúl A. González, and E. Walter Helbling. 2020. "Anthropogenic Pollution of Aquatic Ecosystems: Emerging Problems with Global Implications." *Science of the Total Environment* 713 (April): 136586. doi:10.1016/j.scitotenv.2020.136586.

Hassanshahian, Mehdi. 2014. "Isolation and Characterization of Biosurfactant Producing Bacteria from Persian Gulf (Bushehr Provenance)." *Marine Pollution Bulletin* 86 (1): 361–66. doi:10.1016/j.marpolbul.2014.06.043.

Hatha, A.A.M., Gakpe Edward, and K.S.M. Pattanathu Rahman. 2007. "Microbial Biosurfactants—Review." *Journal of Marine and Atmospheric Research* 3 (2): 1–17.

Hema, T., G. Seghal Kiran, Arya Sajayyan, Amrudha Ravendran, G. Gowtham Raj, and Joseph Selvin. 2019. "Response Surface Optimization of a Glycolipid Biosurfactant Produced by a Sponge Associated Marine Bacterium *Planococcus* sp. MMD26." *Biocatalysis and Agricultural Biotechnology* 18 (December 2018). Elsevier Ltd: 101071. doi:10.1016/j.bcab.2019.101071.

Hentati, Dorra, Meriam Cheffi, Fatma Hadrich, Neila Makhloufi, Francesc Rabanal, Angeles Manresa, Sami Sayadi, and Mohamed Chamkha. 2021. "Investigation of Halotolerant Marine *Staphylococcus* sp. CO100, as a Promising Hydrocarbon-Degrading and Biosurfactant-Producing Bacterium, under Saline Conditions." *Journal of Environmental Management* 277 (January): 111480. doi:10.1016/j.jenvman.2020.111480.

Heryani, Hesty, and Meilana Dharma Putra. 2017. "Kinetic Study and Modeling of Biosurfactant Production Using *Bacillus* Sp." *Electronic Journal of Biotechnology* 27 (May): 49–54. doi:10.1016/j.ejbt.2017.03.005.

Hu, Xiaoke, Caixia Wang, and Peng Wang. 2015. "Optimization and Characterization of Biosurfactant Production from Marine *Vibrio* sp. Strain 3B-2." *Frontiers in Microbiology* 6 (September): 1–13. doi:10.3389/fmicb.2015.00976.

Ibacache-Quiroga, C., J. Ojeda, G. Espinoza-Vergara, P. Olivero, M. Cuellar, and M.A. Dinamarca. 2013. "The Hydrocarbon-Degrading Marine Bacterium *Cobetia* Sp. Strain MM1IDA2H-1 Produces a Biosurfactant That Interferes with Quorum Sensing of Fish Pathogens by Signal Hijacking." *Microbial Biotechnology* 6 (4): 394–405. doi:10.1111/1751-7915.12016.

Jain, D.K., D.L. Collins-Thompson, H. Lee, and J.T. Trevors. 1991. "A Drop-Collapsing Test for Screening Surfactant-Producing Microorganisms." *Journal of Microbiological Methods* 13 (4): 271–79. doi:10.1016/0167-7012(91)90064-W.

Jakinala, Parameshwar, Nageshwar Lingampally, Archana Kyama, and Bee Hameeda. 2019. "Enhancement of Atrazine Biodegradation by Marine Isolate *Bacillus velezensis* MHNK1 in Presence of Surfactin Lipopeptide." *Ecotoxicology and Environmental Safety* 182 (October): 109372. doi:10.1016/j.ecoenv.2019.109372.

Javee, Anand, Ramamoorthy Karuppan, and Nagaraj Subramani. 2020. "Bioactive Glycolipid Biosurfactant from Seaweed *Sargassum myriocystum* Associated Bacteria *Streptomyces* Sp. SNJASM6." *Biocatalysis and Agricultural Biotechnology* 23 (January). Elsevier Ltd: 101505. doi:10.1016/j.bcab.2020.101505.

Kaczorek, Ewa, Amanda Pacholak, Agata Zdarta, and Wojciech Smułek. 2018. "The Impact of Biosurfactants on Microbial Cell Properties Leading to Hydrocarbon Bioavailability Increase." *Colloids and Interfaces* 2 (3). Multidisciplinary Digital Publishing Institute: 35. doi:10.3390/colloids2030035.

Kadam, Deepa, and Devayani Savan. 2019. "Biosurfactant Production from Shrimp Shell Waste by *Pseudomonas stutzeri*." *Indian Journal of Geo Marine Sciences* 48 (9): 1411–18.

Khopade, A., R. Biao, X. Liu, K. Mahadik, L. Zhang, and C. Kokare. 2012. "Production and Stability Studies of the Biosurfactant Isolated from Marine *Nocardiopsis* sp. B4." *Desalination* 285. Elsevier B.V.: 198–204. doi:10.1016/j.desal.2011.10.002.

Khopade, Abhijit, Biao Ren, Xiang-Yang Liu, Kakasaheb Mahadik, Lixin Zhang, and Chandrakant Kokare. 2012. "Production and Characterization of Biosurfactant from Marine *Streptomyces* Species B3." *Journal of Colloid and Interface Science* 367 (1): 311–18. doi:10.1016/j.jcis.2011.11.009.

Kim, Chul Hwan, Dong Wan Lee, Young Mok Heo, Hanbyul Lee, Yeonjae Yoo, Gyu Hyeok Kim, and Jae Jin Kim. 2019. "Desorption and Solubilization of Anthracene by a Rhamnolipid Biosurfactant from *Rhodococcus fascians*." *Water Environment Research* 91 (8): 739–47. doi:10.1002/wer.1103.

Kishimoto, Noriaki, Kenichi Adachi, Shinichi Tamura, Masateru Nishihara, Kenji Inagaki, Tsuyoshi Sugio, and Tatsuo Tano. 1993. "Lipoamino Acids Isolated from *Acidiphilium organovorum*." *Systematic and Applied Microbiology* 16 (1): 17–21. doi:10.1016/S0723-2020(11)80245-6.

Konwar, Bolin Kumar. 2022. *Bacterial Biosurfactants: Isolation, Purification, Characterization, and Industrial Applications.* CRC Press.

Kretschmer, Axel, Hans Bock, and Fritz Wagner. 1982. "Chemical and Physical Characterization of Interfacial-Active Lipids from *Rhodococcus erythropolis* Grown on n-Alkanes." *Applied and Environmental Microbiology* 44 (4). Am Soc Microbiol: 864–70.

Kubicki, Sonja, Alexander Bollinger, Nadine Katzke, Karl-Erich Jaeger, Anita Loeschcke, and Stephan Thies. 2019. "Marine Biosurfactants: Biosynthesis, Structural Diversity and Biotechnological Applications." *Marine Drugs* 17 (7): E408. doi:10.3390/md17070408.

Kügler, Johannes H., Marilize Le Roes-Hill, Christoph Syldatk, and Rudolf Hausmann. 2015. "Surfactants Tailored by the Class Actinobacteria." *Frontiers in Microbiology* 6. www.frontiersin.org/article/10.3389/fmicb.2015.00212.

Lawniczak, Lukasz, Roman Marecik, y Lukasz Chrzanowski. 2013. "Contributions of Biosurfactants to Natural or Induced Bioremediation." *Applied Microbiology and Biotechnology* 97 (6): 2327–39. doi:10.1007/s00253-013-4740-1.

Lee, Dong Wan, Hanbyul Lee, Bong-Oh Kwon, Jong Seong Khim, Un Hyuk Yim, Beom Seok Kim, and Jae-Jin Kim. 2018. "Biosurfactant-Assisted Bioremediation of Crude Oil by Indigenous Bacteria Isolated from Taean Beach Sediment." *Environmental Pollution (Barking, Essex: 1987)* 241 (October): 254–64. doi:10.1016/j.envpol.2018.05.070.

Mabrouk, Mona E.M., Eman M. Youssif, and Soraya A. Sabry. 2014. "Biosurfactant Production by a Newly Isolated Soft Coral-Associated Marine *Bacillus* sp.E34: Statistical Optimization and Characterization." *Life Science Journal* 11 (10): 756–68.

MacDonald, R.M., and Muriel R. Chandler. 1981. "Bacterium-like Organelles in the Vesicular-Arbuscular Mycorrhizal Fungus *Glomus caledonius*." *New Phytologist* 89 (2). Wiley Online Library: 241–46.

Maneerat, Suppasil, Takeshi Bamba, Kazuo Harada, Akio Kobayashi, Hidenori Yamada, and Fusako Kawai. 2006. "A Novel Crude Oil Emulsifier Excreted in the Culture Supernatant of a Marine Bacterium, *Myroides* sp. Strain SM1." *Applied Microbiology and Biotechnology* 70 (2): 254–59. doi:10.1007/s00253-005-0050-6.

Manivasagan, Panchanathan, Palaniappan Sivasankar, Jayachandran Venkatesan, Kannan Sivakumar, and Se-Kwon Kim. 2014. "Optimization, Production and Characterization of Glycolipid Biosurfactant from the Marine Actinobacterium, *Streptomyces* sp. MAB36." *Bioprocess and Biosystems Engineering* 37 (5): 783–97. doi:10.1007/s00449-013-1048-6.

Morikawa, Masaaki, Yoshihiko Hirata, and Tadayuki Imanaka. 2000. "A Study on the Structure—Function Relationship of Lipopeptide Biosurfactants." *Biochimica et Biophysica Acta (BBA)—Molecular and Cell Biology of Lipids* 1488 (3): 211–18. doi:10.1016/S1388-1981(00)00124-4.

Moshtagh, Bahareh, Kelly Hawboldt, and Baiyu Zhang. 2019. "Optimization of Biosurfactant Production by *Bacillus subtilis* N3–1P Using the Brewery Waste as the Carbon Source." *Environmental Technology (United Kingdom)* 40 (25). Taylor & Francis: 3371–80. doi:10.1080/09593330.2018.1473502.

Moshtagh, Bahareh, Kelly Hawboldt, and Baiyu Zhang. 2021. "Biosurfactant Production by Native Marine Bacteria (*Acinetobacter calcoaceticus* P1–1A) Using Waste Carbon Sources: Impact of Process Conditions." *Canadian Journal of Chemical Engineering* 99 (11): 2386–97. doi:10.1002/cjce.24254.

Mukherjee, Soumen, Palashpriya Das, C. Sivapathasekaran, and Ramkrishna Sen. 2008. "Enhanced Production of Biosurfactant by a Marine Bacterium on Statistical Screening of Nutritional Parameters." *Biochemical Engineering Journal* 42 (3): 254–60. doi:10.1016/j.bej.2008.07.003.

Mulligan, Catherine N. 2021. "Sustainable Remediation of Contaminated Soil Using Biosurfactants." *Frontiers in Bioengineering and Biotechnology* 9. www.frontiersin.org/article/10.3389/fbioe.2021.635196.

Muneeswari, R., K.V. Swathi, G. Sekaran, and K. Ramani. 2022. "Microbial-Induced Biosurfactant-Mediated Biocatalytic Approach for the Bioremediation of Simulated Marine Oil Spill." *International Journal of Environmental Science and Technology* 19 (1): 341–54. doi:10.1007/s13762-020-03086-0.

Nakama, Y. 2017. "Chapter 15 — Surfactants." In *Cosmetic Science and Technology*, edited by Kazutami Sakamoto, Robert Y. Lochhead, Howard I. Maibach, and Yuji Yamashita, 231–44. Elsevier. doi:10.1016/B978-0-12-802005-0.00015-X.

Nalini, S., D. Inbakandan, T. Stalin Dhas, and S. Sathiyamurthi. 2021. "Optimization of Biosurfactant Production by Marine *Streptomyces youssoufiensis* SNSAA03: A Comparative Study of RSM and ANN Approach." *Results in Chemistry* 3. Elsevier B.V.: 100223. doi:10.1016/j.rechem.2021.100223.

Nayak, Nisha S., Mamta S. Purohit, Devayani R. Tipre, and Shailesh R. Dave. 2020. "Biosurfactant Production and Engine Oil Degradation by Marine Halotolerant *Bacillus licheniformis* LRK1." *Biocatalysis and Agricultural Biotechnology* 29 (September). Elsevier Ltd: 101808. doi:10.1016/j.bcab.2020.101808.

Nitschke, Marcia, Siddhartha G.V.A.O. Costa, and Jonas Contiero. 2005. "Rhamnolipid Surfactants: An Update on the General Aspects of These Remarkable Biomolecules." *Biotechnology Progress* 21 (6): 1593–1600. doi:10.1021/bp050239p.

Nurfarahin, Abdul Hamid, Mohd Shamzi Mohamed, and Lai Yee Phang. 2018. "Culture Medium Development for Microbial-Derived Surfactants Production-An Overview." *Molecules (Basel, Switzerland)* 23 (5): E1049. doi:10.3390/molecules23051049.

Ohadi, Mandana, Gholamreza Dehghannoudeh, Hamid Forootanfar, Mojtaba Shakibaie, and Majid Rajaee. 2018. "Investigation of the Structural, Physicochemical Properties, and Aggregation Behavior of Lipopeptide Biosurfactant Produced by *Acinetobacter junii* B6." *International Journal of Biological Macromolecules* 112 (June): 712–19. doi:10.1016/j.ijbiomac.2018.01.209.

Patiño, Albert D., Manuela Montoya-Giraldo, Marynes Quintero, Lizbeth L. López-Parra, Lina M. Blandón, and Javier Gómez-León. 2021. "Dereplication of Antimicrobial Biosurfactants from Marine Bacteria Using Molecular Networking." *Scientific Reports* 11 (1): 16286. doi:10.1038/s41598-021-95788-9.

Peele, K. Abraham, V. Ravi Teja Ch., and Vidya P. Kodali. 2016. "Emulsifying Activity of a Biosurfactant Produced by a Marine Bacterium." *3 Biotech* 6 (2): 177. doi:10.1007/s13205-016-0494-7.

Prakash, Arumugam Arul, Natarajan Srinivasa Prabhu, Aruliah Rajasekar, Punniyakotti Parthipan, Mohamad S. AlSalhi, Sandhanasamy Devanesan, and Muthusamy Govarthanan. 2021. "Bio-Electrokinetic Remediation of Crude Oil Contaminated Soil Enhanced by Bacterial Biosurfactant." *Journal of Hazardous Materials* 405 (September 2020). Elsevier B.V.: 124061. doi:10.1016/j.jhazmat.2020.124061.

Primeia, Sandia, Chihiro Inoue, and Mei-Fang Chien. 2020. "Potential of Biosurfactants' Production on Degrading Heavy Oil by Bacterial Consortia Obtained from Tsunami-Induced Oil-Spilled Beach Areas in Miyagi, Japan." *Journal of Marine Science and Engineering* 8 (8). Multidisciplinary Digital Publishing Institute: 577. doi:10.3390/jmse8080577.

Raddadi, Noura, Lucia Giacomucci, Grazia Totaro, and Fabio Fava. 2017. "*Marinobacter* sp. from Marine Sediments Produce Highly Stable Surface-Active Agents for Combatting Marine Oil Spills." *Microbial Cell Factories* 16 (1): 186. doi:10.1186/s12934-017-0797-3.

RamyaDevi, K.C., R.L. Sundaram, D. Asha, V. Sivamurugan, V. Vasudevan, and K.M.E. Gnanambal. 2018. "Demonstration of Bioprocess Factors Optimization for Enhanced Mono-Rhamnolipid Production by a Marine *Pseudomonas guguanensis*." *International Journal of Biological Macromolecules* 108. Elsevier B.V.: 531–40. doi:10.1016/j.ijbiomac.2017.10.186.

Ravindran, Amrudha, Arya Sajayan, Gopal Balasubramian Priyadharshini, Joseph Selvin, and George Seghal Kiran. 2020. "Revealing the Efficacy of Thermostable Biosurfactant in Heavy Metal Bioremediation and Surface Treatment in Vegetables." *Frontiers in Microbiology* 11. www.frontiersin.org/article/10.3389/fmicb.2020.00222.

Ray, Madhurya, Vipin Kumar, and Chiranjib Banerjee. 2022. "Kinetic Modelling, Production Optimization, Functional Characterization and Phyto-Toxicity Evaluation of Biosurfactant Derived from Crude Oil Biodegrading Pseudomonas Sp. IITISM 19." *Journal of Environmental Chemical Engineering* 10 (2). Elsevier Ltd: 107190. doi:10.1016/j.jece.2022.107190.

Romsted, Laurence S. 2014. "Microbially Derived Biosurfactants: Sources, Design, and Structure-Property Relationships." In *Surfactant Science and Technology*, 362–75. CRC Press.

Roy, Suki, Shreta Chandni, Ishita Das, Loganathan Karthik, Gaurav Kumar, and Kokati Venkata Bhaskara Rao. 2015. "Aquatic Model for Engine Oil Degradation by Rhamnolipid Producing Nocardiopsis VITSISB." *3 Biotech* 5 (2): 153–64. doi:10.1007/s13205-014-0199-8.

Samanta, Sudip K., Om V. Singh, and Rakesh K. Jain. 2002. "Polycyclic Aromatic Hydrocarbons: Environmental Pollution and Bioremediation." *Trends in Biotechnology* 20 (6): 243–48. doi:10.1016/S0167-7799(02)01943-1.

Santos, Danyelle Khadydja F., Raquel D. Rufino, Juliana M. Luna, Valdemir A. Santos, and Leonie A. Sarubbo. 2016. "Biosurfactants: Multifunctional Biomolecules of the 21st Century." *International Journal of Molecular Sciences* 17 (3): 401. doi:10.3390/ijms17030401.

Santos-Gandelman, Juliana F., Kimberly Cruz, Sharron Crane, Guilherme Muricy, Marcia Giambiagi-deMarval, Tamar Barkay, and Marinella S. Laport. 2014. "Potential Application in Mercury Bioremediation of a Marine Sponge-Isolated

Bacillus cereus Strain Pj1." *Current Microbiology* 69 (3): 374–80. doi:10.1007/s00284-014-0597-5.

Sar, Nechemia, and Eugene Rosenberg. 1983. "Emulsifier Production By *Acinetobacter calcoaceticus* Strains." *Current Microbiology* 9 (6): 309–13. doi:10.1007/BF01588825.

Satpute, Surekha K., Ibrahim M. Banat, Prashant K. Dhakephalkar, Arun G. Banpurkar, and Balu A. Chopade. 2010. "Biosurfactants, Bioemulsifiers and Exopolysaccharides from Marine Microorganisms." *Biotechnology Advances* 28 (4): 436–50. doi:10.1016/j.biotechadv.2010.02.006.

Satpute, Surekha K., Arun G. Banpurkar, Prashant K. Dhakephalkar, Ibrahim M. Banat, and Balu A. Chopade. 2010. "Methods for Investigating Biosurfactants and Bioemulsifiers: A Review." *Critical Reviews in Biotechnology* 30 (2): 127–44. doi:10.3109/07388550903427280.

Shaligram, Nikhil S., and Rekha S. Singhal. 2010. "Surfactin—A Review on Biosynthesis, Fermentation, Purification and Applications." *Food Technology and Biotechnology* 48 (2): 119–34.

Sharma, Deepika, Shelley Sardul Singh, Piyush Baindara, Shikha Sharma, Neeraj Khatri, Vishakha Grover, Prabhu B. Patil, and Suresh Korpole. 2020. "Surfactin Like Broad Spectrum Antimicrobial Lipopeptide Co-Produced with Sublancin from *Bacillus subtilis* Strain A52: Dual Reservoir of Bioactives." *Frontiers in Microbiology* 11. www.frontiersin.org/article/10.3389/fmicb.2020.01167.

Shekhar, S., A. Sundaramanickam, K. Saranya, M. Meena, S. Kumaresan, and T. Balasubramanian. 2019. "Production and Characterization of Biosurfactant by Marine Bacterium *Pseudomonas stutzeri* (SSASM1)." *International Journal of Environmental Science and Technology* 16: 4697–706. Springer: Berlin Heidelberg. https://doi.org/10.1007/s13762-018-1915-4.

Shukla, Sudhir K., Neelam Mangwani, T. Subba Rao, and Surajit Das. 2014. "8—Biofilm-Mediated Bioremediation of Polycyclic Aromatic Hydrocarbons." In *Microbial Biodegradation and Bioremediation*, edited by Surajit Das, 203–32. Elsevier. doi:10.1016/B978-0-12-800021-2.00008-X.

Silva, Elias J., Priscilla F. Correa, Darne G. Almeida, Juliana M. Luna, Raquel D. Rufino, and Leonie A. Sarubbo. 2018. "Recovery of Contaminated Marine Environments by Biosurfactant-Enhanced Bioremediation." *Colloids and Surfaces B: Biointerfaces* 172 (August): 127–35. doi:10.1016/j.colsurfb.2018.08.034.

Sivapathasekaran, Chandrasekaran, and Ramkrishna Sen. 2013. "Performance Evaluation of Batch and Unsteady State Fed-Batch Reactor Operations for the Production of a Marine Microbial Surfactant." *Journal of Chemical Technology and Biotechnology* 88 (4): 719–26. doi:10.1002/jctb.3891.

Soberón-Chávez, Gloria. 2010. *Biosurfactants: From Genes to Applications*. Springer Science & Business Media.

Solís-González, C.J., and H. Loza-Tavera. 2019. "Alicycliphilus: Current Knowledge and Potential for Bioremediation of Xenobiotics." *Journal of Applied Microbiology* 126 (6): 1643–56. doi:10.1111/jam.14207.

Tripathi, Lakshmi, Victor U. Irorere, Roger Marchant, and Ibrahim M. Banat. 2018. "Marine Derived Biosurfactants: A Vast Potential Future Resource." *Biotechnology Letters* 40 (11): 1441–57. doi:10.1007/s10529-018-2602-8.

Tripathi, Lakshmi, Matthew S. Twigg, Aikaterini Zompra, Karina Salek, Victor U. Irorere, Tony Gutierrez, Georgios A. Spyroulias, Roger Marchant, and Ibrahim M. Banat. 2019. "Biosynthesis of Rhamnolipid by a Marinobacter Species

Expands the Paradigm of Biosurfactant Synthesis to a New Genus of the Marine Microflora." *Microbial Cell Factories* 18 (1). BioMed Central: 1–12. doi:10.1186/s12934-019-1216-8.

Unás, Jorge H., Darlisson de Alexandria Santos, Eduardo Bessa Azevedo, and Marcia Nitschke. 2018. "*Brevibacterium luteolum* Biosurfactant: Production and Structural Characterization." *Biocatalysis and Agricultural Biotechnology* 13 (December 2017). Elsevier Ltd: 160–67. doi:10.1016/j.bcab.2017.12.005.

Velásquez, Anyela Vanessa, Marynés Quintero, Eylin Yaidith Jiménez, Lina Marcela Blandón, and Javier Gómez. 2018. "Microorganismos marinos extremófilos con potencial en bioprospección." *Revista de la Facultad de Ciencias* 7 (2): 9–43. doi:10.15446/rev.fac.cienc.v7n2.67360.

Vigneshwaran, C., V. Sivasubramanian, K. Vasantharaj, N. Krishnanand, and M. Jerold. 2018. "Potential of *Brevibacillus* sp. AVN 13 Isolated from Crude Oil Contaminated Soil for Biosurfactant Production and Its Optimization Studies." *Journal of Environmental Chemical Engineering* 6 (4). Elsevier: 4347–56. doi:10.1016/j.jece.2018.06.036.

Vigneshwaran, C., K. Vasantharaj, N. Krishnanand, and V. Sivasubramanian. 2021. "Production Optimization, Purification and Characterization of Lipopeptide Biosurfactant Obtained from *Brevibacillus* sp. AVN13." *Journal of Environmental Chemical Engineering* 9 (1). Elsevier Ltd: 104867. doi:10.1016/j.jece.2020.104867.

Vilela, W.F.D., S.G. Fonseca, F. Fantinatti-Garboggini, V.M. Oliveira, and M. Nitschke. 2014. "Production and Properties of a Surface-Active Lipopeptide Produced by a New Marine *Brevibacterium luteolum* Strain." *Applied Biochemistry and Biotechnology* 174 (6): 2245–56. doi:10.1007/s12010-014-1208-4.

Walter, Vanessa, Christoph Syldatk, and Rudolf Hausmann. 2013. *Screening Concepts for the Isolation of Biosurfactant Producing Microorganisms.* Madame Curie Bioscience Database 2000–2013. Landes Bioscience. www.ncbi.nlm.nih.gov/books/NBK6189/.

Ward, Alan C., and Nagamani Bora. 2006. "Diversity and Biogeography of Marine Actinobacteria." *Current Opinion in Microbiology* 9 (3): 279–86. doi:10.1016/j.mib.2006.04.004.

White, D.A., L.C. Hird, and S.T. Ali. 2013. "Production and Characterization of a Trehalolipid Biosurfactant Produced by the Novel Marine Bacterium *Rhodococcus* sp., Strain PML026." *Journal of Applied Microbiology* 115 (3): 744–55. doi:10.1111/jam.12287.

Wu, Yanan, Meng Xu, Jianliang Xue, Ke Shi, and Meng Gu. 2019. "Characterization and Enhanced Degradation Potentials of Biosurfactant-Producing Bacteria Isolated from a Marine Environment." *ACS Omega* 4 (1). American Chemical Society: 1645–51. doi:10.1021/acsomega.8b02653.

Xu, Meng, Xinge Fu, Yu Gao, Liangfeng Duan, Congchao Xu, Wenshuang Sun, Yixuan Li, Xianzheng Meng, and Xinfeng Xiao. 2020. "Characterization of a Biosurfactant-Producing Bacteria Isolated from Marine Environment: Surface Activity, Chemical Characterization and Biodegradation." *Journal of Environmental Chemical Engineering* 8 (5). Elsevier: 104277. doi:10.1016/j.jece.2020.104277.

Yalaoui-Guella, Drifa, Samira Fella-Temzi, Salima Djafri-Dib, Khodir Madani, Ibrahim Banat, and Fatiha Brahmi. 2020. "Biodegradation Potential of Crude Petroleum by Hydrocarbonoclastic Bacteria Isolated from Soummam Wadi Sediment and Chemical-Biological Proprieties of Their Biosurfactants." *Journal of Petroleum Science and Engineering* 184 (106554): 1–7. doi:10.1016/j.petrol.2019.106554.

Yao, Shulin, Shengming Zhao, Zhaoxin Lu, Yuqi Gao, Fengxia Lv, and Xiaomei Bie. 2015. "Control of Agitation and Aeration Rates in the Production of Surfactin in Foam Overflowing Fed-Batch Culture with Industrial Fermentation." *Revista Argentina de Microbiología* 47 (4): 344–49. doi:10.1016/j.ram.2015.09.003.

Zenati, Billal, Alif Chebbi, Abdelmalek Badis, Kamel Eddouaouda, Hocine Boutoumi, Mohamed El Hattab, Dorra Hentati, et al. 2018. "A Non-Toxic Microbial Surfactant from *Marinobacter hydrocarbonoclasticus* SdK644 for Crude Oil Solubilization Enhancement." *Ecotoxicology and Environmental Safety* 154 (June): 100–107. doi:10.1016/j.ecoenv.2018.02.032.

Zhang, W., X. Zhang, and H. Cui. 2015. "Isolation, Fermentation Optimization and Performance Studies of a Novel Biosurfactant Producing Strain *Bacillus amyloliquefaciens*." *Chemical and Biochemical Engineering Quarterly* 29 (3): 447.

Zhou, Hanghai, Lijia Jiang, Keliang Li, Chunlei Chen, Xiaoyun Lin, Chunfang Zhang, and Qinglin Xie. 2021. "Enhanced Bioremediation of Diesel Oil-Contaminated Seawater by a Biochar-Immobilized Biosurfactant-Producing Bacteria *Vibrio* Sp. LQ2 Isolated from Cold Seep Sediment." *Science of The Total Environment* 793 (November): 148529. doi:10.1016/j.scitotenv.2021.148529.

Zhu, Zhiwen, Baiyu Zhang, Bing Chen, Qinghong Cai, and Weiyun Lin. 2016. "Biosurfactant Production by Marine-Originated Bacteria *Bacillus subtilis* and Its Application for Crude Oil Removal." *Water, Air, & Soil Pollution* 227 (9): 328. doi:10.1007/s11270-016-3012-y.

4 Rhamnolipids Produced by Marine Microorganisms, A Perspective

Jorge Gracida, Arturo Abreu, Dulce Celeste López Díaz and Evelyn Zamudio Pérez

CONTENTS

INTRODUCTION

The sea as an object of study is a part of the planet that has not been fully investigated. Two factors affect it globally: first, the melting of glaciers, and second, the pollution of its water from different sources. This disturbance in turn causes changes in the diversity of living beings that inhabit it, including the present biota, which we have not finished analyzing. In the case of bacteria, there are some that present biological activities with development potential. Such is the case with the biosynthesis of surfactants. In the chapter, we briefly describe something about it.

MARINE BACTERIA RHAMNOLIPIDS PRODUCERS

Biosurfactants are amphipathic structures, contain hydrophobic and hydrophilic domains, and can reduce surface tension. One major class of biosurfactant is the rhamnolipids. The production of rhamnolipids is found in different phyla, including proteobacteria and Actinobacteria, as shown in Table 4.1. However, *Pseudomonas* spp. have been reported as a major source of rhamnolipids and, due to their pathogenicity, have been of increased interest in the discovery of non-pathogenic rhamnolipid producers. Marine microorganisms possess unique

DOI: 10.1201/9781003307464-4

TABLE 4.1
Marine Bacteria Producers of Rhamnolipid Isolated from Seawater

Organism Producer	Carbon Source
Tetragenococcus koreensis	Hydrocarbons
Pantoea sp.	Hydrocarbons
Acinetobacter calcoaceticus	Petroleum
Cellulomonas cellulans	Hydrocarbon
Myxococcus sp.	Hydrocarbons
Nocardioides sp.	Olefins
Actinoalloteichus hymeniacidonis	Glucose
Pseudomonas aeruginosa	Molasses
Pseudoxanthomonas sp.	Charcoal
Renibacterium salmoninarum	Hexadecane

metabolic and physiological features and are an important source of new bio-molecules, such as biosurfactants (Gudiña et al., 2016). Halotolerant biosurfactants are required for efficient bioremediation of marine oil spills (Kiran et al., 2016). Recent reports show marine microorganisms as a source of rhamnolipid production.

Next, we will talk about some marine species that have been reported as producers of rhamnolipids.

In 2014, Yan et al. isolated *Streptomyces* sp. ISP2–49 (marine actinomycetes) from sediment samples from Galveston Bay, Texas, and screened for production of bioactive metabolites. This was the first report of a rhamnolipid from the genus *Streptomyces*. In 2015 Roy et al. focused on the isolation, screening, characterization, and application of biosurfactant-producing marine actinobacteria. Twenty actinobacteria were isolated from the marine water sample; two isolates were selected (SIS-3 and SIS-20), the biosurfactant produced by SIS-3 was characterized, and the compound detected was rhamnolipid. The isolate (SIS-3) was identified as *Nocardiopsis* (Roy et al., 2015).

Tedesco et al. (2016) applied a biodiscovery pipeline for the identification of anti-Bcc compounds. Antarctic sub-sea sediments were collected from the Ross Sea and used to isolate 25 microorganisms, which were phylogenetically affiliated to three bacterial genera (*Psychrobacter, Arthrobacter*, and *Pseudomonas*) via sequencing and analysis of 16S rRNA genes. Compound structures are shown in Figure 4.1. The molecular formula of compound *1* was established as $C_{28}H_{52}O_9$; data indicated that it is a known rhamnolipid containing two fully saturated lipid chains. The molecular formula of compound *2* was established as $C_{28}H_{50}O_9$, and subsequent dereplication suggested it was new. The molecular formula of compound 3 was established as $C_{30}H_{54}O_9$. Based on the data, compound 3 was similar to 2, the difference being an additional C_2H_4 unit (Tedesco et al., 2016).

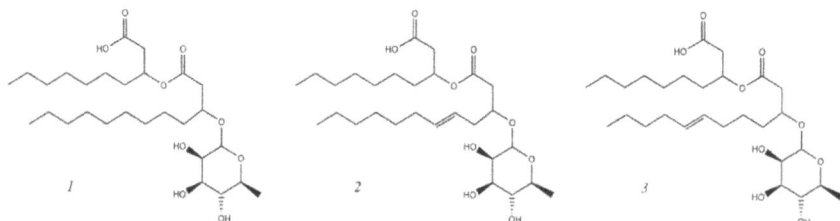

FIGURE 4.1 Structures of rhamnolipids 1–3 isolated from *Pseudomonas* BTN1 (Tedesco, 2016).

In 2017 Cheng et al. isolated a bacterium derived from oil sludge from Zhoushan Islands, China, one of the national strategic oil reservoirs, identified as *P. aeruginosa* ZS1. In this study, they showed that ZS1 isolate was capable of consuming 60% crude oil at 30°C in 14 days and giving yields between 30 and 44 g/L (the highest rhamnolipid yields are reported to range from 70 to 120 g/L by the chemically mutagenized strains DSM 7107 and DSM 7108 derived from an environmental isolate cultivated in bioreactors by Giani et al., 1997). The analysis indicated that the rhamnolipids produced by ZS1 consisted of 7 mono-rhamnolipid and 11 di-rhamnolipid homologs (RL7–11), two of which were novel (Cheng et al., 2017).

A non-pathogenic rhamnolipid-producing marine *Pseudomonas* sp. MCTG214 (3b1) was shown to produce mono- and di-rhamnolipids (87% relative abundance); the analysis of the samples indicated the presence of five different rhamnolipids, all congeners possessing fatty acid moieties consisting of 8–12 carbons, but the more abundant were Rha-Rha-C_{10}-C_{10} (42.75%), as shown in Figure 4.2, and Rha-Rha-C_{10} (23.8%). They postulated that RL synthesis in *Pseudomonas* sp. MCTG214(3b1) is carried out by enzymes expressed from rhlA/B homologs similar to those of *P. aeruginosa*; however, a lack of rhlC potentially indicates the presence of a second novel rhamnosyltransferase responsible for the di-rhamnolipid congeners identified by high-performance liquid chromatography—mass spectrometry (HPLC-MS) (Twigg et al., 2018). These marine bacteria were identified as *Marinobacter* sp. and *Pseudomonas mendocina*, and both species were characterized via HPLC-MS and nuclear magnetic resonance (NMR) (Twigg et al., 2019). An arctic marine bacterium identified as *Pseudomonas fluorescens* species was reported to synthesize five mono-rhamnolipid congeners and the lipid moiety of one of the rhamnolipids (Figure 4.3) (Kristoffersen et al., 2018). Twigg et al. (2019) reported two rhamnolipids produced by marine bacteria. Rhamnolipid synthesis was then confirmed and characterized via HPLC-MS and NMR. Both 16S rDNA and subsequent genomic sequencing revealed these strains to be *Marinobacter* sp. and *Pseudomonas mendocina*, both species where rhamnolipid production was previously unreported.

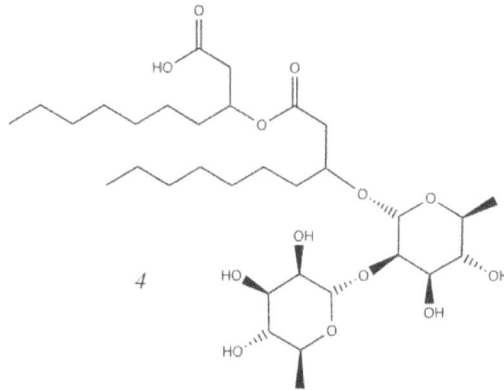

FIGURE 4.2   Structure of rhamnolipid 4, Rha-Rha-C_{10}-C_{10} (Conceição et al., 2020).

Tripathi et al. (2019) investigated biosurfactant production in five bacterial strains isolated from coastal and offshore sites in the United States, Scotland, and Norway. One of these strains, *Marinobacter* sp. MCTG107b, possessed phenotypic traits indicative of biosurfactant synthesis. To analyze the chemical structure, they used high-performance liquid chromatography-mass spectrometry and tandem-MS, and the results confirmed the presence of rhamnolipids. Congener producers of rhamnolipids produced mono- and di-rhamnolipids; however, there was a preference for the synthesis of di-rhamnolipids (95.39% of total rhamnolipids). The most abundantly synthesized rhamnolipids were Rha-Rha C_{10}-C_{10} (52.45%), Rha-Rha C_{10}-C_{10} CH_3 (23.07%), Rha-Rha C_{10} (5.13%), Rha-Rha C_{10}-C_{12} (5.01%), Rha-Rha C_{10}-C_{12} CH_3 (3.26%), and Rha-$C_{14\cdot2}$ (3.18%).

CHARACTERIZATION OF MARINE RHAMNOLIPIDS

The workgroup of Kristoffersen (2018) cultured an Arctic marine strain of *Pseudomonas* sp., strain M10B774, affiliated with the *Pseudomonas fluorescens* group, in four different media broths. Kristoffersen et al. demonstrated the use of MS/MS-based molecular networks as a dereplication strategy to identify known compounds, their analogs, and related compounds. The application of this strategy led to the isolation of a new mono-rhamnolipid, four already known mono-rhamnolipids, and a lipidic fraction of one of the rhamnolipids, which consists of two fatty acid fragments.

In characterization of the compounds isolated by Kristoffersen et al. (2018), lipid 5 and mono-rhamnolipids 6–10 were isolated as viscous liquids, while molecular formulas were calculated using precise mass and isotope distribution by high-resolution electrospray ionization mass spectrometry (HR-ESI-MS). The structures of the isolated compounds are shown in Figure 4.3; these were determined by correlating the information from both the HR-ESI-MS technique and the ¹H and ¹³C NMR and 2D techniques, as well as fragmentation by mass

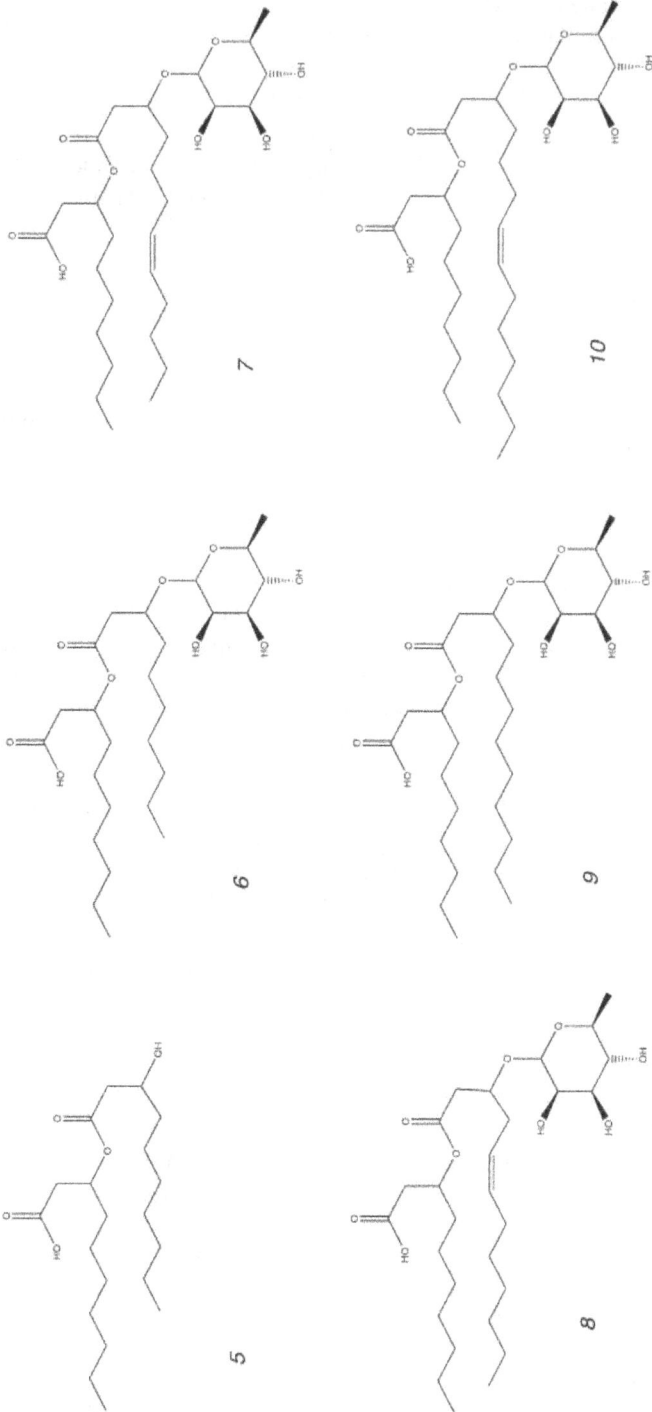

FIGURE 4.3 Structures of compounds 5–10 isolated from *Pseudomonas* sp. (Kristoffersen et al., 2018).

spectrometry tandem (MS/MS). Using the techniques described, the working group determined that compound 5 was the lipid residue of compound 6, while compounds 6–10 were elucidated as mono-rhamnolipids with different fatty acids in their structure; in the case of compounds 7, 8, and 10, they had the presence of a double bond with *cis* configuration.

Using NMR and mass spectrometry techniques, they determined that the lipid chains of compounds 5–10 contained between 10 and 14 carbon atoms. In the ^{1}H NMR spectrum of compound 8, a double bond is found in the 12-carbon chain with *cis* configuration, with a chemical shift around 5.35–5.45 ppm. These data were compared with those described by Tedesco et al. (2016) for a similar compound; however, the double bond configuration is described as a *trans* configuration. It should be noted that these are bacterial strains isolated from the Ross Sea in Antarctica.

In the ^{13}C NMR spectrum of compound 8, signals corresponding to the carbons of an ester (170.0 ppm) and a carboxylic acid (171.6 ppm) can be observed, while the olefinic carbons have a chemical shift of 124.4 and 132.4 ppm and the anomeric carbon at 98.8 ppm. The rest of the carbons are observed in two regions, between 68.8 and 72.5 ppm of the carbons bound to oxygen and between 13.4 and 40.5 ppm of the aliphatic carbons.

The Tedesco workgroup (2016) collected Antarctic submarine sediments from the Ross Sea and managed to isolate 25 microorganisms phylogenetically affiliated to three bacterial genera, *Psychrobacter*, *Arthrobacter*, and *Pseudomonas*, with the ability to inhibit strains of opportunistic human pathogens belonging to the *Burkholderia cepacia* complex (Bcc). The molecular formula of compound 1 was established as $C_{28}H_{52}O_9$ by HR-ESI-MS. For this compound, it was determined that it does not present unsaturation in the lipid chains, which have 10 and 12 carbon atoms, confirmed by tandem mass spectrometry (MS/MS). For molecules 2 and 3, using the same techniques, it was determined that they had unsaturation in the lipid chains in positions five and seven with respect to the carbonyl of the ester, respectively (Figure 4.1).

The ESI MS/MS spectrum of rhamnolipid 1 showed the adduct, the molecular ion plus the sodium cation $[M + Na]^+$, verifying that the lipid fragments of compound 1 do not have double bonds.

As mentioned, compound 2 had unsaturation in the lipid chain, specifically at carbon five with respect to the ester. In the ESI MS/MS spectrum, the $[M + Na]^+$ ion is observed, verifying an additional unsaturation with respect to compound 1 described previously. In its fragmentation pattern, three main fragments can be observed, the signal base, which consists of the lipid fragment; another fragment that contains the rhamnose unit and lipid part of the ester; and finally a fragment that corresponds to the ester. In all cases, sodium is bound to the fragments.

The Tedesco workgroup characterized mono-rhamnolipids by several NMR techniques. In the ^{1}H NMR spectrum, it can be seen that the ^{1}H spectrum is similar to compound 8 described by Kristoffersen et al. (2018). Olefinic hydrogens exhibit a similar chemical shift and multiplicity. However, an important difference is present in the methylene group, which has diastereotopic hydrogens in the 4' position of the ester fragment, and the hydrogens split in an ABCD-type system. This was corroborated by other NMR techniques, for example, the two-dimensional heteronuclear single quantum coherence (HSQC) spectrum. This technique can check the interactions between hydrogens and carbons to a bond; that is, it shows that hydrogens are directly attached to each carbon atom present in the molecules. Unlike the heteronuclear multiple bond connectivity (HMBC) technique, the interactions between hydrogens and carbons can be observed at two or three bonds, that is, at a greater distance. These techniques were used by both research groups.

Tripathi et al. (2019) described the biosynthesis and characterization of glycolipids produced by the strain *Marinobacter* sp. MCTG107b by HPLC-MS. They mention that using this technique, they obtained m/z values that corresponded to already known rhamnolipid values; that is, the MCTG107b strain produced a family (congeners) of rhamnolipids that include mono- and di-rhamnolipids, the latter being the most favored since they represent 95.39% of the production of biosurfactants.

Of all the congeners synthesized by *Marinobacter* sp. MCTG107b, five main biosurfactants stand out, including a-l-rhamnopyranosyl-α-l-rhamnopyranosyl-β-hydroxydecanoyl-b-hydroxydecanoate (Rha-Rha-C_{10}-C_{10}), compound 11, with a molecular weight of 650.79 Da and 52.45% abundance. The next most synthesized congeners, as can be seen in Figure 4.4, were Rha-Rha-C_{10}-$C_{10}CH_3$ (23.07%) compound 12, Rha-Rha-C_{10} (5.13%) compound 13, Rha-Rha-C_{10}-C_{12} (5.01%) compound 14, Rha-Rha-C_{10}-$C_{12}CH_3$ (3.26%) compound 15, and Rha-$C_{14:2}$ (3.18%) compound 16.

The authors describe that the production of rhamnolipids has not been previously observed in strains of the genus *Marinobacter*. To confirm this, they performed tandem mass spectrometry on the main molecular ion, verifying the detection of fragmentation ions of the biosurfactant Rha-Rha-C_{10}-C_{10}.

On the other hand, the group of Twigg et al. (2018) carried out a study to identify and characterize biosurfactants produced by bacteria associated with a marine eukaryotic phytoplankton bloom, a strain of *Pseudomonas* sp. called MCTG214(3b1). They described that it is not related to *P. aeruginosa*, and therefore, it is much less pathogenic. In the study, they were able to identify short-chain di-rhamnolipids biosynthesized by a non-pathogenic marine species of *Pseudomonas*. The main techniques used for characterization were HPLAC-MS and NMR spectroscopy.

The profile of the HPLC-MS technique obtained from the supernatant of the *Pseudomonas* sp. MCTG214(3b1) and purified by solid-phase extraction (SPE) confirmed the presence of five rhamnolipid congeners; one of the congeners is a mono-rhamnolipid, with an abundance of 12.26% (Rha-Rha-C_8-C_{12}), while the remaining four congeners are of the di-rhamnolipid type with a total abundance of 87.74%, with 42.74% Rha-Rha-C_{10}-C_{10} and 23.80% Rha-Rha-C_{10} being the most abundant, followed by 11.42% Rha-C_8-C_{10}/Rha-C_{10}-C_8 and finally Rha-Rha-C_{10}-C_{-12}/Rha-Rha-C_{12}-C_{10} with 9.78%. In Figure 4.5, the value indicated in the parenthesis represents the abundance of each congener.

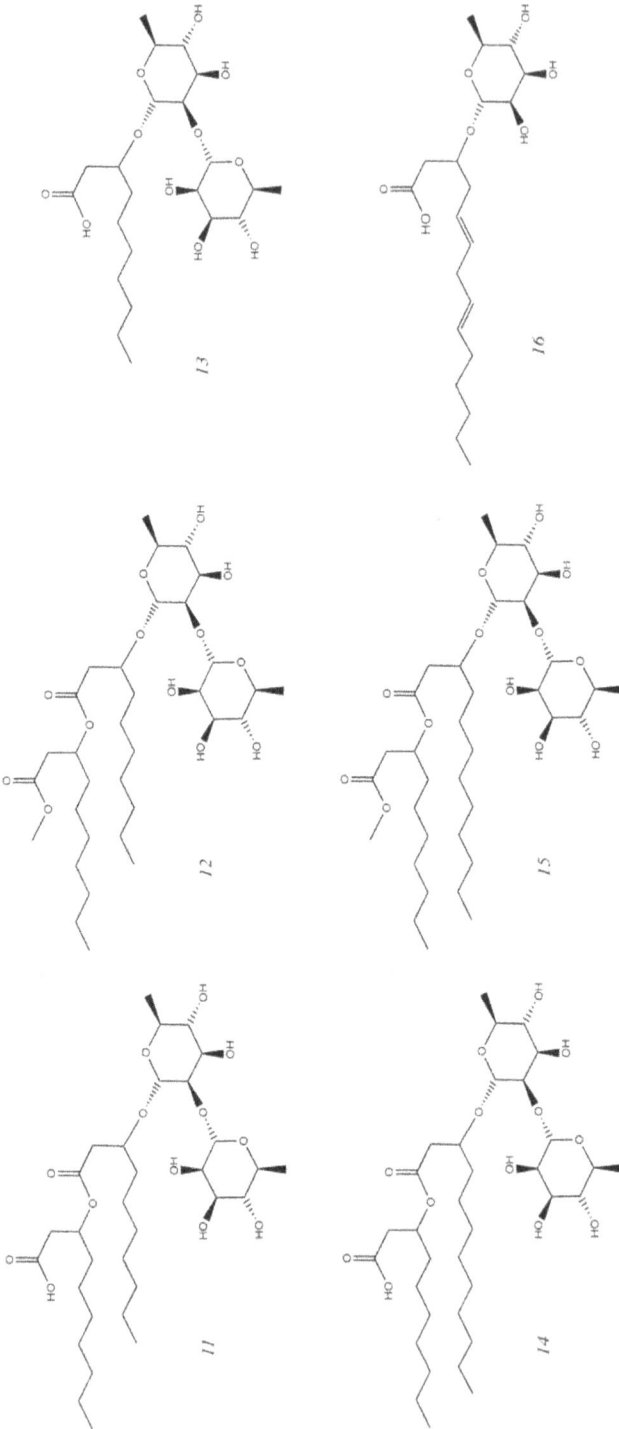

FIGURE 4.4 Structures of mono-rhamnolipid and di-rhamnolipid compounds are more abundant in the isolates described by Tripathi et al. (2019).

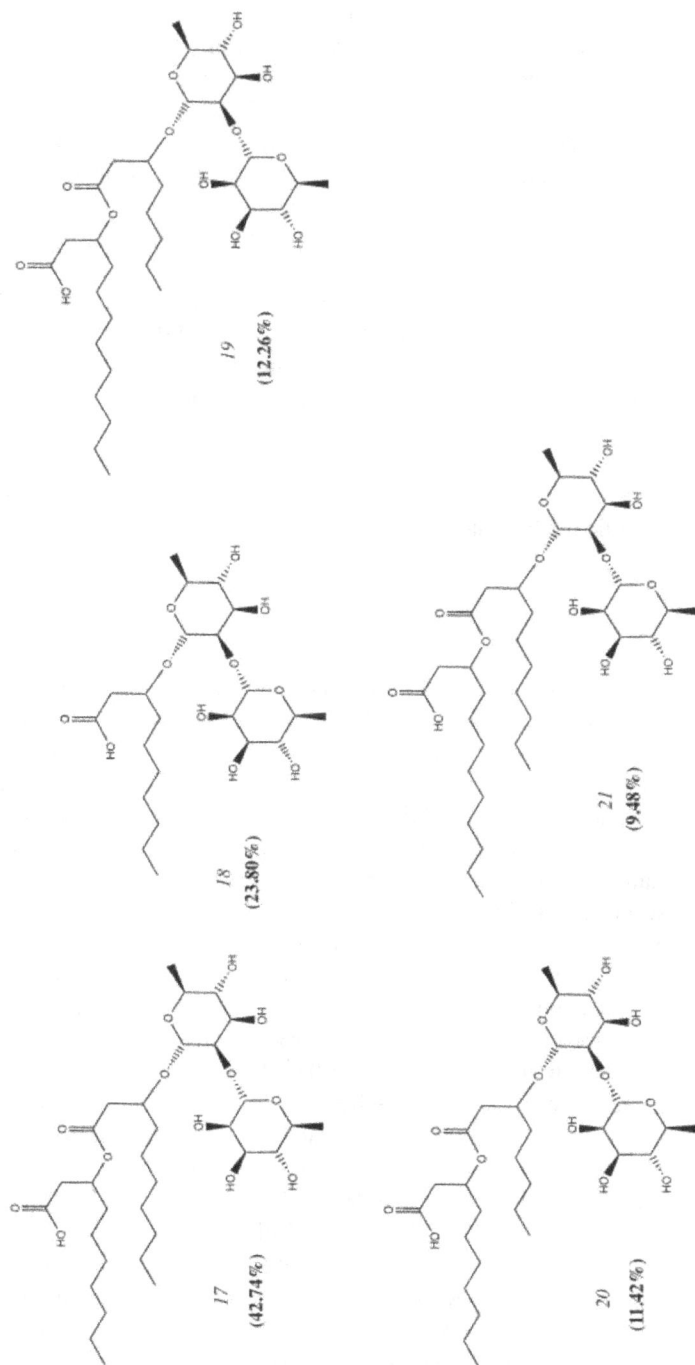

FIGURE 4.5 Structures of five rhamnolipids congeners isolated and characterized by Twigg et al. Congeners 17–19 showed structures of short chains present in lipidic ester.

From the analysis of NMR and MS techniques, it can be concluded that *Pseudomonas* sp. MCTG214(3b1) does not synthesize rhamnolipid congeners that have fatty acid chains of more than 12 carbons. In addition, there is no evidence that the lipid fragments do not present unsaturated, as described by Kristoffersen et al. (2018), Tedesco et al. (2016), and Tripathi et al. (2019).

In recent work, Xu et al. (2020) studied the strain *Paracoccus* sp. MJ9 producer of biosurfactants, isolated from Jiaozhou Bay in Qingdao, Shandong province (China). The results of thin plate chromatography (TLC), Fourier transform infrared spectroscopy (FT-IR), and NMR indicate that the metabolites produced by the strain are rhamnolipids.

Xu et al. (2020) described that FT-IR presents bands that are characteristic of rhamnolipids, and those observed at 3411, 2923, 2852, 1617, 1398, and 1085 cm^{-1} confirm a glycolipid. The adsorption band at 3411 cm^{-1} corresponds to stretching hydrogen bonded of O-H groups. The band at 1617 cm^{-1} is attributed to the ester and carboxylic acid carbonyl groups. Asymmetric and symmetric stretch bands corresponding to the lipid methylene groups appeared at 2923 cm^{-1} and 2852 cm^{-1} (-CH), respectively. The band at 1398 cm^{-1} corresponds to the curved plane of the carboxylic acid, while the band at 1085 cm^{-1} is attributed to the stretching of the C-O ester bond.

MODELING AND RHAMNOLIPIDS

Chemical surface-active or surfactant compounds are used for numerous industrial applications, such as cosmetic, medical, food additives, as well as the removal of contaminants. Due to the ability to interact with hydrophilic compounds and their affinity with this phase, they produce a decrease in surface tension which leads to an increase in solubility (Raiger Iustman and López, 2009). Depending on the industrial, domestic, or natural process where they can be applied, surfactants can be classified by origin, chemical structure, ionization in water, and so on. Regarding their origin, surfactants can be synthesized or biological, known as biosurfactants. Although there are a wide variety of organisms that can produce surfactants, bacteria and fungi are the main producers of these compounds, through a diversity of processing routes, each one has a specific performance depending on the type of microorganism and the conditions of the implemented method (Youssef et al., 2004). These microorganisms have been studied in recent years to obtain the best conditions of pH, nutrients, and temperature to provide the highest yields to scale their production in the future. Typically, rhamnolipids are biomolecules produced mainly by *Pseudomonas aeruginosa* bacteria (Muthusamy et al., 2008).

One of the first works about the alternative of biomolecules in the sea was carried out by Žutić et al. (1981). However, this class of microorganisms lives in an environment with extreme conditions that could affect the production of the desired bio-surfactant, since by trying to be resilient to changes in their environment, their metabolism is affected. The great potential of this alternative is also sustained by being applicable on an industrial scale and its environmental degradation, counteracting its synthetic counterpart in the environmental field concerning contamination by final disposal. Their success in bioremediation also depends on precise physical conditions and the chemical nature of the pollutants in the affected areas. Some studies as presented by Banat et al. (2000) obtained great results using biosurfactants in hydrocarbon production control in marine biotopes considering the effects of these conditions.

However, the advantages of biosurfactants have not been exploited due to low yields and insufficient information about their metabolism to determine adequate conditions. Sometimes sufficient information about biosynthesis is necessary to obtain maximum production, improve cost-efficiency, and realize industrial demands (Tripathi et al., 2018). To mention the importance of this, in the case of surfactants that are produced from marine bacteria, good monitoring using remote sensing techniques can give a better understanding between the ocean and atmosphere (range of wind-wave conditions, organic material as dissolved oil in a column of water, etc.) to define the biophysical processes (Kurata et al., 2016).

Once the optimal growth and processing conditions are defined for the type of marine microorganism selected, the optimal processing routes and reagents (substrates) can be scaled up to evaluate production yields against operational limitations. Determining the cost-benefit in production and implementation of these types of technologies is essential in the economic and technical analysis for the market introduction of a new product, especially when it is sought to have such a great impact at an ecological and political level (Banat et al., 2014). In 2015, Hamilton et al. presented a research project to detect biosurfactant production on the microlayer between the sea surface and bacteria to produce biosurfactants by by monitoring the characteristics in Florida and the Gulf of Mexico (2015).

Due to the importance of production conditions for the growth of bacteria and the connections with their metabolism to increase yields, great efforts in research have been put forth in testing several combinations to find the optimal guidelines, and on many occasions, there is little information about numerous metabolic routes that can impact the results (Suh et al., 2019). Therefore, the use of models to minimize the time and resources to find those combinations of conditions is an essential tool in this area.

Some mathematical models have focused on metabolism networks. In 2016, Satya Eswari and Venkateswarlu proposed a dynamic model to analyze the rhamnolipid production synthesized by *Pseudomonas aeruginosa*, identifying important pathways and reactions to enhance yield through the metabolic engineering problem. For marine rhamnolipids, the bioprocess factors can be enhanced by the production of specific sources using optimization studies, as well as the combination of bioprocess parameters (Asha et al., 2018).

The production of rhamnolipid biosurfactants by *Pseudomonas aeruginosa* is affected by a lack of understanding of the kinetics and relevant interrelation of variables, and current processes are based on heuristic approaches. A model describes the biomass growth, substrates, and products that include kinetic representations of the mandatory mechanisms to understand complex biological processes and metabolic networks (Henkel et al., 2014). Considering production via bio-membranes, Herzog et al. (2020) presented the interaction of rhamnolipids varying the complexity of these units. The mathematical models are not only limited to finding the most suitable metabolic conditions, but they can also determine the behavior of the surfactants in the direct application of contaminant removal to evaluate their efficiency. Specific interaction can be defined as a model of pollutant degradation (Chrzanowski et al., 2011; Oluwaseun et al., 2017). Rhamnolipids can constitute a valid alternative to chemical surfactants in promoting the biodegradation of slow-desorption of polycyclic aromatic hydrocarbons, which is one of the most important problems in bioremediation, but the efficiency depends strongly on the bioremediation stage in which biosurfactant is applied (Posada-Baquero et al., 2019) as well as to be able to provide solid rubrics of the advantage of biological surfactants against synthetics on crude oil spill cleanups and other environmental applications (Dobler et al., 2020). Recently, Rathakumar et al. (2021) managed to produce high purity rhamnolipids using immobilized magnetic biocatalysts, where the application of statistical modeling allowed finding optimal values for parameters of interest to increase the purity and efficiency of the bioremediation using an artificial neural network-coupled genetic algorithm compared with a composite central design.

The models allow variations and have obtained some elementary indexes that are called optimization factors. Optimization factors enhance rhamnolipid production by marine *Pseudomonas* using permutation experiments improved from laboratory-scale production (Asha et al., 2018). These factors can affect generation, biofilm formation, and production by electroactive *Pseudomonas aeruginosa*, and are a good alternative to evaluate the use of green energy sources in some mechanisms. Extraction can implement electrokinetic technologies (Gidudu and Chirwa, 2021) as microbial electrochemical cells (Allam et al., 2021), so this technology offers continuity of the process from upstream to downstream, eliminating complexities involved in conventional methods.

Biosurfactant characterization, production, and purification are processes that appear in a biosurfactant model compiling numerous surfactant properties (Varjani and Upasani, 2017). The surface modeling method enhances rhamnolipid production to develop the production processes, and identity responses are measured, such as biomass and rhamnolipid production (Abalos et al., 2002). Process optimization to produce rhamnolipids is evaluated by the nutritional and environmental conditions necessary for optimal production of rhamnolipids (Arutchelvi et al., 2011).

Another important application of mathematical models in biosurfactant production is the prediction approach, to decrease the number of experiments to evaluate the growth of *Pseudomonas aeruginosa* strain, and the rhamnolipid production data from several carbon sources (i.e. oleic acid as carbon source) (Medina-Moreno et al., 2011). Multiple responses of optimization factors (such as surface tension reduction, substrate concentration, and biomass formation) can be analyzed to apply models such as the Taguchi method, especially when the bacteria are modified to increasingly improve yields and production and purification technologies (Raza et al., 2014).

REFERENCES

Abalos, A., Maximo, F., Manresa, M. A., Bastida, J. 2002. Utilization of response surface methodology to optimize the culture media for the production of rhamnolipids by *Pseudomonas aeruginosa* AT10. *Journal of Chemical Technology & Biotechnology*, 77: 777–784. https://doi.org/10.1002/jctb.637.

Allam, F., Elnouby, M., Sabry, S. A., El-Khatib, K. M., El-Badan, D. E. 2021. Optimization of factors affecting current generation, biofilm formation and rhamnolipid production by electroactive *Pseudomonas aeruginosa* FA17. *International Journal of Hydrogen Energy*, 46: 11419–11432. https://doi.org/10.1016/j.ijhydene.2020.08.070.

Arutchelvi, J., Joseph, C., Doble, M. 2011. Process optimization for the production of rhamnolipid and formation of biofilm by *Pseudomonas aeruginosa* CPCL on polypropylene. *Biochemical Engineering Journal*, 56: 37–45. https://doi.org/10.1016/j.bej.2011.05.004.

Asha, D., Vasudevan, V., Krishnan, M. E. G. 2018. Demonstration of bioprocess factors optimization for enhanced mono-rhamnolipid production by a marine *Pseudomonas guguanensis*. *International Journal of Biological Macromolecules*, 108: 531–540. https://doi.org/10.1016/j.ijbiomac.2017.10.186.

Banat, I. M., Makkar, R. S., Cameotra, S. S. 2000. Potential commercial applications of microbial surfactants. *Applied Microbiology and Biotechnology*, 53: 495–508. https://doi.org/10.1007/s002530051648.

Banat, I. M., Satpute, S. K., Cameotra, S. S., Patil, R., Nyayanit, N. V. 2014. Cost effective technologies and renewable substrates for biosurfactants' production. *Frontiers in Microbiology*, 5: 1–18. https://doi.org/10.3389/fmicb.2014.00697.

Cheng, T., Liang, J., He, J., Hu, X., Ge, Z., Liu, J. 2017. A novel rhamnolipid-producing *Pseudomonas aeruginosa* ZS1 isolate derived from petroleum sludge suitable for bioremediation. *AMB Express*, 7: 1–14. https://doi.org/10.1186/s13568-017-0418-x.

Choi, B. K., Lee, H. S., Kang, J. S., Shin, H. J. (2019). Dokdolipids A–C, hydroxylated rhamnolipids from the marine-derived Actinomycete *Actinoalloteichus hymeniacidonis*. *Marine Drugs*, 17(4): 237–245. https://doi.org/10.3390/md17040237.

Chrzanowski, Ł., Owsianiak, M., Szulc, A., et al. 2011. Interactions between rhamnolipid biosurfactants and toxic chlorinated phenols enhance biodegradation of a model hydrocarbon-rich effluent. *International Biodeterioration & Biodegradation*, 65: 605–611. https://doi.org/10.1016/j.ibiod.2010.10.015.

Conceição, K. S., de Alencar Almeida, M., Sawoniuk, I. C., Marques, G. D., de Sousa Faria-Tischer, P. C., Tischer, C. A., . . . Camilios-Neto, D. (2020). Rhamnolipid production by *Pseudomonas aeruginosa* grown on membranes of bacterial cellulose supplemented with corn bran water extract. *Environmental Science and Pollution Research*, 27(24): 30222–30231. https://doi.org/10.1007/s11356-020-09315-w.

Dobler, L., Ferraz, H. C., de Castilho, L. V. A. (2020). Environmentally friendly rhamnolipid production for petroleum remediation. *Chemosphere*, 252: 126349–126358. https://doi.org/10.1016/j.chemosphere.2020.126349.

Giani, C., Wullbrandt, D., Rothert, R., Meiwes, J. 1997. *Pseudomonas aeruginosa* and its use in a process for the biotechnological preparation of l-rhamnose. *U.S. Patent No. 5,658,793*. Washington, DC: U.S. Patent and Trademark Office.

Gidudu, B., Chirwa, E. M. 2021. Electrokinetic extraction and recovery of biosurfactants using rhamnolipids as a model biosurfactant. *Separation and Purification Technology*, 276: 119327–119338. https://doi.org/10.1016/j.seppur.2021.119327.

Gudiña, E. J., Teixeira, J. A., Rodrigues, L. R. 2016. Biosurfactants produced by marine microorganisms with therapeutic applications. *Marine Drugs*, 14: 38–52. https://doi.org/10.3390/md14020038.

Hamilton, B., Dean, C., Kurata, N., et al. 2015. Surfactant associated bacteria in the sea surface microlayer: Case studies in the Straits of Florida and the Gulf of Mexico. *Canadian Journal of Remote Sensing*, 41: 135–143. https://doi.org/10.1080/07038992.2015.1048849.

Henkel, M., Schmidberger, A., Vogelbacher, M., et al. 2014. Kinetic modeling of rhamnolipid production by *Pseudomonas aeruginosa* PAO1 including cell density-dependent regulation. *Applied Microbiology and Biotechnology*, 98: 7013–7025. https://doi.org/10.1007/s00253-014-5750-3.

Herzog, M., Tiso, T., Blank, L. M., Winter, R. 2020. Interaction of rhamnolipids with model biomembranes of varying complexity. *Biochimica et Biophysica Acta (BBA)-Biomembranes*, 1862: 183431–183444. https://doi.org/10.1016/j.bbamem.2020.183431.

Kiran, G. S., Ninawe, A. S., Lipton, A. N., Pandian, V., Selvin, J. 2016. Rhamnolipid biosurfactants: Evolutionary implications, applications and future prospects from untapped marine resource. *Critical Reviews in Biotechnology*, 36: 399–415. https://doi.org/10.3109/07388551.2014.979758.

Kristoffersen, V., Rämä, T., Isaksson, J., Andersen, J. H., Gerwick, W. H., Hansen, E. 2018. Characterization of rhamnolipids produced by an arctic marine bacterium from the pseudomonas fluorescence group. *Marine Drugs*, 16: 163–181. https://doi.org/10.3390/md16050163.

Kurata, N., Vella, K., Hamilton, B., et al. 2016. Surfactant-associated bacteria in the near-surface layer of the ocean. *Scientific Reports*, 6: 1–8. https://doi.org/10.1038/srep19123.

Medina-Moreno, S. A., Jiménez-Islas, D., Gracida-Rodríguez, J. N., Gutiérrez-Rojas, M., Díaz-Ramírez, I. J. 2011. Modeling rhamnolipids production by *Pseudomonas aeruginosa* from immiscible carbon source in a batch system. *International Journal of Environmental Science & Technology*, 8: 471–482. https://doi.org/10.1007/BF03326233.

Muthusamy, K., Gopalakrishnan, S., Ravi, T. K., Sivachidambaram, P. 2008. Biosurfactants: properties, commercial production and application. *Current Science*: 736–747. www.jstor.org/stable/24100627.

Oluwaseun, A. C., Kola, O. J., Mishra, P., et al. 2017. Characterization and optimization of a rhamnolipid from *Pseudomonas aeruginosa* C1501 with novel biosurfactant activities. *Sustainable Chemistry and Pharmacy*, 6: 26–36. https://doi.org/10.1016/j.scp.2017.07.001.

Posada-Baquero, R., Grifoll, M., Ortega-Calvo, J. J. 2019. Rhamnolipid-enhanced solubilization and biodegradation of PAHs in soils after conventional bioremediation. *Science of the Total Environment*, 668: 790–796. https://doi.org/10.1016/j.scitotenv.2019.03.056.

Raiger Iustman, L. J., López, N. I. 2009. Los biosurfactantes y la industria petrolera. *Química Viva*, 8: 146–161.

Rathankumar, A. K., Saikia, K., Ribeiro, M. H., et al. 2021. Application of statistical modeling for the production of highly pure rhamnolipids using magnetic biocatalysts: Evaluating its efficiency as a bioremediation agent. *Journal of Hazardous Materials*, 412: 125323–125334. https://doi.org/10.1016/j.jhazmat.2021.125323.

Raza, Z. A., Ahmad, N., Kamal, S. 2014. Multi-response optimization of rhamnolipid production using grey rational analysis in Taguchi method. *Biotechnology Reports*, 3: 86–94. https://doi.org/10.1016/j.btre.2014.06.007.

Roy, S., Chandni, S., Das, I., Karthik, L., Kumar, G., Ventata, B. K. 2015. Aquatic model for engine oil degradation by rhamnolipid producing *Nocardiopsis* VITSISB. *3 Biotech*, 5: 153–164. https://doi.org/10.1007/s13205-014-0199-8.

Satya Eswari, J., Venkateswarlu, C. 2016. Dynamic modeling and metabolic flux analysis for optimized production of rhamnolipids. *Chemical Engineering Communications*, 203: 326–338. https://doi.org/10.1080/00986445.2014.996638.

Suh, S. J., Invally, K., Ju, L. K. 2019. Rhamnolipids: Pathways, productivities, and potential. *Biobased Surfactants*, 169–203. https://doi.org/10.1016/b978-0-12-812705-6.00005-8.

Tedesco, P., Maida, I., Palma Esposito, F., et al. 2016. Antimicrobial activity of monoramnholipids produced by bacterial strains isolated from the Ross Sea (Antarctica). *Marine Drugs*, 14: 83–96. https://doi.org/10.3390/md14050083.

Tripathi, L., Irorere, V. U., Marchant, R., Banat, I. M. 2018. Marine derived biosurfactants: A vast potential future resource. *Biotechnol Lett*, 40: 1441–1457. https://doi.org/10.1007/s10529-018-2602-8.

Tripathi, L., Twigg, M. S., Zompra, A., et al. 2019. Biosynthesis of rhamnolipid by a Marinobacter species expands the paradigm of biosurfactant synthesis to a new genus of the marine microflora. *Microb Cell Fact*, 18: 164–175. https://doi.org/10.1186/s12934-019-1216-8.

Twigg, M. S., Tripathi, L., Zompra, A., et al. 2018. Identification and characterization of short chain rhamnolipid production in a previously uninvestigated, non-pathogenic marine pseudomonad. *Applied Microbiology and Biotechnology*, 102: 8537–8549. https://doi.org/10.1007/s00253-018-9202-3.

Twigg, M. S., Tripathi, L., Zompra, A., et al. 2019. Surfactants from the sea: Rhamnolipid production by marine bacteria. *Access Microbiology*, 1: 192–194. https://doi.org/10.1099/acmi.ac2019.po0066.

Varjani, S. J., Upasani, V. N. (2017). Critical review on biosurfactant analysis, purification and characterization using rhamnolipid as a model biosurfactant. *Bioresource Technology*, 232: 389–397. https://doi.org/10.1016/j.biortech.2017.02.047.

Xu, M., Fu, X., Gao, Y., Duan, L., Xu, C., Sun, W., Li, Y., Meng, X., Xiao, X. (2020). Characterization of a biosurfactant-producing bacteria isolated from marine environment: Surface activity, chemical characterization, and biodegradation. *Journal of Environmental Chemical Engineering*, 8: 104277–104283. https://doi.org/10.1016/j.jece.2020.104277.

Yan, X., Sims, J., Wang, B., Hamann, M. T. 2014. Marine actinomycete *Streptomyces* sp. ISP2–49E, a new source of rhamnolipid. *Biochemical Systematics and Ecology*, 55: 292–295. https://doi.org/10.1016/j.bse.2014.03.015.

Youssef, N. H., Duncan, K. E., Nagle, D. P., Savage, K. N., Knapp, R. M., McInerney, M. J. 2004. Comparison of methods to detect biosurfactant production by diverse microorganisms. *Journal of Microbiological Methods*, 56: 339–47. https://doi.org/10.1016/j.mimet.2003.11.001.

Žutić, V., Ćosović, B., Marčenko, E., Bihari, N., Kršinić, F. 1981. Surfactant production by marine phytoplankton. *Marine Chemistry*, 10: 505–520. https://doi.org/10.1016/0304-4203(81)90004-9.

5 Role of Biosurfactants Produced by Marine Microbes in Bioremediation

Ria Desai and Trupti K Vyas

CONTENTS

INTRODUCTION

The marine environment makes up more than 70% of the Earth's surface, harboring a rich biological and chemical diversity (Proksch et al. 2003). The diversity of a marine habitat is attributed to its comprehensive physiological adaptations, unique genetic structures, and metabolic peculiarities compared to its terrestrial homologs. Hence, the marine world promises the discovery of novel bioactive compounds like antibiotics, enzymes, vitamins, lipids, drugs, and biosurfactants, among others.

Through evolution, bacteria have adapted themselves to feed on the water-immiscible substrate by producing and utilizing surface-active compounds

DOI: 10.1201/9781003307464-5

which help them in adsorbing, emulsifying, foaming, wetting, or solubilizing hydrophobic materials in the aqueous phase (Paniagua-Michel et al. 2014). These microbial-derived surface-active metabolites, bio-surfactants (BSs), have received a lot of interest in recent years due to their various advantages over synthetic surfactant ones like low environmental toxicity; higher biodegradability; and stability at extreme conditions of pH, temperature, and salinity. Many synthetic surfactants in the market today require non-renewable petrochemical feedstock, whereas BSs from marine organisms require renewable and economical feedstock, making them a green and clean alternative. This also makes their use suitable in agriculture and therapeutic, food, and cosmetic industries. Microbe-derived BSs can be a good alternative to synthetic chemical surfactants on the market today, which are more toxic in many applications, like battling oil spills, bioremediation amplification, micro-extraction of poly aromatic hydrocarbons, pharmaceutical products, and the detergent industry (Nguyen and Sabatini 2011; Banat et al. 2010).

Considering their wide range of surface activities, researchers have tested marine-derived BSs in bioremediation processes and environmental applications like oil spill cleanup in the sea, microbial-enhanced oil recovery (MEOR), and environmental cleanup processes. With nearly 5.73 million tonnes of oil spilt in the ocean between 1970 and 2016 (ITOPF-International Tanker Owners Pollution Federation 2017), new bioactive chemicals for bioremediation are in high demand, and marine origin BSs can be a good alternative. Also, BSs from marine microbe act as a biocide, protecting them against competitive microbes or predators (Rodrigues et al. 2006) and stimulating antimicrobial defense responses in plants and animals (Vatsa et al. 2010). Many marine microorganisms capable of producing biosurfactants with various structures have been discovered. However, the vast majority of marine microbial diversity has not been extensively studied, owing to the difficulties of isolating and culturing marine microorganisms in the laboratory (Stein et al. 1996). Low yield, insufficient structure elucidation, and lack of culture condition knowledge optimization further make this BS culturing expensive.

This chapter offers an overview of biosurfactants from marine microorganisms, highlighting their structural diversity, their role in bioremediation, and potential biotechnology application with its recent advances.

BIOSURFACTANTS

BSs are surface-active compounds produced by microorganisms as a product of their secondary metabolism exhibiting surface and emulsifying properties. These amphipathic molecules are composed of a hydrophilic moiety containing an acid, peptide, cation, anion, or mono-polysaccharide and a hydrophobic moiety of unsaturated or saturated hydrocarbon chains of fatty acids. These are produced either extracellularly or as part of their cell membrane by hydrocarbon-degrading microbes, including yeasts, filamentous fungi, and bacteria (Chen et al. 2007). These green amphiphilic molecules are likely to interact at the surface level, linking two phases in a heterogeneous system by producing a biofilm that can alter a wide range of surface properties like surface tension and interfacial tension, allowing monomers to form micelles and microemulsions between two phases.

There are two primary categories of BS, one that functions by reducing surface tension at the air-water interface (biosurfactants) and another that acts by reducing the interfacial tension between immiscible liquids or the solid–liquid interface, termed bioemulsifiers (Batista et al. 2006). A change in surface properties by a BS results in greater hydrocarbon mobility and improves biodegradation by increasing organic contaminants' bioavailability, including CO components, causing hydrocarbon elimination (Randhawa and Rahman 2014).

Much of the BS research has concentrated on soil isolates, such as the *Pseudomonas* and *Bacillus* families. In contrast, BS production by marine microbes is a largely unexplored field due to the difficulty of culturing marine microorganisms in the laboratory (Stein et al. 1996). The different hydrocarbon-contaminated water sources are the most studied and exploited marine sources searching for potential BS producers (Dang et al. 2015). These hydrocarbons or oils are known to stimulate the production of BS secondary metabolites. Considering this fact, mainly enrichment culture techniques favor isolation and detection of BS producers (Thavasi et al. 2011). Despite this experimentation, some researchers argue that BS production is associated with hydrocarbon uptake and biological matrices, or filter-feeding organisms as hosts of microbial communities also specialize in BS generation linked with a functional role (Pini et al. 2007).

Currently, the renewable substrate media quantity, the slow-growing rate of organisms on the substrate, low yield, and end product purification step from substrate contaminants are all issues that limit BS industrial production. Polychaetes (Rizzo et al. 2013), sponges (Dhasayan et al. 2015), sea pens, cnidarians (Mabrouk et al. 2014), and fish (Floris et al. 2018) have all been reported as optimal BS producers. Surface-active compounds are known to possess various therapeutic properties like antibacterial, antifungal, and antiviral properties attributed to BSs' ability to decrease surface tension and increase solubility, which can affect the attachment of bacteria, hence making them a global interest for bioactive molecules from marine environments (Rodrigues et al. 2006).

STRUCTURAL DIVERSITY AND CLASSIFICATION

Classification of biosurfactants is primarily based on their chemical structure and microbial origin. The amphiphilic nature of BS usually has a C8 to C10 length chain fatty acid, while its hydrophobic part possesses a small hydroxyl, phosphate, and carboxyl group; carbohydrate (polysaccharide); or peptide moiety contributing to its hydrophilic part. Surfactants based on the nature of a charge on polar moiety can be anionic, non-ionic (polymerization product), cationic, or amphoteric. Anionic and non-ionic compounds make up the majority of biosurfactants. Because cationic biosurfactants are thought to exhibit higher toxicity, they are rarely found (Kosaric et al. 2022). Besides grouping on the charge, BSs are broadly classified into low and high molecular weight (HMW) biosurfactants. Low molecular weight surfactants like rhamnolipids, sophorolipids, trehalose lipids, lipopeptides, phospholipids, fatty acids, and neutral lipids lower surface and interfacial tensions, whereas HMW surfactants like polysaccharides, polysaccharide-protein complexes, lipopolysaccharides, and lipoproteins bind to the surface and are called emulsions (Mulligan et al. 2005). The structural diversity of low molecular mass BSs can be subdivided further into six classes, as described in the following.

GLYCOLIPID

Glycolipids are carbohydrate groups and a lipid part of long-chain aliphatic acid or hydroxyl fatty acid linked with an ester or an ether group. Typical sugars like glucose, mannose, galactose, glucuronic acid, or rhamnose constitute the hydrophilic part. Sophorolipids, mannosylerythritol lipids, trehalose lipids, and rhamnolipids form the well-studied groups of glycolipids. Rhamnolipid biosurfactants have one or two molecules of an α-l rhamnose unit coupled to one or two molecules of β-hydroxy fatty acid moieties with a chain length of C8 to C16 carbon atoms via a glycosidic linkage, produced by *Pseudomonas aeruginosa* (Abdel et al. 2010; Déziel et al. 2000).

Sophorolipids are another type of glycolipid produced primarily by yeasts *Torulopsis bombicola*, *T. petrophilum*, and *T. apicola* and made of di-saccharide sophorose linked to 16 to 18 carbon atom fatty acid chains (predominantly 17-hydroxyoleic acids). They usually form lactones; however, they can also be found in an acidic state, that is, without the ring closure (Nunez et al. 2001). Trehalolipids, another group of BSs from *Mycobacterium*, *Arthrobacter*, and *Rhodococcus* species, is made of the disaccharide trehalose, acylated with long-chained, straight, or α-branched 3-hydroxy fatty acids (mycolic acids) linked at the C6 position (Kügler et al. 2015). Fungal species *Pseudozyma* sp. and *Ustilago maydis* are known to produce mannosylerythritol lipids (MELs). MELs have a carbohydrate moiety of 4-O-β-d-mannopyranosyl-d-erythritol with varied acylation patterns linked with acyl chain lengths (Morita et al. 2009).

LIPOPEPTIDES AND LIPOPROTEINS

This class of cyclic BSs has a lipid moiety linked to a polypeptide or an amino acid chain. *Bacillus*, *Lactobacillus*, *Streptomyces*, *Pseudomonas*, and *Serratia* produce these cyclic BSs. They often have a non-proteinogenic D-enantiomer amino acid attributed to their non-ribosomal origin. Gramicidins, decapeptide antibiotics, and polymyxins, a lipopeptide antibiotic with exceptional surface-active characteristics, are prominent examples of this group (Trimble et al. 2016; Théatre et al. 2021). Another cyclic lipopeptide is surfactin produced by *Bacillus* sp., which has a ring structure of seven amino acids long linked to a fatty acid via lactone. Surfactin is the most effective biosurfactant and can decrease surface tension from 72.5 to 27 mN/m at the air–water interface (Théatre et al. 2021; Yeh et al. 2005).

FATTY ACIDS AND PHOSPHOLIPIDS

Fatty acids from alkane with OH group and alkyl branch-like corynomucolic acid act as surfactants. Most saturated fatty acid chain lengths of C12 to C14 are known to exhibit the most robust surface and interfacial tension of corynomucolic acid. Phospholipid derivatives from yeast and bacteria produce phosphatidyl ethanolamine-rich microemulsions. According to Maneerat and co-workers (2005), *Myroides* sp. SM1, when cultured in marine broth, was observed to produce bile acids, cholic acid, deoxycholic acid, and glycine conjugate (Takeshi et al. 2006).

POLYMERIC BIOSURFACTANTS

Mostly polymeric biosurfactants are heterosaccharide-protein complexes with protein, carbohydrate, and lipid ratios of 35:63:2, respectively (Maneerat et al. 2005). Liposan, emulsan, biodispersan, and alasan are some examples of this group. *Candida lipolytica* produces liposan (83% carbohydrate and 17% protein), and *Saccharomyces cerevisiae* has mannoproteins (44%mannose and 17%protein). Zinjarde and Pant (2002) presented tropical marine strain *Yarrowia lipolytica*–produced emulsifier in alkanes or crude oil presence. In the early stages of growth, an emulsifier (lipid-carbohydrate-lipid) complex was connected with the cell wall, demonstrating extracellular emulsifier activity (Maneerat et al. 2005).

PARTICULATE BIOSURFACTANTS

Extracellular membrane vesicles partition hydrocarbons to form microemulsions, critical for absorption of microbial alkane. *Acinetobacter* sp. vesicles have a diameter of 20–50 nm and a buoyant density of 1.158 cubic g/cm, consisting of protein phospholipids and lipopolysaccharides. Bioemulsifiers or high molecular mass BSs produced by some bacterial strains comprise polysaccharides, proteins, lipopolysaccharides, lipoproteins, or a complex mixture of polymers, lipoheteropolysaccharide. For example, *Acinetobacter calcoaceticus* RAG-1, isolated from the Mediterranean Sea, produces the prototype HMW bioemulsifier emulsion (Rosenberg 2006).

BIOSURFACTANTS FROM MARINE ORGANISMS

Various lipopeptide antibiotics exhibiting significant surface-active character-istics have been reported in microbial BSs from marine sources such as sediments, corals, sponges, sea, and hot water springs (Gandhimathi et al. 2009; Joshi et al. 2013). *Aureobasidium pullulans* YTP6–14, a marine yeast, produces a lactonized hydro fatty acid BS, also used as a fragrance (Luepongpattana et al. 2017). *Serratia rubidaea*, identical to marine bacterium *Serratia marinorubra*, secretes linear three hydroxy fatty acids with surface-active properties named rubiwettin R141.

Glycolipids have been widely explored in the marine system as a diverse group of bacteria makes them that are present in various marine matrices, including animals (annelida, Pteroeides, and fish guts) and contaminated soils such as Arctic and Antarctic sediments (Floris et al. 2018). Some marine bacteria are also known to produce lipoamino acid. For example, a crude oil metabolite was produced by the flavobacterium *Myroides* sp. comprising the non-proteinogenic amino acid ornithine coupled to a 3-hydroxy-13-methyl-tetradecanoic acid diester (Takeshi et al. 2006). *Actinobacter* and *Firmicutes* (*Bacillus* sp.) are significant contributors among marine producers of lipo-peptide BSs. *Rhodococcus* sp., producing actinobacterial rhodofactin, and *N. alba*, producing phenylalanine lipopeptide, are examples of lipooligopeptide (Kiran et al. 2010a). Massetolide A-H, a collection of antibacterial viscosin-like lipopeptides, is known to be made by *Achromobacter* sp. HZ-01, a hydro-carbon-degrading betaproteobacterium, and epiphytic marine *Pseudomonads* (Gerard et al. 1997).

For large-scale industrial production, glycolipids are preferred BSs due to the simple production process of large quantities and their wide applications. *P. aeruginosa* is a well-known human pathogen; it is less appealing for industrial use, and marine *Pseudomonas* sp. MCTG214 (3b1) has been recently charac-terized as a non-pathogenic rhamnolipid producer exhibiting a good alternative strain for industrial production of rhamnolipids (Twigg et al. 2018). Many studies have also been published on the generation of rhamnolipids by marine isolates (Vyas and Dave 2011; Vyas and Dave 2016). It is necessary to highlight the strain *Streptomyces* sp. ISP2–49E that was isolated from Galveston Bay (Texas) marine sediment samples. This strain developed the rhamnolipid biosurfactant l-rhamnosyl-l-rhamnosyl-β-hydroxydecanoyl-β-hydroxydecanoate (Rha-Rha-C10-C10), which is the first identified rhamnolipid-producing *Streptomyces* strain (Yan et al. 2014). Firmicutes are not commonly considered glycolipid producers, and biofilm blocking threose-di-lipid from a snail epiphytic *Staphylococcus lentus* is the only marine case (Hamza et al. 2017).

The number of publications reporting marine biosurfactant producers exceeds the number of molecular structures that have been resolved. To broaden the list of marine biosurfactants, various new forms of surface-active metabolites need to be discovered. Table 5.1 lists some of the studied BS types with the respective marine microbial producers and applicative properties.

TABLE 5.1
BS Types with the Respective Marine Producers and Economic Importance

Name of Microorganism (Marine Source)	Biosurfactant Type (Structure)	Economic importance	References
Pseudoalteromonas sp. strain TG12	Glycolipid	Emulsifying or metal-chelating agent, high emulsifying activities against a range of oil substrates	Gutierrez et al. 2008
Serratia marcescens (hard coral; *Symphyllia* sp.)	Glycolipid (palmitic acid and glucose)	Antibiofilm activity, antimicrobial, antiadhesive against *P. aeruginosa* and *Candida albicans*	Gudiña et al. 2016
Streptomyces sp. B3 (sediment samples)	Glycolipid	Shows antimicrobial activity against *C. albicans*, *E. coli*, and *S. aureus*	Khopade et al. 2012
Brevibacterium casei MSA19 (sponge; *Dendrilla nigra*)	Glycolipid	Works as antibiofilm activity against *E. coli* and *Vibrio sp.* population	Kiran et al. 2010b
Candida antartica (deep sea)	Mannosylerythritol lipid	Has antimicrobial, neurological, and immunological properties	Kitamoto et al. 1993
Aspergillus ustus MSF 3 (sponge; *Fasciospongia cavernosa*)	Glycoprotein	Antimicrobial activity against *E. coli, K. pneumoniae, C. albicans, S. aureus*	Kiran et al. 2009
Bacillus circulans (sea water sample)	Lipopeptide	Antibiofilm, anti-adhesive against *E. coli, S. typhimurium, P. vulgaris, S. marscescens*	Das et al. 2008, 2009
Bacillus circulans DMS-2 (sea water sample)	Lipopeptide	Shows antimicrobial activity against *P. vulgaris* and *C. freundii*	Sivapathasekaran et al. 2009
Nocardia otitidiscaviarum	Glycolipid	Enhances biodegradation of oil	Vyas and Dave 2011
Methylobacterium mesophilicum MTCC 6839	Glycolipid	Enhances biodegradation of oil	Vyas and Dave 2016

BIOSYNTHESIS MECHANISM

In marine organisms, insufficient knowledge of molecular structure correlates with a lack of biosynthetic mechanisms leading to biosurfactant production from marine organisms. Pathways for biosurfactant production have been studied and explored in non-marine organisms, but knowledge of marine strains is limited.

Water-soluble substrates like carbohydrate are used by microorganisms to construct the hydrophilic moiety of biosurfactants, and for the hydrophobic moiety, substrates like oil and fats are used (Desai and Banat 1997). The synthesis of biosurfactant precursor production requires a variety of metabolic pathways that are dependent on the nature of the primary carbon sources used in the culture medium. For instance, the carbon flow is managed when only carbohydrate is the sole carbon source for glycolipid formation. Microbial metabolism will constrain both lipogenic (lipid formation) and glycolytic pathways (hydrophilic moiety formation). A hydrophilic substrate like glucose undergoes a glycolytic pathway, yielding glucose six phosphate (G6P) as one of its intermediates, a major precursor of carbohydrates present in the hydrophilic moiety of BS. Several enzymes catalyze G6P along with the biosurfactant's numerous hydrophilic moieties, including trehalose, sophorose, rhamnose, mannose, and polysaccharide (Nurfarahin et al. 2018). For lipid formation, glucose isoxidized to pyruvate first, then undergoes a tricarboxylic TCA cycle forming malonyl-CoA when combined with oxaloacetate, eventually converting into fatty acid, which acts as a precursor for lipid production (Haritash and Kaushik 2009).

Primary carbon metabolism is responsible for fatty acid and hydroxy fatty acid formation, whereas lipoamino acid BS synthesis from marine bacteria is not yet detailed. It may be similar to N-acyl amino acid biosynthesis from amino acid and activated fatty acids using acyltransferase and N-acyl amino acid synthase performed by microorganisms (Brady et al. 2004). Recently, microbes producing lipopeptides have gained researchers' interest because they are potential biocontrol agents and are a good source of pharmaceutically active chemicals. Lipopeptides such as surfactin are constructed modularly by special non-ribosomal peptide synthetases (NRPSs). Each module adds one amino acid to the peptide chain; additionally, the NRPS can catalyze processes such as lipid incorporation, epimerization, and lactonization (Samel et al. 2006).

One of the most-studied NRPS is the surfactin biosynthetic molecular machinery. This NRPS consists of three multifunctional proteins coded by the srfA operon, made of srfA-A, srfA-B, and srfA-C. Proteins SrfA-A and SrfA-B have three modules, while SrfA-C only has one module with a thioesterase domain. 3-hydroxy fatty acid, a starter molecule, was recognized by the first module of the condensation domain where seven other amino acids (l-Glu, l-Leu, d-Leu, l-Val, l-Asp, d-Leu, and l-Leu) were added consecutively, followed by product release and lactonization catalyzed by the thioesterase domain of SrfA-C (Kohli et al. 2001). Lichenysin, fengycin, plipastatin, fusaricidin, and massetolide/viscosins have all been reported as having similar biosynthetic machinery.

The production of rhamnolipids by *P. aeruginosa* is considered the best-known example of a bacterial glycolipid biosynthetic pathway documented for non-marine and marine strains. Rhamnosyltransferase I and II, two different glycosyltransferases, catalyze rhamnolipid formation. There are two proteins involved, RhlA and RhlB, whose genes are arranged in the bicistronic operon. RhlA converts activated hydroxy fatty acids into 3-(3-hydroxyalkanoyloxy) alkanoic acids (HAAs), while RhlB catalyzes condensation dTDP-l-rhamnose

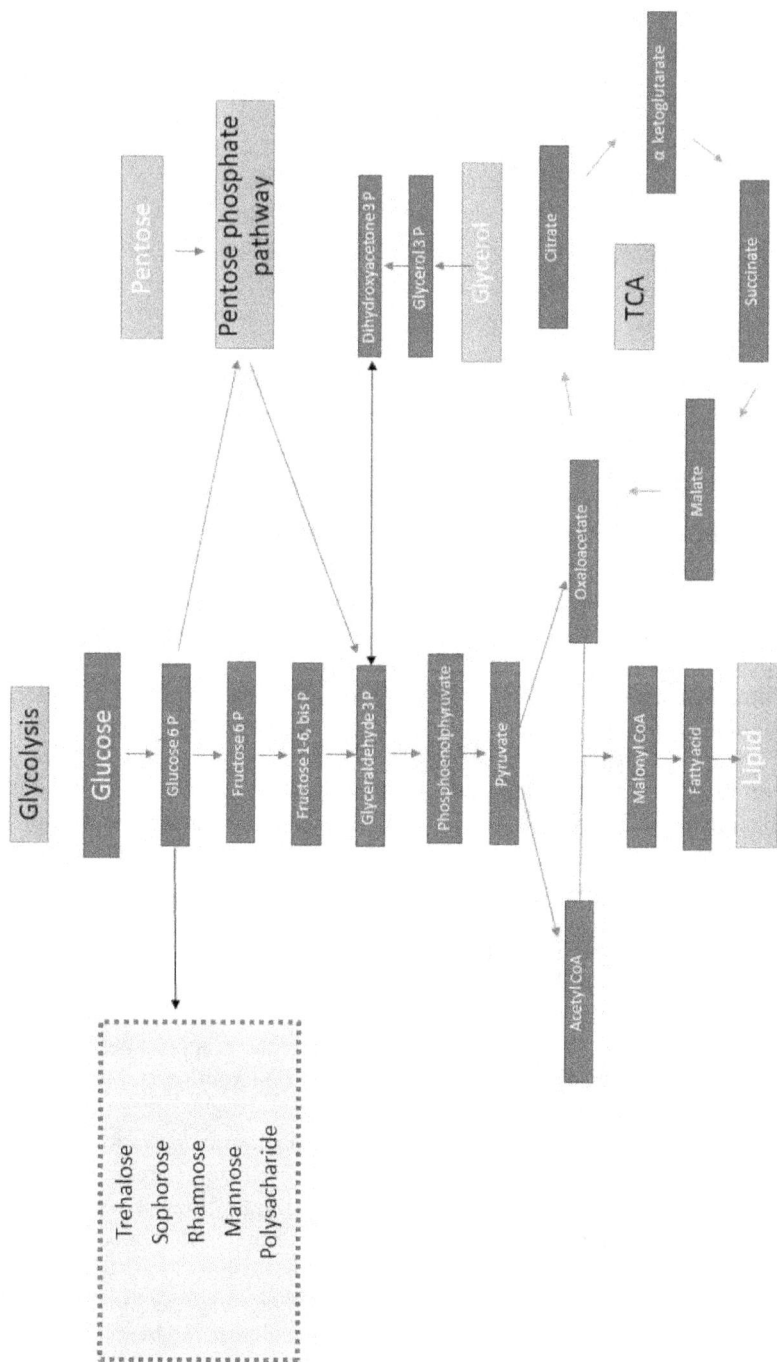

FIGURE 5.1   Metabolic pathways involved synthesizing biosurfactants using the water-soluble substrate.

and the HAA to generate mono-rhamnolipids. Also, HAA is a surface-active compound itself and is secreted in the environment as a BS (Tiso et al. 2017). Other marine glycolipids' biosynthesis has yet to be elucidated. The biosynthesis process of BSs is strictly regulated, due to which they are produced in well-defined conditions like cell density, presence of hydrophobic nutrients, or limited growth conditions. Most of the known biosynthetic pathways for marine biosurfactants come from research on non-marine bacteria. The development of bioinformatics tools and low-cost sequencing methodologies has enabled researchers to discover BS biosynthesis genes from genome sequence data.

BIOSURFACTANT ROLE IN DEGRADATION

The discharge of petroleum hydrocarbons into marine habitats due to oil spills is a significant threat to our marine environment. Various research is going on oil spill bioremediation to address these challenges. Bioremediation involves transforming organic pollutants into a less toxic compound that can be assimilated into the natural biogeochemical cycle performed by living organisms. The biodegradation rates are influenced by environmental parameters like oxygen, pH, the presence of micronutrients, macronutrients, and so on. The dispersion state of a hydrocarbon also affects its biodegradation. Mattei and Bertrand (1985) reported that when the water-insoluble substrate was dissolved, solubilized, or emulsified, it enhanced the rate of biodegradation.

Synthetic dispersants are used worldwide to solve this oil spillage pollution, which combines one or more surfactants and solvents that improve oil dispersion into droplets, resulting in enhanced hydrocarbon mobility and bioavailability. These droplets are further solubilized in aquatic water, enhancing their mobility and hydrocarbon bioavailability. However, synthetic surfactants, on the other hand, are detrimental to marine organisms; thus, replacing them with biological, non-toxic alternatives would be beneficial. Many marine γ-proteobacterial strains produce BSs either extracellularly or on the cell surface that solubilizes aromatic hydrocarbons. When localized on the cell surface, microbial cells adhere to the hydrocarbon, acting as a BS. Biosurfactants can increase the bioavailability of poorly soluble polycyclic aromatic hydrocarbons like phenanthrene and resin (Budzinski et al. 1998). BS development from these bacteria promotes the dispersion of hydrocarbons, allowing non-BS-producing microbes better to degrade them (McGenity et al. 2012).

Biosurfactant activity is influenced by critical micelle concentration (CMC). It is the concentration at which the rate of surface tension reduction causes the formation of micelles and vesicles. This concentration regulates the rate of surface tension reduction of BS. Biosurfactants have CMC values from 1 to 200 mg/L (Filonov et al. 2004) and are believed to have a 10–40× lower CMC than chemical synthesized surfactants, implying that less biosurfactant is needed to reduce surface tension. Biosurfactants enhance the biodegradation of hydrocarbon by mobilization, solubilization, or emulsification mechanisms (Nievas et al. 2008). Below the CMC concentration, the mobilization process

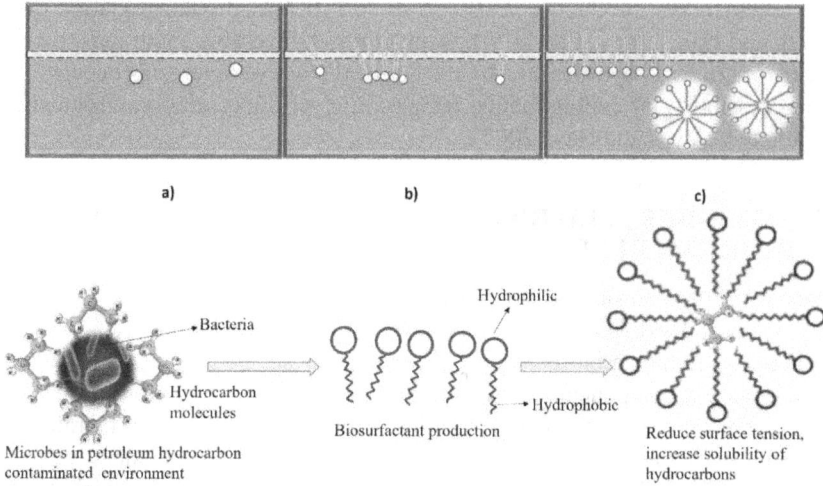

FIGURE 5.2 Mechanism of hydrocarbon solubilization by micelle formation by biosurfactant.

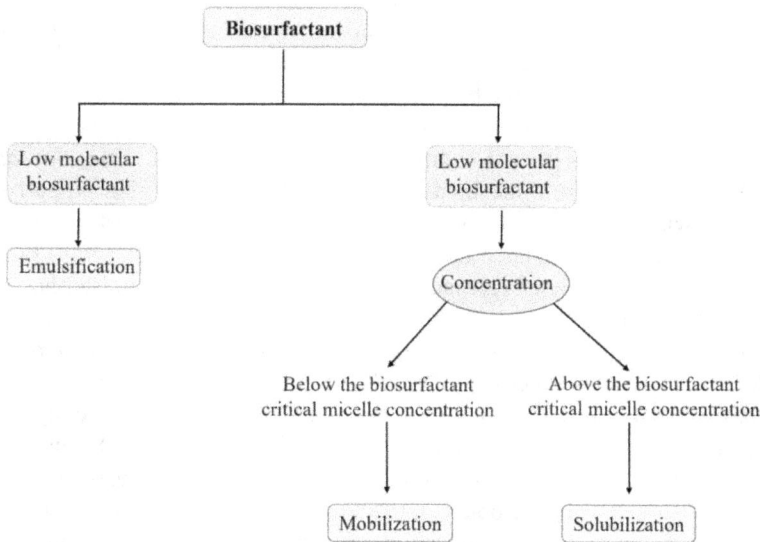

FIGURE 5.3 Mechanism of hydrocarbon removal by biosurfactants.

occurs by reducing surface and interfacial tension by BS. Above the CMC concentration, solubilization takes place by micelle formation, increasing the solubility of hydrocarbon oil with its hydrophobic part in the micelle core. High molecular mass surfactants exhibit an emulsification process (Urum and Pekdemir 2004). In the biostimulation strategy for oil, biodegradation revealed that biosurfactant producing marine microbes increased biodegradation rates (Vyas and Dave 2010, 2011, 2016). Depending on favorable conditions, degradation of crude oil hydrocarbon by metabolically active organisms occurs, like the accessibility of contaminants, temperature, pH, and other environmental parameters (Vyas and Dave 2007).

MARINE-DERIVED BIOSURFACTANTS IN BIOREMEDIATION

Marine microorganisms, such as *Marinobacter, Halomonas, Myroides*, and the marine yeast *Yarrowia lipolytica* can help remove hydrocarbons from polluted regions by producing bio-emulsion. *Acinetobacter* sp. T4 decomposed Arabian light crude oil components, both saturated and aromatic, and created metabolites that enhanced the growth of *Pseudomonas putida* PB4. The PB4 strain was further grown on these metabolites and degraded crude oil aromatic components (Komukai-Nakamura et al. 1996). It was observed that bioaugmentation (introduction of microbial culture from nonindigenous populations) increased the hydrocarbon degradation rate by 30% compared with the treatment without bioaugmentation (Dave et al. 1994).

Halomonas sp. helps eliminate oil spillage by synthesizing bio-emulsifiers. Glycolipid molecules generated on the *Halomonas* cell surface boost the solubility of hydrocarbons, increasing their bioavailability. The surface-active BS produced by *Halomonas* sp. could be employed for improved oil recovery operations in difficult situations, as the emulsifier at low temperatures could improve hydrocarbon bioavailability in cold environments (Dhasayan et al. 2014). *Myroides* sp. SM1 producing ornithine lipid possesses a high emulsifying property for crude oil with good stability at extreme conditions of temperature and pH, exhibiting better performance than synthetic detergent and surfactin (Takeshi et al. 2006). *Acinetobacter* calcoaceticus RAG-1's produced emulsan has been claimed to have promising applications in microbial-enhanced oil recovery and oil spillage cleanup (Belsky et al. 1979). A BS produced from *Alteromonas* sp. 17 could be exploited for effective hydrocarbon breakdown (Al-Mallah et al. 1990). *Marinobacter hydrocarbonoclasticus* strain SdK644 produced glycolipid biosurfactant, which solubilized crude oil two times faster than tween 80, indicating that it could be used in marine bioremediation (Zenati et al. 2018). The bio-emulsifier Yansan, generated by the aerobic yeast *Yarrowia lipolytica*, demonstrated significant emulsifying properties and stability over a pH range of 3–9, with prospective uses in perfluorocarbon (PFC)-based emulsion formation and hydrocarbon degradation (Amaral et al. 2006). The role of BS in pesticide biodegradation has also been explored. When solubility and emulsification increase, it helps in the cellular uptake of contaminants, making biosurfactants quite valuable for the remediation approach comprising biochemical supplements and organisms. BSs from *Pseudomonas* species have been identified as degrading hexachlorocyclohexane and organochlorinated pesticides like DDT and cyclodienes (Karanth et al. 1999).

The particular mode of hydrocarbon adsorption by BS is not completely known. The absorption process of n-hexadecane by rhamnolipids from *P. aeruginosa* was studied, which indicated emulsion formation by rhamnolipid with hexadecane, promoting hydrocarbon and bacteria contact. The absorbed hydrocarbon droplets were masked by BS, proposing the possibility of pinocytosis rather than hydrocarbon capturing by bacteria. However, a negative effect of BS on biodegradation was also observed. A barrier was formed by micelles holding organic hydrocarbon between microbes and hydrocarbon molecules, lowering the substrate availability. Witconol SN70, a non-ionic surfactant derived from alcohol ethoxylate, reduced the hexadecane biodegradation rate and is one of the examples (Colores et al. 2000).

Although there have been various successful applications of BS, the literature shows contradictory reports in terms of bioremediation. For example, rhamnolipid from *P. aeruginosa* encourages the degradation of n-hexadecane, whereas the genus *Rhodococcus* does not, indicating the specificity of microorganisms for degradation. A biosurfactant from *R. erythropolis* 3C-9 has been known to significantly increase the degradation rate of n-hexadecane. As a result, the effect of a biosurfactant on bioremediation is unexpected, and its efficacy must be determined experimentally (Franzetti et al. 2008).

BIOTECHNOLOGICAL APPLICATIONS

Biosurfactants are emerging as attractive molecules because of their structural uniqueness, adaptability, and versatile properties useful for many therapeutic applications. The physical and biological properties make them applicable in various fields of medicine, food, cosmetics, industry, and so on. Marine origin BS have been studied and known to inhibit cell adhesion and the biofilm formation process (Das et al. 2009), exhibiting antimicrobial, anti-adhesive, and biofilm-inhibiting properties against pathogenic organisms. Lipopeptide and glycolipid BSs were reported to selectively interfere with cancer cell growth and disrupt cancer cell membranes through apoptotic pathways (Gudiña et al. 2013). Surprisingly, certain biosurfactants appear to be made up of a mix of molecules with various modes of action. For instance, it was demonstrated that from a natural rhamnolipid mixture, mono-rhamnolipid indicated bacteriostatic activity on *P. aeruginosa*, whereas di-rhamnolipid showed bactericidal activity. BSs were also reported to exhibit a synergistic effect with some antibiotics in some situations, like increasing the antibiotic penetration efficacy into the cell (Diaz De Rienzo et al. 2016).

Biosurfactants like di-rhamnolipid have also been successfully evaluated to treat burn wounds, where they enhanced wound healing and reduced scar formation due to collagen content (Stipcevic et al. 2006). Furthermore, biosurfactants as drug delivery vehicles have both commercial and scientific applications. Due to emulsifying properties of BS like wetting, foaming, dispersing, cleaning, and so on, they have been known to replace synthetic surfactants in the cosmetic industry; as all these properties help the hydrophobic components in the product to solubilize and are an excellent delivery system for permeating the skin barrier (Varvaresou and Iakovou 2015).

The agriculture sector also utilizes the benefits of biosurfactants to increase their productivity, where BSs can aid by increasing healthy microbe plant interaction, improving soil quality, safeguarding from plant pathogens, and stimulating plants for nutrient and fertilizer uptake (Sachdev and Cameotra 2013). Also, the antimicrobial and antibiofilm properties of BSs can help disinfect equipment and halt food spoilage. BSs are used as an anti-adhesive agent in the food processing industry and as a food formulation ingredient, stabilizing the emulsion. BSs have also been studied for a role in pesticide degradation. *Tistrella* sp., a biosurfactant-producing bacterium, completely degraded the pesticide chlorpyrifos with 99.86% efficiency (Ahir et al. 2020). The insecticidal activity was shown by a lipopeptide surfactant from various bacteria against *Drosophila melanogaster*, thus being a potential biopesticide (Mulligan 2005).

In conclusion, biosurfactants are helpful in a wide range of applications. However, for medicinal and therapeutic use, studies have been focused on a small group of well-characterized biosurfactants like sophorolipids, rhamnolipids, mannosylerythritol lipids, and their related compounds. Other biosurfactants will almost certainly bring up new possibilities in the future. Biosurfactants from marine organisms can be of great interest due to their adaptability to extreme conditions of temperature and salinity, which may open up a novel area of its application. The current inability to supply a significant quantity of new compounds to industries necessitates the development of appropriate production

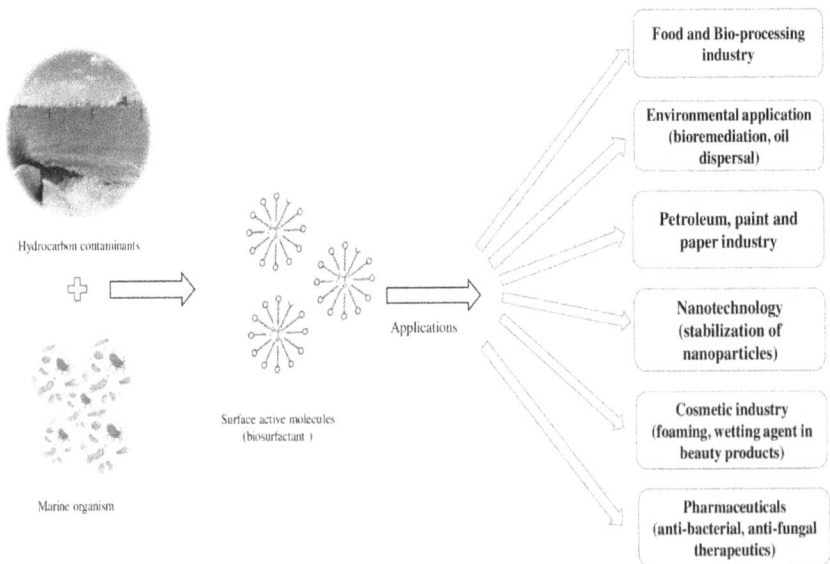

FIGURE 5.4 Biosurfactant production from waste and its application in various sectors.

techniques and methods. Novel applications may result from the structural elucidation and physicochemical characterization of such novel biosurfactants.

CONCLUSION

There has been great interest in studying and exploring marine microorganisms as promising producers of new bioactive compounds for various applications in recent years. The reason is that the marine world represents a rich source of new compounds. Research conducted in the last decade has been encouraging concerning BS producers, effectiveness, role in degradation, and therapeutic applications. The oceans contain a wide range of habitats and should contain many more biosurfactant-producing organisms to be discovered, particularly in promising niches such as highly contaminated water bodies. Various microorganisms are reported as potential marine biosurfactant producers, but *Bacillus, Candida,* and *Pseudomonas* are the most-studied ones. Marine-derived BSs effectively emulsify hydrocarbon, thus increasing the bioavailability of oil and proving a promising agent in bioremediation, especially in oil-polluted marine environments. Because of their low toxicity, biological activity, biocompatibility, and biodegradability, they have the potential to be widely used in personal care and cosmetic goods, and they benefit product efficiency, efficacy, and economics.

Furthermore, sampling and enrichment culture techniques for BSs and function-based metagenomics tools will help develop novel marine biosurfactant producers. Low-cost substrates, ideal conditions for culturing and developing novel characterization, and purification techniques will help make the biosurfactant production industry economical. More marine organisms producing BSs need to be explored, as they have the potential to make a substantial scientific contribution to humanity through their wide range of applications and lower toxicity.

FUTURE PERSPECTIVE

The surface-active compounds obtained from marine microorganisms have a wide range of applications. Biosurfactants can be produced from various natural sources, and their emulsifying and other biodegradation properties can be studied and applied in bioremediation technology. Microorganisms isolated from marine habitats have yielded several biosurfactants with pharmacological and medicinal applications. However, the current use of molecular and genetic tools to examine the variety of aquatic ecosystems has revealed that the vast majority of the marine organisms remain undiscovered, owing to the difficulty of growing most of those microorganisms under laboratory conditions (Kennedy et al. 2010).

Another way to investigate the genetic content of inaccessible marine organisms and explore unknown bioactive natural compounds without the necessity of culturing marine organisms in the lab is by employing culture-independent techniques like metagenomics. Sequence-based metagenomics can be done through random sequencing of metagenomic libraries, analyzing further by bioinformatics tools, and comparing them through the homology-based database.

Some sequences can give an idea about novel genes with unknown functions. Screening of sequences can also be performed for various antimicrobial, antiviral, and other medical properties (Gudiña et al. 2016). Screening functional metagenomic libraries for new antimicrobial activity using *E. coli* and alternative hosts *(Bacillus subtilis, Ralstonia metallidurans,* or *Streptomyces lividans)* simultaneously detected antimicrobial activity in the alternative hosts but not in *E. coli* (Biver et al. 2013). Developing analytical methodologies for measuring anti-inflammatory chemicals and standardizing marine sources is required. As a result, an integrated approach will aid in the development of more effective and safe anti-inflammatory medicines derived from marine sources.

Marine biosurfactants, less explored than their terrestrial ones, have various features that make them powerful agents for various therapeutic, cosmetic, and food applications as alternatives to currently available pharmaceuticals in the market that need to be discovered and investigated. Regarding the biodegradation mechanism of hydrocarbon removal, biosurfactants, particularly in the oil industry, are promising options for replacing synthetic surfactants. Large-scale biosurfactant manufacturing will be aided by investments in strategies to improve the processing of these natural substances.

REFERENCES

Abdel-Mawgoud, Ahmad Mohammad, François Lépine, and Eric Déziel. 2010 Rhamnolipids: Diversity of structures, microbial origins and roles. *Applied Microbiology and Biotechnology* 86 (5): 1323–36. https://doi.org/10.1007/S00253-010-2498-2.

Ahir, U. N., T. K. Vyas, K. D. Gandhi, P. R. Faldu, and K. G. Patel. 2020. In vitro efficacy for chlorpyrifos degradation by novel isolate *Tistrella* Sp. AUC10 isolated from *chlorpyrifos* contaminated field. *Current Microbiology* 77: 2226–32. https://doi.org/10.1007/s00284-020-01998-1.

Al-Mallah, Maha, Madeleine Goutx, Gilbert Mille, and Jean Claude Bertrand. 1990. Production of emulsifying agents during growth of a marine *Alteromonas* in seawater with eicosane as carbon source, a solid hydrocarbon. *Oil and Chemical Pollution* 6 (4): 289–305. https://doi.org/10.1016/S0269-8579(05)80005-X.

Amaral, P. F. F., J. M. da Silva, M. Lehocky, A. M. V. Barros-Timmons, M. A. Z. Coelho, I. M. Marrucho, and J. A. P. Coutinho. 2006. Production and characterization of a bioemulsifier from *Yarrowia lipolytica. Process Biochemistry* 41 (8): 1894–98. https://doi.org/10.1016/J.PROCBIO.2006.03.029.

Banat, Ibrahim M., Andrea Franzetti, Isabella Gandolfi, Giuseppina Bestetti, Maria G. Martinotti, Letizia Fracchia, Thomas J. Smyth, and Roger Marchant. 2010. Microbial biosurfactants production, applications and future potential. *Applied Microbiology and Biotechnology* 87 (2): 427–44. https://doi.org/10.1007/S00253-010-2589-0.

Batista, S. B., A. H. Mounteer, F. R. Amorim, and M. R. Tótola. 2006. Isolation and characterization of biosurfactant/bioemulsifier-producing bacteria from petroleum contaminated sites. *Bioresource Technology* 97 (6): 868–75. https://doi.org/10.1016/J.BIORTECH.2005.04.020.

Belsky, I., D. L. Gutnick, and E. Rosenberg. 1979. Emulsifier of *Arthrobacter* RAG-1: Determination of emulsifier-bound fatty acids. *FEBS Letters* 101 (1): 175–78. https://doi.org/10.1016/0014-5793(79)81320-4.

Biver, Sophie, Sébastien Steels, Daniel Portetelle, and Micheline Vandenbol. 2013. *Bacillus subtilis* as a tool for screening soil metagenomic libraries for antimicrobial activities. *Journal of Microbiology and Biotechnology* 23 (6): 850–55. https://doi.org/10.4014/JMB.1212.12008.

Brady, Sean F., Carol J. Chao, and Jon Clardy. 2004. Long-chain N-acyltyrosine synthases from environmental DNA. *Applied and Environmental Microbiology* 70 (11): 6865–70. https://doi.org/10.1128/AEM.70.11.6865-6870.2004/ASSET/548E106F-115F-4C72-9562-39E623C07CE1/ASSETS/GRAPHIC/ZAM0110449090003.JPEG.

Budzinski, H., N. Raymond, T. Nadalig, M. Gilewicz, P. Garrigues, J. C. Bertrand, and P. Caumette. 1998. Aerobic biodegradation of alkylated aromatic hydrocarbons by a bacterial community. *Organic Geochemistry* 28 (5): 337–48. https://doi.org/10.1016/S0146-6380(98)00002-3.

Chen, Chien Yen, Simon C. Baker, and Richard C. Darton. 2007. The application of a high throughput analysis method for the screening of potential biosurfactants from natural sources. *Journal of Microbiological Methods* 70 (3): 503–10. https://doi.org/10.1016/J.MIMET.2007.06.006.

Colores, Gregory M., Richard E. Macur, David M. Ward, and William P. Inskeep. 2000. Molecular analysis of surfactant-driven microbial population shifts in hydrocarbon-contaminated soil. *Applied and Environmental Microbiology* 66 (7): 2959–64. https://doi.org/10.1128/AEM.66.7.2959-2964.2000/ASSET/C055F9C5-D49B-4971-8BFE-85B54E14360F/ASSETS/GRAPHIC/AM0700043004.JPEG.

Dang, Nga Phuong, Bjarne Landfald, and Nils Peder Willassen. 2015. Biological surface-active compounds from marine bacteria. *Environmental Technology* 37 (9): 1151–58. https://doi.org/10.1080/09593330.2015.1103784.

Das, P., S. Mukherjee, and R. Sen. 2008. Antimicrobial potential of a lipopeptide biosurfactant derived from a marine *Bacillus circulans*. *Journal of Applied Microbiology* 104 (6): 1675–84. https://doi.org/10.1111/J.1365-2672.2007.03701.X.

Das, P., S. Mukherjee, and R. Sen. 2009. Antiadhesive action of a marine microbial surfactant. *Colloids and Surfaces B: Biointerfaces* 71 (2): 183–86. https://doi.org/10.1016/J.COLSURFB.2009.02.004.

Dave, H., C. Ramakrishna, B. D. Bhatt, and J. D. Desai. 1994. Biodegradation of slop oil from a petrochemical industry and bioreclamation of slop oil-contaminated soil. *World Journal of Microbiology and Biotechnology* 10 (6): 653–56. https://doi.org/10.1007/BF00327953.

Desai, Jitendra D., and Ibrahim M. Banat. 1997. Microbial production of surfactants and their commercial potential. *Microbiology and Molecular Biology Reviews* 61 (1): 47–64. https://doi.org/10.1128/MMBR.61.1.47-64.1997.

Déziel, E., F. Lépine, S. Milot, and R. Villemur. 2000. Mass spectrometry monitoring of rhamnolipids from a growing culture of *Pseudomonas aeruginosa* Strain 57RP. *Biochimica et Biophysica Acta (BBA)—Molecular and Cell Biology of Lipids* 1485 (2–3): 145–52. https://doi.org/10.1016/S1388-1981(00)00039-1.

Dhasayan, Asha, G. Seghal Kiran, and Joseph Selvin. 2014. Production and characterisation of glycolipid biosurfactant by *Halomonas* sp. MB-30 for potential application in enhanced oil recovery. *Applied Biochemistry and Biotechnology* 174 (7): 2571–84. https://doi.org/10.1007/S12010-014-1209-3/FIGURES/6.

Dhasayan, Asha, Joseph Selvin, and Seghal Kiran. 2015. Biosurfactant production from marine bacteria associated with sponge *Callyspongia diffusa*. *3 Biotech* 5 (4): 443–54. https://doi.org/10.1007/S13205-014-0242-9/FIGURES/5.

Diaz De Rienzo, M. A., P. S. Stevenson, R. Marchant, and I. M. Banat. 2016. Effect of biosurfactants on *Pseudomonas aeruginosa* and *Staphylococcus aureus* biofilms in a bioflux channel. *Applied Microbiology and Biotechnology* 100 (13): 5773–79. https://doi.org/10.1007/S00253-016-7310-5/FIGURES/3.

Filonov, Andrei E., Irina F. Puntus, Alexander V. Karpov, Irina A. Kosheleva, Konstantin I. Kashparov, Anatoly V. Slepenkin, and Alexander M. Boronin. 2004. Efficiency of naphthalene biodegradation by *Pseudomonas putida* G7 in Soil. *Journal of Chemical Technology & Biotechnology* 79 (6): 562–69. https://doi.org/10.1002/JCTB.998.

Floris, Rosanna, Giuseppe Scanu, Nicola Fois, Carmen Rizzo, Roberta Malavenda, Nunziacarla Spanò, and Angelina lo Giudice. 2018. Intestinal bacterial flora of Mediterranean gilthead sea bream (*Sparus aurata linnaeus*) as a novel source of natural surface active compounds. *Aquaculture Research* 49 (3): 1262–73. https://doi.org/10.1111/ARE.13580.

Franzetti, Andrea, Giuseppina Bestetti, Paolo Caredda, Paolo la Colla, and Elena Tamburini. 2008. Surface-active compounds and their role in the access to hydrocarbons in *Gordonia* strains. *FEMS Microbiology Ecology* 63 (2): 238–48. https://doi.org/10.1111/J.1574-6941.2007.00406.X.

Gandhimathi, R., G. Seghal Kiran, T. A. Hema, Joseph Selvin, T. Rajeetha Raviji, and S. Shanmughapriya. 2009. Production and characterization of lipopeptide biosurfactant by a sponge-associated marine *Actinomycetes nocardiopsis alba* MSA10. *Bioprocess and Biosystems Engineering* 32 (6): 825–35. https://doi.org/10.1007/S00449-009-0309-X/FIGURES/10.

Gerard, Jeff, Richard Lloyd, Todd Barsby, Paul Haden, Michael T. Kelly, and Raymond J. Andersen. 1997. Massetolides A–H, antimycobacterial cyclic depsipeptides produced by two Pseudomonads isolated from marine habitats. *Journal of Natural Products* 60 (3): 223–29. https://doi.org/10.1021/NP9606456.

Giraud, Marie France, and James H. Naismith. 2000. The rhamnose pathway. *Current Opinion in Structural Biology* 10 (6): 687–96. https://doi.org/10.1016/S0959-440X(00)00145-7.

Gudiña, Eduardo J., Vivek Rangarajan, Ramkrishna Sen, and Lígia R. Rodrigues. 2013. Potential therapeutic applications of biosurfactants. *Trends in Pharmacological Sciences* 34 (12): 667–75. https://doi.org/10.1016/J.TIPS.2013.10.002.

Gudiña, Eduardo J., José A. Teixeira, and Lígia R. Rodrigues. 2016. Biosurfactants produced by marine microorganisms with therapeutic applications. *Marine Drugs* 14 (2): 38. https://doi.org/10.3390/md14020038.

Gutierrez, Tony, Tracy Shimmield, Cheryl Haidon, Kenny Black, and David H. Green. 2008. Emulsifying and metal ion binding activity of a glycoprotein exopolymer produced by *Pseudoalteromonas sp.* strain TG12. *Applied and Environmental Microbiology* 74 (15): 4867–76. https://doi.org/10.1128/AEM.00316-08/ASSET/84B1F10F-F47B-465B-82D8-CC10B4881897/ASSETS/GRAPHIC/ZAM0150890500004.JPEG.

Hamza, Faseela, Surekha Satpute, Arun Banpurkar, Ameeta Ravi Kumar, and Smita Zinjarde. 2017. Biosurfactant from a marine bacterium disrupts biofilms of pathogenic bacteria in a tropical aquaculture system. *FEMS Microbiology Ecology* 93 (11). https://doi.org/10.1093/FEMSEC/FIX140.

Haritash, A. K., and C. P. Kaushik. 2009. Biodegradation aspects of polycyclic aromatic hydrocarbons (PAHs): A review. *Journal of Hazardous Materials* 169 (1–3): 1–15. https://doi.org/10.1016/J.JHAZMAT.2009.03.137.

Ibacache-Quiroga, C., J. Ojeda, G. Espinoza-Vergara, P. Olivero, M. Cuellar, and M. A. Dinamarca. 2013. The hydrocarbon-degrading marine bacterium *Cobetia* sp. strain

MM1IDA2H-1 produces a biosurfactant that interferes with quorum sensing of fish pathogens by signal hijacking. *Microbial Biotechnology* 6 (4): 394–405. https://doi. org/10.1111/1751-7915.12016.

Joshi, Sanket J., Harish Suthar, Amit Kumar Yadav, Krushi Hingurao, and Anuradha Nerurkar. 2013. Occurrence of biosurfactant producing *Bacillus* spp. in diverse habitats. *ISRN Biotechnology* 2013 (August): 1–6. https://doi.org/10.5402/2013/652340.

Karanth, N. G. K., P. G. Deo, and N. K. Veenanadig. 1999. Microbial production of biosurfactants and their importance. *Current Science* 77 (1): 116–26.

Kennedy, Jonathan, Burkhardt Flemer, Stephen A. Jackson, David P. H. Lejon, John P. Morrissey, Fergal O'Gara, and Alan D. W. Dobson. 2010. Marine metagenomics: New tools for the study and exploitation of marine microbial metabolism. *Marine Drugs* 8 (3): 608–28. https://doi.org/10.3390/MD8030608.

Khopade, Abhijit, Biao Ren, Xiang Yang Liu, Kakasaheb Mahadik, Lixin Zhang, and Chandrakant Kokare. 2012. Production and characterization of biosurfactant from marine *Streptomyces* species B3. *Journal of Colloid and Interface Science* 367 (1): 311–18. https://doi.org/10.1016/J.JCIS.2011.11.009.

Kiran, G. S., T. Anto Thomas, J. Selvin, B. Sabarathnam, and A. P. Lipton. 2010a. Optimization and characterization of a new lipopeptide biosurfactant produced by marine *Brevibacterium aureum* MSA13 in solid state culture. *Bioresource Technology* 101 (7): 2389–96. https://doi.org/10.1016/J.BIORTECH.2009.11.023.

Kiran, G. S., Balu Sabarathnam, and Joseph Selvin. 2010b. Biofilm disruption potential of a glycolipid biosurfactant from marine *Brevibacterium casei*. *FEMS Immunology and Medical Microbiology* 59 (3): 432–38. https://doi.org/10.1111/J.1574-695X.2010.00698.X.

Kitamoto, Dai, Hiroshi Yanagishita, Toshio Shinbo, Takashi Nakane, Chiyoshi Kamisawa, and Tadaatsu Nakahara. 1993. Surface active properties and antimicrobial activities of mannosylerythritol lipids as biosurfactants produced by *Candida Antarctica*. *Journal of Biotechnology* 29 (1–2): 91–96. https://doi.org/10.1016/0168-1656(93)90042-L.

Kohli, R. M., J. W. Trauger, D. Schwarzer, M. A. Marahiel, and C. T. Walsh. 2001. Generality of peptide cyclization catalyzed by isolated thioesterase domains of nonribosomal peptide synthetases. *Biochemistry* 40 (24): 7099–7108. https://doi.org/10.1021/BI010036J.

Komukai-Nakamura, Syoko, Keiji Sugiura, Yukie Yamauchi-Inomata, Haruhisa Toki, Kasthuri Venkateswaran, Satoshi Yamamoto, Hiroki Tanaka, and Shigeaki Harayama. 1996. Construction of bacterial consortia that degrade Arabian light crude oil. *Journal of Fermentation and Bioengineering* 82 (6): 570–74. https://doi. org/10.1016/S0922-338X(97)81254-8.

Kosaric, Naim, and Fazilet Vardar-Sukan. 2014. *Biosurfactants: Production and utilization–processes, technologies, and economics*. CRC Press, Boca Raton, FL. https:// doi.org/10.1201/b17599.

Kügler, Johannes H., Marilize le Roes-Hill, Christoph Syldatk, and Rudolf Hausmann. 2015. Surfactants tailored by the class *Actinobacteria*. *Frontiers in Microbiology* 6 (March): 212. https://doi.org/10.3389/FMICB.2015.00212/BIBTEX.

Luepongpattana, S., J. Thaniyavarn, and M. Morikawa. 2017. Production of massoia lactone by *Aureobasidium pullulans* YTP6–14 isolated from the Gulf of Thailand and its fragrant biosurfactant properties. *Journal of Applied Microbiology* 123 (6): 1488–97. https://doi.org/10.1111/JAM.13598.

Mabrouk, Mona E. M., Eman M. Youssif, and Soraya A. Sabry. 2014. 756 Biosurfactant production by a newly isolated soft coral-associated marine *Bacillus* sp. E34: Statistical optimization and characterization. *Life Science Journal* 11 (10): 1097–8135. www.lifesciencesite.comwww.lifesciencesite.com.123.

Maneerat, S., T. Nitoda, H. Kanzaki, and F. Kawai. 2005. Bile acids are new products of a marine bacterium, *Myroides* sp. strain SM1. *Applied Microbiology and Biotechnology* 67: 679–83. https://doi.org/10.1007/s00253-004-1777-1.

Mattei, G., and J. C. Bertrand. 1985. Production of biosurfactants by a mixed bacteria population grown in continuous culture on crude oil. *Biotechnology Letters* 7: 217–22.

McGenity, Terry J., Benjamin D. Folwell, Boyd A. McKew, and Gbemisola O. Sanni. 2012. Marine crude-oil biodegradation: A central role for interspecies interactions. *Aquatic Biosystems* 8 (1): 1–19. https://doi.org/10.1186/2046-9063-8-10/FIGURES/3.

Morita, T., M. Kitagawa, M. Suzuki, S. Yamamoto, A. Sogabe, S. Yanagidani, T. Imura, T. Fukuoka, and D. Kitamoto. 2009. A yeast glycolipid biosurfactant, mannosylerythritol lipid, shows potential moisturizing activity toward cultured human skin cells: The recovery effect of MEL-A on the SDS-damaged human skin cells. *Journal of Oleo Science* 58 (12): 639–42.

Mulligan, Catherine N. 2005. Environmental applications for biosurfactants. *Environmental Pollution* 133 (2): 183–98. https://doi.org/10.1016/J.ENVPOL.2004.06.009.

Nguyen, Thu T., and David A. Sabatini. 2011. Characterization and emulsification properties of rhamnolipid and sophorolipid biosurfactants and their applications. *International Journal of Molecular Sciences* 12 (2): 1232–44. https://doi.org/10.3390/IJMS12021232.

Nievas, M. L., M. G. Commendatore, J. L. Esteves, and V. Bucalá. 2008. Biodegradation pattern of hydrocarbons from a fuel oil-type complex residue by an emulsifier-producing microbial consortium. *Journal of Hazardous Materials* 154 (1–3): 96–104. https://doi.org/10.1016/J.JHAZMAT.2007.09.112.

Nunez, A., R. Ashby, T. A. Foglia, and D. K. Y. Solairnan. 2001. Analysis and characterization of sophorolipids by liquid chromatography with atmospheric pressure chemical ionization. *Chromatographia* 53 (11): 673–677.

Nurfarahin, Abdul Hamid, Mohd Shamzi Mohamed, and Lai Yee Phang. 2018. Culture medium development for microbial-derived surfactants production—An overview. *Molecules* 23 (5): 1049. https://doi.org/10.3390/molecules23051049.

Paniagua-Michel, José de Jesús, Jorge Olmos-Soto, and Eduardo Roberto Morales-Guerrero. 2014. Algal and microbial exopolysaccharides: New insights as biosurfactants and bioemulsifiers. *Advances in Food and Nutrition Research* 73 (January): 221–57. https://doi.org/10.1016/B978-0-12-800268-1.00011-1.

Pini, Francesco, Cristina Grossi, Sabrina Nereo, Luigi Michaud, Angelina lo Giudice, Vivia Bruni, Franco Baldi, and Renato Fani. 2007. Molecular and physiological characterisation of psychrotrophic hydrocarbon-degrading bacteria isolated from Terra Nova Bay (Antarctica). *European Journal of Soil Biology* 43 (5–6): 368–79. https://doi.org/10.1016/J.EJSOBI.2007.03.012.

Proksch, Peter, Ruangelie Edrada-Ebel, and Rainer Ebel. 2003. Drugs from the sea—opportunities and obstacles. *Marine Drugs* 1 (1): 5–17. https://doi.org/10.3390/MD101005.

Randhawa, Kamaljeet K. Sekhon, and Pattanathu K. S. M. Rahman. 2014. Rhamnolipid biosurfactants-past, present, and future scenario of lobal market. *Frontiers in Microbiology* 5 (September): 454. https://doi.org/10.3389/FMICB.2014.00454/BIBTEX.

Rizzo, Carmen, Luigi Michaud, Barbara Hörmann, Berna Gerçe, Christoph Syldatk, Rudolf Hausmann, Emilio de Domenico, and Angelina lo Giudice. 2013. Bacteria associated with *Sabellids* (Polychaeta: Annelida) as a novel source of surface active compounds. *Marine Pollution Bulletin* 70 (1–2): 125–33. https://doi.org/10.1016/J. MARPOLBUL.2013.02.020.

Rodrigues, Lígia, Ibrahim M. Banat, José Teixeira, and Rosário Oliveira. 2006. Biosurfactants: Potential applications in medicine. *Journal of Antimicrobial Chemotherapy* 57 (4): 609–18. https://doi.org/10.1093/JAC/DKL024.

Rosenberg, Eugene. 2006. Biosurfactants. *The Prokaryotes*: 834–49. https://doi.org/10.1007/0-387-30741-9_25.

Sachdev, Dhara P., and Swaranjit S. Cameotra. 2013. Biosurfactants in agriculture. *Applied Microbiology and Biotechnology* 97 (3): 1005–1016. https://doi.org/10.1007/s00253-012-4641-8.

Samel, Stefan A., Björn Wagner, Mohamed A. Marahiel, and Lars Oliver Essen. 2006. The thioesterase domain of the fengycin biosynthesis cluster: A structural base for the macrocyclization of a non-ribosomal lipopeptide. *Journal of Molecular Biology* 359 (4): 876–89. https://doi.org/10.1016/J.JMB.2006.03.062.Sivapathasekaran, C., Soumen Mukherjee, Ramapati Samanta, and Ramkrishna Sen. 2009. High-performance liquid chromatography purification of biosurfactant isoforms produced by a marine bacterium. *Analytical and Bioanalytical Chemistry* 395 (3): 845–54. https://doi.org/10.1007/S00216-009-3023-2/TABLES/4.

Stein, Jefferey L., Terence L. Marsh, Ke Ying Wu, Hiroaki Shizuya, and Edward F. Delong. 1996. Characterization of uncultivated prokaryotes: Isolation and analysis of a 40-kilobase-pair genome fragment from a planktonic marine archaeon. *Journal of Bacteriology* 178 (3): 591–99. https://doi.org/10.1128/JB.178.3.591-599.1996.

Stipcevic, Tamara, Ante Piljac, and Goran Piljac. 2006. Enhanced healing of full-thickness burn wounds using di-rhamnolipid. *Burns* 32 (1): 24–34. https://doi.org/10.1016/J.BURNS.2005.07.004.

Takeshi, Suppasil Maneerat, Bamba Kazuo, Harada Akio, Kobayashi Hidenori, and Yamada Fusako Kawai. 2006. Applied microbial and cell physiology a novel crude oil emulsifier excreted in the culture supernatant of a marine bacterium, *Myroides* sp. strain SM1. *Applied Microbiology and Biotechnology* 70: 254–59. https://doi.org/10.1007/s00253-005-0050-6.

Thavasi, Thavasi Renga, Shilpy Sharma, Singaram Jayalakshmi, and Rengathavasi Thavasi. 2011. Evaluation of screening methods for the isolation of biosurfactant producing marine bacteria. *Article in Journal of Petroleum & Environmental Biotechnology* 1: 1. https://doi.org/10.4172/2157-7463.S1-e001.

Théatre, Ariane, Carolina Cano-Prieto, Marco Bartolini, Yoann Laurin, Magali Deleu, Joachim Niehren, Tarik Fida, et al. 2021. The surfactin-like lipopeptides from *Bacillus* spp.: Natural biodiversity and synthetic biology for a broader application range. *Frontiers in Bioengineering and Biotechnology* 9 (March). https://doi.org/10.3389/FBIOE.2021.623701.

Tiso, Till, Rabea Zauter, Hannah Tulke, Bernd Leuchtle, Wing Jin Li, Beate Behrens, Andreas Wittgens, Frank Rosenau, Heiko Hayen, and Lars Mathias Blank. 2017. Designer rhamnolipids by reduction of congener diversity: Production and characterization. *Microbial Cell Factories* 16 (1): 1–14. https://doi.org/10.1186/S12934-017-0838-Y/FIGURES/8.

Trimble, Michael J., Patrik Mlynárčik, Milan Kolář, and Robert E. W. Hancock. 2016. Polymyxin: Alternative mechanisms of action and resistance. *Cold Spring Harbor Perspectives in Medicine* 6 (10). https://doi.org/10.1101/CSHPERSPECT.A025288.

Twigg, Matthew S., L. Tripathi, A. Zompra, K. Salek, V. U. Irorere, T. Gutierrez, G. A. Spyroulias, R. Marchant, and I. M. Banat. 2018. Identification and characterisation of short chain rhamnolipid production in a previously uninvestigated, non-pathogenic marine pseudomonad. *Applied Microbiology and Biotechnology* 102 (19): 8537–49. https://doi.org/10.1007/S00253-018-9202-3.

Urum, Kingsley, and Turgay Pekdemir. 2004. Evaluation of biosurfactants for crude oil contaminated soil washing. *Chemosphere* 57 (9): 1139–50. https://doi.org/10.1016/J.CHEMOSPHERE.2004.07.048.

Varvaresou, A., and K. Iakovou. 2015. Biosurfactants in cosmetics and biopharmaceuticals. *Letters in Applied Microbiology* 61 (3): 214–23. https://doi.org/10.1111/LAM.12440.

Vatsa, Parul, Lisa Sanchez, Christophe Clement, Fabienne Baillieul, and Stephan Dorey. 2010. Rhamnolipid biosurfactants as new players in animal and plant defense against microbes. *International Journal of Molecular Sciences* 11 (12): 5095–5108. https://doi.org/10.3390/IJMS11125095.

Vyas, Trupti K., and B. P. Dave. 2007. Effect of crude oil concentrations, temperature and pH on growth and degradation of crude oil by marine bacteria. *Indian Journal of Marine Sciences* 36 (1): 76–85.

Vyas, Trupti K., and P. Dave. 2010. Effect of addition of nitrogen, phosphorus and potassium fertilizers on biodegradation of crude oil by marine bacteria. *Indian Journal of Marine Sciences* 39 (1): 143–50.

Vyas, T. K., and B. P. Dave. 2011. Production of biosurfactant by *Nocardia Otitidiscaviarum* and its role in biodegradation of crude oil. *International Journal of Environmental Science and Technology* 8 (2): 425–32.

Vyas, T. K., and B. P. Dave. 2016. Biosynthesis of rhamnolipid biosurfactant by newly isolated marine bacterium *Methylobacterium mesophilicum* MTCC 6839 from oil contaminated sites at Alang coast, Gujarat, India. *International Journal of Bio-Resource and Stress Management* 7 (1): 074–79.

Yan, Xia, James Sims, Bin Wang, and Mark T. Hamann. 2014. Marine actinomycete *Streptomyces* sp. ISP2–49E, a new source of rhamnolipid. *Biochemical Systematics and Ecology* 55: 292–295. https://doi.org/10.1016/j.bse.2014.03.015.

Yeh, Mao Sung, Yu Hong Wei, and Jo Shu Chang. 2005. Enhanced production of surfactin from *Bacillus subtilis* by addition of solid carriers. *Biotechnology Progress* 21 (4): 1329–34. https://doi.org/10.1021/BP050040C.

Zenati, Billal, Alif Chebbi, Abdelmalek Badis, Kamel Eddouaouda, Hocine Boutoumi, Mohamed el Hattab, Dorra Hentati, et al. 2018. A non-toxic microbial surfactant from *Marinobacter hydrocarbonoclasticus* SdK644 for crude oil solubilization enhancement. *Ecotoxicology and Environmental Safety* 154 (June): 100–107. https://doi.org/10.1016/J.ECOENV.2018.02.032.

Zinjarde, S. S., and A. Pant. 2002. Emulsifier from a tropical marine yeast, *Yarrowia lipolytica* ncim 3589. *Journal of Basic Microbiology* 42 (1): 67–73. https://doi.org/10.1002/1521-4028(200203)42:1.

6 Diversity of Marine Biosurfactant-Producing Bacteria and Their Role in the Degradation of Heterogenous Petrochemical Hydrocarbons

Brian Gidudu and Evans M.N. Chirwa

CONTENTS

DOI: 10.1201/9781003307464-6

INTRODUCTION

The oil and gas industry remains a major pillar of national economies globally and is the lifeblood of the manufacturing and transport industries due to its usage as a source of energy (Al-Samhan et al. 2022). In addition, the petrochemical industry supports the pharmaceutical industry, food industry and agricultural sectors with products such as medicines, fertilizers, food preservatives, solvents, plastics and pesticides (Khuzwayo and Chirwa 2020). Currently, new international and national policies and laws world over are exerting serious pressure towards moving from fossil fuels to cleaner sources of energy, but the petrochemical sector is continuously getting new industry players, with a reported industry growth exceeding the world's gross product by a factor of more than 2 (Golysheva et al. 2020). The rise in energy and petrochemical product demand has increased the release of petrochemical waste and products produced in offshore/onshore upstream and downstream oil production phases into the environment (Gupta and Pathak 2020). Hence the petrochemical industry is known for discharging regulatory pollutants such as polycyclic aromatic hydrocarbons (PAHs); petroleum hydrocarbons; benzene, toluene, ethylbenzene, and xylene (BTEX) compounds; sulfides; phenols; volatile organic compounds; and heavy metals (Wang et al. 2022). Most of these compounds, especially PAHs, are hazardous to humans and adversely affect the environment because they are mutagenic, carcinogenic and toxic (Gupta and Pathak 2020).

Numerous attempts have been made to use biosurfactants or combine them with other methods for enhanced degradation of petrochemical hydrocarbons in contaminated media (Mulligan 2021). Biosurfactants are preferred because of their low toxicity, high biodegradability, high diversity, high demulsification potential and high selectivity (Mulligan 2009). Biosurfactants can also be used in varying salinity, pH and temperature (Bezza and Chirwa 2015). Biosurfactant production and degradation of hydrophobic compounds are correlated events that have to be studied together (Oliveira et al. 2017). Marine environments have a diverse source of bacteria that efficiently degrade and produce biosurfactants, which are rarely produced by organisms from terrestrial environments (Deng et al. 2016). More surfactation genes are found in water biomes compared to terrestrial biomes, but more degradation genes are found in terrestrial organisms near the equator (Oliveira et al. 2017).

PETROCHEMICALS

The petrochemical industry is a heavy industry that is quite important to most economies globally. However, the sector poses a threat to the environment, as it generates greenhouse gases; consumes a huge share of energy; and produces lots of wastewater, hazardous wastes and noise (Aryanasl et al. 2017). Petrochemical pollutants are introduced into the environment as crude oil, refined products or waste (Gupta and Pathak 2020). The extent of contamination of the environment by petrochemicals is dependent on their crude oil source, refining process and final purpose of the product (Cerqueira et al. 2014). But the major pathway of environmental pollution is the poor disposal of oil sludge and oil spillages (Cerqueira et al. 2014).

Pollutants Found in Petrochemicals

The petrochemical industry produces both organic and inorganic wastes (Tian et al. 2020). Heavy metals are the major inorganic pollutants produced in the petrochemical industry but may also combine with other metals to form secondary compounds such as sulfides and fluorides (Tian et al. 2020). These compounds are found in wastewater, refined products, crude oil and oily waste (Sarma, Bustamante, and Prasad 2019). Depending on the amount of oil an water, oily waste can be referred to as simple oil if it contains less water or as oil sludge if it is highly viscous and has stable solid-water-oil emulsions (Hu, Li, and Zeng 2013).

Oil sludge is the main component of solid waste generated in the petrochemical industry during exploration, refining, transportation and treatment of petrochemical wastewater (Wang et al. 2020). Oil sludge is generally composed of 30–50% water, 30–80% oil and 10–20% residues, making it a critical hazardous waste produced in the petrochemical industry (Chen et al. 2020). In addition, oil sludge may also be composed of aqueous waste, contaminated solids, hydrocarbon wastes, spent catalysts and inorganic wastes (Islam 2015).

The main pollutants of major concern are organic wastes that include polycyclic aromatic hydrocarbons, petroleum hydrocarbons, BTEX compounds and phenols (Wang et al. 2022). Depending on the properties of the hydrocarbons, these could generally be categorized as resins, saturates, aromatics and asphaltenes (Fakher et al. 2020).

Saturates

Saturates are saturated compounds, so they do not contain double bonds since they are bonded to the maximum allowable hydrogen (Obodovskiy 2019). They may also be referred to as alkanes, and examples are methane, ethane, propane, butane and so on (Fakher et al. 2020).

Aromatics

Aromatics are a large component of hydrocarbons with a more complex structure than saturates. They are nonpolar and molecularly stable unsaturated hydrocarbons with multiple carbon-carbon double bonds (Hamilton et al. 2021). Polycyclic aromatic hydrocarbons in particular have two or more fused rings in their structure with various configurations (Premnath et al. 2021). They are classified as light

PAHs if they have fewer than four benzene rings and heavy PAHs if they have more than four benzene rings. Examples of PAHs include naphthalene, acenaphthylene, acenaphthene, fluorene, phenanthrene, anthracene, fluoranthene, pyrene, benz(a)anthracene, chrysene, benzo(b)fluoranthene, benzo(k)fluoranthene, benzo(a)pyrene, benzo(ghi)perylene dibenz(a,h)anthracene and indeno(1,2,3-cd) pyrene (Premnath et al. 2021). Some of them occur as hybrids, but the greater the PAH size and angularity, the more stable and hydrophobic the compound becomes (Lawal 2017). PAHs are highly soluble in lipids so are easily absorbed on the gastro-intestinal tract of mammals, where they are distributed to a variety of tissues in the body (Samanta, Singh, and Jain 2002). In the environment, PAHs often interact with other compounds to increase the potency of the known toxicity of PAHs. They are persistent in the environment and can be transferred through the food chain, posing a threat to flora and fauna (Lawal 2017). PAHs are therefore classified as the priority pollutants for remediation because of their carcinogenicity, mutagenicity and toxicity (Gupta and Pathak 2020).

Asphaltenes

Asphaltenes are nonvolatile and nonpolar compounds that are soluble in aromatics and insoluble in alkanes. They may be made up of carbon, nitrogen, sulfur and oxygen (Azizian and Khosravi 2019). Asphaltenes have a nonpolar part that has a high affinity to oil and a polar part that has an affinity for water and polar compounds (Ramirez-Corredores 2017). Their structure comprises heterocyclic aromatic rings and polyaromatics with side branching and is more complex than saturates, resins and aromatics (Fakher et al. 2020). This makes asphaltenes function as surfactants with solid micelle structures that prevent coalescence and growth of droplets (Ramirez-Corredores 2017). Asphaltenes are characterized as solid, n-alkane insoluble or highly polar based on their heteroatoms (Fakher et al. 2020). Ketones, phenols, porphyrins, fatty acids and esters are some of the major components of asphaltenes (Cheshkova et al. 2019).

Resins

Resins are nonvolatile compounds soluble in alkanes and aromatic solvents but with more complexity than saturates and aromatics (Azizian and Khosravi 2019). These are usually dark brown, adhesive and shiny, with a strong attraction to asphaltenes (Azizian and Khosravi 2019). Their structure is similar to asphaltenes but with fewer heteroatoms and higher hydrogen-to-carbon ratio. Resins link polar hydrocarbons and nonpolar asphaltenes to make more complex petrochemical compounds (Fakher et al. 2020). Resins are mainly made up of amides, pyridines, quinolines and sulfoxides (Cheshkova et al. 2019).

PATHWAYS OF PETROCHEMICAL POLLUTANTS IN THE ENVIRONMENT

The aftermath of disposal or spillage of petrochemical hydrocarbons is the evaporation of hydrocarbons into the atmosphere and the flow of the hydrocarbons into the ground or surface water. In surface water, the oil creates a thin layer on the surface of the water as the toxic volatile compounds evaporate into the atmosphere. In contrast, the low-molecular weight compounds dissolve in water. This is followed by the photolysis of low-molecular weight compounds into phenolic and highly acidic toxic compounds that could be more toxic than the original hydrocarbons (Kingston 2002). The highly hydrophobic and high-pressure PAHs settle in marine sediments of sea beds (Koh et al. 2004). In the air, recalcitrant organic compounds such as PAHs released to the atmosphere are often redeposited to the soil (Premnath et al. 2021). In soil, the non-aqueous hydrophobic compounds adsorb to crystals, organic matter, particles and oily liquid phase (Premnath et al. 2021). In the presence of emulsifying agents such as asphaltenes and resins, the adsorption is very strong, making remediation attempts very difficult (Hu, Li, and Zeng 2013).

EFFECT OF HYDROCARBON POLLUTION ON MARINE ORGANISMS, SOIL, PLANTS AND HUMANS

The pollution of the environment by petrochemicals has perennial effects on the affected ecosystems like salt marshes and mangroves after contamination (Premnath et al. 2021). Pollution by petrochemicals is usually an issue of great concern because they are generally carcinogenic, mutagenic and toxic to mammals (Gupta and Pathak 2020).

Organic compounds in petrochemicals adversely affect the strength, compressibility and structural properties of soil (Gopang et al. 2016). The salinity, morphology and solidity properties of soil are also most likely to increase after contamination (Johnson and Affam 2018). Organic pollutants also end up in the food chain if food is grown in contaminated soils (Lawal 2017). The extent of the effect of soil pollution mainly depends on the properties of the soil and the constituents of the petrochemicals. Depending on the properties of the petrochemicals, the contamination of land by hydrocarbons can make it unusable for plant growth when hydrocarbons are strongly bound with soil (Kisic et al. 2009).

The exposure of plants to oily petrochemicals has detrimental effects on the growth of plants (Bansal and Kim 2015). Lighter oils are reported to have more adverse effects on plants than heavier oils (Pezeshki et al. 2000). PAH pathways into plants are through soil to plant by root uptake or from air to plant by atmospheric deposition through the stomata (Zhang et al. 2017). Studies have reported that aromatics can kill plants at concentrations of 1% (Nelson-Smith 1968). Exposure of plants to oil is also associated with the blocking of the transpiration pathways of plants and their stomata from which the exposed plants experience temperature stress and attenuated transpiration (Alagić, Maluckov, and Radojičić 2015).

Pollutants in petrochemicals have adverse impacts on coral reefs and marine organisms (Guigue et al. 2015). Petrochemicals can inhibit the colonialization of hermatypic coral reefs and decrease the viability of their colonies (Soares et al. 2022). They may also reduce life expectancy and cause the mutilation of coral reefs and planulae (Craveiro et al. 2021). Extensive exposure to PAHs has detrimental effects on coral reefs, such as the destruction of their feeding mechanism and tactile stimuli, extensive bacterial growth due to mucus secretion and the destruction of coral reef tissues and cell structures (Soares et al. 2022). Studies have also reported the accumulation of petrochemical hydrocarbons in the tissues of mussels (Magalhães et al. 2022). The destruction of larvae and fish eggs even after exposure to low concentrations have also been reported (Langangen et al. 2017). It is also common to notice the reduction in growth rates, reduction in feeding and the development of morphological deformities in fish (Langangen et al. 2017).

Respiratory problems, headaches, vomiting, eye irritation, nausea, skin irritation and dizziness are some of the effects experienced by humans after exposure (Lawal 2017). Other effects may include kidney damage, learning disabilities, liver damage, mental disabilities, skin lesions, physical retardation, increased cancer and congenital disabilities (Ossai et al. 2020). The extent of the effect is dependent on the toxicity of the pollutant and the rate of exposure (Premnath et al. 2021). Indeno(1,2,3-cd)pyrene, dibenz(ah)anthracene, benz(a)anthracene, chrysene, benzo(k)fluoranthene, benzo(b)fluoranthene and benzo(a)pyrene are some of the PAH compounds classified as carcinogenic (Lawal 2017).

CHALLENGES OF BIOREMEDIATING PETROCHEMICAL-CONTAMINATED ENVIRONMENTS

Hydrocarbons with C_{20}–C_{30} chains are not easily degraded because of their hydrophobicity, but from the hardest to the easiest to degrade, hydrocarbons are arranged as asphaltenes, PAHs, cyclic alkanes, monoaromatics, low molecular weight n-alkyl aromatics, branched alkenes, branched-chain alkanes and n-alkanes. PAHs are the major pollutants of concern due to their toxicity, yet they are characterized by low bioavailability and solubility (Gidudu and Chirwa 2020). This makes PAHs persistent in the environment and resistant to biodegradation. The bioavailability of hydrocarbons decreases with time but can be increased if they are evaporated or dissolved (Mahjoubi et al. 2017).

Biosurfactants are applied in the presence of microbes or remediation technologies to increase bioavailability and the remediation of petrochemical hydrocarbons (Gidudu and Chirwa 2020). In the combination of biosurfactants and microbes for bioremediation, it is always logical to use microbes that can degrade and produce biosurfactants simultaneously rather than applying biosurfactants to improve the bioavailability of the contaminants to non-biosurfactant-producing organisms. Numerous microorganisms summing up to a total of 9 cyanobacteria species, 14 algae, 103 fungi and 79 bacteria can degrade hydrocarbons (Hassanshahian et al. 2012). Some of these microorganisms, especially bacteria from marine environments, can achieve 100% degradation, while those obtained from soil may easily achieve 50% degradation of hydrocarbons (Das and Chandran 2011).

The efficiency in terms of degradation of each species depends on the nature and the amount of hydrocarbons available to the microbes (Das and Chandran 2011). Hydrocarbons with 5–10 hydrocarbon atoms are not easily degraded because they damage and distort the cell membranes of microbe degraders (Tyagi et al. 2011). Biodegradation also depends on the environment of the soil or water in which the degradation must happen (Jørgensen 2008).

DIVERSITY OF BIOSURFACTANTS PRODUCED BY MARINE ORGANISMS

Biosurfactants are classified into five major groups: glycolipids, lipopeptides, lipopolysaccharides, phospholipids and fatty acids/neutral lipids (Roy 2017). The characteristics and nature of biosurfactants vary depending on the producer organism (Mondal et al. 2019). Various microorganisms, including bacteria, fungi and yeasts, produce biosurfactants (Gudina et al. 2015). *Rhodococcus, Bacillus, Halomonas, Arthrobacter, Pseudomonas, Acinetobacter* and *Enterobacter* are the most widely studied biosurfactant-producing bacteria (Mondal et al. 2019). Bacteria are acknowledged for being great producers of biosurfactants with high and low molecular weight properties (Uzoigwe et al. 2015).

High molecular surface agents such as liposan, alasan, biodispersan and emulsan are excellent bioemulsifiers, whereas low molecular weight metabolites such as glycolipids and lipopeptides are good biosurfactants (Kubicki et al. 2019). Bioemulsifiers are efficient emulsifiers of immiscible liquids but are inefficient in reducing surface tension. On the other hand, biosurfactants have good emulsification capabilities but are also efficient at reducing surface tension and interfacial tension between phases of different polarity (Uzoigwe et al. 2015).

Low molecular weight biosurfactants are composed of glycolipids, sophorolipids and trehalose lipids made up of long chain fatty acids or disaccharides acylated with hydroxy fatty acids (Kubicki et al. 2019). They also comprise carbohydrates attached to long chain lipopeptides or aliphatic acids (Kubicki et al. 2019). High molecular weight biosurfactants are made up of polysaccharides, lipoproteins, lipopolysaccharides, proteins or complex mixtures of these biopolymers (Uzoigwe et al. 2015).

REMEDIATION OF PETROCHEMICAL HYDROCARBONS BY MARINE BACTERIA

Marine organisms have been investigated as potential producers of biosurfactants and hydrocarbon degraders. Pelagic hydrocarbon degraders, also referred to as hydrocarbonoclastic bacteria, degrade recalcitrant hydrocarbons such as aliphatic and aromatic hydrocarbons that are hardly degraded by bacteria (Golyshin et al. 2003). Marine hydrocarbonoclastic bacteria possess low protein counts and dilute cytoplasms with a dry mass of every cell eight times lower than that of heterotrophic bacterium such as *Escherichia coli* that enables them to degrade a few hydrocarbons according to their catabolic genes (Golyshin et al. 2003).

Hydrocarbonoclastic bacteria are found in contaminated and undisturbed water, which is why they are often isolated from contaminated sediments and water samples. The other sources are communities associated with invertebrates like sponges and tunicates (Kubicki et al. 2019). The presence of hydrophobic and insoluble substrates such as oil or hydrocarbons in sediments and water stimulates the production of secondary metabolites that enhance the uptake and degradation of recalcitrant hydrocarbons (Floris, Rizzo, and Giudice 2018). Although hydrocarbon-degrading bacteria constitute less than 0.1% of microbial communities, they often represent 100% of the viable microorganisms in oil-contaminated environments (Oliveira et al. 2017). Some of the best biosurfactant-producing organisms from marine environments are sourced from seawater, marine sediments, oil-spilled seawater samples, the Black Sea, the Red Sea, seawater enrichments and hot water springs (Floris, Rizzo, and Giudice 2018).

In some instances, marine biosurfactant-producing organisms are favored ahead of other organisms because of their ability to produce biosurfactants under extreme conditions such as changing pH, salinity, temperature and limited nutrients (Tripathi et al. 2019). The so-called hydrocarbonoclastic bacteria that grow on petroleum hydrocarbons are often Gram negative, aerobic and affiliated with *Proteobacteria, Marinomonas vaga, Oceanospirillum linum* and *Halomonas elongate* (Golyshin et al. 2003). These belong to the genera *Oleiphilus, Neptunomonas, Marinobacter, Cycloclasticus* or *Alcanivorax* (Golyshin et al. 2003). Strains such as *Marinobacter* sp. CAB, *Marinobacter hydrocarbonoclasticus, Alcanivorax* sp. ST1 and *Alcanivorax borkumensis* degrade linear branched chain aliphatics, whereas *Psychroserpens burtonensis, Cycloclasticus pugetii* and *Cycloclasticus oligotrophus* degrade polycyclic aromatic hydrocarbons (Golyshin et al. 2003).

THE ROLE OF MARINE BIOSURFACTANTS IN BIOREMEDIATION

Biosurfactants increase the bioavailability of hydrophobic pollutants by reducing surface and interfacial tension, thereby increasing the surface area of hydrocarbons, which makes them available to microbes (Gidudu and Chirwa 2020). Biosurfactants improve the bioavailability of pollutants in contaminated media through three major mechanisms: (1) In the presence of nonaqueous-phase liquid organics, the interfacial tension between the aqueous and non-aqueous phases is reduced due to the dispersion of non-aqueous liquid organics. (2) Surface tension of the solid particle-pore water is reduced due to the increase in the solubility of the pollutants. Solubility increases due to the formation of micelles around hydrophobic organic pollutants. (3) Expulsion of the pollutants from the solid matrix due to the interactions between the pollutant and the biosurfactant and the interaction of the solid particles with the biosurfactant.

Degradation rates in the range of 32–88% for PAHs, 41–84% for high molecular weight alkanes (C_{20}–C_{32}) and 99% for low molecular weight alkanes were reported when marine organisms from seawater containing *Cycloclasticus, Alcanivorax* and *Thalassolituus oleivorans* were inoculated (McKew et al. 2007).

Vibrio, Pseudoalteromonas, Marinomonas and *Halomonas* isolated from marine sediments were identified as potential degraders of PAHs such as phenanthrene and chrysene (Melcher, Apitz, and Hemmingsen 2002). Recently, bacteria in the genera *Alcanivorax* and *Ketobacter* were identified as dominant marine organisms with good metabolic capabilities for degradation of linear and branched alkanes and the production of biosurfactants (Yakimov et al. 2019). Table 6.1 shows the diversity of marine organisms, their sources and the biosurfactants produced during hydrocarbon degradation.

GLYCOLIPIDS

Glycolipids are made up of sugars with a long chain of aliphatic hydroxy acids or aliphatic acids (Roy 2017). Examples of glycolipids are rhamnolipids, trehalolipids and sophorolipids.

Rhamnolipids

Rhamnolipids are ionic glycolipid biosurfactants with two or one α-l-rhamnose units linked to two 3-hydroxy fatty acid moieties with 8 to 14 carbon atoms (Varjani and Upasani 2017). The production of rhamnolipid biosurfactants by bacteria is reported in *Pseudomonas aeruginosa, Burkholderia* sp., *Marinobacter* sp., *Pseudomonas chlororaphis* and *Pseudomonas putida* (Upton 2017).

Marinobacter sp. MCTG107b produced a glycolipid biosurfactant with a yield of 150 μg mL^{-1} characterized by 14 separate rhamnolipid congeners using PAHs as the substrate (Tripathi et al. 2019). An *Alcanivorax borkumensis* strain with the ability to scavenge nutrients and organic pollutants and produce biosurfactants contained genes similar to those of *Pseudomonas putida* strains that allowed it to degrade hydrocarbons and genes similar to those of *Pseudomonas aeruginosa* that allowed it to produce glycolipid biosurfactants (Schneiker et al. 2006).

Xu et al. (2020) reported that rhamnolipids with the ability to reduce the surface tension of water from 65.56 to 38.33 mN/m were produced by *Paracoccus* sp. MJ9 strain found in the Jiaozhou Bay in Qingdao, Shandong, China. The strain had great potential for the degradation of hydrophobic compounds, having degraded more than 81% of diesel oil in five days (Xu et al. 2020).

A consortium of *Pseudoalteromonas agarivorans* SDRB-Py1, *Isoptericola chiayiensis* 103-Na4, *Rhodococcus soli* 102-Na5 and *Bacillus algicola* 003-Phe1 isolated from oil-contaminated beach sediments of the Taean coast, Korea, produced rhamnolipid biosurfactants with monorhamnolipid and dirhamnolipid congeners that generally reduced the surface tension of water to 33.9–41.3 mN m^{-1} while growing on crude oil (Lee et al. 2018). The rhamnolipid biosurfactants enhanced the biodegradation of n-alkanes and aromatic hydrocarbons to 48–72%. Most of the hydrocarbons between C9 and C14 were fully degraded, whereas the hydrocarbons from C15–C20 were substantially degraded by all the strains. Of all the strains, *Isoptericola chiayiensis* 103-Na4 and *Pseudoalteromonas agarivorans* SDRB-Py1 were considered the best degraders (Lee et al. 2018).

TABLE 6.1
Biosurfactants Produced by Marine Bacteria

Biosurfactant	Marine Organism	Source of Microorganism	Hydrocarbon	Reference
Rhamnolipid	*Marinobacter* sp. MCTG107b	Sea surface water	PAHs	(Tripathi et al. 2019)
Rhamnolipid	*Pseudomonas* sp. MCTG214(3b1)	Marine eukaryotic phytoplankton bloom	Phenanthrene, anthracene, fluorene and pyrene	(Twigg et al. 2018)
Glycolipid	*Rhodococcus* species BS-15	Deep-sea sediment	None	(Konishi et al. 2014)
Rhamnolipid	*Rhodobacteraceae, Rhodospirillaceae, Halomonadaceae, Oceanospirillaceae, Pseudomonadaceae, Shewanellaceae*	Seawater and sediment samples	Crude oil	(Antoniou et al. 2015)
Sophorolipids	*Rhodobacteraceae, Rhodospirillaceae, Halomonadaceae, Oceanospirillaceae, Pseudomonadaceae, Shewanellaceae*	Seawater and sediment samples	Crude oil	(Antoniou et al. 2015)
Glycolipids	*Rhodococcus, Pseudomonas, Pseudoalteromonas, Idiomarina*	Shoreline sediments	Tetradecane	(Malavenda et al. 2015)
Glycolipids	*Rhodococcus fascians*	Deep sea sediment	Not specified	(Konishi et al. 2014)
Glycolipids, phospholipids	*Alcanivorax borkumensis*	Sea water	Alkanes	(Abraham, Meyer, and Yakimov 1998)
3-hydroxy fatty acids	*Cobetia* sp. strain MM11DA2H-1	Seawater	Dibenzothiophene	(Ibacache-Quiroga et al. 2013)
Rhamnolipids	*Sphingomonas* sp., *Pseudomonas* sp.	Deep sea sediment	PAHs	(Cui et al. 2008)
Trehalose lipids	*Rhodococci*	Sea water	Aliphatic hydrocarbons, volatile fatty acids and biphenyl	(Yakimov et al. 1999)
Trehalose lipids	*Alcanivorax* and *Ketobacter*	Sea water	Aliphatic hydrocarbons, volatile fatty acids and biphenyl	(Yakimov et al. 2019)
Glycolipids	*Alcanivorax borkumensis*	Coastal waters	Alkanes, alkylarenes and alkylcycloalkanes	(Schneiker et al. 2006)

Biosurfactant	Bacteria	Source	Substrate	Reference
Phospholipids	*Alcanivorax borkumensis*	Sea water/sediments	N-alkanes	(Yakimov et al. 1998)
Lipopeptide; proline lipids	*Alcanivorax dieselolei* B-5^T	Oil-contaminated surface sea water	Diesel oil	(Qiao and Shao 2010)
Cyclic lipopeptide	*Achromobacter* sp. HZ01	Crude oil-contaminated seawater	Not specified	(Deng et al. 2016)
Lipopeptide: tauramamide,	*Brevibacillus laterosporus*	Tissues of a tube worm	Not specified	(Desjardine et al. 2007)
Lipopeptide: lichenysin and iturine	*Staphylococcus* sp.	Contaminated sediments	Crude oil and PAHs	(Hentati et al. 2021)
Rhamnolipid	*Paracoccus* sp. MJ9	Sea water	Diesel oil	(Xu et al. 2020)
Rhamnolipid	*Bacillus algicola* (003-Phe1)	Beach sediments	Crude oil and PAHs	(Lee et al. 2018)
Rhamnolipid	*Rhodococcus soli* (102-Na5)	Beach sediments	Crude oil and PAHs	(Lee et al. 2018)
Rhamnolipid	*Isoptericola chiayiensis* (103-Na4)	Beach sediments	Crude oil and PAHs	(Lee et al. 2018)
Rhamnolipid	*Pseudoalteromonas agarivorans* (SDRB-Py1)	Beach sediments	Crude oil and PAHs	(Lee et al. 2018)
Not determined	*Shewanella alga, Shewanella upenei, Vibrio furnissii, Gallaecimonas pentaromativorans, Brevibacterium epidermidis, Psychrobacter namhaensis, Pseudomonas fluorescens*	Sediments and seawater	Diesel oil	(Hassanshahian 2014)
Rhamnolipids	*P. aeruginosa* 110	Roots of *Phragmites australis*	N-alkanes and PAHs	(Wu et al. 2018)
Not determined	*Vibrio* sp. PBN295	Pennatulids	Diesel oil and crude oil	(Graziano et al. 2016)
Surfactin	*Bacillus velezensis* H3	Sea mud	Crude oil	(Liu et al. 2010)
Lipopeptide	*Halomonas* sp., *Pseudomonas* sp., *Aneurinibacillus* sp., *Planomicrobium* sp.	Marine environment	Diesel oil	(Wu et al. 2019)
Lipopeptide	*Bacillus licheniformis* MTCC 5514	Marine samples	Anthracene	(Swaathy et al. 2014)

A *Pseudomonas aeruginosa* L10 strain isolated from the roots of a reed, *Phragmites australis*, in the Yellow River Delta, Shandong, China, efficiently degraded n-alkanes, naphthalene, phenanthrene and pyrene while producing a rhamnolipid biosurfactant (Wu et al. 2018). The strain possessed putative genes encoding aldehyde dehydrogenase, alcohol dehydrogenase, monooxygenase and dioxygenase that enabled it to degrade petrochemical hydrocarbons and *rhlABRI* gene clusters that enabled it to produce rhamnolipids in the presence of petro-chemical substrates (Wu et al. 2018).

Trehalolipids

These are glycolipids containing disaccharide trehalose acylated with mycolic acids (Kubicki et al. 2019). These are produced by *Anthrobacter* sp., *Rhodococcus erythropolis*, *Corynebacterium*, *Norcadia* and *Mycobacterium* (Floris, Rizzo, and Giudice 2018). *Rhodococcus, Pseudomonas, Pseudoalteromonas* and *Idiomarina* isolated from shoreline sediments were reported to produce treha-lolipid glycolipids during the biodegradation of tetradecane as the hydrophobic substrate (Malavenda et al. 2015). Yakimov et al. (1999) reported that marine *Rhodococci* isolated from seawater produced trehalose lipids while degrading aliphatic hydrocarbons, volatile fatty acids and biphenyl (Yakimov et al. 1999).

Sophorolipids

Sophorolipids are ionic glycolipid biosurfactants containing di-saccharide sopho-rose and 17-hydroxyoleic acids (Kubicki et al. 2019). These are mainly produced by *Candida* spp. and the yeast *Starmerella bombicola* (Floris, Rizzo, and Giudice 2018). A mixed consortium of hydrocarbon degraders sourced from seawater and sediment samples dominated by *Alcanivorax borkumensis* SK2 and *Paracoccus marcusii* produced sophorolipids and rhamnolipids while utilizing crude oil as a carbon source (Antoniou et al. 2015). Of these strains, *Paracoccus marcusii*, a strain isolated from marine sediments, exhibited properties of remaining trapped in the carbon phase and hence showed the potential to degrade hydrocarbons in marine environments without getting diluted by sea currents (Antoniou et al. 2015).

The production of glycolipid biosurfactants without advanced identification of the biosurfactant has also been done. For instance, glycolipid biosurfactants have also been produced by a *Alcanivorax dieselolei* B-5[T] strain obtained from oil con-taminated seawater while growing on diesel oil as the carbon and energy source (Qiao and Shao 2010). A direct correlation was also reported between hydrocarbon degradation and the production of biosurfactants by rare species *Brevibacterium* and *Vibrio* isolated from sea pen *Pteroeides spinosum* (Graziano et al. 2016).

LIPOPEPTIDES

These are made up of lipids connected to a polypeptide chain. The main explain example of lipopeptides is surfactin containing gramicidins or polymyxins (Roy 2017).

Surfactin

Surfactin is an anionic lipopeptide biosurfactant containing a heptapeptide chain linked to a hydroxyl fatty acid by a lactone bond (Patel et al. 2019). It is commonly produced by *Bacillus* species such as *B. pumilus, B. mojavensis, B. licheniformis, B. amyloliquefaciens* and *B. subtilis* (Patel et al. 2019).

Bacillus species such as *Bacillus velezensis* H3 isolated from sea mud collected from the Huanghai and Bohai Seas, Dalian, China, produced a surfactin biosurfactant composed of *anteiso*C15-surfactin and *n*C14-surfactin using crude oil as the sole carbon source (Liu et al. 2010). The surfactin biosurfactant reduced the surface tension of a phosphate buffer from 71.8 to 24.8 mN/m and had a critical micelle concentration of 2.03×10^{-5} and 3.06×10^{-5} mol/L for *anteiso*C15-surfactin and *n*C14-surfactin, respectively (Liu et al. 2010).

Alvarez et al. (2020) produced a surfactin biosurfactant using a *Bacillus subtilis* AB2.0 strain obtained from sea sediments from Abraão Beach, Rio de Janeiro, Brazil (Alvarez et al. 2020). The biosurfactant was identified as a good demulsifier, having reduced the surface tension of a light crude oil from 36.4 to 3.8 mN/m, hexadecane from 30.1 to 7.5 mN/m, crude oil from 42.3 to 6.9 mN/m and water from 72.0 to 24.7 mN/m (Alvarez et al. 2020).

A novel *Alcanivorax dieselolei* B-5^{T} strain produced crude proline lipids with the utilization of diesel oil as the hydrophobic substrate (Qiao and Shao 2010). Furthermore, Deng et al. (2016) used a hydrocarbon-degrading bacteria *Achromobacter* sp. HZ01 to produce a cyclic lipopeptide with a critical micelle concentration of 48 mg L^{-1} under a wide range of temperatures, pH values and salinities.

More than 72% degradation of aliphatic hydrocarbons was achieved by a halotolerant species *Staphylococcus* sp. with the aid of a lipopeptide biosurfactant classified as lichenysin and iturine with critical micelle concentrations of 65 and 750 mg/L, respectively (Hentati et al. 2021). *Staphylococcus* sp. also degraded recalcitrant PAHs such as phenanthrene, fluoranthene and pyrene in the presence of surface-active secondary metabolites (Hentati et al. 2021). Swaathy et al. (2014) used a marine *Bacillus licheniformis* MTCC with a licA3 gene and catechol 2,3 dioxygenase (C23O) gene to produce a lipopeptide biosurfactant that aided in the degradation of 95% of anthracene in 22 days.

PHOSPHOLIPIDS

These are categorized as fatty acid and phospholipid biosurfactants produced by *Rhodococcus erythropolis, Anthrobacter* sp., *Penicillium spiculisporum, Corynebacterium* sp., *Thiobacillus thio-oxidan* and *Capnocytophaga* sp. (Floris, Rizzo, and Giudice 2018). An *Alcanivorax borkumensis* strain produced a phospholipid ester-linked fatty acid that reduces the surface tension of water from 72 to 29 mNm^{-1} as it efficiently degraded n-alkanes (Yakimov et al. 1998). Phospholipids and lipid-organic acids were also produced by *Alcanivorax dieselolei* B-5^{T} strain alongside other dominant biosurfactants such as lipopeptides with diesel oil as the main carbon and energy source (Qiao and Shao 2010).

CONCLUSION

Marine environments such as oceans and seas are a diverse source of efficient biosurfactant producing and hydrocarbon-degrading bacteria. Novel and unique biosurfactants are produced by *Proteobacteria, Marinomonas vaga, Oceanospirillum linum* and *Halomonas elongate*. The biosurfactants produced aid in the degradation of hydrocarbons by improving solubilization and desorption of hydrophobic compounds. However, hydrocarbonoclastic bacteria from marine environments have a low substrate spectrum compared to heterotrophic bacteria because of their unique physiology. Still, the ability of pelagic bacteria to function under extreme conditions shows the opportunities in exploring marine environments for efficient degraders and biosurfactant-producing bacteria for bioremediation of contaminated environments.

REFERENCES

Abraham, Wolf-Rainer, Holger Meyer, and Misha Yakimov. 1998. "Novel glycine containing glucolipids from the alkane using bacterium *Alcanivorax borkumensis.*" *Biochimica et Biophysica Acta (BBA)—Lipids and Lipid Metabolism* 1393 (1):57–62. https://doi.org/10.1016/S0005-2760(98)00058-7.

Alagić, Slađana Č., Biljana S. Maluckov, and Vesna B. Radojičić. 2015. "How can plants manage polycyclic aromatic hydrocarbons? May these effects represent a useful tool for an effective soil remediation? A review." *Clean Technologies and Environmental Policy* 17 (3):597–614. https://doi.org/10.1007/s10098-014-0840-6.

Al-Samhan, Meshal, Jamal Al-Fadhli, Ahmad M. Al-Otaibi, Fatma Al-Attar, Rashed Bouresli, and Mohan S. Rana. 2022. "Prospects of refinery switching from conventional to integrated: An opportunity for sustainable investment in the petrochemical industry." *Fuel* 310:122161. https://doi.org/10.1016/j.fuel.2021.122161.

Alvarez, Vanessa Marques, Carolina Reis Guimarães, Diogo Jurelevicius, Livia Vieira Araujo de Castilho, Joab Sampaio de Sousa, Fabio Faria da Mota, Denise Maria Guimarães Freire, and Lucy Seldin. 2020. "Microbial enhanced oil recovery potential of surfactin-producing *Bacillus subtilis* AB2.0." *Fuel* 272:117730. https://doi.org/10.1016/j.fuel.2020.117730.

Antoniou, Eleftheria, Stilianos Fodelianakis, Emmanouela Korkakaki, and Nicolas Kalogerakis. 2015. "Biosurfactant production from marine hydrocarbon-degrading consortia and pure bacterial strains using crude oil as carbon source." *Frontiers in Microbiology* 6:274. https://doi.org/10.3389/fmicb.2015.00274.

Aryanasl, Amir, Jamal Ghodousi, Reza Arjmandi, and Nabiollah Mansouri. 2017. "Components of sustainability considerations in management of petrochemical industries." *Environmental Monitoring and Assessment* 189 (6):274. https://doi.org/10.1007/s10661-017-5962-y.

Azizian, Saeid, and Maryam Khosravi. 2019. "Chapter 12—Advanced oil spill decontamination techniques." In *Interface Science and Technology*, edited by George Z. Kyzas and Athanasios C. Mitropoulos, 283–332. Elsevier.

Bansal, Vasudha, and Ki-Hyun Kim. 2015. "Review of PAH contamination in food products and their health hazards." *Environment International* 84:26–38. https://doi.org/10.1016/j.envint.2015.06.016.

Bezza, Fisseha Andualem, and Evans M. Nkhalambayausi Chirwa. 2015. "Biosurfactant from *Paenibacillus dendritiformis* and its application in assisting polycyclic aromatic hydrocarbon (PAH) and motor oil sludge removal from contaminated soil and sand media." *Process Safety and Environmental Protection* 98:354–364. https://doi.org/10.1016/j.bej.2015.05.007.

Cerqueira, Vanessa S., Maria do Carmo R. Peralba, Flávio A. O. Camargo, and Fátima M. Bento. 2014. "Comparison of bioremediation strategies for soil impacted with petrochemical oily sludge." *International Biodeterioration & Biodegradation* 95:338–345. https://doi.org/10.1016/j.ibiod.2014.08.015.

Chen, Guanyi, Jiantao Li, Kai Li, Fawei Lin, Wangyang Tian, Lei Che, Beibei Yan, Wenchao Ma, and Yingjin Song. 2020. "Nitrogen, sulfur, chlorine containing pollutants releasing characteristics during pyrolysis and combustion of oily sludge." *Fuel* 273:117772. https://doi.org/10.1016/j.fuel.2020.117772.

Cheshkova, Tatiana V., Valery P. Sergun, Elena Yu Kovalenko, Natalya N. Gerasimova, Tatiana A. Sagachenko, and Raisa S. Min. 2019. "Resins and asphaltenes of light and heavy oils: Their composition and structure." *Energy & Fuels* 33 (9):7971–7982. https://doi.org/10.1021/acs.energyfuels.9b00285.

Craveiro, Nykon, Rodrigo Vinícius de Almeida Alves, Juliana Menezes da Silva, Edson Vasconcelos, Flavio de Almeida Alves-Junior, and José Souto Rosa Filho. 2021. "Immediate effects of the 2019 oil spill on the macrobenthic fauna associated with macroalgae on the tropical coast of Brazil." *Marine Pollution Bulletin* 165:112107. https://doi.org/10.1016/j.marpolbul.2021.112107.

Cui, Zhisong, Qiliang Lai, Chunming Dong, and Zongze Shao. 2008. "Biodiversity of polycyclic aromatic hydrocarbon-degrading bacteria from deep sea sediments of the Middle Atlantic Ridge." *Environmental Microbiology* 10 (8):2138–2149. https://doi.org/10.1111/j.1462-2920.2008.01637.x.

Deng, M.-C., J. Li, Y.-H. Hong, X.-M. Xu, W.-X. Chen, J.-P. Yuan, J. Peng, M. Yi, and J.-H. Wang. 2016. "Characterization of a novel biosurfactant produced by marine hydrocarbon-degrading bacterium *Achromobacter* sp. HZ01." *Journal of Applied Microbiology* 120 (4):889–899. https://doi.org/10.1111/jam.13065.

Desjardine, Kelsey, Alban Pereira, Helen Wright, Teatulohi Matainaho, Michael Kelly, and Raymond J. Andersen. 2007. "Tauramamide, a lipopeptide antibiotic produced in culture by *Brevibacillus laterosporus* isolated from a marine habitat: Structure elucidation and synthesis." *Journal of natural products* 70 (12):1850–1853. https://doi.org/10.1021/np070209r

Fakher, Sherif, Mohamed Ahdaya, Mukhtar Elturki, and Abdulmohsin Imqam. 2020. "Critical review of asphaltene properties and factors impacting its stability in crude oil." *Journal of Petroleum Exploration and Production Technology* 10 (3):1183–1200. https://doi.org/10.1007/s13202-019-00811-5.

Floris, Rosanna, Carmen Rizzo, and Angelina Lo Giudice. 2018. "Biosurfactants from marine microorganisms." In *Metabolomics-New Insights into Biology and Medicine*. IntechOpen.

Gidudu, Brian, and Evans M. Nkhalambayausi Chirwa. 2020. "The combined application of a high voltage, low electrode spacing, and biosurfactants enhances the bioelectrokinetic remediation of petroleum contaminated soil." *Journal of Cleaner Production* 276. https://doi.org/10.1016/j.jclepro.2020.122745.

Golysheva, E. A., O. V. Zhdaneev, V. V. Korenev, A. S. Lyadov, and A. S. Rubtsov. 2020. "Petrochemical industry in Russia: State of the art and prospects for development."

Russian Journal of Applied Chemistry 93 (10):1596–1603. https://doi.org/10.1134/s107042722010158.

Golyshin, Peter N., Vitor A. P. Martins Dos Santos, Olaf Kaiser, Manuel Ferrer, Yulia S. Sabirova, H. Lünsdorf, Tatyana N. Chernikova, Olga V. Golyshina, Michail M. Yakimov, Alfred Pühler, and Kenneth N. Timmis. 2003. "Genome sequence completed of *Alcanivorax borkumensis*, a hydrocarbon-degrading bacterium that plays a global role in oil removal from marine systems." *Journal of Biotechnology* 106 (2):215–220. https://doi.org/10.1016/j.jbiotec.2003.07.013.

Gopang, I.A., H. Mahar, A.S. Jatoi, K.S. Akhtar, M. Omer, and M.S. Azeem. 2016. "Characterization of the sludge deposits in crude oil." *Journal of Faculty of Engineering & Technology* 23 (1):57–64.

Graziano, Marco, Carmen Rizzo, Luigi Michaud, Erika Maria Diletta Porporato, Emilio De Domenico, Nunziacarla Spanò, and Angelina Lo Giudice. 2016. "Biosurfactant production by hydrocarbon-degrading *Brevibacterium* and *Vibrio* isolates from the sea pen *Pteroeides spinosum* (Ellis, 1764)." *Journal of Basic Microbiology* 56 (9):963–974. https://doi.org/10.1002/jobm.201500701.

Gudina, Eduardo J., Ana I. Rodrigues, Eliana Alves, M. Ròsário Domingues, José A. Teixeira, and Lígia R. Rodrigues. 2015. "Bioconversion of agro-industrial by-products in rhamnolipids toward applications in enhanced oil recovery and bioremediation." *Bioresource Technology* 177:87–93. https://doi.org/10.1016/j.biortech.2014.11.069.

Guigue, Catherine, Lionel Bigot, Jean Turquet, Marc Tedetti, Nicolas Ferretto, Madeleine Goutx, and Pascale Cuet. 2015. "Hydrocarbons in a coral reef ecosystem subjected to anthropogenic pressures (La Réunion Island, Indian Ocean)." *Environmental Chemistry* 12 (3):350–365. https://doi.org/10.1071/EN14194.

Gupta, Shalini, and Bhawana Pathak. 2020. "Chapter 6—Mycoremediation of polycyclic aromatic hydrocarbons." In *Abatement of Environmental Pollutants*, edited by Pardeep Singh, Ajay Kumar and Anwesha Borthakur, 127–149. Elsevier.

Hamilton, Jerry, Yousef Sadat, Matthew Dwyer, Pierre Ghali, and Bhupendra Khandelwal. 2021. "Chapter 8 — Thermal stability and impact of alternative fuels." In *Aviation Fuels*, edited by Bhupendra Khandelwal, 149–218. Academic Press.

Hassanshahian, Mehdi. 2014. "Isolation and characterization of biosurfactant producing bacteria from Persian Gulf (Bushehr provenance)." *Marine Pollution Bulletin* 86 (1):361–366. https://doi.org/10.1016/j.marpolbul.2014.06.043.

Hassanshahian, Mehdi, Giti Emtiazi, and Simone Cappello. 2012. "Isolation and characterization of crude-oil-degrading bacteria from the Persian Gulf and the Caspian Sea." *Marine Pollution Bulletin* 64 (1):7–12. https://doi.org/10.1016/j.marpolbul.2011.11.006.

Hentati, Dorra, Meriam Cheffi, Fatma Hadrich, Neila Makhloufi, Francesc Rabanal, Angeles Manresa, Sami Sayadi, and Mohamed Chamkha. 2021. "Investigation of halotolerant marine *Staphylococcus* sp. CO100, as a promising hydrocarbon-degrading and bio-surfactant-producing bacterium, under saline conditions." *Journal of Environmental Management* 277:111480. https://doi.org/10.1016/j.jenvman.2020.111480.

Hu, G., J. Li, and G. Zeng. 2013. "Recent development in the treatment of oily sludge from petroleum industry: A review." *Journal of Hazardous Materials* 261:470–90. https://doi.org/10.1016/j.jhazmat.2013.07.069.

Ibacache-Quiroga, C., J. Ojeda, G. Espinoza-Vergara, P. Olivero, M. Cuellar, and M. A. Dinamarca. 2013. "The hydrocarbon-degrading marine bacterium *Cobetia* sp. strain MM1IDA2H-1 produces a biosurfactant that interferes with quorum sensing of fish pathogens by signal hijacking." *Microbial biotechnology* 6 (4):394–405. https://doi.org/10.1111/1751-7915.12016.

Islam, Badrul. 2015. "Petroleum sludge, its treatment and disposal: A review." *International Journal of Chemical Sciences* 13 (4):1584–1602.

Johnson, Olufemi Adebayo, and Augustine Chioma Affam. 2018. "Petroleum sludge treatment and disposal: A review." *Environmental Engineering Research* 24 (2):191–201. https://doi.org/10.4491/eer.2018.134.

Jørgensen, S. E., 2008. "Biodegradation." In *Encyclopedia of Ecology*, edited by S. E. Jørgensen and B. D. Fath, 366–367. Academic Press.

Khuzwayo, Zakhele Zack, and Evans Chirwa. 2020. "Chapter 8 — Photocatalysis as a clean technology for the degradation of petrochemical pollutants." In *Emerging Eco-friendly Green Technologies for Wastewater Treatment*, edited by R. N. Bharagava, 171–191. Springer.

Kingston, Paul F. 2002. "Long-term environmental impact." *Spill Science & Technology Bulletin* 7 (1–2):53–61.

Kisic, Ivica, Sanja Mesic, Ferdo Basic, Vladislav Brkic, Milan Mesic, Goran Durn, Zeljka Zgorelec, and Lidija Bertovic. 2009. "The effect of drilling fluids and crude oil on some chemical characteristics of soil and crops." *Geoderma* 149 (3–4):209–216. https://doi.org/10.1016/j.geoderma.2008.11.041.

Koh, C. H., J. S. Khim, K. Kannan, D. L. Villeneuve, K. Senthilkumar, and J. P. Giesy. 2004. "Polychlorinated dibenzo-p-dioxins (PCDDs), dibenzofurans (PCDFs), biphenyls (PCBs), and polycyclic aromatic hydrocarbons (PAHs) and 2,3,7,8-TCDD equivalents (TEQs) in sediment from the Hyeongsan River, Korea." *Environmental Pollution* 132 (3):489–501. https://doi.org/10.1016/j.envpol.2004.05.001.

Konishi, Masaaki, Shinro Nishi, Tokuma Fukuoka, Dai Kitamoto, Tomo-o Watsuji, Yuriko Nagano, Akinori Yabuki, Satoshi Nakagawa, Yuji Hatada, and Jun-ichi Horiuchi. 2014. "Deep-sea *Rhodococcus* sp. BS-15, lacking the phytopathogenic fas genes, produces a novel glucotriose lipid biosurfactant." *Marine Biotechnology* 16 (4):484–493. http://doi.org/10.1007/s10126-014-9568-x.

Kubicki, Sonja, Alexander Bollinger, Nadine Katzke, Karl-Erich Jaeger, Anita Loeschcke, and Stephan Thies. 2019. "Marine biosurfactants: Biosynthesis, structural diversity and biotechnological applications." *Marine Drugs* 17 (7):408. https://doi.org/10.3390/md17070408.

Langangen, O., E. Olsen, L. C. Stige, J. Ohlberger, N. A. Yaragina, F. B. Vikebo, B. Bogstad, N. C. Stenseth, and D. O. Hjermann. 2017. "The effects of oil spills on marine fish: Implications of spatial variation in natural mortality." *Marine Pollution Bulletin* 119 (1):102–109. http://doi.org/10.1016/j.marpolbul.2017.03.037.

Lawal, Abdulazeez T. 2017. "Polycyclic aromatic hydrocarbons. A review." *Cogent Environmental Science* 3 (1):1339841. http://doi.org/10.1080/23311843.2017.1339841.

Lee, Dong Wan, Hanbyul Lee, Bong-Oh Kwon, Jong Seong Khim, Un Hyuk Yim, Beom Seok Kim, and Jae-Jin Kim. 2018. "Biosurfactant-assisted bioremediation of crude oil by indigenous bacteria isolated from Taean beach sediment." *Environmental Pollution* 241:254–264. https://doi.org/10.1016/j.envpol.2018.05.070.

Liu, Xiangyang, Biao Ren, Ming Chen, Haibin Wang, Chandrakant R. Kokare, Xianlong Zhou, Jidong Wang, Huanqin Dai, Fuhang Song, Mei Liu, Jian Wang, Shujin Wang, and Lixin Zhang. 2010. "Production and characterization of a group of bioemulsifiers from the marine *Bacillus velezensis* strain H3." *Applied Microbiology and Biotechnology* 87 (5):1881–1893. http://doi.org/10.1007/s00253-010-2653-9.

Magalhães, Karine Matos, Renato Silva Carreira, José Souto Rosa Filho, Pedro Palmeira Rocha, Francisco Marcante Santana, and Gilvan Takeshi Yogui. 2022. "Polycyclic aromatic hydrocarbons (PAHs) in fishery resources affected by the 2019 oil spill in Brazil: Short-term environmental health and seafood safety." *Marine Pollution Bulletin* 175:113334. https://doi.org/10.1016/j.marpolbul.2022.113334.

Mahjoubi, Mouna, Simone Cappello, Yasmine Souissi, A. tef Jaouani, and Ameur Cherif. 2017. Microbial bioremediation of petroleum hydrocarbon–contaminated marine environments. In *Recent Insights in Petroleum Science and Engineering*, edited by Mansoor Zoveidavianpoor. IntechOpen.

Malavenda, Roberta, Carmen Rizzo, Luigi Michaud, Berna Gerçe, Vivia Bruni, Christoph Syldatk, Rudolf Hausmann, and Angelina Lo Giudice. 2015. "Biosurfactant production by Arctic and Antarctic bacteria growing on hydrocarbons." *Polar Biology* 38 (10):1565–1574. https://doi.org/10.1007/s00300-015-1717-9.

McKew, Boyd A., Frédéric Coulon, A. Mark Osborn, Kenneth N. Timmis, and Terry J. McGenity. 2007. "Determining the identity and roles of oil-metabolizing marine bacteria from the Thames estuary, UK." *Environmental Microbiology* 9 (1):165–176. https://doi.org/10.1111/j.1462-2920.2006.01125.x.

Melcher, Rebecca J., Sabine E. Apitz, and Barbara B. Hemmingsen. 2002. "Impact of irradiation and polycyclic aromatic hydrocarbon spiking on microbial populations in marine sediment for future aging and biodegradability studies." *Applied and Environmental Microbiology* 68 (6):2858–2868. https://doi:10.1128/AEM.68.6.2858-2868.2002.

Mondal, Madhumanti, Gopinath Halder, Gunapati Oinam, Thingujam Indrama, and Onkar Nath Tiwari. 2019. "Chapter 17 — Bioremediation of organic and inorganic pollutants using microalgae." In *New and Future Developments in Microbial Biotechnology and Bioengineering*, 223–235. Elsevier.

Mulligan, C.N. 2009. "Recent advances in the environmental applications of biosurfactants." *Current Opinion in Colloid & Interface Science* 14:372–378. https://doi.org/10.1016/j.cocis.2009.06.005.

Mulligan, Catherine N. 2021. "Sustainable remediation of contaminated soil using biosurfactants." *Frontiers in Bioengineering and Biotechnology* 9:635196. http://doi.org/10.3389/fbioe.2021.635196.

Nelson-Smith, Anthony. 1968. "The effects of oil pollution and emulsifier cleansing on shore life in south-west Britain." *Journal of applied Ecology*:97–107. https://doi.org/10.2307/2401276.

Obodovskiy, Ilya. 2019. "Chapter 33 — Basics of biochemistry." In *Radiation*, edited by Ilya Obodovskiy, 399–427. Elsevier.

Oliveira, Jorge S., Wydemberg J. Araújo, Ricardo M. Figueiredo, Rita C. B. Silva-Portela, Alaine de Brito Guerra, Sinara Carla da Silva Araújo, Carolina Minnicelli, Aline Cardoso Carlos, Ana Tereza Ribeiro de Vasconcelos, Ana Teresa Freitas, and Lucymara F. Agnez-Lima. 2017. "Biogeographical distribution analysis of hydrocarbon degrading and biosurfactant producing genes suggests that near-equatorial biomes have higher abundance of genes with potential for bioremediation." *BMC Microbiology* 17 (1):168. http://doi.org/10.1186/s12866-017-1077-4.

Ossai, Innocent Chukwunonso, Aziz Ahmed, Auwalu Hassan, and Fauziah Shahul Hamid. 2020. "Remediation of soil and water contaminated with petroleum hydrocarbon: A review." *Environmental Technology & Innovation* 17:100526. http://doi.org/10.1016/j.eti.2019.100526.

Patel, Seema, Ahmad Homaei, Sangram Patil, and Achlesh Daverey. 2019. "Microbial biosurfactants for oil spill remediation: Pitfalls and potentials." *Applied Microbiology and Biotechnology* 103 (1):27–37. http://doi.org/10.1007/s00253-018-9434-2.

Pezeshki, S. R., M. W. Hester, Q. Lin, and J. A. Nyman. 2000. "The effects of oil spill and clean-up on dominant US Gulf coast marsh macrophytes: A review." *Environmental Pollution* 108 (2):129–139. https://doi.org/10.1016/S0269-7491(99)00244-4.

Premnath, N., K. Mohanrasu, R. Guru Raj Rao, G. H. Dinesh, G. Siva Prakash, V. Ananthi, Kumar Ponnuchamy, Govarthanan Muthusamy, and A. Arun. 2021. "A crucial review on polycyclic aromatic hydrocarbons—Environmental occurrence and strategies for microbial degradation." *Chemosphere* 280:130608. https://doi. org/10.1016/j.chemosphere.2021.130608.

Qiao, N., and Z. Shao. 2010. "Isolation and characterization of a novel biosurfactant produced by hydrocarbon-degrading bacterium *Alcanivorax dieselolei* B-5." *Journal of Applied Microbiology* 108 (4):1207–1216. https://doi.org/10.1111/j.1365-2672.2009. 04513.x.

Ramirez-Corredores, Maria Magdalena. 2017. "Chapter 2 — Asphaltenes." In *The Science and Technology of Unconventional Oils*, edited by Maria Magdalena Ramirez-Corredores, 41–222. Amsterdam: Academic Press.

Roy, Arpita. 2017. "Review on the biosurfactants: Properties, types and its applications." *Journal of Fundamentals of Renewable Energy and Applications* 8:1–14. https:// doi.org/10.4172/2090-4541.1000248.

Samanta, Sudip K., Om V. Singh, and Rakesh K. Jain. 2002. "Polycyclic aromatic hydrocarbons: Environmental pollution and bioremediation." *Trends in Biotechnology* 20 (6):243–248. https://doi.org/10.1016/S0167-7799(02)01943-1.

Sarma, Hemen, Karla Liliana Tarango Bustamante, and Majeti Narasimha Vara Prasad. 2019. "6 — Biosurfactants for oil recovery from refinery sludge: Magnetic nanoparticles assisted purification." In *Industrial and Municipal Sludge*, edited by Majeti Narasimha Vara Prasad, Paulo Jorge de Campos Favas, Meththika Vithanage and S. Venkata Mohan, 107–132. Butterworth-Heinemann.

Schneiker, Susanne, Vítor A. P. Martins dos Santos, Daniela Bartels, Thomas Bekel, Martina Brecht, Jens Buhrmester, Tatyana N. Chernikova, Renata Denaro, Manuel Ferrer, Christoph Gertler, Alexander Goesmann, Olga V. Golyshina, Filip Kaminski, Amit N. Khachane, Siegmund Lang, Burkhard Linke, Alice C. McHardy, Folker Meyer, Taras Nechitaylo, Alfred Pühler, Daniela Regenhardt, Oliver Rupp, Julia S. Sabirova, Werner Selbitschka, Michail M. Yakimov, Kenneth N. Timmis, Frank-Jörg Vorhölter, Stefan Weidner, Olaf Kaiser, and Peter N. Golyshin. 2006. "Genome sequence of the ubiquitous hydrocarbon-degrading marine bacterium *Alcanivorax borkumensis*." *Nature Biotechnology* 24 (8):997–1004. http://doi. org/10.1038/nbt1232.

Soares, Marcelo Oliveira, Carlos Eduardo Peres Teixeira, Luis Ernesto Arruda Bezerra, Emanuelle Fontenele Rabelo, Italo Braga Castro, and Rivelino Martins Cavalcante. 2022. "The most extensive oil spill registered in tropical oceans (Brazil): The balance sheet of a disaster." *Environmental Science and Pollution Research* 29 (13):19869–19877. http://doi.org/10.1007/s11356-022-18710-4.

Swaathy, Sreethar, Varadharajan Kavitha, Arokiasamy Sahaya Pravin, Asit Baran Mandal, and Arumugam Gnanamani. 2014. "Microbial surfactant mediated degradation of anthracene in aqueous phase by marine *Bacillus licheniformis* MTCC 5514." *Biotechnology Reports* 4:161–170. https://doi.org/10.1016/j.btre.2014.10.004.

Tian, Xiangmiao, Yudong Song, Zhiqiang Shen, Yuexi Zhou, Kaijun Wang, Xiaoguang Jin, Zhenfeng Han, and Tao Liu. 2020. "A comprehensive review on toxic petrochemical wastewater pretreatment and advanced treatment." *Journal of Cleaner Production* 245:118692. https://doi.org/10.1016/j.jclepro.2019.118692.

Tripathi, Lakshmi, Matthew S. Twigg, Aikaterini Zompra, Karina Salek, Victor U. Irorere, Tony Gutierrez, Georgios A. Spyroulias, Roger Marchant, and Ibrahim M. Banat. 2019. "Biosynthesis of rhamnolipid by a *Marinobacter* species expands the paradigm of biosurfactant synthesis to a new genus of the marine microflora." *Microbial Cell Factories* 18 (1):164. http://doi.org/10.1186/s12934-019-1216-8.

Twigg, Matthew S., Lakshmi Tripathi, Aikaterini Zompra, Karina Salek, V. U. Irorere, Tony Gutierrez, G. A. Spyroulias, Roger Marchant, and Ibrahim M. Banat. 2018. "Identification and characterisation of short chain rhamnolipid production in a previously uninvestigated, non-pathogenic marine pseudomonad." *Applied Microbiology and Biotechnology* 102 (19):8537–8549. https://doi.org/10.1007/s00253-018-9202-3.

Tyagi, Meenu, M. Manuela R. da Fonseca, and Carla CCR de Carvalho. 2011. "Bioaugmentation and biostimulation strategies to improve the effectiveness of bioremediation processes." *Biodegradation* 22 (2):231–241. https://doi.org/10.1007/s10532-010-9394-4.

Upton, Charles R. 2017. "Biosurfactants." Hauppauge Nova Science Publishers, Inc. https://search.ebscohost.com/login.aspx?direct=true&scope=site&db=nlebk&db=nlabk&AN=1512010.

Uzoigwe, C., J. G. Burgess, C. J. Ennis, and P. K. Rahman. 2015. "Bioemulsifiers are not biosurfactants and require different screening approaches." *Frontiers in Microbiology* 6:245. http://doi.org/10.3389/fmicb.2015.00245.

Varjani, Sunita J., and Vivek N. Upasani. 2017. "Critical review on biosurfactant analysis, purification and characterization using rhamnolipid as a model biosurfactant." *Bioresource Technology* 232:389–397. https://doi.org/10.1016/j.biortech.2017.02.047.

Wang, Meng, Xue Li, Mei Lei, Lunbo Duan, and Huichao Chen. 2022. "Human health risk identification of petrochemical sites based on extreme gradient boosting." *Ecotoxicology and Environmental Safety* 233:113332. https://doi.org/10.1016/j.ecoenv.2022.113332.

Wang, Zhentong, Zhiqiang Gong, Wei Wang, and Zhe Zhang. 2020. "Study on combustion characteristics and the migration of heavy metals during the co-combustion of oil sludge char and microalgae residue." *Renewable Energy* 151:648–658. https://doi.org/10.1016/j.renene.2019.11.056.

Wu, Tao, Jie Xu, Wenjun Xie, Zhigang Yao, Hongjun Yang, Chunlong Sun, and Xiaobin Li. 2018. "*Pseudomonas aeruginosa* L10: A hydrocarbon-degrading, biosurfactant-producing, and plant-growth-promoting endophytic bacterium isolated from a reed (Phragmites australis)." *Frontiers in Microbiology* 9. http://doi.org/10.3389/fmicb.2018.01087.

Wu, Yanan, Meng Xu, Jianliang Xue, Ke Shi, and Meng Gu. 2019. "Characterization and enhanced degradation potentials of biosurfactant-producing bacteria isolated from a marine environment." *ACS Omega* 4 (1):1645–1651. https://doi.org/10.1021/acsomega.8b02653.

Xu, Meng, Xinge Fu, Yu Gao, Liangfeng Duan, Congchao Xu, Wenshuang Sun, Yixuan Li, Xianzheng Meng, and Xinfeng Xiao. 2020. "Characterization of a biosurfactant-producing bacteria isolated from marine environment: Surface activity, chemical characterization and biodegradation." *Journal of Environmental Chemical Engineering* 8 (5):104277. https://doi.org/10.1016/j.jece.2020.104277.

Yakimov, M. M., L. Giuliano, V. Bruni, S. Scarfì, and P. N. Golyshin. 1999. "Characterization of Antarctic hydrocarbon-degrading bacteria capable of producing bioemulsifiers." *The New Microbiologica* 22 (3):249–256.

Yakimov, Michail M., Peter N. Golyshin, Francesca Crisafi, Renata Denaro, and Laura Giuliano. 2019. "Marine, aerobic hydrocarbon-degrading gammaproteobacteria: The family Alcanivoracaceae." In *Taxonomy, Genomics and Ecophysiology of Hydrocarbon-Degrading Microbes*, 167–179. Springer. https://doi.org/10.1007/978-3-030-14796-9_24.

Yakimov, Michail M., Peter N. Golyshin, Siegmund Lang, Edward R. B. Moore, Wolf-Rainer Abraham, Heinrich Lünsdorf, and Kenneth N. Timmis. 1998. "*Alcanivorax borkumensis* gen. nov., sp. nov., a new, hydrocarbon-degrading and surfactant-producing marine bacterium." *International Journal of Systematic and Evolutionary Microbiology* 48 (2):339–348. https://doi.org/10.1099/00207713-48-2-339.

Zhang, Shichao, Hong Yao, Yintao Lu, Xiaohua Yu, Jing Wang, Shaobin Sun, Mingli Liu, Desheng Li, Yi-Fan Li, and Dayi Zhang. 2017. "Uptake and translocation of polycyclic aromatic hydrocarbons (PAHs) and heavy metals by maize from soil irrigated with wastewater." *Scientific Reports* 7 (1):12165. http://doi.org/10.1038/s41598-017-12437-w.

7 Production and
Characterization
of Biosurfactants
to Improve the
Effectiveness and
Bioavailability of
Insoluble Antibiotics

C. Elizabeth Rani and Mithrambigai

CONTENTS

INTRODUCTION

A biosurfactant is defined as a surface-active molecule produced by living cells, mainly by microorganisms [1]. These amphiphilic compounds contain a

DOI: 10.1201/9781003307464-7

hydrophobic and a hydrophilic moiety and have the capability to reduce interfacial tension between different fluid phases. These superior biomolecules are found to occur in a variety of chemical structures, such as glycolipids, lipopeptides and lipoproteins, fatty acids, neutral lipids, phospholipids, and polymeric and particulate structures [1]. By reducing the interfacial tension, biosurfactants increase solubility and the contact area of insoluble compounds and encourage the flexibility, bioavailability, and biodegradation efficiency of insoluble compounds.

Compared with chemical surfactants, the superior biomolecules, biosurfactants, have advantages such as minimal toxicity, high biodegradability, good environmental compatibility, high foaming ability, and high environmental tolerance [2]. Biosurfactants such as surfactin, fengycin, iturin, and bacillomycins possess wide-spectrum bioactive compounds with applications in both pharmaceutical and biotechnological industries. When most drugs are administered, they are first absorbed systematically and then transported to their site(s) of action [3].

The two dynamic methods involved are dissolution followed by absorption. Biosurfactants have intrinsic capabilities to improve drug dissolution and solubility of drugs by lowering the melting point owing to their capability to lower interfacial tension and micellar formation [4]. The rate of solubility is the measure of its saturation to remain undissolved while adding solute with a solvent.

The solubility of constituents basically depends on the solvent used as well as temperature and pressure. The oral route of drug intake is the most highly adapted method and the ideal method of delivery because of its suitability and comfort of absorption, but for many drugs, it can be difficult due to poor solubility, which will lead to low bioavailability.

The solubility of drug constituents is tied to the dissolution rate per the Noyes-Whitney equation, and solubility is a significant aspect for evaluation of the bioavailability. The therapeutic efficiency of a drug hinges on the dissolution, bioavailability, and ultimately solubility of drug molecules [5].

METHODS

Soil samples were collected. The samples were serially diluted and the decimal dilutions spread on nutrient agar medium containing 5% glycerol.

Bacterial colonies surrounded by an emulsified area were identified as biosurfactant-producing bacteria after the incubation at 37°C incubation for 24 hrs. The biosurfactant-synthesizing bacterial isolates were picked up and inoculated in nutrient broth containing 5% of glycerol. After 24 h, the bacterial cells were harvested by centrifugation at 10,000 rpm for 30 min [6,7,8,9,10].

EXTRACTION OF BIOSURFACTANTS

Liquid–liquid extraction procedure using chloroform: Methanol solvent was preferred in the ratio of 2:1. The supernatants were acidified for pH 2 by adding 6M HCl. After acidification, the cell-free extract was left overnight at 4°C for the acidification-induced precipitation of biosurfactants. An equal quantity of solvent was added to the acidified broth.

The combination was mixed vigorously for 30 min and placed in a resting position for the parting of polaric and non-polaric compounds overnight. After the separation of phases, the biosurfactant containing the middle phase was collected and kept for evaporation [11,12,13,14,15,16].

SYNTHESIS OF BIOSURFACTANTS OF MICROBIAL ORIGIN

As natural surface-active agents, biosurfactants can be attained from microorganism action (such as from *Pseudomonas* and *Bacillus*), through enzyme-substrate–based synthesis and aerobic and facultative anaerobic fermentation processes are usually carried out extracellularly by means of biocatalyst enzymes.

Both the hydrophobic and hydrophilic moieties of biosurfactants can be created in two self-determining paths: both parts can be substrate dependent, or one can be synthesized *de novo* while the other is produced by the substrate [17,18,19,20].

GLYCOLIPID BIOSURFACTANTS

Biosurfactants may be lipopeptides, glycolipids, phospholipids, and so on, but glycolipids are the most general type of biosurfactants. Some of the common glycolipid biosurfactants, such as rhamnolipids, trehalolipids, sophorolipids, and mannosylerythritol lipids (MELs), encompass mono- and disaccharides shared with long-chain aliphatic acids or hydroxy-aliphatic acids [21,22].

Rhamnolipids are one of the most significant glycolipids and are recognized for their exceptional physicochemical assets [23,24]. Rhamnolipids are mostly produced by *Pseudomonas* and *Burkholderia* species. The initial stage in the production of rhamnolipids comprises synthesis of the sugar part encompassing rhamnose from D-glucose and the hydrophobic acid part from fatty acids [25]. The enzymes essential for this first phase are generally present in most bacteria, but the specific enzymes desirable for the biosynthesis of rhamnolipids are present almost exclusively in *P. aeruginosa* and *Burkholderia* species. Five diverse enzymes, RhlA, RhlB, RhlC, RhlG, and RhlI, have been described to be related to the production of rhamnolipids in *P. aeruginosa* [26]. Microbial fermentation can result in diverse kinds of rhamnolipids.

Mono-rhamnolipids and di-rhamnolipids vary in the number of rhamnose groups present in the molecular structure. Rhamnolipids also vary with respect to chain length, degree of branching, and degree of unsaturation in the fatty acid chains, all dependent on various factors related to environmental and internal growth conditions [27].

Around 60 diverse rhamnolipid congeners and homologues have been described [28] by means of a mixture of *Pseudomonas* and other bacterial species. Numerous *Burkholderia* species have been shown to produce longer alkyl chain rhamnolipids compared to those produced by *P. aeruginosa* [29]. Different substrates, such as alkenes, citrates, glucose, fructose, glycerol, and olive oil, have also been used as a substrates for producing biosurfactants with diverse characteristics [30]. Numerous studies have been carried out to reveal the best way to produce rhamnolipids both in terms of safety and affordability.

Genetically engineered *P. aeruginosa* to reduce its pathogenicity is one such method, while other methods include the expression of the key genes accountable for rhamnolipid production in non-pathogenic strains [30].

The cost of producing rhamnolipids can be reduced by selecting suitable substrates, such as vegetable oils, and optimizing the fermentation process [31]. Sophorolipids can be produced by several non-pathogenic yeast species, the genus *Candida* being the most common, and *C. bombicola* and *C. apicola* (encompassing enzymes essential for the termina oxidation of alkanes to produce fatty acids) are the greatest used species [32].

Sophorolipids can originate in two dissimilar formulae, the lactonic and acidic forms. Similar to rhamnolipids, the price of manufacturing sophorolipids (compared to chemically manufactured surfactants) is actually high, which limits their industrial creation [33]. The usage of oil byproducts or food waste has been suggested as an alternative to decrease the manufacturing costs [34].

LIPOPEPTIDE BIOSURFACTANTS

The biosynthesis of surfactin ensues via non-ribosomal machinery, which is catalyzed using surfactin synthetase, which is a protein complex. The compound encompasses four enzymatic subunits, among which the subunit SrfD is vital to recruit the synthesis [35]. Supplementary lipopeptide biosurfactants, iturin, lichenysin, and arthrofactin, are also produced by comparable synthase complexes [36].

Bacillus subtilis is the key bacteria used for surfactin manufacture. Genetic modification using recombinant DNA technology of the wild-type strain to improve low harvests has been described. Wu et al. [37] adapted 53 dissimilar genes in *B. subtilis* to spread a harvest of around 42% of the theoretical yield. Apart from the usual submerged fermentation, surfactin can also be synthesized by adapting the method of solid state fermentation (SSF). This is a kind of fermentation in which microorganisms will be cultivated on or within solid substrates or supports in the absence of free water.

Lipopeptide biosurfactants have also been described as being synthesized by *Pseudomonas aeruginosa* by means of renewable resources, such as greasing oil and peanut oil [38].

HIGH MOLECULAR WEIGHT BIOSURFACTANTS/BIOEMULSIFIERS

Bioemulsifiers (BEs) are complexes with high molecular weight synthesized by bacteria, yeast, and fungi. They are usually produced as compound mixtures of heteropolysaccharides, lipopolysaccharides, lipoproteins, and also proteins, which can be anchored to the cell surface or unconfined [39]. Bioemulsifiers display an extensive diversity of physicochemical features exhibited by variable microorganisms that produce them. *Acinetobacter* spp. is one of the oldest known bacteria producing BEs. Emulsan and alasan are the best examples of commercially available bioemulsifiers produced by *Acinetobacter* spp. [40]. Emulsan is a lipoheteropolysaccharide polymer containing D-galactose-amine and is the secondary metabolite synthesized throughout the idio phase. Maximum yield will be produced when culture media comprising 12 carbon-containing fatty acids are used as the source of carbon. Emulsan production is possible with fermentation methods such as batch, chemo-stat, immobilized cell system, and self-cycling fermentation. Additional bioemulsifiers such as mannose sugar containing protein have been described, and it was reported to be synthesized inside the cell wall of *Saccharomyces* spp. and *Kluyveromyces marxianus* and liberated from the cell wall of yeast by means of pressurized heat [41].

PHYSICOCHEMICAL CHARACTERIZATION

Various microorganisms can synthesize a extensive variety of comparable bio-surfactants, but with diverse biological action. Also regarding their sources, the physicochemical characteristics of these biosurfactants are predisposed by synthesis and purification processes.

Considering these characteristics is significant for their industrial application [42]. This section explicates some important characteristics of biosurfactants that affect their use as emulsifiers in pharmaceutical industries.

SURFACE AND INTERFACIAL TENSION

The greatest significant property of a bio-emulsifier is its capability to reduce of surface and interfacial tension between two molecules. This is an essential purpose of the significant functional property of an amphiphilic molecule for the development of kinetically alleviated emulsions. These molecules adhere on interfaces such as air/liquid, liquid/liquid and solid/liquid, replacing water or oil particles along the boundary and lowering intermolecular forces among solvent molecules and superficial or interfacial tension. Compared to synthetically pre-pared surfactants, microbial sources of biosurfactants are capable of decreasing the interfacial tension extra-efficiently. For instance, surfactin is one of the most active surface tension-reducing biosurfactants.

Surfactin shows sensitive surface action from 72 to 27 ± 2 mN/m [43], and interfacial tension values were estimated as 3.79 ± 0.27 mN/m and 0.32 ± 0.02 mN/m below favorable conditions of both intrinsic and extrinsic physical and chemical conditions [44]. For instance, lipopeptide biosurfactin, also called arthrofactin, was synthesized by *Arthrobacter* sp. strain MIS38 [45], and another important example, the biosurfactant synthesized by *Candida lipolytica* UCP 0988 [46], presented low surface action. These kinds of exclusive surface characteristics are related to the more complex chemical structure of biosurfactants. Dissimilar artificially prepared surfactants do not posses a strong delivery of polarity and also cover divided or circular assemblies [47].

Lipopeptide surfactin is capable of forming spherical structures to facilitate close packing at interfaces and the formation of structures with a low aggregation number [48]. Unusual surface properties of saponins have been observed. Their behavior was explained by dense molecular packing at the interface of phases and a strong hydrogen bond between saccharide groups in the interfacial layer [49]. The very compact surface layers formed are denser than those observed in most common amphiphiles.

The aforementioned characteristics of biosurfactants govern their requisite mechanisms for biomolecules and cell membranes. A comprehensive evaluation of properties, surface activity and critical micellization concentration (CMC) values, is presented in [50]. In some cases, low CMC values and surface activity can be influenced with the occurrence of impure compounds with surfactant compositions.

SELF-ASSEMBLY

Micellization is a process which results in thermodynamically steady nanostructures. Surfactants impulsively form micelles in the aqueous phase at concentrations just above the CMC. Then the surfactant monomers are assembled in micelles [51]. It is a natural process of biosurfactants, which have a propensity to impulsively self-assemble through hydrophobia and weak Van der Waals exchanges.

The efficiency of the biosurfactant is defined by its capability to decrease surface tension. Water has a surface tension of 72 mN/m, and a perfect biosurfactant can decrease this value to 30 mN/m [52]. There are several factors will influence the size and shape of the micelle and are reported to be based on changes in the concentration of biosurfactant, temperature, pH, pressure, and salts, among others.

The main important repulsive forces of the head groups limit the amount of related biosurfactants in the development of micelles. Elevating the surfactant concentration just above the CMC value impacts the development of a larger number of micelles. The development of micelles with a low accumulation number is typical of both rhamnolipids and surfactin [53,54,55,56,57].

SOLUBILIZATION

Amphiphilic compounds that are self-assembled in aqueous phases can promote the solubilization of hydrophobic composites (oil, for example), which specially exist in the specific hydrophobic domains of the amphiphilic structural molecule. The ability of biosurfactants to improve the solubility of compounds with hydrophobic groups depends on the amount of biosurfactant used and also other external factors such as pH and the mixture of adjuvants and electrolytes. They can especially alter the magnitude of micelles [58]. The rhamnolipid kind of biosurfactant can influence and elevate the ability of hydrophobic compounds to dissolve by elevating the hydrophobicity of amphiphilic molecules.

Frequently biosurfactants have a propensity to form vesicles and micelles with optimum pH, which confines the dissolution of other molecules. The characteristics of biosurfactants are exact to the substrate, and solubilization or emulsification diversifies hydrocarbons at different rates [59]. Biosurfactants are more active solubilizing agents than synthetic surfactants [60].

A biosurfactant attained from *Rhodococcus erythropolis* HX-2 showed advanced solubilization for petroleum and compounds like polycyclic aromatic hydrocarbons when compared to synthetic surfactants like sodium dodecyl sulfate (SDS), polysorbate (Tween 80), Triton X-100, and also rhamnolipid [61]. It is thought-provoking to note that rhamnolipids can solubilize n-alkanes not only at a rate above CMC [62] but also below CMC [63]. Their solubilization efficacy of n-alkanes is 3–4 times more efficient below CMC [64]. There was a synergistic development in solubility of polycyclic aromatic hydrocarbons detected in the case of contact between two biosurfactants (rhamnolipids and sophorolipids) associated with one glycolipid [65].

EMULSIFYING ACTION

Suspensions are kinetically alleviated schemes but deprived of balance. Their construction, stability, and appearance are based on the composition of the aqueous phases (oil and water), emulsifiers (chemical structure and physicochemical properties), and conditions of the preparation (temperature and pressure) and process (input energy, mixing time, and the kind of equipment) [66]. The mixture can be fragmented by several mechanisms, such as skimming, flocculation coagulation, a specific Ostwald maturation, and coalescence. Blending may happen by alteration in concentrations between the oil and aqueous phases, wherein the suspension dews travel as a function of the natural force called gravitational force, subsequent to phase separation. The usual emulsifier must be quickly adsorbed on the superficial of oil droplets and thus quickly decreases the interfacial tension to enable droplet failure and formation of small droplets [67]. The well-studied and utilized biosurfactant applied as an emulsification agent in the food industry is taken from quillaja saponin extract.

There were several biosurfactants with emulsification activity that was found to be the cyclic lipopeptide pseudofactin II and was synthesized by *P. fluorescens* BD5, which is improved in action when compared with synthetic surfactants like Tween 20 and Triton X-100. Pseudofactin II was found to be more effective at emulsifying aromatic and aliphatic hydrocarbons and also some vegetable oils [68]. Rhamnolipids have a better effect on suspension droplet size decrease than lecithin and monoglycerides and also protect the thermal steadiness of the emulsion [69].

PHARMACEUTICAL APPLICATIONS OF BIOSURFACTANTS

Biosurfactants can be used for various functions in the pharmaceutical industry; they have numerous activities such as antimicrobial, anti-adhesive, antiviral, anticancer, spermicidal, hemolytic, anti-inflammatory, and immunomodulatory [70]. Antimicrobial application is one of the most anticipated usages mentioned in pharmaceutical literature. Of several successful biosurfactants, surfactin has numerous stimulating properties, in addition to its antimicrobial, antiviral, antitumor, hypocholesterolemic, anti-adhesive, biopesticidal, apoptotic, and hemolytic action. Among other functions, surfactin prevents fibrin clotting progression and shows antitumor action against Ehrlich's ascites carcinoma cells, inhibiting cyclic adenosine 3'5'-monophosphate phosphodiesterase, and antifungal activities [71].

According to the research outcome, a kind of lipopeptide biosurfactant exhibited various in vitro activities such as antimicrobial, antibiofilm, and cytotoxic. These kinds of biosurfactants were synthesized by the bacteria *Acinetobacter junii* (AjL), displaying inhibition to *Candida utilis* and aiding a possible new drug. In this way, Fernandes et al. [72,73] utilized kitchen waste oil as a substrate for *Wickerhamomyces anomalus* CCMA 0358, and the fungi yielded a biosurfactant with larvicidal action against the mosquito *Aedes aegypti* larvae. The biomolecule also shows effective antibacterial activities against pathogens (*Bacillus cereus, Salmonella enteritidis, Staphylococcus aureus*, and *Escherichia coli*) and fungal strains such as Aspergillus, Cercospora, Colletotrichum, and Fusarium. The key advantages were the low cost of production and ease of industrial scale-up processes.

Jiang et al. [74] developed a biosurfactant using Gram-positive probiotic bacteria *Lactobacillus helveticus* strains and showed anti-adhesive and inhibited effects of biofilm formation, and a similar result was obtained in research carried out by Ashitha et al. [75] using *Burkholderia* sp. WYAT7 strains to obtain a biosurfactant with drug likeliness against *Pseudomonas aeruginosa* (MTCC 2453), *Escherichia coli* (MTCC 1610), *Salmonella paratyphi*, and *Bacillus subtilis* strains. Moreover, biosurfactants are very useful in cosmetic applications. According to a current report, they have lower toxicity, providing biocompatibility and hydrating effects on skin [76]. Ferreira et al. [77] clearly explained a claim for green cosmetics to replace petroleum based. The investigators established a biosurfactant created from *Lactobacillus paracasei* by oil-in-water (o/w) emulsification. These reports were associated with the conventional surfactant, sodium dodecyl sulfate (SDS), used in the cosmetics industry, and they attained comparable results [73].

No toxic result was detected in fibroblasts; otherwise, SDS presented a robust inhibitory effect. Our present investigation on its capability to improve the drug delivery of the commercial antibiotics with drug resistance against pathogens has proved its ability to be used in the pharmaceutical industry as an adjuvant to increase the functionality of marketable antibiotics against clinical isolates, particularly an alarming drug resistant *Staphylococcus aureus* (Figures 7.1–7.3). The biosurfactants were developed from seven different *Bacillus* sp., and the chemical characteristics were found using the standard methods and conferred glycolipid nature. There were five commercial antibiotics selected and used for the antibiotic sensitivity assay by the Kirby Bauer method, and the antibiotics selected

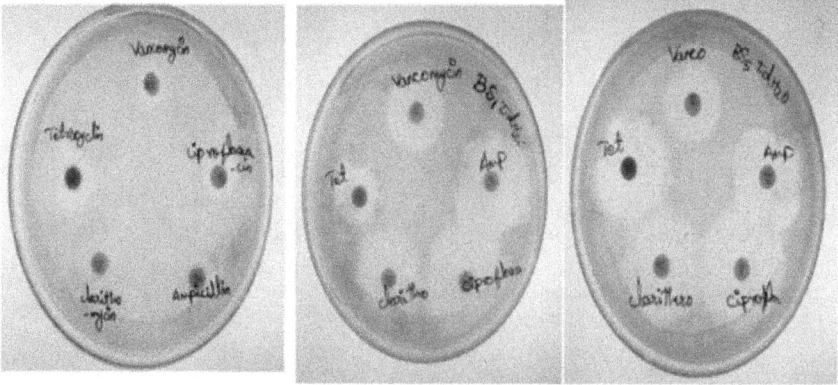

FIGURE 7.1 Comparative analysis of antibiotic sensitivity of discs impregnated with commercial antibiotics and biosurfactants against *S. aureus.*

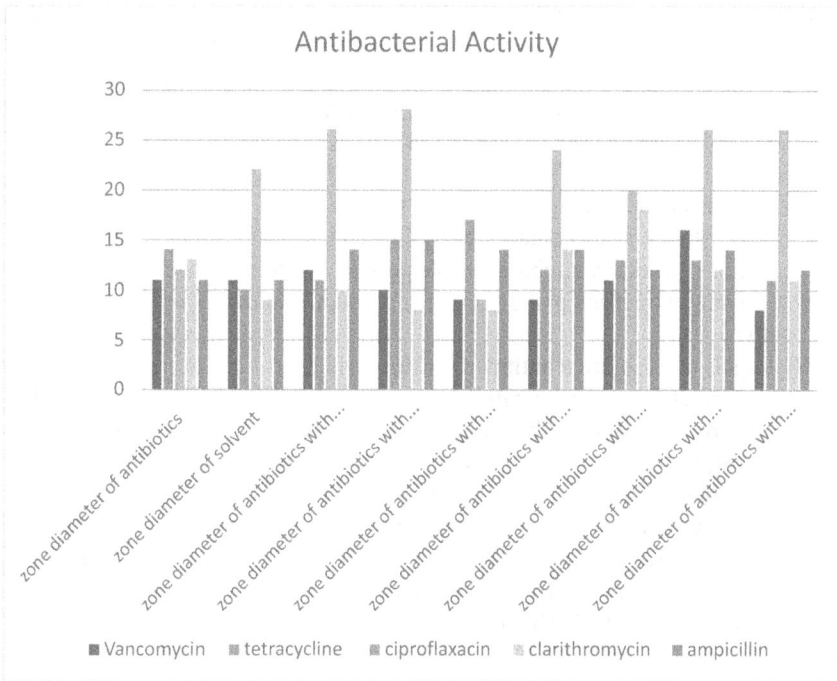

FIGURE 7.2 Synergistic activity of solvent-based biosurfactants with commercial antibiotics.

Antibacterial Activity

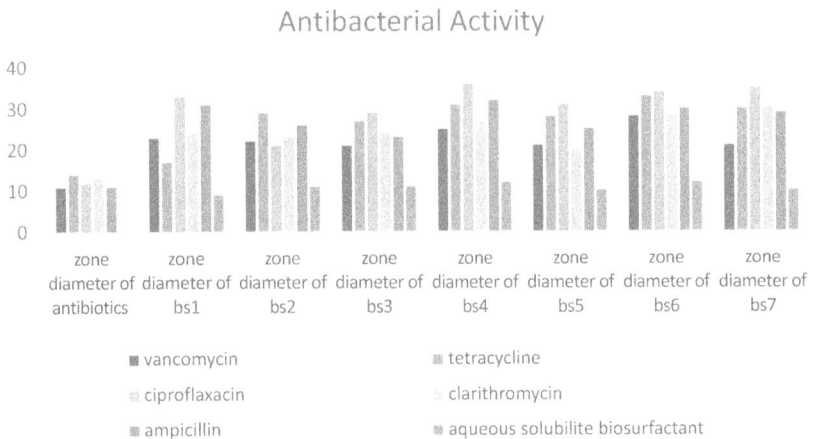

FIGURE 7.3   Efficacy improvement using aqueous biosurfactant–antibiotics complex.

were vancomycin, tetracycline, ciprofloxacin, clarithromycin, and ampicillin. We obtained a significant improvement in antibacterial activity against *S. aureus*.

Biosurfactants have been found to be active as a drug delivery system (DDS) to increase the oral bioavailability of a huge quantity of drugs that exhibit low aqueous solubility, and this has been a chief task in the arena of the pharmaceutical industry. There are countless possibilities for these biomolecules in auto assembly and emulsification activities [78].

Microemulsion drug delivery systems (MDDSs) have been considered accepted and have accomplished improvement of the oral bioavailability of hydrophobic drugs. They contain lipids, surfactants, cosurfactants, and/or cosolvents and are generally globular in shape [79]. Currently, MDDSs are being explored using numerous paths such as ingestion, inhalation, ophthalmic, topical, and venous [80].

Hydrogels incorporating antimicrobial biosurfactants have been recommended as wound curative arrangements against skin drug-resistant infections [81]. Hydrogels from poly(vinyl alcohol) (PVA), polyethylene oxide (PEO), and poly (acrylic acid) (PAA) polymers can be utilized as semisolid vehicles for superficial application-based products including antimicrobial surfactants.

Instances of commercially available products containing antimicrobial biosurfactants are the moisturizer Kanebo (Kanebo Cosmetics, Tokyo, Japan), the facial cleaner Sopholiance (Givaudan Active Beauty, Paris, France), and the body moisturizer Relipidium (BASF, Monheim, Germany) [82,83].

CONCLUSIONS

Many systematic publications have reported the extraordinary properties of biosurfactants (e.g. antimicrobial, emulsifying, and anti-adhesive), synthesized either by chemical synthesis or by microbial activity, and have proposed possible applications of these compounds in a range of industries. New biosurfactants (glycolipids, surfactin, and high molecular weight biosurfactants) produced by microorganisms (bacteria, yeasts, fungi) may be helpful in the discovery of diverse molecules in relation to their structure and physicochemical characteristics. Their amphipathic nature and capacity for self-assembly open perspectives in the development of new inventions with great potential for the delivery of bioactives of pharmaceutical interest.

REFERENCES

1. Burilova E.A., Pashirova T.N., Lukashenko S.S., Sapunova A.S., Voloshina A.D., Zhiltsova E.P., Campos J.R., Souto E.B., Zakharova L.Y. Synthesis, biological evaluation and structure-activity relationships of self-assembled and solubilization properties of amphiphilic quaternary ammonium derivatives of quinuclidine. *J. Mol. Liq.* 2018;272:722–730. doi:10.1016/j.molliq.2018.10.008.
2. Pashirova T.N., Sapunova A.S., Lukashenko S.S., Burilova E.A., Lubina A.P., Shaihutdinova Z.M., Gerasimova T.P., Kovalenko V.I., Voloshina A.D., Souto E.B., et al. Synthesis, structure-activity relationship and biological evaluation of tetracationic gemini Dabco-surfactants for transdermal liposomal formulations. *Int. J. Pharm.* 2019;575:118953. doi:10.1016/j.ijpharm.2019.118953.
3. Zakharova L.Y., Pashirova T.N., Doktorovova S., Fernandes A.R., Sanchez-Lopez E., Silva A.M., Souto S.B., Souto E.B. Cationic surfactants: Self-assembly, structure-activity correlation and their biological applications. *Int. J. Mol. Sci.* 2019;20:5534. doi:10.3390/ijms20225534
4. Subramaniam M.D., Venkatesan D., Iyer M., Subbarayan S., Govindasami V., Roy A., Narayanasamy A., Kamalakannan S., Gopalakrishnan A.V., Thangarasu R., et al. Biosurfactants and anti-inflammatory activity: A potential new approach towards COVID-19. *Curr. Opin. Environ. Sci. Health.* 2020;17:72–81. doi:10.1016/j.coesh.2020.09.002.
5. Liu L. Penetration of surfactants into skin. *J. Cosmet. Sci.* 2020;71:91–109.
6. Johnson P., Trybala A., Starov V., Pinfield V.J. Effect of synthetic surfactants on the environment and the potential for substitution by biosurfactants. *Adv. Colloid Interface Sci.* 2021;288:102340. doi:10.1016/j.cis.2020.102340.
7. Bhadani A., Kafle A., Ogura T., Akamatsu M., Sakai K., Sakai H., Abe M. Current perspective of sustainable surfactants based on renewable building blocks. *Curr. Opin. Colloid Interface Sci.* 2020;45:124–135. doi:10.1016/j.cocis.2020.01.002.
8. Foley P., Kermanshahi Pour A., Beach E.S., Zimmerman J.B. Derivation and synthesis of renewable surfactants. *Chem. Soc. Rev.* 2012;41:1499–1518. doi:10.1039/C1CS15217C.
9. Van Hamme J.D., Singh A., Ward O.P. Physiological aspects. Part 1 in a series of papers devoted to surfactants in microbiology and biotechnology. *Biotechnol. Adv.* 2006;24:604–620. doi:10.1016/j.biotechadv.2006.08.001.
10. Nitschke M., Silva S.S.E. Recent food applications of microbial surfactants. *Crit. Rev. Food Sci. Nutr.* 2018;58:631–638. doi:10.1080/10408398.2016.1208635.
11. Singh P., Patil Y., Rale V. Biosurfactant production: Emerging trends and promising strategies. *J. Appl. Microbiol.* 2019;126:2–13. doi:10.1111/jam.14057.
12. Inès M., Dhouha G. Glycolipid biosurfactants: Potential related biomedical and biotechnological applications. *Carbohydr. Res.* 2015;416:59–69. doi:10.1016/j.carres.2015.07.016.

13. De Graeve M., De Maeseneire S.L., Roelants S., Soetaert W. *Starmerella bombicola*, an industrially relevant, yet fundamentally underexplored yeast. *FEMS Yeast Res.* 2018;18:72. doi:10.1093/femsyr/foy072.

14. Jahan R., Bodratti A.M., Tsianou M., Alexandridis P. Biosurfactants, natural alternatives to synthetic surfactants: Physicochemical properties and applications. *Adv. Colloid Interface Sci.* 2020;275:102061. doi:10.1016/j.cis.2019.102061.

15. Ramos da Silva A., Manresa M., Pinazo A., García M.T., Pérez L. Rhamnolipids functionalized with basic amino acids: Synthesis, aggregation behavior, antibacterial activity and biodegradation studies. *Colloids Surf. B Biointerfaces.* 2019;181:234–243. doi:10.1016/j.colsurfb.2019.05.037.

16. Morya V.K., Ahn C., Jeon S., Kim E.K. Medicinal and cosmetic potentials of sophorolipids. *Mini Rev. Med. Chem.* 2013;13:1761–1768. doi:10.2174/13895575113139990002.

17. Haque F., Sajid M., Cameotra S.S., Battacharyya M.S. Anti-biofilm activity of a sophorolipid-amphotericin B niosomal formulation against *Candida albicans.* *Biofouling.* 2017;33:768–779. doi:10.1080/08927014.2017.1363191.

18. Wang J., Zhang J., Liu K., He J., Zhang Y., Chen S., Ma G., Cui Y., Wang L., Gao D. Synthesis of gold nanoflowers stabilized with amphiphilic daptomycin for enhanced photothermal antitumor and antibacterial effects. *Int. J. Pharm.* 2020;580:119231. doi:10.1016/j.ijpharm.2020.119231.

19. Oliveira D.M.L., Rezende P.S., Barbosa T.C., Andrade L.N., Bani C., Tavares D.S., da Silva C.F., Chaud M.V., Padilha F., Cano A., et al. Double membrane based on lidocaine-coated polymyxin-alginate nanoparticles for wound healing: In vitro characterization and in vivo tissue repair. *Int. J. Pharm.* 2020;591:120001. doi:10.1016/j.ijpharm.2020.120001.

20. Severino P., Chaud M.V., Shimojo A., Antonini D., Lancelloti M., Santana M.H., Souto E.B. Sodium alginate-cross-linked polymyxin B sulphate-loaded solid lipid nanoparticles: Antibiotic resistance tests and HaCat and NIH/3T3 cell viability studies. *Colloids Surf. B Biointerfaces.* 2015;129:191–197. doi:10.1016/j.colsurfb.2015.03.049.

21. Severino P., Silveira E.F., Loureiro K., Chaud M.V., Antonini D., Lancellotti M., Sarmento V.H., da Silva C.F., Santana M.H.A., Souto E.B. Antimicrobial activity of polymyxin-loaded solid lipid nanoparticles (PLX-SLN): Characterization of physicochemical properties and in vitro efficacy. *Eur. J. Pharm. Sci.* 2017;106:177–184. doi:10.1016/j.ejps.2017.05.063.

22. Karas J.A., Carter G.P., Howden B.P., Turner A.M., Paulin O.K.A., Swarbrick J.D., Baker M.A., Li J., Velkov T. Structure-activity relationships of daptomycin lipopeptides. *J. Med. Chem.* 2020;63:13266–13290. doi:10.1021/acs.jmedchem.0c00780.

23. Arima K., Kakinuma A., Tamura G. Surfactin, a crystalline peptidelipid surfactant produced by *Bacillus subtilis*: Isolation, characterization and its inhibition of fibrin clot formation. *Biochem. Biophys. Res. Commun.* 1968;31:488–494. doi:10.1016/0006-291X(68)90503-2.

24. Drakontis C.E., Amin S. Biosurfactants: Formulations, properties, and applications. *Curr. Opin. Colloid Interface Sci.* 2020;48:77–90. doi:10.1016/j.cocis.2020.03.013.

25. Abdel-Mawgoud A.M., Lépine F., Déziel E. A stereospecific pathway diverts β-oxidation intermediates to the biosynthesis of rhamnolipid biosurfactants. *Chem. Biol.* 2014;21:156–164. doi:10.1016/j.chembiol.2013.11.

26. Kiss K., Ng W.T., Li Q. Production of rhamnolipids-producing enzymes of Pseudomonas in *E. coli* and structural characterization. *Front. Chem. Sci. Eng.* 2017;11:133–138. doi:10.1007/s11705-017-1637-z.

27. Ławniczak Ł., Marecik R., Chrzanowski Ł. Contributions of biosurfactants to natural or induced bioremediation. *Appl. Microbiol. Biotechnol.* 2013;97:2327–2339. doi:10.1007/s00253-013-4740-1.
28. Chrzanowski Ł., Ławniczak Ł., Czaczyk K. Why do microorganisms produce rhamnolipids? *World J. Microbiol. Biotechnol.* 2012;28:401–419. doi:10.1007/s11274-011-0854-8.
29. Nitschke M., Costa S.G., Contiero J. Rhamnolipids and PHAs: Recent reports on Pseudomonas-derived molecules of increasing industrial interest. *Process Biochem.* 2011;46:621–630. doi:10.1016/j.procbio.2010.12.012
30. Dobler L., Vilela L.F., Almeida R.V., Neves B.C. Rhamnolipids in perspective: Gene regulatory pathways, metabolic engineering, production and technological forecasting. *New Biotechnol.* 2016;33:123–135. doi:10.1016/j.nbt.2015.09.005.
31. Ndlovu T., Rautenbach M., Vosloo J.A., Khan S., Khan W. Characterisation and antimicrobial activity of biosurfactant extracts produced by *Bacillus amyloliquefaciens* and *Pseudomonas aeruginosa* isolated from a wastewater treatment plant. *AMB Express.* 2017;7:108. doi:10.1186/s13568-017-0363-8.
32. Van Bogaert I.N., Saerens K., De Muynck C., Develter D., Soetaert W., Vandamme E.J. Microbial production and application of sophorolipids. *Appl. Microbiol. Biotechnol.* 2007;76:23–34. doi:10.1007/s00253-007-0988-7.
33. Daverey A., Pakshirajan K. Production, characterization, and properties of sophorolipids from the yeast *Candida bombicola* using a low-cost fermentative medium. *Appl. Biochem. Biotechnol.* 2009;158:663–674. doi:10.1007/s12010-008-8449-z.
34. Kaur G., Wang H., To M.H., Roelants S.L., Soetaert W., Lin C.S.K. Efficient sophorolipids production using food waste. *J. Clean. Prod.* 2019;232:1–11. doi:10.1016/j.jclepro.2019.05.326.
35. Ndlovu T., Rautenbach M., Khan S., Khan W. Variants of lipopeptides and glycolipids produced by *Bacillus amyloliquefaciens* and *Pseudomonas aeruginosa* cultured in different carbon substrates. *AMB Express.* 2017;7:109. doi:10.1186/s13568-017-0367-4.
36. Das P., Mukherjee S., Sen R. Genetic regulations of the biosynthesis of microbial surfactants: An overview. *Biotechnol. Genet. Eng. Rev.* 2008;25:165–186. doi:10.5661/bger-25-165.
37. Wu Q., Zhi Y., Xu Y. Systematically engineering the biosynthesis of a green biosurfactant surfactin by *Bacillus subtilis* 168. *Metab. Eng.* 2019;52:87–97. doi:10.1016/j.ymben.2018.11.004.
38. Thavasi R., Subramanyam Nambaru V.R.M., Jayalakshmi S., Balasubramanian T., Banat I.M. Biosurfactant production by *Pseudomonas aeruginosa* from renewable resources. *Indian J. Microbiol.* 2011;51:30–36. doi:10.1007/s12088-011-0076-7.
39. Uzoigwe C., Burgess J.G., Ennis C.J., Rahman P.K.S.M. Bioemulsifiers are not biosurfactants and require different screening approaches. *Front. Microbiol.* 2015;6:245. doi:10.3389/fmicb.2015.00245.
40. Mujumdar S., Joshi P., Karve N. Production, characterization, and applications of bioemulsifiers (BE) and biosurfactants (BS) produced by *Acinetobacter* spp.: A review. *J. Basic Microbiol.* 2019;59:277–287. doi:10.1002/jobm.201800364.
41. Alizadeh-Sani M., Hamishehkar H., Khezerlou A., Azizi-Lalabadi M., Azadi Y., Nattagh-Eshtivani E., Fasihi M., Ghavami A., Aynehchi A., Ehsani A. Bioemulsifiers derived from microorganisms: Applications in the drug and food industry. *Adv. Pharm. Bull.* 2018;8:191. doi:10.15171/apb.2018.023.
42. Pereira J.F.B., Gudiña E.J., Costa R., Vitorino R., Teixeira J.A., Coutinho J.A.P., Rodrigues L.R. Optimization and characterization of biosurfactant production by

Bacillus subtilis isolates towards microbial enhanced oil recovery applications. *Fuel.* 2013;111:259–268. doi:10.1016/j.fuel.2013.04.040.

43. Alvarez V.M., Jurelevicius D., Marques J.M., de Souza P.M., de Araújo L.V., Barros T.G., de Souza R.O., Freire D.M., Seldin L. *Bacillus amyloliquefaciens* TSBSO 3.8, a biosurfactant-producing strain with biotechnological potential for microbial enhanced oil recovery. *Colloids Surf. B Biointerfaces.* 2015;136:14–21. doi:10.1016/j.colsurfb.2015.08.046.

44. Al-Wahaibi Y., Joshi S., Al-Bahry S., Elshafie A., Al-Bemani A., Shibulal B. Biosurfactant production by *Bacillus subtilis* B30 and its application in enhancing oil recovery. *Colloids Surf. B Biointerfaces.* 2014;114:324–333. doi:10.1016/j.colsurfb.2013.09.022.

45. Morikawa M., Daido H., Takao T., Murata S., Shimonishi Y., Imanaka T. A new lipopeptide biosurfactant produced by *Arthrobacter* sp. strain MIS38. *J. Bacteriol.* 1993;175:6459–6466. doi:10.1128/jb.175.20.6459-6466.1993.

46. Rufino R.D., de Luna J.M., de Campos Takaki G.M., Sarubbo L.A. Characterization and properties of the biosurfactant produced by *Candida lipolytica* UCP 0988. *Electron. J. Biotechnol.* 2014;17:34–38. doi:10.1016/j.ejbt.2013.12.006.

47. Kumar P.S., Ngueagni P.T. A review on new aspects of lipopeptide biosurfactant: Types, production, properties and its application in the bioremediation process. *J. Hazard. Mater.* 2021;407:124827. doi:10.1016/j.jhazmat.2020.124827.

48. Shen H.H., Thomas R.K., Chen C.Y., Darton R.C., Baker S.C., Penfold J. Aggregation of the naturally occurring lipopeptide, surfactin, at interfaces and in solution: An unusual type of surfactant? *Langmuir ACS J. Surf. Colloids.* 2009;25:4211–4218. doi:10.1021/la802913x.

49. Penfold J., Thomas R.K., Tucker I., Petkov J.T., Stoyanov S.D., Denkov N., Golemanov K., Tcholakova S., Webster J.R.P. Saponin adsorption at the air-water interface-neutron reflectivity and surface tension study. *Langmuir ACS J. Surf. Colloids.* 2018;34:9540–9547. doi:10.1021/acs.langmuir.8b02158.

50. Otzen D.E. Biosurfactants and surfactants interacting with membranes and proteins: Same but different? *Biochim. Biophys. Acta Biomembr.* 2017;1859:639–649. doi:10.1016/j.bbamem.2016.09.024.

51. Lee Y.S., Woo K.W. Micellization of aqueous cationic surfactant solutions at the micellar structure transition concentration—Based upon the concept of the pseudophase separation. *J. Colloid Interface Sci.* 1995;169:34–38. doi:10.1006/jcis.1995.1003.

52. Akbari S., Abdurahman N.H., Yunus R.M., Fayaz F., Alara O.R. Biosurfactants—A new frontier for social and environmental safety: A mini review. *Biotechnol. Res. Innov.* 2018;2:81–90. doi:10.1016/j.biori.2018.09.001.

53. Desai J.D., Banat I.M. Microbial production of surfactants and their commercial potential. *Microbiol. Mol. Biol. Rev.* 1997;61:47–64. doi:10.1128/.61.1.47-64.1997.

54. Aveyard R., Binks B.P., Clint J.H. Emulsions stabilised solely by colloidal particles. *Adv. Colloid Interface Sci.* 2003;100–102:503–546. doi:10.1016/S0001-8686(02)00069-6.

55. Wu L.-M., Lai L., Lu Q., Mei P., Wang Y.-Q., Cheng L., Liu Y. Comparative studies on the surface/interface properties and aggregation behavior of mono-rhamnolipid and di-rhamnolipid. *Colloids Surf. B Biointerfaces.* 2019;181:593–601. doi:10.1016/j.colsurfb.2019.06.012.

56. Liu K., Sun Y., Cao M., Wang J., Lu J.R., Xu H. Rational design, properties, and applications of biosurfactants: A short review of recent advances. *Curr. Opin. Colloid Interface Sci.* 2020;45:57–67. doi:10.1016/j.cocis.2019.12.005.

57. Cui H., Cheetham A.G., Pashuck E.T., Stupp S.I. Amino acid sequence in constitutionally isomeric tetrapeptide amphiphiles dictates architecture of one-dimensional nanostructures. *J. Am. Chem. Soc.* 2014;136:12461–12468. doi:10.1021/ja507051w.

58. Nagarajan R., Ruckenstein E. Theory of surfactant self-assembly: A predictive molecular thermodynamic approach. *Langmuir ACS J. Surf. Colloids.* 1991;7:2934–2969. doi:10.1021/la00060a012.

59. Shao B., Liu Z., Zhong H., Zeng G., Liu G., Yu M., Liu Y., Yang X., Li Z., Fang Z., et al. Effects of rhamnolipids on microorganism characteristics and applications in composting: A review. *Microbiol. Res.* 2017;200:33–44. doi:10.1016/j.micres.2017.04.005.

60. López-Prieto A., Moldes A.B., Cruz J.M., Pérez Cid B. Towards more ecofriendly pesticides: Use of biosurfactants obtained from the corn milling industry as solubilizing agent of copper oxychloride. *J. Surfactants Deterg.* 2020;23:1055–1066. doi:10.1002/jsde.12463.

61. Hu X., Qiao Y., Chen L.-Q., Du J.-F., Fu Y.-Y., Wu S., Huang L. Enhancement of solubilization and biodegradation of petroleum by biosurfactant from *Rhodococcus erythropolis* HX-2. *Geomicrobiol. J.* 2020;37:159–169. doi:10.1080/01490451.2019.1678702.

62. Satpute S.K., Mone N.S., Das P., Banat I.M., Banpurkar A.G. Inhibition of pathogenic bacterial biofilms on PDMS based implants by *L. acidophilus* derived biosurfactant. *BMC Microbiol.* 2019;19:39. doi:10.1186/s12866-019-1412-z.

63. Yang X., Tan F., Zhong H., Liu G., Ahmad Z., Liang Q. Sub-CMC solubilization of n-alkanes by rhamnolipid biosurfactant: The influence of rhamnolipid molecular structure. *Colloids Surf. B Biointerfaces.* 2020;192:111049. doi:10.1016/j.colsurfb.2020.111049.

64. Zhang Y., Miller R.M. Enhanced octadecane dispersion and biodegradation by a *Pseudomonas rhamnolipid* surfactant (biosurfactant). *Appl. Environ. Microbiol.* 1992;58:3276–3282. doi:10.1128/aem.58.10.3276-3282.1992.

65. Percebom A.M., Towesend V.J., de Paula Silva de Andrade Pereira M., Pérez Gramatges A. Sustainable self-assembly strategies for emerging nanomaterials. *Curr. Opin. Green Sustain. Chem.* 2018;12:8–14. doi:10.1016/j.cogsc.2018.04.004.

66. Kaizu K., Alexandridis P. Effect of surfactant phase behavior on emulsification. *J. Colloid Interface Sci.* 2016;466:138–149. doi:10.1016/j.jcis.2015.10.016.

67. McClements D.J., Gumus C.E. Natural emulsifiers—Biosurfactants, phospholipids, biopolymers, and colloidal particles: Molecular and physicochemical basis of functional performance. *Adv. Colloid Interface Sci.* 2016;234:3–26. doi:10.1016/j.cis.2016.03.002.

68. Janek T., Łukaszewicz M., Rezanka T., Krasowska A. Isolation and characterization of two new lipopeptide biosurfactants produced by *Pseudomonas fluorescens* BD5 isolated from water from the Arctic Archipelago of Svalbard. *Bioresour. Technol.* 2010;101:6118–6123. doi:10.1016/j.biortech.2010.02.109.

69. Russell C., Zompra A.A., Spyroulias G.A., Salek K., Euston S.R. The heat stability of Rhamnolipid containing egg-protein stabilised oil-in-water emulsions. *Food Hydrocoll.* 2021;116:106632. doi:10.1016/j.foodhyd.2021.106632.

70. Jimoh A.A., Lin J. Biosurfactant: A new frontier for greener technology and environmental sustainability. *Ecotoxicol. Environ. Saf.* 2019;184:109607. doi:10.1016/j.ecoenv.2019.109607.

71. Vecino X., Cruz J.M., Moldes A.B., Rodrigues L.R. Biosurfactants in cosmetic formulations: Trends and challenges. *Crit. Rev. Biotechnol.* 2017;37:911–923. doi:10.1080/07388551.2016.1269053.

72. Marchant R., Funston S., Uzoigwe C., Rahman P., Banat I.M. Production of biosurfactants from nonpathogenic bacteria. *Biosurfactants Prod. Util. Process. Technol. Econ.* 2014;159:73–82.

73. Chakraborty J., Das S. *Microbial Biodegradation and Bioremediation, Chapter 7.* Elsevier; Amsterdam, The Netheralands: 2014. Biosurfactant-Based Bioremediation of Toxic Metals.

74. Xinpeng Jiang, Xin Yan, Shanshan Gu, Yan Yang, Biosurfactants of Lactobacillus helveticus for biodiversity inhibit the biofilm formation of Staphylococcus aureus and cell invasion, 2019, *Future Microbiology* 14(13):1133–1146.

75. Ashitha A., Mathew EKR, *J. Biotechnol*, 2020; Characterization of biosurfactant produced by the endophyte Burkholderia sp. WYAT7 and evaluation of its antibacterial and antibiofilm potentials, *J. Biotechnol*, 313:1–10.

76. Santos D.K.F., Rufino R.D., Luna J.M., Santos V.A., Sarubbo L.A. Biosurfactants: Multifunctional biomolecules of the 21st century. *Int. J. Mol. Sci.* 2016;17:401. doi:10.3390/ijms17030401.

77. Campos J.M., Montenegro Stamford T.L., Sarubbo L.A., de Luna J.M., Rufino R.D., Banat I.M. Microbial biosurfactants as additives for food industries. *Biotechnol. Prog.* 2013;29:1097–1108. doi:10.1002/btpr.1796.

78. Silva A., Santos P., Silva T., Andrade R., Campos-Takaki G. Biosurfactant production by fungi as a sustainable alternative. *Arq. Inst. Biológico.* 2018;85:e0502017. doi:10.1590/1808-1657000502017.

79. Meneses D.P., Gudiña E.J., Fernandes F., Gonçalves L.R.B., Rodrigues L.R., Rodrigues S. The yeast-like fungus *Aureobasidium thailandense* LB01 produces a new biosurfactant using olive oil mill wastewater as an inducer. *Microbiol. Res.* 2017;204:40–47. doi:10.1016/j.micres.2017.07.004.

80. Gudiña E.J., Teixeira J.A., Rodrigues L.R. Biosurfactants produced by marine microorganisms with therapeutic applications. *Mar. Drugs.* 2016;14:38. doi:10.3390/md14020038.

81. Marchant R., Banat I.M. Microbial biosurfactants: Challenges and opportunities for future exploitation. *Trends Biotechnol.* 2012;30:558–565. doi:10.1016/j.tibtech.2012.07.003.

82. Chen W.-C., Juang R.-S., Wei Y.-H. Applications of a lipopeptide biosurfactant, surfactin, produced by microorganisms. *Biochem. Eng. J.* 2015;103:158–169. doi:10.1016/j.bej.2015.07.009.

83. Rodrigues L.R. Inhibition of bacterial adhesion on medical devices. *Adv. Exp. Med. Biol.* 2011;715:351–367. doi:10.1007/978-94-007-0940-9_22.

8 Marine Algal and Cyanobacterial Surface-Active Compounds

J.J. Mehjabin and T. Okino

CONTENTS

INTRODUCTION

Biosurfactants have recently emerged as a diverse group of surface-active amphipathic compounds with distinct chemical structures produced by a wide variety of microorganisms and are potentially helpful for many therapeutic applications (Gudina et al. 2013; Banat et al. 1995). Biosurfactants and bioemulsifiers are both amphiphilic molecules; however, there is a difference between their physicochemical properties. Biosurfactants are mainly formed as secondary metabolites, generally extracellularly excreted or localized on microbial cell surfaces, and play important roles in the survival of their producing microorganisms by facilitating nutrient transport, interfering in microbe-host interactions and quorum-sensing mechanisms, or acting as biocide agents (Marchant and Banat 2012; Fracchia et al. 2015). Biosurfactants are usually classified as glycolipids, lipopeptides, fatty acids, and polymers (Gudina et al. 2013; Satpute et al. 2010; Shekhar et al. 2015). Biosurfactants have attracted much attention as a potential replacement for synthetic surfactants in industrial and environmental applications and the discovery of several therapeutic agents (Fracchia et al. 2014; Banat et al. 2000; Fracchia et al. 2012). On the other hand, bioemulsifiers are high molecular weight biomolecules (exopolysaccharides, lipopolysaccharides, and lipoprotein) produced by a diverse group of microorganisms that show potential applications in the food, cosmetics, and pharmaceutical industries (Rahman et al. 2019).

DOI: 10.1201/9781003307464-8

ALGAL AND CYANOBACTERIAL EXOPOLYSACCHARIDES AND PROTEIN AS BIOSURFACTANT AND BIOEMULSIFIER

Marine algae and cyanobacteria are promising sources of biomolecules and could be sustainable sources to produce biosurfactants and bioemulsifiers. Many studies have been done regarding the characterization of exopolysaccharides (EPSs) from algae and cyanobacteria; nevertheless, the studies focusing on the surfactant/emulsifying properties are still unexploited. Algal extracellular polymeric substances include polysaccharides, protein, nucleic acids, and lipids; among them, proteins exhibit good emulsifying properties, and polar lipid layers of cell membranes and cellular organelles are natural surfactants (Law et al. 2018). The benthic cyanobacterium *Phormidium* J-1 produces excellent extracellular polymeric substances known as emulcyans that have emulsifying properties (Fattom and Shilo 1985). The marine microalgae *Dunaliella salina, Nannochloropsis salina, Tetraselmis tetrathele*, red alga *Porphyridium cruentum* and marine cyanobacterium *Phormidium* sp. have been reported for their surfactant and emulsifying properties (Mishra et al. 2011; Law et al. 2018; Farahin et al. 2019; Guil-Guerrero et al. 2004; Morales and Paniagua-Michel 2013). Table 8.1 includes the marine algal and cyanobacterial strains reported to produce emulsifiers and surfactants.

TABLE 8.1
Algal/Cyanobacterial EPSs/Proteins as Surfactants and Emulsifiers and Their Applications

Algae/ Cyanobacteria	EPSs/Protein	Functions/Applications	References
Dunaliella salina	EPSs: aldohexoses (glucose and galactose), ketohexose (fructose) and pentose (xylose), uronic acid, primary amine, aromatic compounds, halides, aliphatic alkyl, and sulfides	Surfactant and/or emulsifier	Mishra and Jha 2009 Mishra et al. 2011
Nannochloropsis salina	EPSs, intracellular components: proteins, lipids: neutral lipids (91–93% w/w), polar lipids (5–7% w/w glycolipid, 1–2% w/w phospholipid)	Surfactant and emulsifier	Law et al. 2018
Porphyridium cruentum	EPSs: pentoses, hexoses, glucoronic acid	Emulsifier and functional food properties	Guil-Guerrero et al. 2004
Cyanothece sp. CCY 0110	Neutral sugars (56.8%), uronic acids (13.7%), sulfated residue (11%), protein (4%)	Emulsifier and skin wound–healing properties	Mota et al. 2020 Costa et al. 2021
Synechococcus elongates BDUI30911	total sugars (22%), uronic acid (8%), sulfate (5%), protein (6%)	Surfactant and emulsifier, uranium adsorption properties	Rashmi et al. 2021
Phormidium sp.	EPSs: carbohydrates	Surfactant/emulsifying activity and bioremediation of hexadecane and diesel oil	Morales and Paniagua-Michel 2013

ALGAL AND CYANOBACTERIAL SECONDARY METABOLITES WITH SURFACTANT STRUCTURE

Marine algae and cyanobacteria are prolific producers of secondary metabolites that exhibit various structural properties, including natural surfactants with enormous potential in different industries. A few surfactant structures were reported from cyanobacteria, for example, somocystinamide A (1), apratoxin A (2) from marine cyanobacteria, and 2-acyloxyethylphosphonate (3) from the freshwater cyanobacterium strain *Aphanizomenon flos-aquae* (Gudiña et al. 2016; Kaya et al. 2006). Besides these compounds, several glycosidic compounds and glycolipids are reported exhibiting surfactant structures (Dembitsky 2004; Dembitsky 2005a, 2005b). A 16-membered brominated macrolide lyngbyaloside (4) from *Lyngbya bouillonii* collected in Laing Island, Papua New Guinea, is the first glycosidic macrolide from marine cyanobacteria (Klein et al. 1997). The structural analogs of lyngbyaloside, lyngbyaloside B (5) acquired from the Palauan collection of *Lyngbya* sp. exhibited weak cytotoxicity against KB cells (IC$_{50}$ value of 4.3 μM) and LoVo cells (IC$_{50}$ value of 15 μM) (Luesch et al. 2002). Later in 2010, analogs of lyngbyaloside, that is, 2-*epi*-lyngbyaloside (6), 18*E*- and 18*Z*-lyngbyalosides (7, 8) with weak cytotoxicity against HT29 colorectal adenocarcinoma and HeLa cervical carcinoma cells were reported from *L. bouillonii* collected in Apra Harbor, Guam (Matthew et al. 2010). Another glycosidic macrolide, lyngbouilloside (9) from the same species, *L. bouillonii*, collected in Papua New Guinea, showed moderate cytotoxicity (IC$_{50}$ value of 17 μM) against neuroblastoma cells (Tan et al. 2002). Fatal toxins and glycosidic macrolides, polycavernoside A (10) and structural analogs polycavernoside A2 (11), A3 (12), B (13), and B2 (14), are produced by the marine red alga *Polycavernosa tsudai* collected from Tanguisson Beach, Guam (Yotsu-Yamashita et al. 1993, 1995). Water-soluble alkaloid glycosides are a class of natural compounds found in microorganisms, plants, and marine invertebrates and exhibit a wide range of biological activities. Glycosidic quinoline alkaloid (15) was afforded in low yield from the lipid fraction of Caribbean cyanobacterium *Lyngbya majuscula* (Orjala et al. 1997). A series of bioactive malyngamide compounds, typically containing fatty acid side chain and hydrophilic unusual vinyl chloride functionality, were isolated from different collections of marine cyanobacteria *Lyngbya* sp. Among them, fatty acid amide glycoside malyngamide J (16) was isolated from *Lyngbya majuscula* collected in Curaçao island (Wu et al. 1997). Filamentous cyanobacterium *Nodularia harveyana* is reported to produce heterocyst glycolipids capable of N$_2$ fixation against the penetration of O$_2$. Based on spectroscopic analysis, the structure has been established to be 1-(*O*-α-D-glucopyranosyl)-3*R*,25*R*-hexacosanediol (17), 1-(*O*-α-D-glucopyranosyl)-3*S*,25*R*-hexacosanediol (18), and 1-(*O*-α-D-glucopyranosyl)-3-*keto*-25*R*-hexacosanol (19) (Soriente et al. 1992). These glycolipids (17, 19) with another two similar heterocyst glycolipids, 1-(*O*-α-D-glucopyranosyl)-3-*keto*-25*S*,27*R*-octacosanediol (20) and 1-(*O*-α-D-glucopyranosyl)-3*R*,25*S*,27*R*-octacosanetriol (21), are detected in cyanobacteria *Anabaena cylindrical* and *Anabaena torulosa* (Soriente et al. 1995). Figure 8.1 shows the structures of natural surfactants from algae and cyanobacteria.

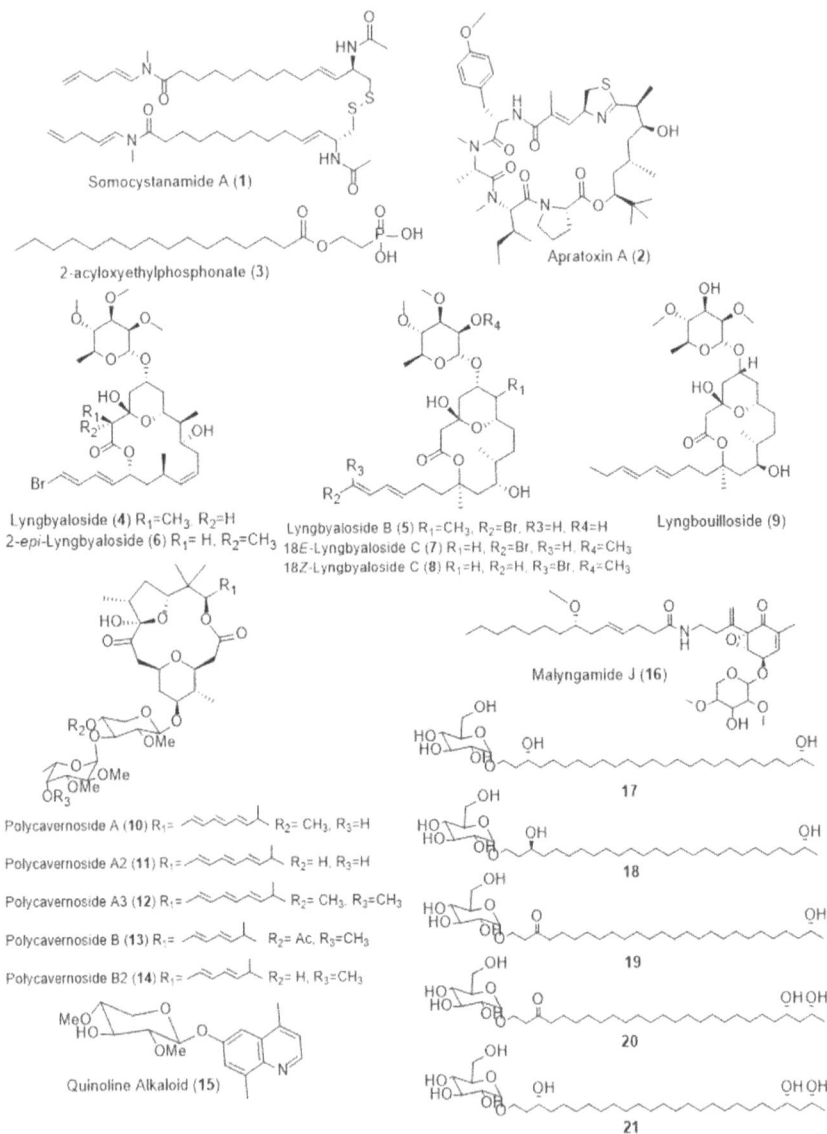

FIGURE 8.1   Structures of natural surfactants from algae and cyanobacteria.

CASE STUDIES

In our ongoing studies to search for bioactive secondary metabolites from Malaysian marine cyanobacteria, several compounds with intriguing bioactivity, such as columbamides for surfactant properties, homohydroxydolabellin for anti-malarial activity, and cytotoxicities of crude extracts against the MCF7 breast cancer cell lines were investigated (Mehjabin et al. 2020; Fathoni et al. 2020; Krisridwany and Okino 2020). In this study, marine cyanobacterial samples were collected by scuba at different locations in Kota Kinabalu, Sabah, Malaysia, in September 2016 and November 2017. The marine samples were identified by molecular phylogenetic analysis based on 16S rRNA gene sequence data (Nübel et al. 1997). Cyanobacterial samples preserved in methanol (MeOH) were homogenized three times with MeOH. Then the dried MeOH extract was further partitioned with ethyl acetate (EtOAc), n-butanol (BuOH), and water (H_2O). These crude extracts were tested for their surfactant activity by performing an oil displacement assay (Morikawa et al. 2000). For this assay, 10 µL of crude oil were added to the surface of 40 mL of distilled H_2O in a 15-cm glass petri dish to form a thin oil layer. The crude oil used in this experiment was manufactured by Tokyo Chemical Industry Co., LTD (S0432). Crude extracts were dissolved in H_2O for hydrophilic fractions and ethanol (EtOH) for lipophilic fractions to make the 10-mg/mL final concentration. Then, 10 µL of extracted samples were gently placed on the center of the oil layer. Biosurfactant activity was observed for the extracts with the capabilities to remove oil, and the diameter of this clearing zone on the oil surface correlates to the surfactant activity, also called oil displacement activity. Each fraction (ethyl acetate, butanol, and water) from the solvent partition were tested, and the preliminary screening results showed a higher oil displacement area for EtOAc fractions compared to other fractions because of the presence of surface-active compounds in these fractions. Table 8.2 shows the results of the oil displacement assay for selected marine cyanobacteria samples.

The cyanobacterial sample (M1705) collected from Rocky Point, Mantanani Island, Sabah, Malaysia (06°42′16″ N; 116°19′23″ E), in November 2017 by using scuba diving at a depth of 5–10 m was chosen for the isolation of surface-active compounds based on the amount and chemical profiling of active EtOAc fraction. The chemical profiling of EtOAc and BuOH fractions revealed the presence of known compounds, including cytotoxic apratoxins, wewakazole, kanamienamide, lyngbyabellins, and possible new halogenated compounds according to

TABLE 8.2
Oil Displacement Assay Results of EtOAc, BuOH and H_2O Fractions

Species Name	Oil Displacement Area (mm)		
	EtOAc Fr.	BuOH Fr.	H_2O Fr.
M1620 (*Moorea producens*)	139	88	1
M1622 (*Moorea bouilloni*)	102	0	5
M1623 (*Okeania* sp.)	79	101	12
M1629 (*Moorea bouilloni*)	91	0	5
M1701 (*Moorea bouilloni*)	100	40	25
M1705 (*Moorea bouilloni*)	110	60	35
M1704 (*Moorea bouilloni*)	98	55	40

the MarinLit database. The EtOAc extract (600 mg) was applied onto a silica gel (size 0.063–0.200 mm) column for further separation and eluted with a stepwise gradient starting with 100% *n*-hexane to 100% EtOAc (98:2, 95:5, 90:10, 80:20, 50:50-v/v) followed by 100% EtOAc to 100% MeOH (95:5, 90:10, 80:20, 50:50) to yield 12 fractions. The solvents were dried *in vacuo*, and the samples were placed in a clean vial. The 50:50 (v/v) hexane-EtOAc fraction was subjected to semi-preparative reversed-phase HPLC (gradient conditions: 0–30 min 50–80% acetonitrile (MeCN); 30–40 min 80–100% MeCN, Cosmosil cholester column, 10 × 250 mm, flow rate: 3 mL/min, and UV detection at 210 nm) using HPLC-grade MeCN (Wako) and MilliQ H_2O to obtain di- and tri-chlorinated compounds, columbamides F (22) (5 mg, t_R = 43.3 min) and G (23) (2 mg, t_R = 44.7 min), respectively. The silica gel fraction eluted with 100% EtOAc was subjected to RP-HPLC with similar conditions to obtain the mono-chlorinated compound columbamide H (24) (3.5 mg, t_R = 39.1 min). Then compounds 22 and 24 were further purified using RP-HPLC (gradient 0–30 min, 80–100% MeCN, Cosmosil cholester column, 4.6 × 250 mm, flow rate: 1 mL/min, and UV detection at 210 nm) to yield 22 (3.2 mg, t_R = 13.9 min) and 24 (2 mg, t_R = 12 min). Figure 8.2 represents the HPLC profile of hexane/EtOAc (50:50-v/v) (A) and 100% EtOAc fraction (B) fractions.

The planar structures of 22–24 were established by analyzing NMR and MS data. An *E* configuration of the double bond for compounds 22–24 was

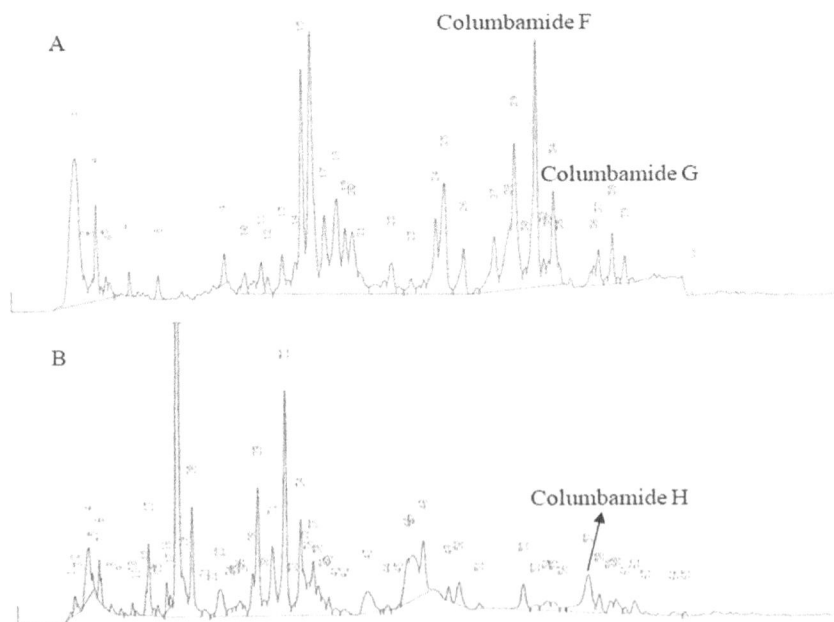

FIGURE 8.2 HPLC chromatogram of the hexane/EtOAc (50:50-v/v) (A) and 100% EtOAc fractions (B).

obtained by measuring the coupling constant ($^3J_{H,H}$ = 16–17 Hz) from the ^{13}C satellites observed by non-decoupled HSQC analysis. R-configurations for N,O-dimethylserinol residues were established by comparing the retention times of Marfey's derivatives of hydrolysates (22–24) and synthetic standards. The absolute configuration of columbamide F (22) was determined to be (10R,20R) by Marfey's analysis and chiral-phase HPLC analysis of synthetic standards and columbamide F (22) after derivatization with Ohrui's acid (Mehjabin et al. 2020). Figure 8.3 shows the structure of fatty acid amides from marine cyanobacteria *M. bouillonii* and *Okeania* sp.

The biosurfactant properties of columbamide F (22) were measured by oil displacement assay only due to the limited amount of sample and the diameter of the clear zone was approximately 90 mm for the concentration of 10 mg/mL, which was higher than the oil displacement area of the synthetic surfactant SDS (~84 mm) and pluronic F-68 (~54 mm) for the same concentration. Because of the presence of fatty acid amide columbamide D (Lopez et al. 2017) in the active fraction with the surfactant properties, it was checked by measuring the surface tension–lowering activity along with structurally similar serinolamide C (Petitbois et al. 2017). Both fatty acid amides showed lower critical micelle concentrations (CMCs) than the synthetic surfactant SDS and were comparable to pluronic F-68 and triton X-100 CMC values. Table 8.3 shows the surfactant activity of fatty acid amides from marine cyanobacteria and synthetic surfactants. The antifouling property (IC$_{50}$ value of 0.88 μg/mL) of serinolamide C against barnacle larvae *Amphibalanus amphitrite* was reported, and the possible reason behind this activity could be the biosurfactant properties that reduce the surface tension of the seawater in the microtitre plate and reduce the adherence.

The first reported columbamides were isolated from chemical extracts from the cultured *M. bouillonii*. A combination of mass spectrometric metabolic profiling and genomic analysis resulted in the discovery of the unique acyl amides columbamide (A-C, 25–27) from the laboratory culture of the *M. bouillonii* sample, which were found to be potent ligands for the cannabinoid receptors CB$_1$ and CB$_2$ (Kleigrewe et al. 2015). Later, related compounds columbamide

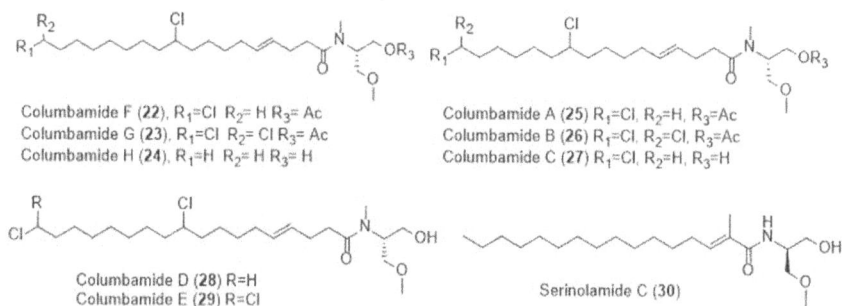

Columbamide F (22), R$_1$=Cl R$_2$= H R$_3$= Ac
Columbamide G (23), R$_1$=Cl R$_2$= Cl R$_3$= Ac
Columbamide H (24), R$_1$=H R$_2$= H R$_3$= H

Columbamide A (25) R$_1$=Cl, R$_2$=H, R$_3$=Ac
Columbamide B (26) R$_1$=Cl, R$_2$=Cl, R$_3$=Ac
Columbamide C (27) R$_1$=Cl, R$_2$=H, R$_3$=H

Columbamide D (28) R=H
Columbamide E (29) R=Cl

Serinolamide C (30)

FIGURE 8.3 Structure of fatty acid amides from marine cyanobacteria.

TABLE 8.3
Biosurfactant Activity of Pure Compounds

	Oil Displacement Area (mm) Concentration 10 mg/mL	Critical Micelle Concentration (CMC)	Reference
Columbamide D (**28**)	~110 mm	0.33 mM	Mehjabin et al. 2020
Columbamide F (**22**)	~ 90 mm	—	Mehjabin et al. 2020
Serinolamide C (**30**)	~ 54 mm	0.78 mM	Mehjabin et al. 2020
SDS	~ 84 mm	8.2 mM	Baldeweg et al. 2019
Pluronic F-68	~ 54 mm	0.07–0.35 mM	Samith et al. 2013
Triton X-100	Not tested	0.33 mM	Crook et al. 1964

D and E (28 and 29) were isolated from the *M. bouillonii* sample collected from Mantanani Island, and the absolute configuration was solved by Marfey's analysis and total synthesis of four diastereomers (Lopez et al. 2017). In this study, biosurfactant-assay guided isolation led to the isolation of three new columbamides (22–24), along with columbamides D and E from *M. bouillonii* samples collected in Mantanani Island, Sabah, Malaysia. Columbamide F (22) was checked for the biosurfactant properties by oil displacement assay, which is the indirect measurement of surface activity, and it was found that columbamide F can displace oil from the surface of water. Then it was decided to check the biosurfactant activity of related fatty acid amides columbamide D (28) and serinolamide C (30) by measuring the critical micelle concentration value. Biosurfactants have gained attention in recent years and possess similar or enhanced performance compared to synthetic surfactants. For example, KANEKA surfactin has been widely used as a cosmetic ingredient since 2001 and shows low critical micelle concentration, easy biodegradation, and less irritation than synthetic surfactant SDS. The CMC values of columbamide D and serinolamide C showed comparable value with synthetic standards. These results revealed the biosurfactant properties of fatty acid amides and their increased application in different fields.

FUTURE PERSPECTIVES

Besides the enormous potential of marine algal and cyanobacterial metabolites as surfactants/emulsifiers and their applications in diverse fields, this area is still untapped, and extensive research should be done in the future for identification of suitable species and metabolites capable of exhibiting surfactant properties. The main drawbacks of biosurfactants are the production cost on an industrial scale, and extensive research should be done to improve the conditions for increasing the yield.

REFERENCES

Baldeweg, F., P. Warncke, D. Fischer, and M. Gressler. 2019. Fungal biosurfactants from *Mortierella alpine. Organic Letters*, 21:1444–48.

Banat, I. M. 1995. Characterization of biosurfactants and their use in pollution removal— State of the art (Review). *Acta Biotechnoogica* 15:251–67.

Banat, I. M., R. S. Makkar, and S. S. Cameotra. 2000. Potential commercial applications of microbial surfactants. *Applied Microbiology and Biotechnology* 53:495–508.

Costa, R., L. Costa, I. Rodrigues, et al. 2021. Biocompatibility of the biopolymer cyanoflan for applications in skin wound healing. *Marine Drugs* 19:147.

Crook, E. H., G. F. Trebbi, and D. B. Fordyce. 1964. Thermodynamic properties of solutions of homogeneous *p,t*-octylphenoxyethoxyethanols (OPE_{1-10}). *The Journal of Physical Chemistry* 68:3592–99.

Dembitsky, V. M. 2004. Astonishing diversity of natural surfactants: 1. Glycosides of fatty acids and alcohols. *Lipids* 39:933–953.

Dembitsky, V. M. 2005a. Astonishing diversity of natural surfactants: 6. Biologically active marine and terrestrial alkaloid glycosides. *Lipids* 40:1081–1105.

Dembitsky, V. M. 2005b. Astonishing diversity of natural surfactants: 2. Polyether glycosidic ionophores and macrocyclic glycosides. *Lipids* 40:219–248.

Farahin, A. W., F. M. Yusoff, M. Basri, N. Nagao, and M. Shariff. 2019. Use of microalgae: *Tetraselmis tetrathele* extract in formulation of nanoemulsions for cosmeceutical application. *Journal of Applied Phycology* 31:1743–52.

Fathoni, I., J. G. Petitbois, W. M. Alarif, et al. 2020. Bioactivities of lyngbyabellins from cyanobacteria of *Moorea* and *Okeania* genera. *Molecules* 25:3986.

Fattom, A. and M. Shilo. 1985. Production of emulcyan by *Phormidium* J-1: Its activity and function. *FEMS Microbiology Letters* 31:3–9.

Fracchia, L., J. J. Banat, M. Cavallo, and I. M. Banat. 2015. Potential therapeutic applications of microbial surface-active compounds. *AIMS Bioengineering* 2:144–62.

Fracchia, L., M. Cavallo, M. G. Martinotti, and I. M. Banat. 2012. Biosurfactants and bioemulsifiers, biomedical and related applications-present status and future potentials. *Biomedical Science, Engineering and Technology*, ed. D. N. Ghista, 325–370. Rijeka: InTech.

Fracchia, L., C. Ceresa, A. Franzetti, et al. 2014. Industrial applications of biosurfactants. *Biosurfactant: Production and Utilization—Processes, Technologies, and Economics*, ed. N. Kosaric and F. V. Sukan, 245–267. Boca Raton: CRS Press/ Taylor & Francis.

Gudina, E. J., V. Rangarajan, R. Sen, and L. R. Rodrigues. 2013. Potential therapeutic applications of biosurfactants. *Trends in Pharmacological Sciences* 34:667–75.

Gudiña, E. J., J. A. Teixeira, and L. R. Rodrigues. 2016. Biosurfactants produced by marine microorganisms with therapeutic applications. *Marine Drugs* 14:38.

Guil-Guerrero, J. L., R. Navarro-Juárez, J. C. Lopez-Martinez, P. Campra-Madrid, M. M. Rebolloso-Fuentes. 2004. Functional properties of the biomass of three microalgal species. *Journal of Food Engineering* 65:511–517.

Kaya, K., L. F. Morrison, G. A. Codd, et al. 2006. A novel biosurfactant, 2 acyloxyethylphosphonate, isolated from waterblooms of *Aphanizomenon flos-aquae. Molecules* 11:539–548.

Kleigrewe, K., J. Almaliti, I. Y. Tian, et al. 2015. Combining mass spectrometric metabolic profiling with genomic analysis: A powerful approach for discovering natural products from cyanobacteria. *Journal of Natural Products* 78:1671–82.

Klein, D., J. C. Braekman, D. Daloze, L. Hoffmann, and V. Demoulin. 1997. Lyngbyaloside, a novel 2, 3, 4-tri-*O*-methyl-6-deoxy-α-mannopyranoside macrolide from *Lyngbya bouillonii* (Cyanobacteria). *Journal of Natural Products* 60:1057–59.

Krisridwany, A., and T. Okino. 2020. The cytotoxic activities of the ethyl acetate and butanol crude extracts of marine cyanobacteria collected from Udar Island, Malaysia. *Pharmaciana* 10:23–34.

Law, S. Q. K., S. Mettu, M. Ashokkumar, P. J. Scales, and G. J. O. Martin. 2018. Emulsifying properties of ruptured microalgae cells: Barriers to lipid extraction or promising biosurfactants? *Colloids and Surfaces B: Biointerfaces* 170:438–46.

Lopez, J. A. V., J. G. Petitbois, C. S. Vairappan, T. Umezawa, F. Matsuda, and T. Okino. 2017. Columbamides D and E: Chlorinated fatty acid amides from the marine cyanobacterium *Moorea bouillonii* collected in Malaysia. *Organic Letters* 19:4231–34.

Luesch, H., W. Y. Yoshida, G. G. Harrigan, J. P. Doom, R. E. Moore, and V. J. Paul. 2002. Lyngbyaloside B, a new glycoside macrolide from a Palauan marine cyanobacterium, *Lyngbya* sp. *Journal of Natural Products* 65:1945–48.

Marchant, R., and I. M. Banat. 2012. Microbial biosurfactants: Challenges and opportunities for future exploitation. *Trends in Biotechnology* 30:558–65.

Matthew, S., L. A. Salvador, P. J. Schupp, V. J. Paul, and H. Luesch. 2010. Cytotoxic halogenated macrolides and modified peptides from the apratoxin-producing marine cyanobacterium *Lyngbya bouillonii* from Guam. *Journal of Natural Products* 73:1544–52.

Mehjabin, J. J., L. Wei, J. G. Petitbois, et al. 2020. Biosurfactants from marine cyanobacteria collected in Sabah, Malaysia. *Journal of Natural Products* 83:1925–30.

Mishra, A., and B. Jha. 2009. Isolation and characterization of extracellular polymeric substances from micro-algae *Dunaliella salina* under salt stress. *Bioresource Technology* 100:3382–86.

Mishra, A., K. Kavita, and B. Jha. 2011. Characterization of extracellular polymeric substances produced by micro-algae *Dunaliella salina*. *Carbohydrate Polymers* 83:852–57.

Morales, A. R., and J. Paniagua-Michel. 2013. Bioremediation of hexadecane and diesel oil is enhanced by photosynthetically produced marine biosurfactants. *Journal of Bioremediation and Biodegradation* S4:005.

Morikawa, M., Y. Hirata, and T. Imanaka. 2000. A study on the structure-function relationship of lipopeptide biosurfactants. *Biochimica et Biophysica Acta (BBA)* 1488:211–18.

Mota, R., R. Vidal, C. Pandeirada, et al. 2020. Cyanoflan: A cyanobacterial sulfated carbohydrate polymer with emulsifying properties. *Carbohydrate Polymers* 229:115525.

Nübel, U., F. Garcia-Pichel, and G. Muyzer. 1997. PCR primers to amplify 16S rRNA genes from cyanobacteria. *Applied and Environmental Microbiology* 63:3327–32.

Orjala, J., and W. H. Gerwick. 1997. Two quinoline alkaloids from the Caribbean cyanobacterium *Lyngbya majuscula*. *Phytochemistry* 45:1087–90.

Petitbois, J. G., L. O. Casalme, J. A. V. Lopez, et al. 2017. Serinolamides and lyngbyabellins from an *Okeania* sp. cyanobacterium collected from the Red Sea. *Journal of Natural Products* 80:2708–15.

Rahman, P. K. S. M., A. Mayat, J. G. H. Harvey, K. S. Randhawa, L. E. Relph, and M. C. Armstrong. 2019. Biosurfactants and bioemulsifiers from marine algae. *The Role of Microalgae in Wastewater Treatment*, ed. L. Sukla, E. Subudhi, and D. Pradhan, 169–188. Singapore: Springer.

Rashmi, V., A. Darshana, T. Bhuvaneshwari, S. K. Saha, L. Uma, and D. Prabaharan. 2021. Uranium adsorption and oil emulsification by extracellular polysaccharides

(EPS) of a halophilic unicellular marine cyanobacterium *Synechococcus elongatus* BDUI30911. *Current Research in Green and Sustainable Chemistry* 4:100051.

Samith, V. D., G. Miño, E. Ramos-Moore, and N. Arancibia-Miranda. 2013. Effects of pluronic F68 micellization on the viability of neuronal cells in culture. *Journal of Applied Polymer Science* 138:2159–64.

Satpute, S. K., I. M. Banat, P. K. Dhakephalkar, A. G. Banpurkar, and B. A. Chopade. 2010. Biosurfactants, bioemulsifiers and exopolysaccharides from marine microorganisms. *Biotechnology Advances* 28:436–50.

Shekhar, S., A. Sundaramanickam, and T. Balasubramanian. 2015. Biosurfactant producing microbes and their potential applications: a review. *Critical Reviews in Environmental Science and Technology* 45:1522–54.

Soriente, A., T. Bisogno, A. Gambacorta, I. Romano, C. Sili, A. Trincone, and G. Sodano. 1995. Reinvestigation of heterocyst glycolipids from the cyanobacterium, *Anabaena cylindrical*. *Phytochemistry* 38:641–45.

Soriente, A., G. Sodano, A. Cambacorta, and A. Trincone 1992. Structure of the "heterocyst glycolipids" of the marine cyanobacterium *Nodularia harveyana*. *Tetrahedron* 48:5375–84.

Tan, L. T., B. L. Márquez, and W. H. Gerwick. 2002. Lyngbouilloside, a novel glycosidic macrolide from the marine cyanobacterium *Lyngbya bouillonii*. *Journal of Natural Products* 65:925–28.

Wu, M., K. E. Milligan, and W. H. Gerwick. 1997. Three new malyngamides from the marine cyanobacterium *Lyngbya majuscula*. *Tetrahedron* 53:15983–990.

Yotsu-Yamashita, M., R. L. Haddock, and T. Yasumoto. 1993. Polycavernoside A: a novel glycosidic macrolide from the red alga *Polycavernosa tsudai* (*Gracilaria edulis*). *Journal of American Chemical Society* 115:1147–48.

Yotsu-Yamashita, M., T. Seki, V. J. Paul, H. Naoki, and T. Yasumoto. 1995. Four new analogs of Polycavernoside A. *Tetrahedron Letters* 36:5563–66.

9 The Exploitation of Marine Biosurfactants in India

Camelia Bhattacharyya and Sumitra Datta

CONTENTS

INTRODUCTION

Biosurfactants are anionic or neutral, amphiphilic compounds possessing both hydrophobic and hydrophilic domains, which act on the surface of a growth medium to decrease the surface tension and interfacial tension in it (Nayarisseri et al. 2018; Santos et al. 2016; Cameotra et al. 2010). They form a micelle structure between two materials with their hydrophilic head and hydrophobic tail to finally form emulsions (Costa et al. 2010). Biosurfactants can be categorized into glycolipids, lipoproteins, lipopeptides, fatty acids, phospholipids, and neutral lipids (low molecular mass molecules), as well as polymeric biosurfactants (high molecular mass molecules) (Desai and Banat 1997; Petrikov et al. 2013) and have high affinity for polar as well as non-polar compounds (Das et al. 2014). They have been exploited by several industries, starting from the food industry to the cosmetic industry as well as the drug industry (Kiran et al. 2017). These compounds are planned to be used in bioremediation as an alternative to synthetic methods in the near future. This would make the process less toxic and more environmentally friendly (Das et al. 2014) since biosurfactants are produced as a metabolic by-product by specific microorganisms either on the surface or excreted extracellularly. The process for the degradation of biological products by biosurfactants takes place through several oxidative processes like the tricarboxylic cycle. Oxygenases and peroxidases take part in the incorporation of the hydrogen molecule and step by step take the whole moiety to a level of biomass, thus increasing the efficacy of the degradation (Behera and Prasad 2020).

The very first evidence of biosurfactants isolated from India can be traced back to the 1990s when Das et al. published papers on the production of biosurfactants by *Micrococcus* sp. on n-alkanes and sugars (Das 1993; Das et al. 1990; Das et al. 1998). Our present chapter focuses on the advances in this field of biotechnology since then. We have mentioned the different coastal regions from where biosurfactants for several applications and studies have been isolated and also the roles these biosurfactants play in establishing a more modern and green way of problem-solving in certain areas of the civilized world.

MARINE BIOSURFACTANTS FROM COASTAL AREAS OF INDIA

India, with its vast coastline and a variety of climatic features differing from east to west, has a great diversity of flora and fauna, thus allowing the growth of several microorganisms. Several of these microorganisms have been found to thrive in extreme climatic conditions like salt lakes, lagoons, biodiversity hotspots, and so on, thus receiving the name extremophiles (Thombre and Mangrola 2022). All these microbes, both among extremophiles as well as mesophiles, are thus used for several studies, including the study of biosurfactants (Citarasu 2021). These biosurfactants isolated from the coastal areas of the country can thus be classified into three groups, based on the three well-known coastlines of the country. Mangroves are a rich breeding ground for several faunae and cover a major area of India—the Sundarbans mangroves, the Mahanadi mangroves, the Krishna-Godavari mangroves, the Cauvery deltaic mangroves, the Andaman and Nicobar Island mangroves, the Goa mangroves, the Ratnagiri mangroves, and the mangroves of Gujarat. These faunae contain several microbes with valuable potential for human exploitation. Several halophilic and halotolerant microbes are collected, isolated, and studied from the mangroves of the Western and Eastern ghats of India and have been found to be exploitable for several microbial and bioremediation studies (Keerthi et al. 2018). The biosurfactants produced from such microbes would be capable of bioremediation of oil spills in extreme conditions of temperature, humidity, and salinity.

BIOSURFACTANTS ISOLATED FROM THE WESTERN COAST OF INDIA

The tropical humid climate of the western ghats is home to several microbe flora and fauna. This supports much research on the organisms of the area. In one study, a marine bacterium, *Bacillus licheniformis* LRK1, was collected and isolated from a waterlogged region at Bhavnagar in Gujarat by Nayak et al. (2020) and was studied for biosurfactant production and bioremediation. The bacteria could produce biosurfactants and was able to perform bioremediation on marine oil spills, showing new hope for a greener way of cleaning oil spills and protecting life under the sea.

Mumbai is the industrial capital of India, mostly because of its harbor. Mohanram et al. (2016), in their study, isolated 19 bacterial strains from oil-polluted areas of the Mumbai harbor. These strains isolated using a modified Bushnell–Haas medium were found to be capable of being surface-active agents (SAAs). This shows the prevalence of using these different organisms in bioremediation. Marine sponges too can be great sources of biosurfactants. *Planococcus* sp. MMD26 isolated from the sponges present on the Vizhinjam coast has exhibited ability for environment cleaning and bioremediation (Hema et al. 2019). This opens the gateway of science to a new area of environmental management.

BIOSURFACTANTS ISOLATED FROM THE EASTERN COAST OF INDIA

Lipopeptide biosurfactants can be used in several applications, one of them being food production. A recent study by Ravindran et al. (2022) shows the use of *Bacillus licheniformis* MS48 for the isolation of a lipopeptide biosurfactant from Pondicherry that has the ability to improve the survival rate of the probiotics used in the production of yoghurt, thus improving its flavor, shelf life, sustainability, extracellular polymeric substance (EPS) production and texture while lowering its syneresis. A similar study of *Pseudomonas aeruginosa* PBSC1 isolated from the Pichavaram mangroves of Tamil Nadu was found to serve the purpose of producing stable biosurfactants under minimal salt medium (MSM). These biosurfactants further showed hydrocarbon-degrading properties with high tolerance through a range of high and low temperatures, pH and salinity (Anna and Parthasarathi 2014). Markande and group, in their study, isolated the AM1 strain of the bacterial strain of *S. silvestris* from the Vellar estuary of Tamil Nadu. This strain showed high bio-emulsifying properties against a broad range of hydrocarbons. This strain was also tested against the strain of *A. calcoaceticus* RAG-1, and AM1 was established to be a better emulsifier than RAG-1 (2013).

The southeastern coast of India experiences eastern winds, thus ensuring a moist climate that favors the growth of several microorganisms. In a study on a yeast strain isolated from the southern coasts of India, Srinivasan et al. proved for the very first time the capability of *Wickerhamomyces anomalus* strain MSD1 as a non-*Saccharomyces* yeast to be used as a biofertilizer and biostimulant due to its ability for nitrogen fixation, ammonia production and nutrient maintenance, thus allowing proper growth of a crop while also reducing stress (2021). In a study by Keerthi and group, water samples of different salt concentrations were collected from the southeastern coast of India (Bheemunipatnam, Parawada, Kakinada, and Iskapalli) and checked for different types of heterotrophic co-inhabitants of the halophilic microbes *Dunaliella salina* (2018). The metabolic pathways of these microbes were studied, and the pigmentations were checked with the biochemical components of the organism. This study further enhances the possibilities of using more of such extremophiles in the production and industrial exploitation of biosurfactants.

The increase in the concentration of mercury through different levels of the food web has been observed. This increase in the level of mercury at each stage of the web leads to increases in toxicity in the environment, mostly the marine environment. Certain bacteria have been found to transform and detoxify mercury by oxidation, reduction, methylation, and demethylation. These biotransformations are able to happen in humans as well, so a better understanding of the process is required to design a process that can help reduce Hg^{2+} toxicity inside the human body. De (2014) showed the importance of *B. cereus* BW-03 in bioremediation in a similar process so that mercury-resistant bacteria could absorb the mercury from ocean water, rendering the water clean. *Pseudomonas aeruginosa* KVD-HR42 isolated from a sample from the Krishna estuary of Andhra Pradesh showed very high emulsifying capability. This suggests a future approach toward remediation (Deepika et al. 2016).

While mentioning the vast store and availability of microorganisms on the coasts, the Andaman and Nicobar Islands cannot be ignored; the land comprises ecosystems of unique characteristics due to its geographic location in the Bay of Bengal in the northeast where the influence of the western Pacific is still felt. Sponges and several other marine species from the islands are under research to produce biosurfactants, thus establishing a proper way to exploit these great natural resources (Vinithkumar et al. 2008).

APPLICATIONS OF MARINE BIOSURFACTANTS IN INDIA

Biosurfactants cover a large area of the industrial application when it comes to finding a greener and cleaner way of serving the needs of daily life as well as the luxury needs of the population as a whole. From bioremediation to agriculture, food, textiles, cosmetics, and more, biosurfactants have been studied in every field and are yet to be exploited in the rest. Figure 9.1 shows the different areas where biosurfactants are exploited.

Polycyclic aromatic hydrocarbons (PAHs), produced due to incomplete combustion of organic substances, are harmful to the environment yet are difficult to remove due to their low water solubility and high hydrophobicity, which encourage the biochemical to easily adsorb on soils and sediments (Kumar et al. 2021; Dai et al. 2022). Bioremediation using *Pseudomonas* sp., *Rhodococcus* sp., *Paenibacillus* sp., *Mycobacterium* sp., and *Haemophilus* sp. has been performed to purify such soil successfully (Premnath et al. 2021). Heavy metals (HMs) are also very harmful and result in pollution; concern about the same has resulted in studies on cyanobacteria and its ability for the bioremediation of HMs. Cyanobacteria have been genetically modified for a greener and cleaner method of environment cleaning for both PAHs and HMs (Mandal et al. 2021; Chakdar et al. 2022; Chakraborty and Das 2014). Heavy metals are harmful to

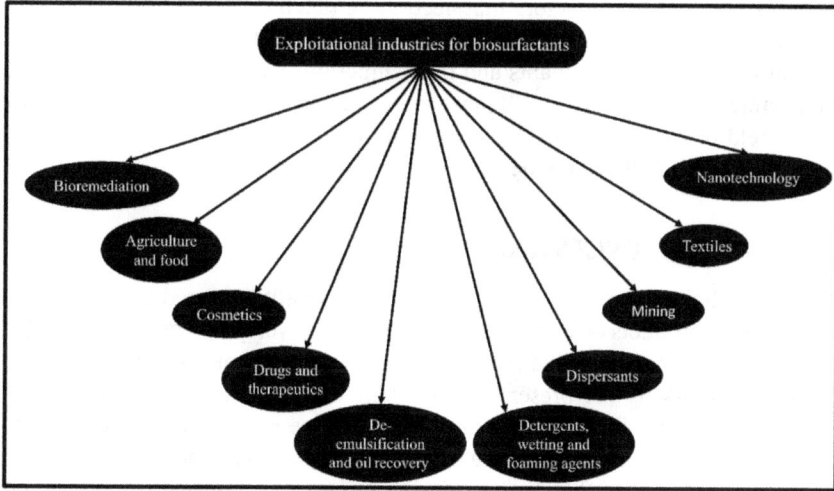

FIGURE 9.1 The various industries that use biosurfactants.

the environment; biosurfactants are successfully used to clean this type of pollution too through micelle formation, thus reducing the energy in the interface (Yadav et al. 2019). Soil remediation and plant pathogen removal through biosurfactants are now used in agriculture for better yields (Figure 9.1). Several other studies have shown the use of these biosurfactants in cosmetics, mining, textiles, and so on.

JOURNEY OF MARINE BIOSURFACTANTS FROM LAB TO INDUSTRY

The exponential increase in the rate of population has resulted in an increase in the demand for energy. This demand cannot be fulfilled unless the wastage and misuse of the already available resources are limited. Thus, the recovery of oil is much needed. This is when microbial-enhanced oil recovery (MEOR) comes to light in the form of biosurfactants, biomass, biopolymers, bio-acids, bio-solvents, biogases, and bio-emulsions to enhance the production of biofuels from depleted reservoirs. The recovery efficiency of this process can be calculated as the product of microscopic oil displacement efficiency and volumetric sweep efficiency (Tatar 2018). This process is highly preferred in a fast-moving world like ours; it is now being implemented in several developing countries like India. Oil pollution can be hazardous for the lives of marine species, ranging from the simplest to the most complicated diseases of their systems. Biotransformation is thus used for the purpose of bioremediation. Currently, the process gets approval on a very small scale, but scientists and industrialists are looking forward to using this economical technique on a large scale soon

(Speight 2018). Certain factors that need to be taken care of while implementing biosurfactants in bioremediation are adsorption, reabsorption, diffusion, and dissolution of the contaminants and pH, temperature, and aeration levels of the environment (Dhanya 2021). All these factors, alongside the never-ending study in the field of biosurfactants, have made improvement in this field even easier and far better than it used to be.

FUTURE CHALLENGES AND PROSPECTS

The main challenge now is to construct proper quality-assured laboratories to manufacture products on an industrial scale. For a developing country like India, a one-time investment in the proper laboratory equipment for bulk production would mean risking the present of the country for a better future. Thus, production and proper handling would take time to be established in the years to come. For now, research in the area of biosurfactants has already paved the way; the implementation of this research in the proper direction with cost-effective planning is the most vital procedure now. In these facilities, once created, biosurfactants would be the most useful microbe or safety component to be used in research.

REFERENCES

Anna, J. P., and R. Parthasarathi. 2014. Production and Characterization of Biosurfactant from *Pseudomonas aeruginosa* PBSC1 Isolated from Mangrove Ecosystem. *African Journal of Biotechnology* 13, no. 33 (August): 3394–3401. https://academicjournals.org/journal/AJB/article-full-text-pdf/6BEFAD946623.

Behera, B. K. 2020. Aqueous-Phase Conservation and Management. In *Environmental Technology and Sustainability*. Ed. R. Prasad, 73–141. Elsevier.

Cameotra, S. S., and R. S. Makkar. 2010. Biosurfactant-Enhanced Bioremediation of Hydrophobic Pollutants. *Pure and Applied Chemistry* 82, no. 1 (January): 97–116. www.degruyter.com/document/doi/10.1351/PAC-CON-09-02-10/html?lang=en.

Chakdar, H., S. Thapa, A. Srivastava, and P. Shukla. 2022. Genomic and Proteomic Insights into the Heavy Metal Bioremediation by Cyanobacteria. *Journal of Hazardous Materials* 424 (February): 127609. www.sciencedirect.com/science/article/abs/pii/S0304389421025772.

Chakraborty, J. 2014. Biosurfactant-Based Bioremediation of Toxic Metals. In *Microbial Biodegradation and Bioremediation*. Ed. S. Das, 167–201. Elsevier.

Citarasu, T. 2021. Biosurfactants from Halophilic Origin and Their Potential Applications. In *Green Sustainable Process for Chemical and Environmental Engineering and Science*. Ed. E. Thirumalaikumar, P. Abinaya, M. M. Babu, and G. Uma, 489–521. Elsevier.

Costa, S. G. V. A. O., M. Nitschke, F. Lépine, E. Déziel, and J. Contiero. 2010. Structure, Properties and Applications of Rhamnolipids Produced by *Pseudomonas aeruginosa* L2–1 from Cassava Wastewater. *Process Biochemistry* 45, no. 9 (September): 1511–16. https://reader.elsevier.com/reader/sd/pii/S1359511310002 14X?token=7957F892AED7EE84E2718987C9977CDE0B9078011BCFC638E1 2C01E7BC863F717CBB2820DCC6BB802ACD514ECF63B294&originRegion =eu-west-1&originCreation=20220406160055.

Dai, C., Y. Han, Y. Duan, et al. 2022. Review on the Contamination and Remediation of Polycyclic Aromatic Hydrocarbons (PAHs) in Coastal Soil and Sediments. *Environmental Research* 205 (April): 112423. www.sciencedirect.com/science/article/abs/pii/S0013935121017242?via%3Dihub.

Das, M. 1993. Measurement of Surface Tension of a Microbial Culture Producing Biosurfactants on n-Alkanes: A Critical Analysis of the Current Methodology. *Journal of Surface Science and Technology* 9, no. 1–4: 47–55. www.informatics-journals.com/index.php/jsst/article/view/1821.

Das, M., S. K. Das, and R. K. Mukherjee. 1998. Surface Active Properties of the Culture Filtrates of a Micrococcus Species Grown on N-Alkanes and Sugars. *Bioresource Technology* 63, no. 3 (March): 231–35. www.sciencedirect.com/science/article/abs/pii/S0960852497001338.

Das, M., R. K. Mukherjee, and M. M. Chakrabarty. 1990. Production of Surfactants by a Micrococcus Species. *Journal of Surface Science and Technology* 6: 259–264.

Das, P., X.-P. Yang, and L. Z. Ma. 2014. Analysis of Biosurfactants from Industrially Viable *Pseudomonas* Strain Isolated from Crude Oil Suggests How Rhamnolipids Congeners Affect Emulsification Property and Antimicrobial Activity. *Frontiers in Microbiology* 5 (December). www.frontiersin.org/articles/10.3389/fmicb.2014.00696/full.

De, J. 2014. Mercury Pollution and Bioremediation—A Case Study on Biosorption by a Mercury-Resistant Marine Bacterium. In *Microbial Biodegradation and Bioremediation*. Ed. H. R. Dash and S. Das, 137–66. Elsevier.

Deepika, K.V., S. Kalam, P. R. Sridhar, A. R. Podile, and P. V. Bramhachari. 2016. Optimization of Rhamnolipid Biosurfactant Production by Mangrove Sediment Bacterium *Pseudomonas aeruginosa* KVD-HR42 Using Response Surface Methodology. *Biocatalysis and Agricultural Biotechnology* 5 (January): 38–47. www.sciencedirect.com/science/article/abs/pii/S1878818115300189.

Desai, J. D., and I. M. Banat. 1997. Microbial Production of Surfactants and Their Commercial Potential. *Microbiology and Molecular Biology Reviews* 61, no. 1 (March): 47–64. www.ncbi.nlm.nih.gov/pmc/articles/PMC232600/.

Dhanya, M. S. 2021. Biosurfactant-Enhanced Bioremediation of Petroleum Hydrocarbons: Potential Issues, Challenges, and Future Prospects. In *Bioremediation for Environmental Sustainability*, 215–50. Elsevier.

Hema, T., G. S. Kiran, A. Sajayyan, A. Ravendran, G. G. Raj, and J. Selvin. 2019. Response Surface Optimization of a Glycolipid Biosurfactant Produced by a Sponge Associated Marine Bacterium *Planococcus* Sp. MMD26. *Biocatalysis and Agricultural Biotechnology* 18 (March): 101071. www.sciencedirect.com/science/article/abs/pii/S1878818118309460.

Keerthi, S., U. D. Koduru, S. S. Nittala, and N. R. Parine. 2018. The Heterotrophic Eubacterial and Archaeal Co-Inhabitants of the Halophilic Dunaliella Salina in Solar Salterns Fed by Bay of Bengal along South Eastern Coast of India. *Saudi Journal of Biological Sciences* 25, no. 7 (November): 1411–19. www.sciencedirect.com/science/article/pii/S1319562X15002600.

Kiran, G. S., S. Priyadharsini, A. Sajayan, G. B. Priyadharsini, N. Poulose, and J. Selvin. 2017. Production of Lipopeptide Biosurfactant by a Marine *Nesterenkonia* Sp. and Its Application in Food Industry. *Frontiers in Microbiology* 8 (June): 1138. www.frontiersin.org/articles/10.3389/fmicb.2017.01138/full.

Kumar, M., N. S. Bolan, S. A. Hoang, et al. 2021. Remediation of Soils and Sediments Polluted with Polycyclic Aromatic Hydrocarbons: To Immobilize, Mobilize, or

Degrade? *Journal of Hazardous Materials* 420 (October): 126534. www.sciencedirect.com/science/article/abs/pii/S0304389421014990.

Mandal, A., A. Dutta, Reshmi Das, and J. Mukherjee. 2021. Role of Intertidal Microbial Communities in Carbon Dioxide Sequestration and Pollutant Removal: A Review. *Marine Pollution Bulletin* 170 (September): 112626. www.sciencedirect.com/science/article/abs/pii/S0025326X21006603?via%3Dihub.

Markande, A. R., S. R. Acharya, and A. S. Nerurkar. 2013. Physicochemical Characterization of a Thermostable Glycoprotein Bioemulsifier from *Solibacillus silvestris* AM1. *Process Biochemistry* 48, no. 11 (November): 1800–1808. www.sciencedirect.com/science/article/abs/pii/S1359511313004789.

Mohanram, R., C. Jagtap, and P. Kumar. 2016. Isolation, Screening, and Characterization of Surface-Active Agent-Producing, Oil-Degrading Marine Bacteria of Mumbai Harbor. *Marine Pollution Bulletin* 105, no. 1 (April): 131–38. www.sciencedirect.com/science/article/abs/pii/S0025326X16300984.

Nayak, N. S., M. S. Purohit, D. R. Tipre, and S. R. Dave. 2020. Biosurfactant Production and Engine Oil Degradation by Marine Halotolerant *Bacillus licheniformis* LRK1. *Biocatalysis and Agricultural Biotechnology* 29 (October): 101808. www.sciencedirect.com/science/article/abs/pii/S1878818120304916.

Nayarisseri, A., P. Singh, and S. K. Singh. 2018. Screening, Isolation and Characterization of Biosurfactant Producing *Bacillus subtilis* Strain ANSKLAB03. *Bioinformation* 14, no. 06 (June): 304–314. www.ncbi.nlm.nih.gov/pmc/articles/PMC6137570/pdf/97320630014304.pdf.

Petrikov, K., Y. Delegan, A. Surin, et al. 2013. Glycolipids of *Pseudomonas* and *Rhodococcus* Oil-Degrading Bacteria Used in Bioremediation Preparations: Formation and Structure. *Process Biochemistry* 48, no. 5–6 (May): 931–35. https://pubag.nal.usda.gov/catalog/1140179.

Premnath, N., K. Mohanrasu, R. G. R. Rao, et al. 2021. A Crucial Review on Polycyclic Aromatic Hydrocarbons—Environmental Occurrence and Strategies for Microbial Degradation. *Chemosphere* 280 (October): 130608. www.sciencedirect.com/science/article/abs/pii/S0045653521010791?via%3Dihub.

Ravindran, A., G. S. Kiran, and J. Selvin. 2022. Revealing the Effect of Lipopeptide on Improving the Probiotics Characteristics: Flavor and Texture Enhancer in the Formulated Yogurt. *Food Chemistry* 375 (May): 131718. www.sciencedirect.com/science/article/abs/pii/S0308814621027242.

Santos, D., R. Rufino, J. Luna, V. Santos, and L. Sarubbo. 2016. Biosurfactants: Multifunctional Biomolecules of the 21st Century. *International Journal of Molecular Sciences* 17, no. 3 (March): 401. www.mdpi.com/1422-0067/17/3/401.

Speight, J. G. 2018. Bioremediation of Marine Oil Spills. In *Introduction to Petroleum Biotechnology*. Ed. N. S. El-Gendy, 419–70. Elsevier.

Srinivasan, R., S. R. Krishnan, K. S. Ragunath, et al. 2021. Prospects of Utilizing a Multifarious Yeast (MSD1), Isolated from South Indian Coast as an Agricultural Input. *Biocatalysis and Agricultural Biotechnology* 39 (January): 102232. www.sciencedirect.com/science/article/abs/pii/S1878818121003285.

Tatar, A. 2018. Microbial Enhanced Oil Recovery. In *Fundamentals of Enhanced Oil and Gas Recovery from Conventional and Unconventional Reservoirs*, 291–508. Elsevier.

Thombre, R. S. 2022. Microbial Diversity and Ecology of Saline Environments from India. In *Microbial Diversity in Hotspots*. Ed. A. V. Mangrola, 45–59. Elsevier.

Vinithkumar, N. V. 2008. Marine Ecosystems of Andaman and Nicobar Islands—Species Abundance and Distribution. In *Biodiversity and Climate Change Adaptation in Tropical Islands*. Ed. T. Sathish, A. K. Das, et al., 217–56. Elsevier.

Yadav, A. N. 2019. Metabolic Engineering to Synthetic Biology of Secondary Metabolites Production. In *New and Future Developments in Microbial Biotechnology and Bioengineering*. Ed. D. Kour, K. L. Rana, et al., 279–320. Elsevier.

10 Advances in Biosurfactant Production from Marine Waste and Its Potential Application in the Marine Environment

Chiamaka Linda Mgbechidinma, Xiaoyan Zhang, Guiling Wang, Edidiong Okokon Atakpa, Lijia Jiang and Chunfang Zhang

CONTENTS

DOI: 10.1201/9781003307464-10

INTRODUCTION

Surfactants are compounds with a diverse and significant role in various industries, including petroleum, agricultural, food, pharmaceutical, and the environment (Martins and Martins, 2018; Zhu et al., 2020). Over the past century, the
global market of surfactants in both production and consumption has been on
a continuous increase (Farias et al., 2021). However, most of these surfactants
are petrochemical based with characteristic properties like partial biodegradability, detrimental environmental impact, and toxicity to living organisms
(Kumar et al., 2021). Accordingly, it is recommended to investigate equally
efficient but eco-friendly alternatives. Thus, there are numerous studies on
biosurfactant derivation from microorganisms, mainly bacteria, filamentous
fungi, and yeasts (Geetha et al., 2018; Moshtagh et al., 2021; Narciso-Ortiz
et al., 2020).

Microbial surfactants are a structurally diverse assemblage of surface-active
agents with amphipathic moieties. These green surface-active agents are categorized by their chemical structures into five major groups: glycolipids, lipopeptides, phospholipids, fatty acids, and polymers (Baskaran et al., 2021; Farias
et al., 2021). There are also grouped based on the microbial types involved as bacterial and fungi-synthesized biosurfactants (Kumar et al., 2021). Commercially,
fungi-produced sophorolipids constitute a significant market in industries such
as Ecover (Belgium), Akzo Nobel, Saraya (Germany), and Jeneil Biotech Inc.
(United States) (Gaur et al., 2022b). The nutrient source and environmental factors such as temperature, pH, and salinity also determine the biosurfactant type
produced. Several techno-economic analyses on biosurfactant synthesis have
demonstrated that waste usage in minimizing production costs is more sustainable due to its relative availability (Kumar et al., 2021; Olasanmi and Thring,
2018). As such, recent research attempts are focused on replacing conventional
nutrient sources, including glucose, glycerol, olive oil, yeast extract, and urea,
with renewable substrates such as agro, food, side-stream, and marine wastes to
reduce production costs (Baskaran et al., 2021; Gaur et al., 2022b). Nevertheless,
marine wastes are less explored despite being generated in high volume and
accounting for about 50–70% of total production from aquatic sources (Dauda
et al., 2019; Sana et al., 2017; Venugopal, 2021). Examples of marine waste are
aquacultural by-products, seafood shells, leachates, fish by-catches, marine animal heads, viscera, and bones (Mgbechidinma et al., 2022; Venugopal, 2021).
These wastes are rich in minerals, vitamins, protein, and lipids, with nutritional
contents aligning with most biosurfactant producers' cultural medium requirements (Nurfarahin et al., 2018).

Biosurfactants are 21st-century molecules with excellent environmental compatibility, as they are biodegradable and non-ecotoxic (Gaur et al., 2022a). Several investigations on the use of biosurfactants at extreme pH, temperature, and salinity have proven effective even at low concentrations (Kumar et al., 2021). These characteristics have attracted significant attention to biosurfactants as a competitive alternative to synthetic surfactants produced by organic processes. Also, the ecological application of biosurfactants in marine environments can be attributed to their wide variety of functions, including emulsification, foaming, detergency, wetting, dispersing, solubilization, and surface modification (Feng et al., 2019; Ravindran et al., 2020; Wang et al., 2022). Biosurfactants play physiological roles in increasing the bioavailability of hydrophobic molecules involved in cellular signaling or differentiation processes, thus facilitating microbial carbon consumption (Coelho et al., 2015; Dierickx et al., 2022). This suggests that biosurfactants can improve microbial communities in diverse ecosystems by increasing the bioavailability of the substrates as an energy source in biosurfactant-microbial interactions (Galitskaya et al., 2021; Lu et al., 2019; Van Landuyt et al., 2020; Zhou et al., 2020). However, the mode of action of biosurfactants during remediation in different marine ecosystems is not fully known.

The remediation of contaminated marine ecosystems using marine biosurfactants is a means of attaining sustainability and achieving the United Nations Sustainable Development Goals (UN-SDGs) (https://sdgs.un.org/goals) linked to the environment. Nevertheless, biosurfactant supply is limited compared to demand despite its numerous applications in the marine environment and potential links to the UN-SDGs, hence the importance of researching alternative substrates and processes for biosurfactant production. This chapter describes biosurfactant production with marine waste as nutrient substrates. Recent advances in biosurfactant upstream and downstream production processes were reviewed with possible links to incorporating marine waste. The improved strategy for enhancing biosurfactant production through engineered processes with possible applications in the marine environment is comprehensively discussed. We have also identified the bottlenecks and presented future perspectives. To the best of our knowledge, a compilation study of the production and possible advancements in the derivation of marine biosurfactants as a potential linkage for achieving environmental protection has not been previously reported.

MARINE BIOSURFACTANT PRODUCTION

Marine biosurfactant production is greatly influenced by the upstream (microorganisms used, fermentable substrate, operational parameters) and downstream processes (extraction mechanisms), as shown in Figure 10.1. Marine biosurfactants are amphipathic molecules with hydrophilic compounds having positive, negative, or amphoteric charged ions, whereas hydrophobic compounds consist of long-chain fatty acids (Kumar et al., 2021; Moshtagh et al., 2021). Marine biosurfactants can be broadly classified based on their molecular weight, either low or high. Those with low molecular mass are glycolipids (rhamnolipids,

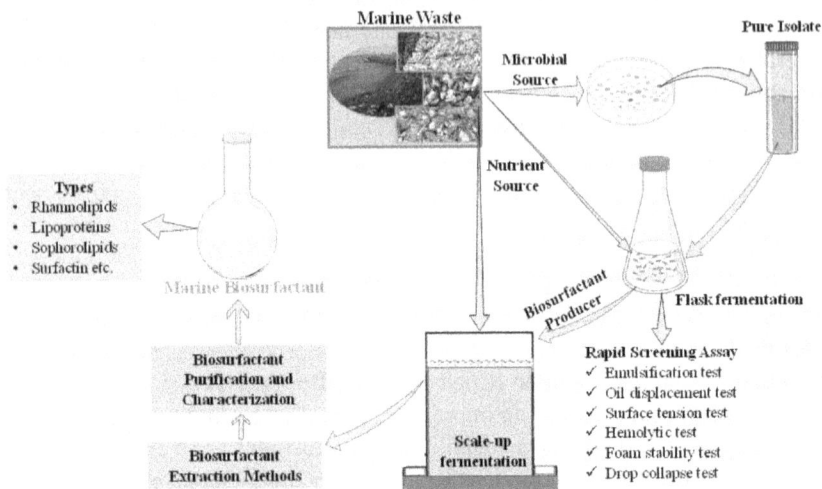

FIGURE 10.1 A typical systematic illustration of marine biosurfactant production.

sophorolipids, trehalose lipids), phospholipids (phospholipids, corinomiocolic acid, fatty acids), and lipopeptides (surfactin, wisconsin, gramicidin, subtilisin, peptide lipid, lichenysin). Meanwhile, biosurfactants with high molecular mass are polymeric (liposan, emulsan, biodispersion, mannan-lipid protein, carbohydrate lipid-protein) and particulate (vesicles).

MARINE MICROORGANISMS FOR BIOSURFACTANT PRODUCTION

Although biosurfactants are produced from several organisms, those from microorganisms are more widely studied. These microbial strains consist of bacteria, filamentous fungi, and yeasts, as described in Table 10.1, for efficient biosurfactant production. The quality and quantity of the produced biosurfactants depend on several intrinsic and extrinsic factors, including the microbial type, media supplements, and substrate constituent (Dobler et al., 2020; Jiang et al., 2021; Kumar et al., 2021). Most biosurfactant-producing microbial strains have been isolated from petroleum/oil-contaminated terrestrial ecosystems, but nowadays, these microbial isolates are also screened from several aquatic sources (Alvarez et al., 2015; Amaral et al., 2006; Balan et al., 2019, 2017, 2016; Coronel-león et al., 2015; El-Sersy, 2012; Fooladi et al., 2018; Janek et al., 2012; Khopade et al., 2012; Kiran et al., 2017, 2013, 2010, 2009; Konishi et al., 2010; Liu et al., 2010; Mani et al., 2016; Moshtagh et al., 2021; Narciso-Ortiz et al., 2020; Pele et al., 2019; Saimmai et al., 2013; Sharma et al., 2019; Shekhar et al., 2018; Wei et al., 2020a; Xu et al., 2020; Zinjarde and Pant, 2002). Moreover, biosurfactants are mainly synthesized (either intracellularly or extracellularly) during nutrient limitation at

TABLE 10.1

Marine Biosurfactant Producers and Their Fermentation Conditions

Marine Biosurfactant Producers	Isolation Source	Fermentation Condition			Biosurfactant Type	Yield	References
		Bioreactor Size/ Fermentation Volume	Inoculum Size	Condition (pH, Temp, Salinity)			
Marine Bacteria							
Pontibacter korlensis strain SBK-47	Coastal waters of Karaikal, Puducherry, India	2.1 L	10^8 CFU mL^{-1}	pH 8.0, 37°C, NaCl (0.001%)	Pontifactin	0.56 g/L	(Balan et al., 2016)
Bacillus licheniformis	Deception Island (Antarctica)	500 mL baffled flask	2% (v/v) cell suspension	pH 7.0, 30°C	Lipopeptides	NA	(Coronel-león et al., 2015)
Paracoccus sp. MJ9	Jiaozhou Bay in Qingdao, Shandong Province	NA	NA	pH 7.2–7.5, 30°C	Rhamnolipid	NA	(Xu et al., 2020)
Pseudomonas fluorescens BD5	Freshwater from the Arctic Archipelago of Svalbard	NA	NA	pH 5.5, 28°C, 10 g/L NaCl	Pseudofactin II	NA	(Janek et al., 2012)
Bacillus velezensis strain H3	Huanghai and Bohai Seas, Dalian, China	250 mL flasks	2% (v/v) inocula level	pH 7.2, 35°C, 2% NaCl	Lipopeptide	NA	(Liu et al., 2010)
Brevibacterium aureum MSA13	Southwest coast of India by SCUBA	250 mL flasks	1.7×10^5 spores	pH 7.0, 30°C, 0.9% NaCl	Lipopeptide	NA	(Kiran et al., 2010)
Nocardiopsis sp. B4	Mumbai coastal region of India	2 L	2.0 ml of spore suspension	pH 7.0, 30°C, 3% NaCl	Rhamnolipid	NA	(Khopade et al., 2012)
Bacillus subtilis N10	NA	250 mL flasks	0.5ml of stock culture	pH 7.2, 30°C, 5% NaCl	Rhamnolipid	4.2 g/L	(El-Sersy, 2012)
Inquilinus limosus KB3	South of Thailand	250 mL flasks	NA	30±3°C, 12% NaCl	Lipopeptide	5.3 g/L	(Saimmai et al., 2013)

(Continued)

TABLE 10.1

Marine Biosurfactant Producers and Their Fermentation Conditions. Continued

Marine Biosurfactant Producers	Isolation Source	Fermentation Condition			Biosurfactant Type	Yield	References
		Bioreactor Size/ Fermentation Volume	Inoculum Size	Condition (pH, Temp, Salinity)			
Marine Bacteria							
Staphylococcus saprophyticus SBPS 15	Puducherry, India	3 L	1×10^8 cfu/mL	pH 8.0, 37°C, 34 ppt NaCl	Glycolipid	1.345 ± 0.056g/L	(Mani et al., 2016)
Aneurinibacillus aneurinilyticus SBP-11	Coastal sites of Gulf of Mannar, Pamban, Tamilnadu, India	3 L	10^8 CFU mL^{-1}	pH 8.0, 37°C, 34 ppt NaCl	Lipopeptide	NA	(Balan et al., 2017)
Brevibacterium casei MSA19	Southwest coast of India	250-mL flasks	1.7×10^5 spores	pH 7.0, 30°C, 2% NaCl	Glycolipid	18 g/L	(Kiran et al., 2010)
Actinobacterial strain MSA31	Southwest coast of India	NA	1 L	pH 7.0, 28°C, 1 g NaCl	Lipopeptide	NA	(Kiran et al., 2017)
Streptomyces sp. B3	West coast of India	1 L	$2.5–3.0 \times 10^6$ cfu/ml	pH 7.0 30°C 4% (w/v) of NaCl	Glycolipid	NA	(Khopade et al., 2012)
Bacillus pumilus 2IR	Oil field	5 L	NA	pH 6–8, 40–45°C 7% (w/v) of NaCl	Rhammolipid	0.98–1.06 g/L	(Fooladi et al., 2018)
Pseudomonas stutzeri (SSASM1)	Pondicherry harbor region	250 ml	NA	pH 9.0, 35°C, 15 psu of salinity	Rhammolipid	NA	(Shekhar et al., 2018)
Acinetobacter calcoaceticus P1–1A	Off-shore oil and gas platform in Atlantic Canada	250 ml	0.5 OD 660nm	pH 7.7, 34.8°C 13.5g/L NaCl	Lipopeptide	862mg/L	(Moshtagh et al., 2021)
Bacillus amyloliquefaciens TSBSO 3.8	Virgin offshore field in the Atlantic Ocean, Rio de Janeiro, Brazil	3 L	NA	pH 8.0 28–42 C 2–3.5% NaCl	Lipopeptide	NA	(Alvarez et al., 2015)

Marine Fungi

Organism	Source	Volume	Inoculum	Conditions	Product	Yield	Reference
Rhizopus arrhizus UCP 1607	Semi-arid region of Caatinga biome in the state of Rio Grande do Norte, northeast of Brazil	100 mL in 250 mL	5% (v/v) of the spore suspension	pH 5.5, 28°C, 2% NaCl	Glycoprotein	1.74 g/L	(Pele et al., 2019)
Aspergillus ustus MSF3	Peninsular coast of India	1000 ml	NA	pH 7.0, 20°C, 3% NaCl	Glycolipoprote	NA	(Kiran et al., 2009)
Aspergillus sp. MSF1	Bay of Bengal region of the Indian peninsular coast	250 ml	NA	pH 7.0, 30°C, 2% NaCl	Rhamnolipid	NA	(Kiran et al., 2013)
Cyberlindnera saturnus SBPN-27	Coastal regions of Tamil Nadu state, India, including Porto Novo	3 L	NA	pH 6.5, 30°C, 34 ppt	Glycolipid	NA	(Balan et al., 2019)
Pseudozyma hubeiensis SY62	Calyptogena soyoae (deep-sea cold-seep clam, Shirouri-gai) at 1156 m in Sagami Bay	NA	2 mL	pH 6.0, 25°C	Mannosylerythritol lipids	30 g/l	(Konishi et al., 2010)
Meyerozyma guilliermondii YK32	Hydrocarbon-polluted locations of Hisar, Haryana	NA	0.25 g	pH 5.5, 30 $\pm$ 2°C	Glycolipids	9.1 g/L	(Sharma et al., 2019)
Yarrowia lipolytica NCIM 3589	Seawater near Mumbai, India	250 ml	2×10^9 cells/ml	pH 8.0 30°C, 2 to 3% NaCl	Bioemulsifier	0.72 mg/mL	(Zinjarde and Pant, 2002)
Yarrowia lipolytica IMUFRJ50682	Guanabara Bay in Rio de Janeiro	500 ml	NA	28°C	Carbohydrate protein complex	NA	(Amaral et al., 2006)

the late exponential or stationary growth phase. The observable properties at this stage of microbial growth, such as surface tension reduction, oil displacement, drop collapse, emulsification, and cell hydrophobicity, are the basis for the development of several rapid methods for the isolation and screening of biosurfactant-producing microbes (Huang et al., 2020; Zhou et al., 2020).

MARINE SUBSTRATES FOR BIOSURFACTANT PRODUCTION

The high production costs of biosurfactants mainly result from the high prices of feedstocks used (Baskaran et al., 2021; Dierickx et al., 2022; Farias et al., 2021; Kumar et al., 2021; Martins and Martins, 2018; Roelants et al., 2019; Shi et al., 2021). Featuring Agae Technologies, Biotensidon, Ecover United Kingdom, Evonik, Jeneil Biotech, Logos Technologies, Saraya, and Soliance, the global biosurfactant market is estimated to reach a worth of US$6.5 billion by 2027, growing at a 5.3% compound annual growth rate (CAGR) from 2020–2027, despite the COVID-19 crisis (www.researchandmarkets.com/). However, Agae Technologies (www.agaetech.com/) reported a market price of pure rhamnolipids greater than the price of chemical surfactants. This is similar to the claim by Olasanmi and Thring (2018) that biosurfactants are 20–30% more expensive than synthetic surfactants. Therefore, considerable efforts are required to develop an economical and sustainable production process to increase biosurfactants' cost competitiveness to encourage their large-scale applicability.

In previous reports, alternative low-cost feedstocks such as agro-industrial/food wastes have been widely explored and contributed massively to global biosurfactant production (Gaur et al., 2022b; Radzuan et al., 2017; Rane et al., 2017). However, nutrient-rich marine wastes (Dauda et al., 2019; Mgbechidinma et al., 2022; Sana et al., 2017; Venugopal, 2021) are yet to receive substantial attention in the global biosurfactant market. Figure 10.2 reveals a potential mechanism of crude fish oil utilization as a nutrient source by biosurfactant producers. This can provide a new frontier for utilizing other marine waste as a carbon and nitrogen source for microbial growth and biorefinery.

Carbon Sources

The aquacultural (seafood and fishery) industries are highly valued in the food sector. However, they generate over 30–80% of their entire products as waste during processing (Mgbechidinma et al., 2022; Zhu et al., 2020), thus posing significant environmental and health problems. These wastes are a rich carbon source consisting mainly of oil and fat, which can serve as nutrients for biosurfactant producers since hydrophobic substrates are reportedly known to favor surfactant yield (Baskaran et al., 2021; Pele et al., 2019; Saimmai et al., 2013). The number of carbons strongly influences the type of biosurfactant produced and the associated properties (Nurfarahin et al., 2018). Biosurfactant-producing microorganisms usually show varying susceptibility to carbon groups used for growth and metabolite secretion. According to Nurfarahin et al. (2018), the

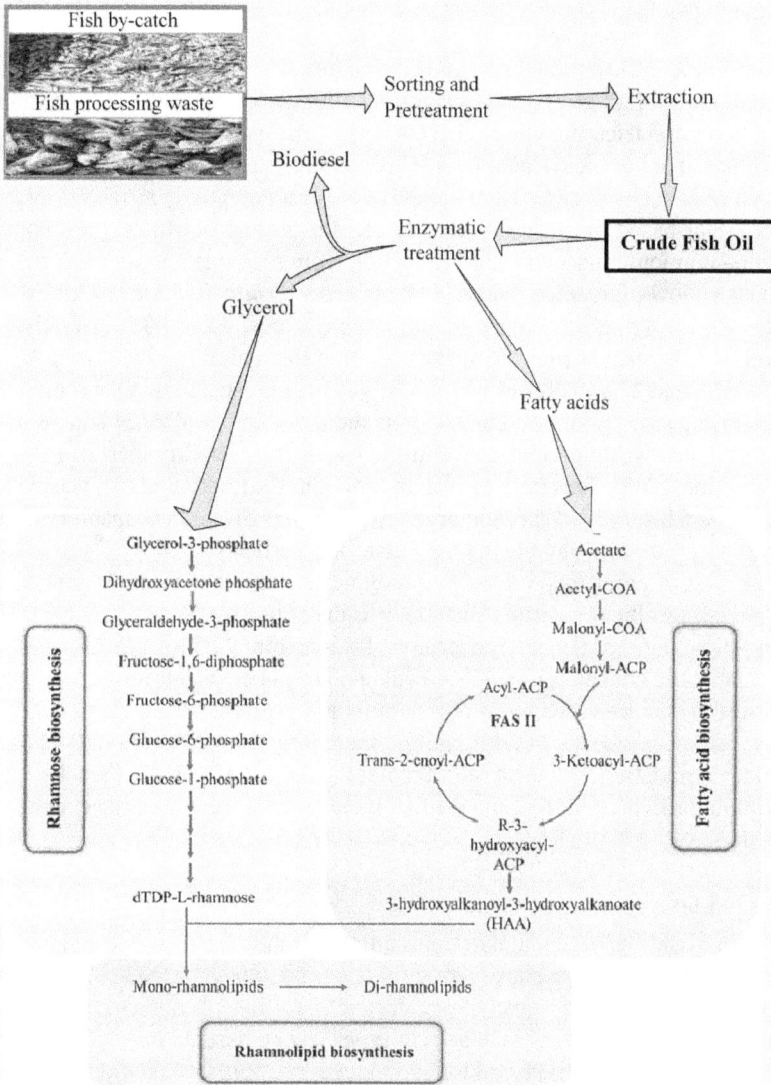

FIGURE 10.2 Potential mechanism of fish oil utilization as a nutrient source by biosurfactant producers.

utilization of hydrocarbons by biosurfactant producers follows a specific pattern of linear alkanes > branched alkanes > small aromatics > cyclic alkanes.

Unlike the sugar-based substrates that undergo the glycolysis pathway, the fatty acid content of marine waste can feed into the central carbon metabolism at the level of glyceraldehyde-3-phosphate without employing the pentose phosphate pathway (Figure 10.2). However, during cell maintenance, the pentose phosphate pathway is activated through a redox cofactor synthesis leading to carbon dioxide (CO_2) production and carbon loss. CO_2 production is also experienced during a high growth rate associated with complete oxidation through the acetyl-CoA and tricarboxylic acid (TCA) cycle. Pathways requiring more energy contribute to lower biosurfactant yield (Nurfarahin et al., 2018). Thus, a marine-derived lipid such as crude fish oil could serve as a potentially attractive alternative to sugar-based substrates (Figure 10.2). Martins and Martins (2018) reported significant anionic biosurfactant production from a convex, non-hemolytic, and slightly yellow-pigmented bacteria (*Corynebacterium aquaticum*) using fish waste with 13.8% of fats and 0.5% of carbohydrates. After 48 h of cultivation with 3 and 5% fish waste, the surface tension decreased by 25% from 37.4 to 36.9 mN/m and 27.8 to 28.1 mN/m. The study also presented that the high initial emulsifying activity followed by a drastic reduction to 0% after 24 and 48 h was due to the fish waste protein concentration with functional properties (Martins and Martins, 2018). This implies that an individual extraction of fish waste components (lipid, carbohydrate, and protein) will better reveal the capability of biosurfactant producers while using fish waste as a carbon source.

The first attempt to produce rhamnolipid, a prevalent biosurfactant, using *Catla catla* fish fat as a cheap substrate, was described by (Sana et al., 2017). The rhamnolipid reduced the surface tension of water from 72.5 to 32.7 mN/m, with a critical micelle concentration (CMC) value of 46 mg/L. The low CMC observed indicated the hydrophobicity of the rhamnolipid produced from *Catla catla* fish fat by *Pseudomonas* sp. In addition, the rhamnolipid concentration (0.421 g/L), fish fat degradation (38.56%), and biomass growth (0.0758 g L) reached their maximum level after 72 h of incubation, which was at the end of the exponential growth phase.

Nitrogen Sources

Nitrogen is another essential nutrient required for microbial growth and metabolism. Aquacultural wastes are rich in proteins, peptides, amino acids, and their co-products, which are mostly organic and significantly support cell growth (Nurfarahin et al., 2018). These protein types can be derived through isoelectric solubilization precipitation, mild acid-induced gelation, flocculation, and enzymatic hydrolysis (Venugopal, 2021). However, during biosurfactant production, nitrates are more easily reduced to nitrite before turning into ammonium, which is assimilated to form glutamate by glutamate dehydrogenase or to form glutamine by glutamine synthetase. Glutamine and α-ketoglutarate are converted to glutamine by l-glutamine 2-oxoglutarate aminotransferase (GOGAT). These nitrate reduction and glutamine-glutamate processes have inhibitory effects on microbial metabolism (Nurfarahin et al., 2018), leading to the formation of the lipid moiety rather than the sugar moiety. This explains why biosurfactants such as rhamnolipids are mainly produced at high yields during nitrogen-limiting conditions. Thus, nitrogen is a rate-determining factor for biosurfactant production as linked to high yield during the late exponential and stationary growth phase of the producing cell.

Biosurfactants from fish proteins and peptides exhibit great potential in production cost reduction. Zhu et al. (2020) produced a lipopeptide using protein hydrolysate derived from fish cod liver and head wastes. The critical micellar dilutions of the produced lipoprotein from the peptones of the fish liver and head waste were 54.72 and 47.59, respectively. Lipoprotein generated by fish liver peptone had a lower CMC of 0.18g/L and reduced the surface tension of water to 27.9 mN/m. The study revealed the biosurfactant production capability of *Bacillus subtilis* N3–1P and reflected fish waste-based peptones as a suitable nitrogen source for bacteria growth (Zhu et al., 2020). However, aside from the potential use of marine leachate as a carbon source due to its large amount of organic contaminants and hydrocarbons suspended solids, leachate is also rich in inorganic salts and ammonia (Oliva et al., 2021). Hence, yeast extract, urea, and other biosurfactant production media compositions can likely be replaced with marine leachates.

Other Nutrient Sources

Apart from the carbon and nitrogen sources, phosphorus, multivalent ions (Mg^{2+}, Ca^{2+}), trace element salts, and the carbon to nitrogen (C/N) ratio are important parameters in formulating biosurfactant cultivation medium (Jiang et al., 2020). The latter is more relevant in the use of marine waste and describes the proportion of carbon and nitrogen required by different biosurfactant producers (Kumar et al., 2021) depending on the microbial type used, the culture conditions, and the desired product type (Nurfarahin et al., 2018). Studies have shown that nitrogen limitation conditions cause microorganisms to produce higher biosurfactants and alter the product's composition (Nurfarahin et al., 2018). Also, high C/N ratios (i.e., low nitrogen levels) restrict bacterial growth, favoring cell metabolism to produce metabolites. Although several optimal C/N values have been reported for conventional and agro-waste, there is limited information on the C/N kinetics of marine wastes during biosurfactant production.

ADVANCES IN BIOSURFACTANT PRODUCTION

Besides selecting over-producers and using marine waste streams for biosurfactant production, optimizing the fermentation parameters is also a sustainable means of enhancing the upstream process. For example, Zhu et al. (2020) optimized feedstock (fish waste peptones) utilization for lipopeptide production using response surface methodology (RSM) with a central composite design (CCD). Similarly, Sana et al. (2017) determined the most suitable C/N, C/P, and $MgSO_4$ concentrations for rhamnolipid production from fish fat using RSM. Other common efforts that have been employed to improve biosurfactant production over the past few decades include the preparation of high-yield strains, medium optimization, and the development of diverse cultivation strategies (Jiang et al., 2021).

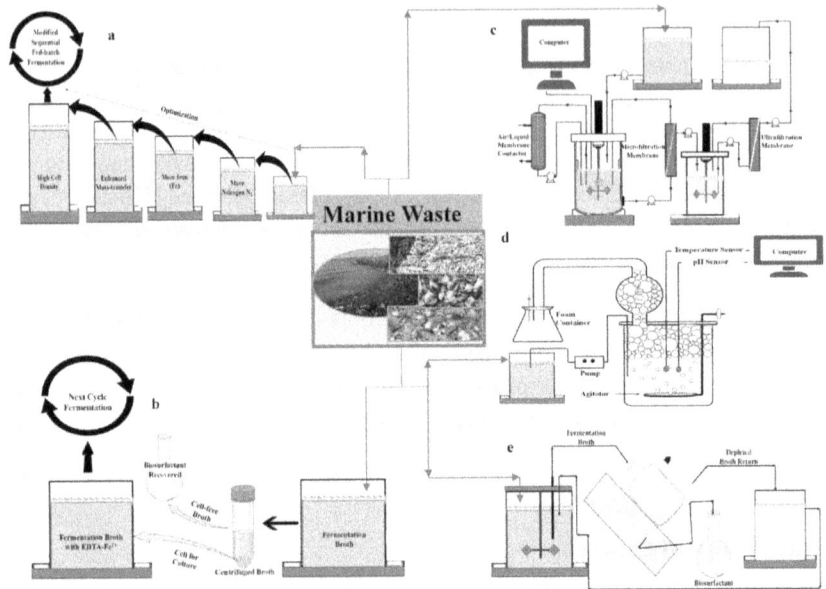

FIGURE 10.3 Advanced processes in biosurfactant production. (a) Advanced sequential fed-batch fermentation. (b) Bubbleless membrane bioreactor coupled with external hollow fiber membrane (air/liquid contactor), microfiltration membrane and ultrafiltration membrane for continuous biosurfactant production. (c) Foam fractionation. (d). Integrated gravity separation. (e) Repeated fed-batch fermentation.

Separation, extraction, and purification are cost-limiting downstream processes that account for about 60% of the total production cost of biosurfactants (Gaur et al., 2022a; Kumar et al., 2021). Most downstream processes include salt precipitation, chromatography, solvent extraction, gravity separation, acid precipitation, and ultra-filtration. Although these processes are used individually, combining them can improve the yield and purity of derived biosurfactants. Also, the biosurfactant's physicochemical properties, such as the hydrophilic-lipophilic balance (HLB), solubility, critical micelle concentration, and ionic charge, should be considered when selecting the suitable downstream process (Gaur et al., 2022a). As described in the following, advances in the production processes of biosurfactants are mostly through the development of integrated systems due to the challenges involved in recovering these metabolites, such as the negative effect of cell biomass, foam, and second phase formation (Figure 10.3).

Advanced Sequential/Repeated Fed-Batch Fermentation

This process is based on enhancing high cell density culture as an efficient way to achieve high volumetric productivity of biosurfactants during fermentation (Figure 10.3a). According to Jiang et al. (2021), a total rhamnolipid yield of 665 g/L was achieved after 18 days of cultivation, showing that $NaNO_3$, $FeSO_4·7H_2O$, healthy cell density, enhanced mass-transfer, and high dissolved oxygen content are essential parameters for the development of an advanced sequential fed-batch fermentation process. Although regular fed-batch fermentation does not improve biosurfactant production, its recovery through repeated fed-batch fermentation is a more suitable method (Figure 10.3b). The critical parameters to consider in a repeated fed-batch fermentation are the EDTA–Fe^{2+} level (low) and the cell density (high) (Mei et al., 2021). Worthy of mention is that excess $FeSO_4·7H_2O$ induces biofilm formation and inhibits biosurfactant synthesis. Also, steel bioreactors are preferable to glass flasks, and sequential/repeated fed-batch fermentation methods cannot sustain biosurfactant production indefinitely due to strain degeneration and aging of producing microbes during fermentation. However, advanced sequential/repeated fed-batch fermentation methods can be improved by using cell immobilization within magnetic particles to prevent microbial contamination during cell transfer.

Integrated Membranes/Filtration Separation

The membrane separation process relies on the selective permeability of biosurfactants in a fermentation broth and a pressure gradient to drive flow across the membrane (Figure 10.3c). Herein, microbial cells are recovered by the microfiltration membrane, while the biosurfactant-rich cell-free supernatant is retained as retentate in the ultrafiltration membrane (Dolman et al., 2019). Lipopeptides, glycolipids, and sophorolipids in scaled-up production have been explored with membrane separation techniques. Although limited to the laboratory scale, advances in this technique are channeled towards the recovery of non-phase-separating, soluble biosurfactants that are harder to recover with other methods. In addition, membrane separation is straightforward, as both fermentation and separation can be entirely decoupled after an initial cell separation with a microfiltration membrane. Even if the cell separation phase is not cost effective, the membrane separation process is suitable for soluble biosurfactants and production improvement. Roelants et al. (2019) reported using integrated ultrafiltration-based methods for selective cell removal to improve sophorolipid production with greater than 500 g/L solubilities with increased yield from 0.37 to 0.63 g/L after 1 h. Likewise, optimal lipopeptide production from 110 mg/L to 3.3 g/L per hour by *Bacillus subtilis* was achieved with 95% purity using an ultrafiltration coupled bubbleless membrane bioreactor, a strategic continuous fermentation process (Coutte et al., 2013).

Foam Fractionation

The foam fractionation process mitigates biosurfactant separation hindrance due to severe emulsification of the multiphase fermentation system while reducing the mass transfer efficiency of carbon sources (Jia et al., 2020). A coupled foam fractionation and fermentation system enables continuous biosurfactant production and separation. High rhamnolipid yield (25.8 g/L) and recovery rate (93.4%) have been reported with the foam fractionation process using *Pseudomonas aeruginosa* D1 under the conditions of the feed flow rate of animal fat hydrolysate (2.0 mL/min), gas volumetric flow rate (90 mL/min), and inoculation scale (6.0%) (Jia et al., 2020). Aside from rhamnolipids, this process has been explored in producing lipopeptides, sophorolipids, and mannosylerythritol lipids (Dolman et al., 2019). Figure 10.3d shows a typical foam fractionation setup.

Integrated Gravity Separation

Gravity separation techniques for improved biosurfactant production result in highly viscous second phase formation (Dolman et al., 2019). Key advances in using gravity separation for biosurfactant production include the recirculating system and gravity-based bioreactors (Figure 10.3e). However, there are limited reports on biosurfactant production through gravity separation.

Others

Recently, electrokinetic extraction and high-purity recovery of biosurfactants in the anode compartment were detailed by Gidudu and Chirwa (2021), with a voltage-dependent extraction efficiency. Likewise, improved sophorolipid recovery of 73.8 g/L from *Candida bombicola* ATCC 22214 was obtained using an ultrasound cell separation strategy, although a suboptimal feeding strategy for cell separation was required (Palme et al., 2010). In addition, high-efficiency rhamnolipid production (over 150 g/L) can be achieved through a sequential fed-batch fermentation approach based on a fill-and-draw strategy coupled with high cell density, nutrient replenishment, and dilution of toxic by-products (He et al., 2017). This process has a more substantial yield efficiency of biosurfactant production than the traditional batch and fed-batch processes.

ENGINEERED BIOSURFACTANT BIOSYNTHESIS

Limited by the physiological functions and genetic background of biosurfactant producers, some optimization processes described in the previous section still face some bottlenecks in yield improvement, such as transformation efficiency, fermentation substrate selection, and product types. Meanwhile, the microbial synthesis pathways of biosurfactants involve multi-step enzyme-catalyzed reactions mainly regulated by quorum sensing (QS), substrate types, and other factors at the transcription and metabolism levels that influence the fermentation process. Strain improvement by genetic engineering is being increasingly explored and expected to improve further the safety, yield, and product performance associated with biosurfactant synthesis. With the development of metabolic engineering and synthetic biology technologies, various transformation strategies have been applied to optimize biosurfactant production, such as metabolic pathway engineering, genetic engineering, fermentation engineering, enzyme engineering, directed evolution, chassis engineering, laboratory adaptive evolution, and so on.

Biosurfactant Synthesis by Genetically Engineered Microorganisms

The reference biosurfactant for this section is rhamnolipid, an extensively studied surface-active molecule in the glycolipids class. This biosurfactant consists of one or two rhamnose molecules with one or two β-hydroxyl fatty acid groups containing 8–14 carbons (Baskaran et al., 2021; Chen et al., 2021; Jiang et al., 2020; Radzuan et al., 2017; Shi et al., 2021). The industrial production of rhamnolipids mainly utilizes bacteria in the genus *Pseudomonas*. The pathogenic nature and possible limited rhamnolipid yield by *P. aeruginosa* reported in previous studies have strongly encouraged genetic engineering approaches in optimal direct synthesis of specific rhamnolipid homologs (Dierickx et al., 2022; Roelants et al., 2019). The most recent research on marine biosurfactants production has mainly reported on rhamnolipids; thus, understanding the biosynthetic pathway and regulation of rhamnolipid-related genes and enzymes is vital.

Biosynthetic Pathway and Regulation of Rhamnolipids

In 1963, Burger et al. first proposed the *de novo* biosynthetic pathway of rhamnolipids using radiolabeled precursor substances to trace enzyme extracts, the dTDP-L-rhamnose, and β-hydroxy fatty acids were considered the main precursors. Nowadays, three biosynthetic pathways have been revealed in *P. aeruginosa*. As shown in Figure 10.2, the precursor dTDP-L-rhamnose provides glycosyl moieties for rhamnolipids, while β-hydroxy fatty acid is responsible for the lipid chain in the rhamnolipids. RhlA and RhlB constitute a complex enzyme, rhamnosyltransferase 1, which can synthesize the precursor dTDP-L-rhamnose and HAA into mono-rhamnolipids, which are further transformed into di-rhamnolipids by rhamnosyltransferase 2 (RhlC) (Müller et al., 2010; Nurfarahin et al., 2018). RhlAB and RhlC catalyze three key reactions in the rhamnolipid biosynthesis. RhlAB is encoded in the operon rhlABRI with a QS-regulated expression under the control of the promoter P_rhl (Baskaran et al., 2021; Jiang et al., 2021; Kumar et al., 2021). There are four QS networks in *P. aeruginosa*, Las, Rhl, PQS, and IQS (Lee and Zhang, 2015). The encoding gene of the RhlR/I system exists as the operon rhlABRI, which directly regulates the transcriptional expression of rhlAB. QS is a cell density-based intercellular communication system that links to the microbial synthesis of rhamnolipid at the gene transcription level. Zhao et al. (2015) constructed an engineered strain *P. aeruginosa* PoprAB by replacing the original promoter of the rhl AB gene with a strong constitutive promoter of P_orpL, and the copy number of the fusion gene Popr-rhl AB was increased to improve rhamnolipid production. The engineered strain PoprAB can produce 1094.6 mg/L of rhamnolipids under anaerobic conditions, which is 60.2% more than the yield of the original strain SG (Zhao et al., 2015).

Although cell density-dependent rhamnolipid synthesis is crucial for microbes to adapt to complex environments and perform population behaviors, it does not favor industrial production since QS results in a long fermentation cycle and low conversion rate. In addition, QS-associated complex-related gene expression regulation hinders genetic engineering practice. There are nonpathogenic strains of *P. fluorescens*, *P. putida*, and *E. coli* capable of producing rhamno-lipids through heterologous approaches; these microbes are widely used as emi-nent model strains for cell factory chassis due to their better biosafety (Bator et al., 2020; Du et al., 2017). Suitable biosurfactant producers can synthesize precursors efficiently and have a high tolerance to the overexpression of RhlA, RhlB, and RhlC proteins required for rhamnolipid production (Bator et al., 2020; Gaur et al., 2022a; Jiang et al., 2021; Müller et al., 2010). Wittgens et al. (2011) found that *P. putida* can produce up to 90 g/L of rhamnolipids through a non-QS-regulated system, making the strain an effective rhamnolipid producer that is not growth dependent. This discovery has a good prospect for driving efficient industrial production of rhamnolipids. Du et al. (2017) optimized the expression of rhlAB-rhlC in *E. coli* and simultaneously introduced the rfbD gene to improve the supply of the precursor D-TDP-L-rhamnose, resulting in a 43% increase in rhamnolipid production.

APPLICATION OF BIOSURFACTANTS IN THE MARINE ENVIRONMENT

Biosurfactant properties like mobilization, solubilization, reduced surface ten-sion, and emulsification significantly support the bioremediation phenomenon in polluted marine ecosystems (Table 10.2). Studies on marine ecosystem remedia-tion using biosurfactants still remain imminent (Chen et al., 2013; Dell'Anno et al., 2018; Du et al., 2016; Jadhav et al., 2013; Kuyukina et al., 2005; Long et al., 2013; Narciso-Ortiz et al., 2020; Saeki et al., 2009; Wei et al., 2020b). Promising applications of biosurfactants include oil spill clean-up, hydrocarbon degradation, heavy metal removal, and remediation of other pollutants (Gaur et al., 2022b; Cai et al., 2021). This section narrates the role of biosurfactants in enhancing the degradation and solubilization of inorganic and organic marine pollutants.

MARINE HYDROCARBON REMEDIATION

Hydrocarbons are hydrophobic compounds categorized as saturated, unsaturated, and aromatic based on their structure (whether single bonds, double bonds, or aromatic rings). Furthermore, hydrocarbons are classified as aliphatic (alkanes, alkynes, and alkenes) and aromatic (benzene, ethylbenzene, toluene, and other polycyclic aromatic hydrocarbons) based on their origin. Hydrocarbon pollution is a common occurrence in marine ecosystems (Kumar et al., 2019; Rodrigues et al., 2020), emanating from processes such as exploration, drilling, mining, refining, transportation, and storage of crude oil (Atakpa et al., 2022; Bezza and Chirwa, 2017; Huang et al., 2020; Jemil et al., 2018; Muriel-Millán et al., 2019; Varjani et al., 2017; Wang et al., 2022; Zhou et al., 2020). These pollutants fun-damentally disrupt the aquatic food chain; pose severe health and environmental risks to humans, plants, and animals inhabiting these niches; and impede the penetration of light necessary for phytoplankton productivity. Thus, microbial

TABLE 10.2

Marine Ecosystems Prone to Pollutants and Their Degradation Mechanisms

Marine Ecosystems	Pollutant	Biosurfactant Type/ Amount	Mechanism of Action/ Functioning	Remark/Outcome	References
Intertidal zone	Lubricant oil, crude oil, diesel, kerosene	Glycolipoprotein/ 1.0mg/ml	Emulsification and biodegradation	Emulsification order was kerosene > lubricant oil > diesel > crude oil. The emulsification activity (D_{610}) of biosurfactant was 0.64 as compared to 0.31 with SDS. 56 and 90% crude oil degradation after 9 and 27 days, respectively.	(Jadhav et al., 2013)
	Kerosene	Crude rhamnolipid (glycolipid)/25 g/L	Demulsification of waste crude oil	Over 90% of dewatering efficiency on refractory waste crude oil.	(Long et al., 2013)
Wetland	Polysaccharides, proteins, nucleic acid, and lipids	Rhamnolipid (glycolipid)/0.12 g/L	Improved solubilization	Enhanced extracellular polymeric substance (EPS) dissolution and dispersion, resulting in enhanced pollutant removal.	(Du et al., 2016)
	Crude oil	Rhamnolipid (glycolipid)/ 1.4%(v/v)	Improved solubilization	Enhanced solubilization and bioavailability of crude oil from soil pores with increased contact with soil microbes for degradation.	(Wei et al., 2020b)
Marshes	Crude oil	Rhamnolipid (glycolipid)/0.1%(v/v)	Increased solubilization	80.9% TPH removal.	(Wei et al., 2020a)
	Crude glycerin	Rhamnolipid (glycolipid)/1 g/L	Increased oil remediation	70% PAH degradation.	(Dobler et al., 2020)
Coastal shorelines/ sediment	South Louisiana crude oil (SLC), Prudhoe Bay crude oil (PBC), Upper Zakum crude oil (UZKC)	Glycolipid/0.5%, 1%, 2%, 3% (v/v)	Oil dispersion, dissolution, localization at the oil/water interface and emulsification activity	Dispersant and the removal of oil from the surface of contaminated sea sand as a washing agent.	(Saeki et al., 2009)
	Crude oil from Chashkinskoe oilfield	Lipopeptide/0.72g/ml	Micelle formation and surfactant-enhanced soil washing	Enhanced oil mobilization and degradation.	(Kuyukina et al., 2005)
	Pyrene	Rhamnolipid (glycolipid) /5g/L	Enhanced contaminants desorption	70% pyrene removal.	(Dell'Anno et al., 2018)
Deep-sea	The crude oil was obtained from Shengli Oilfield of China	Rhamnolipid (glycolipid) /0.1g/L	Increased solubilization and hydrocarbon emulsification	Rhamnolipids enhanced total oil biodegradation efficiency by 5.63%.	(Chen et al., 2013)
Coral reefs and watersheds	Gulf of Mexico in Veracruz	Biosurfactant by Acenitobacter sp.	Emulsifying activity	The hydrocarbon droplet diameter decreased.	(Narciso-Ortiz et al., 2020)

remediation of polluted marine ecosystems is urgently required to revamp and restore these ecosystems to their pristine state (Table 10.3). The mechanism of petroleum hydrocarbon bioremediation in the marine environment (water and soil systems) is presented in Figure 10.4(a and b). The major bioremediation mechanisms include microbial adherence to the hydrophobic substrate, emulsification activity of biosurfactants to form micelles, hydrocarbon solubilization through reduced surface tension, segregation through repulsive forces created by biosurfactants, mobilization, decrease in the capillary force which binds the soil to hydrocarbon (specific to the soil system), and biodegradation.

MICROBIAL-ENHANCED OIL RECOVERY

In recent years, using nontoxic and biodegradable biosurfactants to enhance oil recovery has been gaining more attention in the upstream sector of the oil industry (Table 10.3). Oil as a major energy source cannot naturally be fully extracted from the subsurface reservoirs. As such, microbial-enhanced oil recovery technology has assisted engineers in extracting unrecovered oil using biosurfactants due to its extensive properties (interfacial tension, wettability, and emulsification) and environmentally friendly nature (Ambaye et al., 2021; Liu et al., 2021; Saravanan et al., 2020). Applying biosurfactant-mediated microbial-enhanced oil recovery can be achieved through two approaches (*in-situ* and *ex-situ* methods) (Geetha et al., 2018). The *in-situ* method involves the production of the biosurfactant in the reservoir through nutritive stimulation and/or injection of indigenous producers. Meanwhile, the *ex-situ* methods entails fermentative biosurfactant production in a bioreactor prior to its injection through the well to the reservoir. The *ex-situ* method is more widely employed since the effect of reservoir conditions on the biosurfactant-producing strain is sidetracked. Although biosurfactants for microbial-enhanced oil recovery are mainly screened to reduce the interfacial tension between crude oil and water, wettability alteration and emulsifying ability are more potent strategies (Kryachko, 2018).

REMOVAL OF HEAVY METALS

Anthropogenic activities and perturbations like smelting, mining, industrial and domestic waste discharge, burning of fossil fuels, and fertilizer and pesticide usage in agriculture have significantly contributed to the contamination of the marine environment by heavy metals (Yang et al., 2018). These metals pose serious environmental and health concerns due to their persistence, virulence, and toxicity, even at low concentrations. However, diverse conventional techniques have been adopted to decontaminate and remove these contaminants in the environment. Still, due to their high cost of operation, high energy, and post-disposal issues, they have become unsuitable and unfriendly in the environment (da Rocha Junior et al., 2019; Sarubbo et al., 2018). To solve this environmental problem, Table 10.3 shows the use of biosurfactants produced by strains of microorganisms in heavy metal removal (da Rocha Junior et al., 2019;

TABLE 10.3

Application of Biosurfactants for the Remediation of Potent Marine Environmental Pollutants

Marine Environmental Pollutant	Microbe-Producing Biosurfactant	Biosurfactant Produced	Effects	References
(a) Organic hydrocarbon	P. aeruginosa CH1	Glycolipid	Effective n-alkane biodegradation (272.21 to 56.93 mg/L) within 7 days.	(Huang et al., 2020).
	Enteobacter cloacae C3	Lipopeptides	Good emulsifying abilities with diesel degradation (2.0% v/v) reaching 48% after incubating for 15 days.	(Jemil et al., 2018)
	P. aeruginosa GOM1	Rhamnolipid	Degradation of long and medium chain n-alkanes (C_{12}–C_{36}).	(Muriel-Millán et al., 2019)
	Bacillus species XT-2	Lipopeptides	High emulsifying ability.	(Wang et al., 2022)
	Combination of bacteria and fungi producing biosurfactant	Lipopeptides	Petroleum hydrocarbon degradation from 23.36 to 58.61% after incubating for 7 days. Degradation of n-alkanes (3789.27 to 940.33 mg/L) and PAHs (1667.33 to 661.5 µg/L).	(Atakpa et al., 2022)
	B. cereus SPL-4	Lipopeptide	Effective in the removal of 4-ring (51.2%), 5-ring (55%), and 6-ring (55%) PAHs in soil.	(Bezza and Chirwa, 2017)
(b) Microbial-enhanced oil recovery	B. licheniformis L20	Lipopeptides	Altered wettability of the hydrophobic surface of core slices with different permeability. Stable emulsification activity in pH 4–11, 85°C, 25 wt% NaCl or 17.5 wt% $CaCl_2$ solution; 19.58% increased oil recovery rate through core flooding experiments.	(Liu et al., 2021)
	C. tropicalis	Glycolipids and lipopeptide	Stable emulsion formation, reduces tensions and capillary forces that impede oil flow through the rock pores.	(Ambaye et al., 2021)
	B. licheniformis JF-2	Not mentioned	Effective in thermotolerant and anaerobic conditions when compared to synthetic surfactants.	(Ambaye et al., 2021)
	B. subtilis	Not mentioned	Effective oil recovery from saturated sand oil column using in-situ experiments.	(Ambaye et al., 2021)
	B. subtilis	Lipopeptide	Enhanced oil recovery yield, oil immobilization and storage tanks, and oil immobilization due to stable environmental conditions.	(Ambaye et al., 2021)
	Luteimonas huabeiensis sp.	Lipopeptides	11% improved oil recovery in laboratory-based core column flooding.	(Saravanan et al., 2020)

(Continued)

TABLE 10.3

Application of Biosurfactants for the Remediation of Potent Marine Environmental Pollutants. Continued

Marine Environmental Pollutant	Microbe-Producing Biosurfactant	Biosurfactant Produced	Effects	References
	C. tropicalis	Surfactin	39.80% improved oil recovery in sand packed method.	(Saravanan et al., 2020)
	Pseudomonas sp.	Not mentioned	24.4% enhanced oil recovery in core flooding.	(Saravanan et al., 2020)
	Rhodococcus ruber	Not mentioned	8.88–25.78% enhanced oil recovery in sand package technology.	(Saravanan et al., 2020)
	B. subtilis	Surfactin isoforms	14.21% enhanced oil recovery in sand pack test.	(Saravanan et al., 2020)
	B. safensis	Not mentioned	13% enhanced oil recovery in core flood experiment.	(Saravanan et al., 2020)
	B. amyloliquefaciens	Surfactin	43% enhanced oil recovery in microbial-enhanced oil recovery column experiment.	(Saravanan et al., 2020)
	Acinetobacter baylyi	Lipopeptide	28% enhanced oil recovery in core displacement experiment.	(Saravanan et al., 2020)
	B. licheniformis and C. albicans	Sophorolipids	16.6 and 8.6% enhanced oil recovery in sand packed column technique.	(Saravanan et al., 2020)
(c) Removal of heavy metals	Bacillus sp. MSI 54	Lipopeptide	Cd (99.93%), Mn (89.5%), and Hg (75%) removal.	(Ravindran et al., 2020)
	Pseudomonas sp. CQ2	Rhamnolipid	Pb (59%), Cu (65.7%), and Cd (78.7%) removal.	(Sun et al., 2021)
	P. aeruginosa	Rhamnolipid	Zn (97%), Cd (95%), and As (50%) removal.	(Lopes et al., 2021)
	Rahnella sp. RM	Rhamnolipid	Effective removal of Cr, Pb, and Cu.	(Govarthanan et al., 2017)
	Burkholderia sp. Z-90	Not mentioned	Mn (52.2%), Zn (44.0%), Cd (37.7%), Pb (32.5%), As (31.6%), and Cu (24.1%) removal within 5 days.	(Yang et al., 2018)
	Pseudomonas sp.	Rhamnolipid	85.5% of vanadium was removed using 240 mg/L rhamnolipid.	(San Martin et al., 2021)
	C. guilliemondii UCP 0992	Not mentioned	Pb (89.1%), Fe (89.3%), and Zn (98%) removal	(Sarubbo et al., 2018)
	Starmerella bombicola CGMCC 1576.	Sophorolipid	44.5% and 83.6% removal of Pb and Cd.	(Qi et al., 2018)
	C. tropicalis UCP0996	Not mentioned	Cu and Zn removal.	(da Rocha Junior et al., 2019)

Govarthanan et al., 2017; Lopes et al., 2021; Qi et al., 2018; Ravindran et al., 2020; San Martín et al., 2021; Sarubbo et al., 2018; Sun et al., 2021; Yang et al., 2018). The mechanism for heavy metal removal by biosurfactants is shown in Figure 10.4c. It entails three major steps: binding and sorption of biosurfactant to the metal and soil particles, metal separation from the soil, and heavy metal association with micelles.

POTENTIAL NANOTECHNOLOGY OF BIOSURFACTANTS IN THE MARINE ENVIRONMENT

Nanotechnology is used to describe structure design, production, and application at the nanoscale level that can be used in various fields (Christopher et al., 2019). The wide interest in the utilization of nanomaterials for the manufacture of new products as well as investigations for their synthesis in green and eco-friendly techniques has gained geometric popularity in the past decades. However, nano-material synthesis entails adopting physical and chemical techniques that pro-duce by-products that are hazardous and toxic to the environment and human health (Christopher et al., 2019). Due to their toxicity in the environment, an alternative technique for a sustainable approach has been devised. This involves using biomolecules derived from microorganisms to aid the synthesis of nano-materials. In this regard, biosurfactants have emerged as an environmentally friendly option.

Biosurfactants are utilized in nanotechnology to synthesize nanoparticles (both metallic and organic), which are valuable for the restoration of polluted marine ecosystems as follows:

i. Surfactin from biosurfactants is used as a stabilizer in formulating gold and silver nanoparticles.

ii. Glycolipid produced by a bacterial strain (*Brevibacterium casei*) is used to prepare silver nanoplastics by a microemulsion technique using reverse micelles.

iii. A biosurfactant from *Pseudomonas aeruginosa* is also used in enhanc-ing stability in the synthesis of silver nanoparticles (Farias et al., 2014).

iv. Biosurfactants are utilized as reducing and capping agents in the syn-thesis of nanoparticles.

v. Lipopeptide produced by *Bacillus subtilis* acts as a stabilizer and allows the synthesis of nanoparticles without reducing agents (chemicals) (Rane et al., 2017).

vi. Biosurfactants are also used in optimizing polyethylene nanoparticle synthesis.

vii. A blend of biosurfactant and nanoparticles serves dual functions in the oil recovery mechanism. This approach is competitively advantageous as the chemical-based for the remedial biosurfactant would be nanopar-ticles (Sircar et al., 2021).

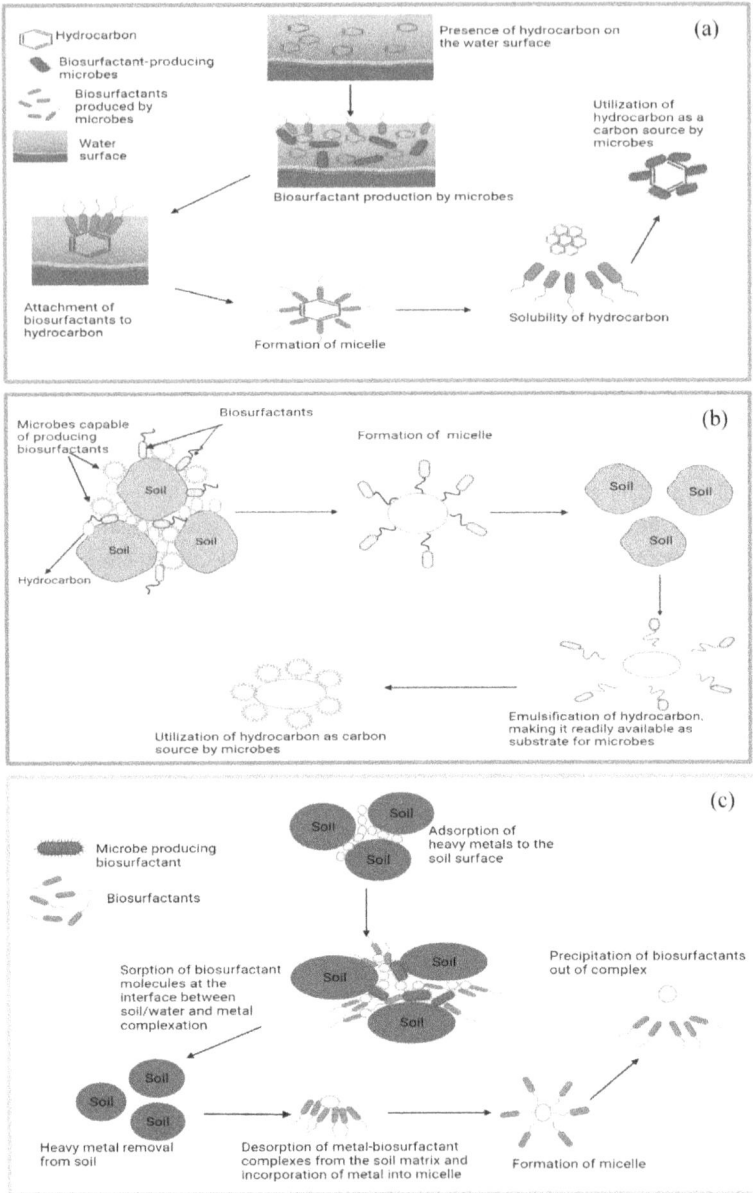

FIGURE 10.4 Application of biosurfactants in the marine environment. (a) Mechanism for bioremediation of petroleum hydrocarbon by biosurfactants in marine environment. (b) Mechanism for bioremediation of petroleum hydrocarbon by biosurfactants in soil of marine environment (typical in the intertidal zone, wetlands, marshes). (c) Mechanism of action of biosurfactants in heavy metal removal. It is important to note that the "water system" is a representation of the deep-sea, coastal shorelines, coral reefs, and watersheds, while the "soil system" is a representation of the intertidal zone, wetlands, and marshes.

viii. The synthesis of silver nanoparticles for lipopeptide biosurfactant produced by *Bacillus paramycoides* improves the biodegradation efficiency
of low-density and high-density polyethylene. This process is superior
to biogenic synthesis for polyethylene degradation by *B. paramycoides*
(Nehal and Singh, 2021).

FUTURE PROSPECTS

The use of biosurfactants produced by microorganisms has recorded huge suc-
cesses in the remediation of heavy metals and aliphatic and aromatic hydrocarbons. Despite this, a lacuna still exists in their utilization or application in various
areas of environmental remediation. For instance, in saline habitats, studies on
the use of biosurfactants in improving such conditions are not documented or
explored. This is a prospective, novel, and emerging area that should be explored
to solve salinity problems in various biota in marine ecosystems. Unlike metal
removal from the marine environment, studies on the effect of biosurfactants
during microplastic degradation/removal are limited despite their prevalence as
a pollutant of public concern (Akan et al., 2021; Okeke et al., 2022).

The commercial success of biosurfactants has been hindered due to their high
production cost. In this regard, efforts should also be channeled towards utilizing
diverse marine waste as cheap alternative substrates for biosurfactant production. In addition, the use of leachate, crude fish oil, and fracturing flow-back
fluid as a substrate for producing biosurfactants should be fully explored and
considered an option.

Furthermore, most studies on bioremediation using biosurfactants have been
conducted in controlled laboratory settings, with very few field experiments.
More field trials are necessary to assess their bioremediation success since
they possess high instability and complexity under field conditions. Thus, more
experimental studies should be directed towards *in situ* bioremediation using
large-scale production of marine biosurfactants. In addition, more efforts toward
evaluating and expanding information concerning the modification of structural
features of biosurfactants are needed to improve their productivity and efficiency.

Information about biomolecules and accurate monitoring and testing methods
for screening the best producer of biosurfactants is not well known. Therefore,
more research is required to understand the pathways of marine biosurfactants
at the level of species and genes using -omics technology (genomics, transcrip-
tomics, and proteomics).

CONCLUSION

This chapter showed the prospects of biosurfactant production by microorganisms with the use of marine waste and its application in the marine environment
for remediation of pollutants. Utilizing marine waste is a sustainable approach to
achieving biosurfactants at a reduced cost and, as such, a possible means to meet

the high market demand. Marine biosurfactants are natural, greener, and eco-friendly substitutes for synthetic surfactants. The production of marine biosurfactants, as discussed, can be improved through the use of engineered microbes and integrated systems (upstream and downstream processes). This would help to produce a more profitable biosurfactant. Furthermore, knowing the social and economic benefits of marine biosurfactants, the optimal conditions for their preparation need to be further investigated. Also, there is a need to explore more kinds of marine wastes in biosurfactant production.

REFERENCES

Akan, O.D., Udofia, G.E., Okeke, E.S., Mgbechidinma, C.L., Okoye, C.O., Zoclanclounon, Y.A.B., Atakpa, E.O., Adebanjo, O.O., 2021. Plastic waste: Status, degradation and microbial management options for Africa. *J. Environ. Manage.* 292, 112758. https://doi.org/10.1016/j.jenvman.2021.112758

Alvarez, V.M., Jurelevicius, D., Marques, J.M., de Souza, P.M., de Araujo, L.V., Barros, T.G., de Souza, R.O.M.A., Freire, D.M.G., Seldin, L., 2015. *Bacillus amyloliquefaciens* TSBSO 3.8, a biosurfactant-producing strain with biotechnological potential for microbial enhanced oil recovery. *Colloids Surf. B Biointerfaces* 136, 14–21. https://doi.org/10.1016/j.colsurfb.2015.08.046

Amaral, P.F.F., Silva, J.M., Lehocky, M., Barros-timmons, A.M. V, Coelho, M.A.Z., Marrucho, I.M., Coutinho, J.A.P., 2006. Production and characterization of a bio-emulsifier from *Yarrowia lipolytica*. *Process Biochem.* 41, 1894–1898. https://doi.org/10.1016/j.procbio.2006.03.029

Ambaye, T.G., Vaccari, M., Prasad, S., Rtimi, S., 2021. Preparation, characterization and application of biosurfactant in various industries: A critical review on progress, challenges and perspectives. *Environ. Technol. Innov.* 24, 102090. https://doi.org/10.1016/j.eti.2021.102090

Atakpa, E.O., Zhou, H., Jiang, L., Ma, Y., Liang, Y., Li, Y., Zhang, D., Zhang, C., 2022. Improved degradation of petroleum hydrocarbons by co-culture of fungi and bio-surfactant-producing bacteria. *Chemosphere* 290, 133337. https://doi.org/10.1016/j.chemosphere.2021.133337

Balan, S.S., Kumar, C.G., Jayalakshmi, S., 2016. Pontifactin, a new lipopeptide biosurfactant produced by a marine *Pontibacter korlensis* strain SBK-47: Purification, characterization and its biological evaluation. *Process Biochem.* https://doi.org/10.1016/j.procbio.2016.09.009

Balan, S.S., Kumar, C.G., Jayalakshmi, S., 2017. Aneurinifactin, a new lipopeptide biosurfactant produced by a marine *Aneurinibacillus aneurinilyticus* SBP-11 isolated from Gulf of Mannar: Purification, characterization and its biological evaluation. *Microbiol. Res.* 194, 1–9. https://doi.org/10.1016/j.micres.2016.10.005

Balan, S.S., Kumar, C.G., Jayalakshmi, S., 2019. Physicochemical, structural and biological evaluation of Cybersan (trigalactomargarate), a new glycolipid biosurfactant produced by a marine yeast, *Cyberlindnera saturnus* strain SBPN-27. *Process Biochem.* 80, 171–180. https://doi.org/10.1016/j.procbio.2019.02.005

Baskaran, S.M., Zakaria, M.R., Mukhlis Ahmad Sabri, A.S., Mohamed, M.S., Wasoh, H., Toshinari, M., Hassan, M.A., Banat, I.M., 2021. Valorization of biodiesel side stream waste glycerol for rhamnolipids production by *Pseudomonas aeruginosa* RS6. *Environ. Pollut.* 276. https://doi.org/10.1016/j.envpol.2021.116742

Bator, I., Karmainski, T., Tiso, T., Blank, L.M., 2020. Killing two birds with one stone—Strain engineering facilitates the development of a unique rhamnolipid production process. *Front. Bioeng. Biotechnol.* 8, 899. https://doi.org/10.3389/fbioe.2020.00899

Bezza, F.A., Chirwa, E.M.N., 2017. The role of lipopeptide biosurfactant on microbial remediation of aged polycyclic aromatic hydrocarbons (PAHs)-contaminated soil. *Chem. Eng. J.* 309, 563–576. https://doi.org/10.1016/j.cej.2016.10.055

Cai, Q., Zhu, Z., Chen, B., Lee, K., Nedwed, T.J., Greer, C., Zhang, B., 2021. A cross-comparison of biosurfactants as marine oil spill dispersants: Governing factors, synergetic effects and fates. *J. Hazard. Mater.* 416, 126122. https://doi.org/10.1016/j.jhazmat.2021.126122

Chen, Q., Bao, M., Fan, X., Liang, S., Sun, P., 2013. Rhamnolipids enhance marine oil spill bioremediation in laboratory system. *Mar. Pollut. Bull.* 71, 269–275. https://doi.org/10.1016/j.marpolbul.2013.01.037

Chen, Q., Li, Y., Liu, M., Zhu, B., Mu, J., Chen, Z., 2021. Removal of Pb and Hg from marine intertidal sediment by using rhamnolipid biosurfactant produced by a *Pseudomonas aeruginosa* strain. *Environ. Technol. Innov.* 22, 101456. https://doi.org/10.1016/j.eti.2021.101456

Christopher, F.C., Ponnusamy, S.K., Ganesan, J.J., Ramamurthy, R., 2019. Investigating the prospects of bacterial biosurfactants for metal nanoparticle synthesis—A comprehensive review. *IET Nanobiotechnol.* 13, 243–249. https://doi.org/10.1049/iet-nbt.2018.5184

Coelho, L.M., Rezende, H.C., Coelho, L.M., de Sousa, P.A.R., Melo, D.F.O., Coelho, N.M.M., 2015. Bioremediation of polluted waters using microorganisms. In: *Advances in Bioremediation of Wastewater and Polluted Soil*. InTech, pp. 172–209. https://doi.org/10.5772/60770

Coronel-león, J., Grau, G. De, Grau-campistany, A., Farfan, M., Rabanal, F., Manresa, A., Marques, A.M., 2015. Biosurfactant production by AL 1.1, a *Bacillus licheniformis* strain isolated from Antarctica: Production, chemical characterization and properties. *Ann. Microbiol.* 65, 2065–2078. https://doi.org/10.1007/s13213-015-1045-x

Coutte, F., Lecouturier, D., Leclère, V., Béchet, M., Jacques, P., Dhulster, P., 2013. New integrated bioprocess for the continuous production, extraction and purification of lipopeptides produced by *Bacillus subtilis* in membrane bioreactor. *Process Biochem.* 48, 25–32. https://doi.org/10.1016/j.procbio.2012.10.005

da Rocha Junior, R.B., Meira, H.M., Almeida, D.G., Rufino, R.D., Luna, J.M., Santos, V.A., Sarubbo, L.A., 2019. Application of a low-cost biosurfactant in heavy metal remediation processes. *Biodegradation* 30, 215–233. https://doi.org/10.1007/s10532-018-9833-1

Dauda, A.B., Ajadi, A., Tola-Fabunmi, A.S., Akinwole, A.O., 2019. Waste production in aquaculture: Sources, components and managements in different culture systems. *Aquac. Fish.* 4, 81–88. https://doi.org/10.1016/j.aaf.2018.10.002

Dell'Anno, F., Sansone, C., Ianora, A., Dell'Anno, A., 2018. Biosurfactant-induced remediation of contaminated marine sediments: Current knowledge and future perspectives. *Mar. Environ. Res.* 137, 196–205. https://doi.org/10.1016/j.marenvres.2018.03.010

Dierickx, S., Castelein, M., Remmery, J., De Clercq, V., Lodens, S., Baccile, N., De Maeseneire, S.L., Roelants, S.L.K.W., Soetaert, W.K., 2022. From bumblebee to bioeconomy: Recent developments and perspectives for sophorolipid biosynthesis. *Biotechnol. Adv.* 54, 107788. https://doi.org/10.1016/j.biotechadv.2021.107788

Dobler, L., Ferraz, H.C., de Castilho, L.V.A., Sangenito, L.S., Pasqualino, I.P., Santos, A.L.S. dos, Neves, B.C., Oliveira, R.R., Freire, D.M.G., Almeida, R.V., 2020. Environmentally friendly rhamnolipid production for petroleum remediation. *Chemosphere* 252, 126349. https://doi.org/10.1016/j.chemosphere.2020.126349

Dolman, B.M., Wang, F., Winterburn, J.B., 2019. Integrated production and separation of biosurfactants. *Process Biochem.* 83, 1–8. https://doi.org/10.1016/j.procbio.2019.05.002

Du, J., Zhang, A., Jing, H., 2017. Biosynthesis of di-rhamnolipids and variations of congeners composition in genetically-engineered *Escherichia coli*. *Biotechnol. Lett.* 39, 1041–1048. https://doi.org/10.1007/s10529-017-2333-2

Du, M., Xu, D., Trinh, X., Liu, S., Wang, M., Zhang, Y., Wu, J., Zhou, Q., Wu, Z., 2016. EPS solubilization treatment by applying the biosurfactant rhamnolipid to reduce clogging in constructed wetlands. *Bioresour. Technol.* 218, 833–841. https://doi.org/10.1016/j.biortech.2016.07.040

El-Sersy, N.A., 2012. Plackett-Burman design to optimize biosurfactant production by marine *Bacillus subtilis* N10. *Rom. Biotechnol. Lett.* 17, 7049–7064.

Farias, C.B.B., Almeida, F.C.G., Silva, I.A., Souza, T.C., Meira, H.M., Soares da Silva, R. de C.F., Luna, J.M., Santos, V.A., Converti, A., Banat, I.M., Sarubbo, L.A., 2021. Production of green surfactants: Market prospects. *Electron. J. Biotechnol.* 51, 28–39. https://doi.org/10.1016/j.ejbt.2021.02.002

Farias, C.B.B., Silva, A.F., Rufino, R.D., Luna, J.M., Souza, J.E.G., Sarubbo, L.A., 2014. Synthesis of silver nanoparticles using a biosurfactant produced in low-cost medium as stabilizing agent. *Electron. J. Biotechnol.* 17, 122–125. https://doi.org/10.1016/j.ejbt.2014.04.003

Feng, J.-Q., Gang, H.-Z., Li, D.-S., Liu, J.-F., Yang, S.-Z., Mu, B.-Z., 2019. Characterization of biosurfactant lipopeptide and its performance evaluation for oil-spill remediation. *RSC Adv.* 9, 9629–9632. https://doi.org/10.1039/C9RA01430F

Fooladi, T., Abdeshahian, P., Moazami, N., Reza, M., 2018. Enhanced biosurfactant production by *Bacillus pumilus* 2IR in fed-batch fermentation using 5-L bioreactor. *Iran. J. Sci. Technol. Trans. A Sci.* 4, 5994. https://doi.org/10.1007/s40995-018-0599-4

Galitskaya, P., Biktasheva, L., Blagodatsky, S., Selivanovskaya, S., 2021. Response of bacterial and fungal communities to high petroleum pollution in different soils. *Sci. Rep.* 11, 164–182. https://doi.org/10.1038/s41598-020-80631-4

Gaur, V.K., Sharma, P., Gupta, S., Varjani, S., Srivastava, J.K., Wong, J.W.C., Hao, H., 2022a. Opportunities and challenges in omics approaches for biosurfactant production and feasibility of site remediation: strategies and advancements. *Environ. Technol. Innov.* 25, 102132. https://doi.org/10.1016/j.eti.2021.102132

Gaur, V.K., Sharma, P., Sirohi, R., Varjani, S., Taherzadeh, M.J., Chang, J.-S., Yong Ng, H., Wong, W.C.J., Kim, S.-H., 2022b. Production of biosurfactants from agro-industrial waste and waste cooking oil in a circular bioeconomy: An overview. *Bioresour. Technol.* 343, 126059. https://doi.org/10.1016/j.biortech.2021.126059

Geetha, S.J., Banat, I.M., Joshi, S.J., 2018. Biosurfactants: Production and potential applications in microbial enhanced oil recovery (MEOR). *Biocatal. Agric. Biotechnol.* 14, 23–32. https://doi.org/10.1016/j.bcab.2018.01.010

Gidudu, B., Chirwa, E.M.N., 2021. Electrokinetic extraction and recovery of biosurfactants using rhamnolipids as a model biosurfactant. *Sep. Purif. Technol.* 276, 119327. https://doi.org/10.1016/j.seppur.2021.119327

Govarthanan, M., Mythili, R., Selvankumar, T., Kamala-Kannan, S., Choi, D., Chang, Y.-C., 2017. Isolation and characterization of a biosurfactant-producing heavy metal resistant *Rahnella* sp. RM isolated from chromium-contaminated soil. *Biotechnol. Bioprocess Eng.* 22, 186–194. https://doi.org/10.1007/s12257-016-0652-0

He, N., Wu, T., Jiang, J., Long, X., Shao, B., Meng, Q., 2017. Toward high-efficiency production of biosurfactant rhamnolipids using sequential fed-batch fermentation based on a fill-and-draw strategy. *Colloids Surfaces B Biointerfaces* 157, 317–324. https://doi.org/10.1016/j.colsurfb.2017.06.007

Huang, X., Zhou, H., Ni, Q., Dai, C., Chen, C., Li, Y., Zhang, C., 2020. Biosurfactant-facilitated biodegradation of hydrophobic organic compounds in hydraulic fracturing flowback wastewater: A dose—effect analysis. *Environ. Technol. Innov.* 19, 100889. https://doi.org/10.1016/j.eti.2020.100889

Jadhav, V.V., Yadav, A., Shouche, Y.S., Aphale, S., Moghe, A., Pillai, S., Arora, A., Bhadekar, R.K., 2013. Studies on biosurfactant from *Oceanobacillus* sp. BRI 10 isolated from Antarctic sea water. *Desalination* 318, 64–71. https://doi.org/10.1016/j.desal.2013.03.017

Janek, T., Marcin, Ł., Krasowska, A., 2012. Antiadhesive activity of the biosurfactant pseudofactin II secreted by the Arctic bacterium *Pseudomonas fluorescens* BD5. *BMC Microbiol.* 12, 24. https://doi.org/10.1186/1471-2180-12-24

Jemil, N., Hmidet, N., Ayed, H. Ben, Nasri, M., 2018. Physicochemical characterization of Enterobacter cloacae C3 lipopeptides and their applications in enhancing diesel oil biodegradation Physicochemical characterization of *Enterobacter cloacae* C3 lipopeptides and their applications in enhancing diesel oil. *Process Saf. Environ. Prot.* 117, 399–407. https://doi.org/10.1016/j.psep.2018.05.018

Jia, L., Zhou, J., Cao, J., Wu, Z., Liu, W., Yang, C., 2020. Foam fractionation for promoting rhamnolipids production by *Pseudomonas aeruginosa* D1 using animal fat hydrolysate as carbon source and its application in intensifying phytoremediation. *Chem. Eng. Process.—Process Intensif.* 158, 108177. https://doi.org/10.1016/j.cep.2020.108177

Jiang, J., Zhang, D., Niu, J., Jin, M., Long, X., 2021. Extremely high-performance production of rhamnolipids by advanced sequential fed-batch fermentation with high cell density. *J. Clean. Prod.* 326, 129382. https://doi.org/10.1016/j.jclepro.2021.129382

Jiang, J., Zu, Y., Li, X., Meng, Q., Long, X., 2020. Recent progress towards industrial rhamnolipids fermentation: Process optimization and foam control. *Bioresour. Technol.* 298, 122394. https://doi.org/10.1016/j.biortech.2019.122394

Khopade, A., Biao, R., Liu, X., Mahadik, K., Zhang, L., Kokare, C., 2012. Production and stability studies of the biosurfactant isolated from marine *Nocardiopsis* sp. B4. *DES* 285, 198–204. https://doi.org/10.1016/j.desal.2011.10.002

Kiran, G.S., Hema, T.A., Gandhimathi, R., Selvin, J., Thomas, T.A., Ravji, T.R., Natarajaseenivasan, K., 2009. Optimization and production of a biosurfactant from the sponge-associated marine fungus *Aspergillus ustus* MSF3. *Colloids Surfaces B Biointerfaces* 73, 250–256. https://doi.org/10.1016/j.colsurfb.2009.05.025

Kiran, G.S., Priyadharsini, S., Sajayan, A., Priyadharsini, G.B., 2017. Production of lipopeptide biosurfactant by a marine *Nesterenkonia* sp. and its application in food industry. *Front Microbiol.* 8, 1–11. https://doi.org/10.3389/fmicb.2017.01138

Kiran, G.S., Sabu, A., Selvin, J., 2010. Synthesis of silver nanoparticles by glycolipid biosurfactant produced from marine *Brevibacterium casei* MSA19. *J. Biotechnol.* 148, 221–225. https://doi.org/10.1016/j.jbiotec.2010.06.012

Kiran, G.S., Thajuddin, N., Hema, T.A., Idhayadhulla, A., Kumar, R.S., Selvin, J., 2013. Optimization and characterization of rhamnolipid biosurfactant from sponge associated marine fungi. *Desalin. Water Treat.* 24, 37–41. https://doi.org/10.5004/dwt.2010.1569

Konishi, M., Fukuoka, T., Nagahama, T., Morita, T., Imura, T., Kitamoto, D., Hatada, Y., 2010. Biosurfactant-producing yeast isolated from *Calyptogena soyoae* (deep-sea cold-seep clam) in the deep sea. *J. Biosci. Bioeng.* 110, 169–175. https://doi.org/10.1016/j.jbiosc.2010.01.018

Kryachko, Y., 2018. Novel approaches to microbial enhancement of oil recovery. *J. Biotechnol.* 266, 118–123. https://doi.org/10.1016/j.jbiotec.2017.12.019

Kumar, A., Singh, S.K., Kant, C., Verma, H., Kumar, D., Singh, P.P., Modi, A., Droby, S., Kesawat, M.S., Alavilli, H., Bhatia, S.K., Saratale, G.D., Saratale, R.G., Chung, S.M., Kumar, M., 2021. Microbial biosurfactant: A new frontier for sustainable agriculture and pharmaceutical industries. *Antioxidants* 10, 1472. https://doi.org/10.3390/antiox10091472

Kumar, G.A., Mathew, N.C., Sujitha, K., Kirubagaran, R., Dharani, G., 2019. Genome analysis of deep sea piezotolerant *Nesiotobacter exalbescens* COD22 and toluene degradation studies under high pressure condition. *Sci. Rep.* 9, 18724. https://doi.org/10.1038/s41598-019-55115-9

Kuyukina, M.S., Ivshina, I.B., Makarov, S.O., Litvinenko, L.V., Cunningham, C.J., Philp, J.C., 2005. Effect of biosurfactants on crude oil desorption and mobilization in a soil system. *Environ Int.* 31, 155–161. https://doi.org/10.1016/j.envint.2004.09.009

Lee, J., Zhang, L., 2015. The hierarchy quorum sensing network in *Pseudomonas aeruginosa*. *Protein Cell* 6, 26–41. https://doi.org/10.1007/s13238-014-0100-x

Liu, Q., Niu, J., Yu, Y., Wang, C., Lu, S., Zhang, S., Lv, J., Peng, B., 2021. Production, characterization and application of biosurfactant produced by *Bacillus licheniformis* L20 for microbial enhanced oil recovery. *J. Clean. Prod.* 307, 127193. https://doi.org/10.1016/j.jclepro.2021.127193

Liu, X., Ren, B., Chen, M., Wang, H., Kokare, C.R., Zhou, X., Wang, J., Dai, H., Song, F., Liu, M., Wang, J., 2010. Production and characterization of a group of bioemulsifiers from the marine *Bacillus velezensis* strain H3. *Appl. Microbiol. Biotechnol.*, 1881–1893. https://doi.org/10.1007/s00253-010-2653-9

Long, X., Zhang, G., Shen, C., Sun, G., Wang, R., Yin, L., Meng, Q., 2013. Application of rhamnolipid as a novel biodemulsifier for destabilizing waste crude oil. *Bioresour. Technol.* 131, 1–5. https://doi.org/10.1016/j.biortech.2012.12.128

Lopes, C.S.C., Teixeira, D.B., Braz, B.F., Santelli, R.E., de Castilho, L.V.A., Gomez, J.G.C., Castro, R.P.V., Seldin, L., Freire, D.M.G., 2021. Application of rhamnolipid surfactant for remediation of toxic metals of long- and short-term contamination sites. *Int. J. Environ. Sci. Technol.* 18, 575–588. https://doi.org/10.1007/s13762-020-02889-5

Lu, C., Hong, Y., Liu, J., Gao, Y., Ma, Z., Yang, B., Ling, W., Waigi, M.G., 2019. A PAH-degrading bacterial community enriched with contaminated agricultural soil and its utility for microbial bioremediation. *Environ. Pollut.* 251, 773–782. https://doi.org/10.1016/j.envpol.2019.05.044

Mani, P., Dineshkumar, G., Jayaseelan, T., Deepalakshmi, K., Kumar, C.G., Balan, S.S., 2016. Antimicrobial activities of a promising glycolipid biosurfactant from a novel marine *Staphylococcus saprophyticus* SBPS 15. *3 Biotech* 6, 163. https://doi.org/10.1007/s13205-016-0478-7

Martins, P.C., Martins, V.G., 2018. Biosurfactant production from industrial wastes with potential remove of insoluble paint. *Int. Biodeterior. Biodegrad.* 127, 10–16. https://doi.org/10.1016/j.ibiod.2017.11.005

Mei, Y., Yang, Z., Kang, Z., Yu, F., Long, X., 2021. Enhanced surfactin fermentation via advanced repeated fed-batch fermentation with increased cell density stimulated by EDTA—Fe (II). *Food Bioprod. Process.* 127, 288–294. https://doi.org/10.1016/j. fbp.2021.03.012

Mgbechidinma, C.L., Zheng, G., Baguya, E.B., Zhou, H., Okon, S.U., Zhang, C., 2022. Fatty acid composition and nutritional analysis of waste crude fish oil obtained by optimized milder extraction methods. *Environ. Eng. Res.* In Press. https://doi. org/10.4491/eer.2022.034

Moshtagh, B., Hawboldt, K., Zhang, B., 2021. Biosurfactant production by native marine bacteria (*Acinetobacter calcoaceticus* P1–1A) using waste carbon sources: Impact of process conditions. *Can. J. Chem. Eng.* 1–12. https://doi.org/10.1002/cjce.24254

Müller, M.M., Hörmann, B., Syldatk, C., Hausmann, R., 2010. *Pseudomonas aeruginosa* PAO1 as a model for rhamnolipid production in bioreactor systems. *Appl. Microbiol. Biotechnol.* 87, 167–174. https://doi.org/10.1007/s00253-010-2513-7

Muriel-Millán, L.F., Rodríguez-Mejía, J.L., Godoy-Lozano, E.E., Rivera-Gómez, N., Gutierrez-Rios, R.-M., Morales-Guzmán, D., Trejo-Hernández, M.R., Estradas-Romero, A., Pardo-López, L., 2019. Functional and genomic characterization of a *Pseudomonas aeruginosa* strain isolated from the Southwestern Gulf of Mexico reveals an enhanced adaptation for long-chain alkane degradation. *Front. Mar. Sci.* 6, 572. https://doi.org/10.3389/fmars.2019.00572

Narciso-Ortiz, L., Vargas-García, K.A., Vázquez-Larios, A.L., Quiñones-Muñoz, T.A., Hernández-Martínez, R., Lizardi-Jiménez, M.A., 2020. Coral reefs and watersheds of the Gulf of Mexico in Veracruz: Hydrocarbon pollution data and bioremediation proposal. *Reg. Stud. Mar. Sci.* 35, 101155. https://doi.org/10.1016/j.rsma.2020. 101155

Nehal, N., Singh, P., 2021. Role of nanotechnology for improving properties of biosurfactant from newly isolated bacterial strains from Rajasthan. *Mater. Today Proc.* In Press. https://doi.org/10.1016/j.matpr.2021.05.682

Nurfarahin, A.H., Mohamed, M.S., Phang, Y.L., 2018. Culture medium development for microbial-derived surfactants production—An overview. *Molecules* 23, 1049. https://doi.org/10.3390/molecules23051049

Okeke, E.S., Okoye, C.O., Atakpa, E.O., Ita, R.E., Nyaruaba, R., Mgbechidinma, C.L., Akan, O.D., 2022. Microplastics in agroecosystems—Impacts on ecosystem functions and food chain. *Resour. Conserv. Recycl.* 177, 105961. https://doi.org/10.1016/j. resconrec.2021.105961

Olasanmi, I.O., Thring, R.W., 2018. The role of biosurfactants in the continued drive for environmental sustainability. *Sustainability* 10, 1–12. https://doi.org/10.3390/ su10124817

Oliva, M., Marchi, L. De, Cuccaro, A., Pretti, C., 2021. Bioassay-based ecotoxicological investigation on marine and freshwater impact of cigarette butt littering. *Environ. Pollut.* 288, 117787. https://doi.org/10.1016/j.envpol.2021.117787

Palme, O., Comanescu, G., Stoineva, I., Radel, S., Benes, E., Develter, D., Wray, V., Lang, S., 2010. Sophorolipids from *Candida bombicola*: Cell separation by ultrasonic particle manipulation. *Eur. J. Lipid Sci. Technol.* 112, 663–673. https://doi.org/10.1002/ ejlt.200900163

Pele, M.A., Ribeaux, D.R., Vieira, E.R., Souza, A.F., Luna, M.A.C., Rodríguez, D.M., Andrade, R.F.S., Alviano, D.S., Alviano, C.S., Barreto-Bergter, E., Santiago, A.L.C.M.A., Campos-Takaki, G.M., 2019. Conversion of renewable substrates for biosurfactant production by *Rhizopus arrhizus* UCP 1607 and enhancing the removal of diesel oil from marine soil. *Electron. J. Biotechnol.* 38, 40–48. https://doi.org/10.1016/j.ejbt.2018.12.003

Qi, X., Xu, X., Zhong, C., Jiang, T., Wei, W., Song, X., 2018. Removal of cadmium and lead from contaminated soils using sophorolipids from fermentation culture of *Starmerella bombicola* CGMCC 1576 fermentation. *Int. J. Environ. Res. Public Health* 15, 2334. https://doi.org/10.3390/ijerph15112334

Radzuan, N.M., Banat, I.M., Winterburn, J., 2017. Production and characterization of rhamnolipid using palm oil agricultural refinery waste. *Bioresour. Technol.* 225, 99–105. https://doi.org/10.1016/j.biortech.2016.11.052

Rane, A.N., Baikar, V.V., Ravi Kumar, V., Deopurkar, R.L., 2017. Agro-industrial wastes for production of biosurfactant by *Bacillus subtilis* ANR 88 and its application in synthesis of silver and gold nanoparticles. *Front. Microbiol.* 8, 492. https://doi.org/10.3389/fmicb.2017.00492

Ravindran, A., Sajayan, A., Priyadharshini, G.B., Selvin, J., Kiran, G.S., 2020. Revealing the efficacy of thermostable biosurfactant in heavy metal bioremediation and surface treatment in vegetables. *Front. Microbiol.* 11. https://doi.org/10.3389/fmicb.2020.00222

Rodrigues, E.M., Cesar, D.E., Santos de Oliveira, R., de Paula Siqueira, T., Tótola, M.R., 2020. Hydrocarbonoclastic bacterial species growing on hexadecane: Implications for bioaugmentation in marine ecosystems. *Environ. Pollut.* 267, 115579. https://doi.org/10.1016/j.envpol.2020.115579

Roelants, S., Renterghem, L. Van, Maes, K., Everaert, B., Demaeseneire, S., Microbial, W.S., 2019. Microbial biosurfactants: from lab to market. In: *Microbial Biosurfactants and their Environmental and Industrial Applications.* CRC Press. https://doi.org/10.1201/b21950

Saeki, H., Sasaki, M., Komatsu, K., Miura, A., Matsuda, H., 2009. Oil spill remediation by using the remediation agent JE1058BS that contains a biosurfactant produced by *Gordonia* sp. strain JE-1058. *Bioresour. Technol.* 100, 572–577. https://doi.org/10.1016/j.biortech.2008.06.046

Saimmai, A., Udomsilp, S., Maneerat, S., 2013. Production and characterization of bio-surfactant from marine bacterium *Inquilinus limosus* KB3 grown on low-cost raw materials. *Ann. Microbiol.* 1–13. https://doi.org/10.1007/s13213-012-0592-7

Sana, S., Datta, S., Biswas, D., Bhattacharya, M., 2017. Production kinetics of rham-nolipid using fish fat: A step towards environmental hazard control of sewage. *Environ. Technol. Innov.* 8, 299–308. https://doi.org/10.1016/j.eti.2017.07.004

San Martín, Y.B., Toledo León, H.F., Rodríguez, A.Á., Marqués, A.M., López, M.I.S., 2021. Rhamnolipids application for the removal of vanadium from contaminated sediment. *Curr. Microbiol.* 78, 1949–1960. https://doi.org/10.1007/s00284-021-02445-5

Saravanan, A., Kumar, P.S., Vardhan, K.H., Jeevanantham, S., Karishma, S.B., Yaashikaa, P.R., Vellaichamy, P., 2020. A review on systematic approach for micro-bial enhanced oil recovery technologies: Opportunities and challenges. *J. Clean. Prod.* 258, 120777. https://doi.org/10.1016/j.jclepro.2020.120777

Sarubbo, L.A., Brasileiro, P.P.F., Silveira, G.N.M., Juliana, M., Rufino, R.D., Valdemir, A., 2018. Application of a low cost biosurfactant in the removal of heavy metals in soil. *Chem. Eng. Trans.* 64, 433–438. https://doi.org/10.3303/CET1864073

Sharma, P., Sangwan, S., Kaur, H., 2019. Process parameters for biosurfactant production using yeast *Meyerozyma guilliermondii* YK32. *Environ. Monit. Assess.* 191.

Shekhar, S., Sundaramanickam, A., Saranya, K., Meena, M., Kumaresan, S., Balasubramanian, T., 2018. Production and characterization of biosurfactant by marine bacterium *Pseudomonas stutzeri* (SSASM1). *Int. J. Environ. Sci. Technol.* 1–11. https://doi.org/10.1007/s13762-018-1915-4

Shi, J., Chen, Y., Liu, X., Li, D., 2021. Rhamnolipid production from waste cooking oil using newly isolated halotolerant *Pseudomonas aeruginosa* M4. *J. Clean. Prod.* 278, 123879. https://doi.org/10.1016/j.jclepro.2020.123879

Sircar, A., Rayavarapu, K., Bist, N., Yadav, K., Singh, S., 2021. Applications of nanoparticles in enhanced oil recovery. *Pet. Res.* In Press. https://doi.org/10.1016/j.ptlrs.2021.08.004

Sun, W., Zhu, B., Yang, F., Dai, M., Sehar, S., Peng, C., Ali, I., Naz, I., 2021. Optimization of biosurfactant production from *Pseudomonas* sp. CQ2 and its application for remediation of heavy metal contaminated soil. *Chemosphere* 265, 129090. https://doi.org/10.1016/j.chemosphere.2020.129090

Van Landuyt, J., Cimmino, L., Dumolin, C., Chatzigiannidou, I., Taveirne, F., Mattelin, V., Zhang, Y., Vandamme, P., Scoma, A., Williamson, A., Boon, N., 2020. Microbial enrichment, functional characterization and isolation from a cold seep yield piezotolerant obligate hydrocarbon degraders. *FEMS Microbiol. Ecol.* 96. https://doi.org/10.1093/femsec/fiaa097

Varjani, S.J., Gnansounou, E., Pandey, A., 2017. Comprehensive review on toxicity of persistent organic pollutants from petroleum refinery waste and their degradation by microorganisms. *Chemosphere* 188, 280–291. https://doi.org/10.1016/j.chemosphere.2017.09.005

Venugopal, V., 2021. Valorization of seafood processing discards: Bioconversion and biorefinery approaches. *Front. Sustain. Food Syst.* 5, 611835. https://doi.org/10.3389/fsufs.2021.611835

Wang, X.-T., Liu, B., Li, X.-Z., Lin, W., Li, D.-A., Dong, H., Wang, L., 2022. Biosurfactants produced by novel facultative-halophilic *Bacillus* sp. XT-2 with biodegradation of long chain n-alkane and the application for enhancing waxy oil recovery. *Energy* 240, 122802. https://doi.org/10.1016/j.energy.2021.122802

Wei, Z., Wang, J.J., Gaston, L.A., Li, J., Fultz, L.M., Delaune, R.D., Dodla, S.K., 2020a. Remediation of crude oil-contaminated coastal marsh soil: Integrated effect of biochar, rhamnolipid biosurfactant and nitrogen application. *J. Hazard. Mater.* 396, 122595. https://doi.org/10.1016/j.jhazmat.2020.122595

Wei, Z., Wang, J.J., Meng, Y., Li, J., Gaston, L.A., Fultz, L.M., Delaune, R.D., 2020b. Potential use of biochar and rhamnolipid biosurfactant for remediation of crude oil-contaminated coastal wetland soil: Ecotoxicity assessment. *Chemosphere* 253, 126617. https://doi.org/10.1016/j.chemosphere.2020.126617

Wittgens, A., Tiso, T., Arndt, T.T., Wenk, P., Hemmerich, J., Müller, C., Rosenau, F., Blank, L.M., 2011. Growth independent rhamnolipid production from glucose using the non-pathogenic *Pseudomonas putida* KT2440 Growth independent rhamnolipid production from glucose using the non-pathogenic *Pseudomonas putida* KT2440. *Microb. Cell Fact.* 10, 80. https://doi.org/10.1186/1475-2859-10-80

Xu, M., Fu, X., Gao, Y., Duan, L., Xu, C., Sun, W., Li, Y., Meng, X., Xiao, X., 2020. Characterization of a biosurfactant-producing bacteria isolated from marine environment: Surface activity, chemical characterization and biodegradation. *J. Environ. Chem. Eng.* 8, 104277. https://doi.org/10.1016/j.jece.2020.104277

Yang, Z., Liang, L., Yang, W., Shi, W., Tong, Y., Chai, L., Gao, S., Liao, Q., 2018. Simultaneous immobilization of cadmium and lead in contaminated soils by hybrid bio-nanocomposites of fungal hyphae and nano-hydroxyapatites. *Environ. Sci. Pollut. Res.* 25, 11970–11980. https://doi.org/10.1007/s11356-018-1492-6

Zhao, F., Cui, Q., Han, S., Dong, H., Zhang, J., 2015. Enhanced rhamnolipid production of *Pseudomonas aeruginosa* SG by increasing copy number of rhlAB genes with modified promoter. *RSC Adv.* 5, 70546–70552. https://doi.org/10.1039/C5RA13415C

Zhou, H., Huang, X., Liang, Y., Li, Y., Xie, Q., Zhang, C., You, S., 2020. Enhanced bioremediation of hydraulic fracturing flowback and produced water using an indigenous biosurfactant-producing bacteria *Acinetobacter* sp. Y2. *Chem. Eng. J.* 397, 125348. https://doi.org/10.1016/j.cej.2020.125348

Zhu, Z., Zhang, B., Cai, Q., Ling, J., Lee, K., Chen, B., 2020. Fish waste based lipopeptide production and the potential application as a bio-dispersant for oil spill control. *Front. Bioeng. Biotechnol.* 8, 1–16. https://doi.org/10.3389/fbioe.2020.00734

Zinjarde, S.S., Pant, A., 2002. Emulsifier from a tropical marine yeast, *Yarrowia lipolytica* NCIM 3589. *J. Basic Microbiol.* 42, 67–73.

11 Marine Biosurfactants
Applications in Agriculture

*Shashank Reddy, Vartika Verma
and Nidhi Srivastava*

CONTENTS

INTRODUCTION

The worldwide population is growing at an exponential rate, with about 200,000 people added to the world's food need every day, and this trend is expected to continue into the next century. By the year 2050, population growth is expected to follow the United Nations medium prediction, resulting in a population of around 10 billion people. One of the most important questions is how global food supply can be raised to meet future population growth. To provide most people with an appropriate diet, current levels of food production would need to be increased more than proportionally to population growth (Kindall and Pimentel

DOI: 10.1201/9781003307464-11

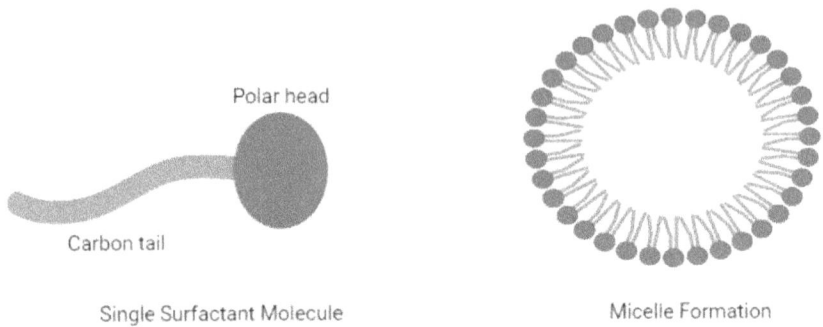

Polar head

Carbon tail

Single Surfactant Molecule   Micelle Formation

FIGURE 11.1   Schematic representation of surfactant molecule (single) and micelle.

1994). Plant health and crop yield must be closely maintained in order to preserve food quality and availability in order to fulfil the world's expanding population demands (Bee et al. 2019). Agronomic sciences, as well as its various fields, should be able to provide solutions for increasing food production. Furthermore, many elements such as irrigation, drainage, tillage, and fertilizer applied to the soil have an impact on crop yields. These treatments change the physical properties of soil, whereas fertilizer use improves the chemical qualities of soil. Finally, the use of pesticides to combat pests and weeds is critical in modern agriculture (Castro et al. 2013). The ability of agricultural output to meet the expanding demands of the human population is a major concern. Biosurfactants can assist in overcoming this problem, and surfactants are used in a variety of agricultural applications.

Biosurfactants have seen a considerable increase in research and development, as well as commercialization of biological agents, in recent years. Microorganisms make a wide variety of amphiphilic metabolites, many of which are structurally distinct. Structural similarities, diameters, moieties, hydrophobicities, degree of change, and other physical and chemical criteria are utilized to categorize microbial biosurfactants in addition to traditional surfactant categorization procedures. In addition to minimizing surface stress, microbial surfactants may offer a number of other advantages. Because of the combined effect of bioactivity and interfacial activity, which is highly dependent on the structure and composition of each molecule, the vast majority of biological structures have unique opportunities for pharmaceutical, agricultural, and environmental applications that have yet to be discovered (Bustamante et al. 2012; Das et al. 2022)

MARINE BIOSURFACTANTS

Marine bacteria have lately emerged as a rich source of these natural compounds, which have surface-active qualities and can be used as detergents, wetting and foaming agents, solubilizers, emulsifiers, and dispersants. Biosurfactants are

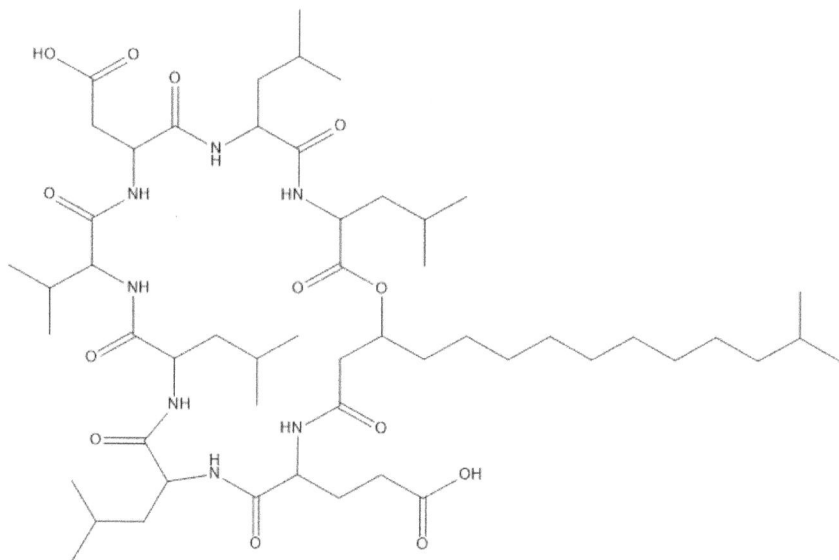

FIGURE 11.2 Structure of biosurfactant molecule (surfactin) isolated from *Bacillus* sp.

microbial molecules that have a high degree of surface activity. Because their complex structures are made up of a hydrophilic and hydrophobic half, biosurfactants have unique amphipathic characteristics (Maneerat 2005). Because they have both hydrophobic and hydrophilic domains in the same molecule, they partition at liquid-liquid interfaces (Desai et al. 1997; Satpute et al. 2010).

DIVERSITY OF BIOSURFACTANTS

Biosurfactants produced by various microorganisms come in a wide variety of chemical structures. The hydrophobic part of biosurfactants usually consists of saturated or unsaturated fatty acids, hydroxy fatty acids, or fatty alcohols with a chain length between 8 and 18 carbon atoms, while the hydrophilic part usually consists of saturated or unsaturated fatty acids, hydroxy fatty acids, or fatty alcohols with a chain length between 8 and 18 carbon atoms. Small hydroxyl, phosphate, or carboxyl groups, as well as carbohydrate (such as mono-, oligo-, or polysaccharides) or (poly-) peptide moieties, make up the hydrophilic components. Anionic and non-ionic chemicals make up the majority of biosurfactants (Kubicki *et al.* 2019)

GLYCOLIPIDS

Glycolipids are microbial surface-active molecules made up of a carbohydrate moiety linked to fatty acids that are produced by a wide range of bacteria. Rhamnose lipids, trehalose lipids, sophorose lipids, cellobiose lipids,

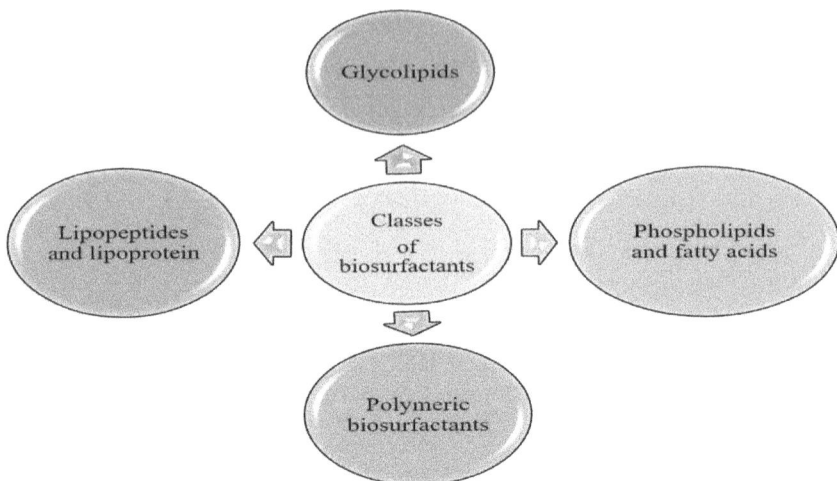

FIGURE 11.3   Various classes of biosurfactants.

mannosylerythritol lipids, lipomannosyl-mannitols, lipomannans and lipo-
arabinomannanes, diglycosyl diglycerides, monoacylglycerol, and galactosyl-
diglyceride are some of the glycolipids (Kitamoto et al. 2002). They work as
biopesticides, preventing plant diseases and extending the shelf life of stored
items. Because of their antifungal effect against phytopathogenic fungi, as well
as their larvicidal and mosquitocidal capabilities, glycolipid biosurfactants protect plants and plant crops from pest capture (Mnif et al. 2016). *Planococcus* sp.
XW-1, a cold-adapted strain, was extracted from the Yellow Sea. The strain may
produce biosurfactant using petroleum as the source for carbon at low temperatures (4°C). The biosurfactant was recognized as a glycolipid-type biosurfactant
species using thin-layer chromatography (TLC) and Fourier transform infrared
spectroscopy (FTIR) (Guo et al. 2022).

LIPOPEPTIDES AND LIPOPROTEINS

Lipopeptides are the most well-known antimicrobial biosurfactant class.
Surfactin is the first and most well-known lipopeptide, produced by *Bacillus
subtilis*. Antibacterial lipopeptides produced by *B. subtilis* include fengycin, iturin, bacillomycins, and mycosubtilins. Lichenysin and pumilacidin are antimicrobial lipopeptides produced by *Bacillus licheniformis* and *Bacillus pumilus*,
respectively.

FATTY ACID AND PHOSPHOLIPID DERIVATIVES

Low molecular weight (LMW) biosurfactants include simple free fatty acids and
phospholipids and amino acids which were coupled to lipids, lipopeptides, and

glycolipids. For example, branched fatty acids studied as corynomycolic acids with chain lengths of C12–C14 showed a significant reduction in surface and interfacial tensions. Lipoamino acid biosurfactants are formed when (hydroxy) fatty acids are bound to non-proteinogenic or proteinogenic amino acids, such as ornithine lipids, lysine lipids, N-acyltyrosines, or cerilipin containing ornithine and taurine obtained by *Myroides* sp., *Gluconobacter cerinus*, and *Nitrosomonas europaea*.

POLYMERIC BIOSURFACTANTS

The well-known polymeric biosurfactants are emulsan, lipomanan, alasan, liposan, and various protein complexes of polysaccharide. Emulsan is a hydrocarbon emulsifier that can emulsify hydrocarbons in water at lowest concentrations of 0.001% to 0.1%. Liposan is an extracellular water miscible emulsifier made up of 83% carbohydrates and 17% proteins that is produced by *C. lipolytica*. Chakrabarti talks about how liposan is used as an emulsifier in the food and cosmetics sectors (Santos et al. 2016).

FACTORS FOR THE PRODUCTION OF MARINE BIOSURFACTANTS

Optimal growth conditions are critical for maximum marine biosurfactant (M-BS) production. The productivity and composition of BSs are altered by factors such as carbon availability, pH, temperature, salinity, nitrogen, and agitation. Many marine species need salt to thrive; some moderate halophiles need 15% (w/v) NaCl for optimal growth, while extreme halophiles need 25% (w/v) NaCl for good growth (Margesin and Schinner 2001). Carbon sources in the growing medium alter the configuration of BS synthesis. Carbon sources for M-BS production have included glycerol, glucose, crude oil, sucrose, and diesel. M-BSs can be produced by marine bacteria using hydrocarbons as a substrate. Nitrogen is another essential source; the kind and concentration of nitrogen supply are critical elements in optimizing M-BS synthesis. M-BSs can be created from ammonium sulfate, sodium nitrate, urea, peptone, ammonium nitrate, and yeast extract, among other nitrogen sources (Davis et al. 1999). Growth circumstances (temperature, pH, agitation speed, and oxygen) are also crucial factors that influence cell proliferation and M-BS synthesis. Candida species produces maximum amounts of biosurfactant throughout a wide pH range, including pH 5.7 for *Candida glabrata*, pH 7.8 for *Candida* sp., pH 5.1 for *Candida Lipolytica*, and pH 6.0 for *Candida batistae*. Furthermore, at pH 5.5 and 7.0, *Pichia anamola* and *Aspergillus ustus* produce the highest biosurfactant yields. For several species of Candida, such as *Candida* sp. SY16, *C. bombicola*, *T. bombicola*, and *C. batistae*, the most favorable temperature for the synthesis of biosurfactants by diverse fungus is 30°C. The optimal temperature for *C. lipolytica* has been discovered to be 27°C. The duration of incubation has a considerable impact on biosurfactant generation (Santos et al. 2016). Furthermore, increasing the agitation speed allowed *P. aeruginosa* UCP 0992 cultured in glycerol to accumulate a bio-surfactant. The industrial world prefers M-BSs produced by thermophilic micro-organisms because of their high thermos ability at temperatures above 40°C; however, M-BSs produced by mesophilic microorganisms are also very thermo-stable, and psychrophilic marine bacteria competent of producing M-BS could be used for bioremediation in cool environments (Tripathi et al. 2018).

S. No	Source Type	Source	Examples	References
1.	Carbon	Crude oil, diesel, glucose, sucrose, glycerol.	*Halomonas* sp. strain C2SS100	(Mnif et al. 2009)
			Brevibacterium luteolum	(Vilela et al. 2014).
			Brevibacillus strain	(Reddy et al. 2010)
			Alteromonas sp. *17*	(Al-Mallah et al. 1990)
			Corynebacterium kutscheri	(Thavasi et al. 2007)
2.	Nitrogen	Sodium nitrate, yeast extract, urea, peptone, ammonium nitrate, ammonium sulfate	*Streptomyces* species B3	(Khopade et al. 2012b).
			B. subtilis N3–4P	(Zhu et al. 2016)
			Nocardiopsis B4	(Khopade et al. 2012a)

PROPERTIES OF BIOSURFACTANTS

Biosurfactants were found to have superior characteristics to chemically manu-factured counterparts, as well as a wide range of substrate availability, making them acceptable for commercial use. Surface movement, resilience to tempera-ture, pH and ionic quality, biodegradability, low toxic quality, demulsifying and emulsifying capacity, and antimicrobial action are all characteristics of microbial surfactants.

ANTI-ADHESIVE PROPERTY

Biosurfactants are surface-active products of microbial metabolism that can reduce surface and interfacial tension. The hydrophobicity of a substratum sur-face is affected by biosurfactant adsorption, interfering with microbial adhesion and desorption processes. In addition to their antifungal, antibacterial, and anti-viral capabilities, these compounds have been proven to be potent inhibitors of microbial adhesion and as well as biofilm formation (Das et al. 2008). *E. coli, M. flavus*, and *P. vulgaris* were used to assess the biosurfactant's antiadhesive effectiveness against a variety of potentially hazardous opportunistic microor-ganisms. As a consequence, 10 g L1 pure biosurfactant could inhibit up to 89% of microbe adherence (Das et al. 2009)

Major properties

Anti-adhesive agents

Biodegradability and Low toxicity

Temperature and pH tolerance

Surface and interface activity

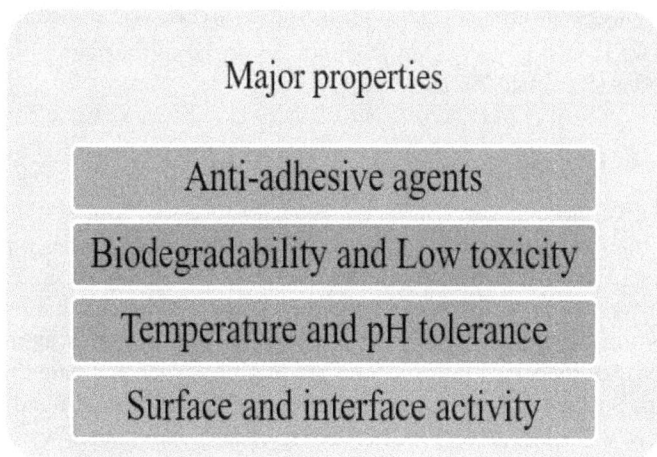

FIGURE 11.4 Properties of biosurfactants.

BIODEGRADABILITY AND LOW TOXICITY

Biosurfactants are environmentally beneficial since they have high biodegradability; have lower toxicity compared with synthetic surfactants; have their own distinct activity; are effective at high variations of temperature, pH, and salinity; and are safe to use. The maritime environment is recognized as a viable source for biosurfactants with low toxicity, environmental compatibility, and biodegradability when compared to their synthetic equivalents. Rhamnolipids were produced by *Marinobacter* MCTG107b and *Pseudomonas* MCTG214(3b1) cultures that had previously been identified and described, and their toxic profile was investigated using well-established techniques. *In vitro* models of human skin and also liver were tested with alamar blue and propidium iodide and demonstrated minimal cytotoxicity (Voulgaridou et al. 2021).

TEMPERATURE AND pH TOLERANCE

Microorganisms from the sea can be found in a range of environments, including the ocean and severe environments. The pH, salinity, and temperature ranges in the ocean are very narrow, whereas volcanic eruptions are subjected to extremes. These bacteria have metabolic and physiological modifications that enable them to survive under very extreme temperatures, pressures, salinity, and pH (Das et al. 2009). Similarly, when the C/N ratio was 2:1 and the best bioprocess parameters were pH 7.0, 30°C, and 3% salt content, *Nocardiopsis* produced the greatest biosurfactant (Khopade et al. 2012a). *Brevundimonas* sp. IITISM 11, *Pseudomonas* sp. IITISM 19, and *Pseudomonas* sp. IITISM 24 produced biosurfactants that were found to be stable throughout a wide range of temperature (0–100°C), pH (4–12), and salinity (up to 20% NaCl). In addition, the strains displayed endurance to high concentrations of anthracene and fluorene (275 mg/L), as well as a high level of hydrophobicity on the cell surface with different hydrocarbons (Ray et al. 2021).

SURFACE AND INTERFACE ACTIVITY

Surface-active biosurfactants have the capacity to lower both surface tension (ST) and interfacial tension (IT). Surface behavior, which entails a reduction of ST and IT, is one of the BSs' fundamental features. BSs build up at the fluid surface or interface, weakening the intermolecular interactions that hold liquid molecules together. When their bulk is lowered, biosurfactants exhibit lower surface tension, increased potency, and productivity when compared to regular surfactants. Surface and interfacial tension (oil/water) are around 1 and 30 ml/m, respectively, while biosurfactant CMC (competence measurement) varies from 1 to 2000 mg/l (Shakeri et al. 2020; Joshi et al. 2016).

AGRICULTURE-RELATED APPLICATIONS OF BIOSURFACTANTS

Biosurfactants have greater advantages than chemically manufactured surfactants due to their dual hydrophobic and hydrophilic characteristics. These biosurfactants can be used in a variety of agricultural applications (Ranjan and Sow 2021).

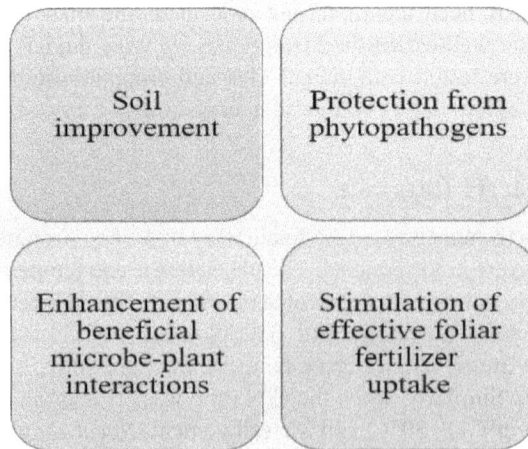

Soil improvement	Protection from phytopathogens
Enhancement of beneficial microbe-plant interactions	Stimulation of effective foliar fertilizer uptake

FIGURE 11.5   Applications of biosurfactants in agriculture.

IMPROVEMENT OF SOIL QUALITY

The presence of various organic and inorganic pollutants that cause abiotic stress to farmed crop plants reduces the productivity of agricultural land. Bioremediation is essential to improve the quality of soil contaminated by hydrocarbons and heavy metals. Different technologies, such as soil washing techniques and clean-up combination technology, use biosurfactants to effectively remove hydrocarbons and metals, respectively, because biosurfactants are known to improve bioavailability and also achieve biodegradation of hydrophobic substances (Kang et al. 2010). Biosurfactants can also help to break down certain chemical insecticides that have built up in agricultural soil. Rhamnolipids have been discovered to be effective at removing pentachlorophenol and polyaromatic hydrocarbons from soil. As a result, biosurfactants can be used to improve the quality of agricultural soil. However, the expensive expense of producing biosurfactants keeps these green surfactants from being used for bioremediation of crude oil and/or petroleum contamination in soil (Moldes et al. 2011). The use of agro-industrial waste materials to produce green surfactants that may then be utilized to biodegrade hydrocarbons from soil must be further investigated. With a new method of foaming surfactant technology, biosurfactants such as surfactin and rhamnolipid are known to remove heavy metals such as Ni, Ba, Li, Mg, Mn, Ca, Cd, Cu, and Zn (ions) from soil. As a result, rather than using damaging synthetic surfactants, biosurfactant overproducers may be the most beneficial for bioremediation (Ranjan and Sow 2021).

PLANT PATHOGEN ELIMINATION

Several other biosurfactants derived from bacteria exhibit antimicrobial activity against various plant diseases, making them a prospective biocontrol molecule for achieving long-term agricultural sustainability. Rhizobacteria biosurfactants are known to exhibit antagonistic characteristics (Nihorimbere et al. 2011). Parasitism, antibiosis, competition, induced systemic resistance, and hypo virulence are some of the biocontrol mechanisms used by plant growth-promoting bacteria when chemical surfactants and biosurfactants are used in agriculture (Singh et al. 2007). These surfactants are employed in agriculture to boost the antagonistic actions of microorganisms and microbial products in large quantities (Jazzar and Hammad 2003; Kim et al. 2004). Surfactants have been shown to improve the insecticidal activities of other systems in a number of *in vitro* and *in situ* experiments (Jazzar and Hammad 2003; Gronwald et al. 2002). Furthermore, along with fungus (*Myrothecium verrucaria*), these surfactants are employed in conjunction to remove weed species that have a negative impact on land production as well as the proliferation of such weed species (Boyette et al. 2002). Aflatoxin synthesis by *Aspergillus* sp., which infects crops such as peanuts, cotton seed, and corn during storage as well as in agricultural fields, has been observed to be inhibited by surfactants. Thus, surfactants (both synthetic and biological) have a variety of roles in plant pathogen removal, both directly and indirectly, as well as in several agricultural processes (Rodriguez and Mahoney 2006).

ASSETS FOR BENEFICIAL PLANT MICROBE INTERACTION

It is critical for rhizobacteria to have contact with plant surfaces such as roots in order to give favorable effects to plants (Nihorimbere et al. 2011). To build an association with the plant, microbial features such as motility, the ability of biofilm formation on the root surface, and the production of quorum-sensing molecules are necessary. Quorum-sensing molecules like acyl homoserine lactone (AHL) are essential for rhizobacteria to synthesize antifungal chemicals. According to studies, the concentration of these molecules in the rhizosphere is higher than in bulk soil (soil away from plant roots), implying that AHL and AHL-like molecules play a role in colony competence in rhizosphere (ability of beneficial microorganisms to colonize the root surface). These AHLs are linked to the control of exopolysaccharide, which is required for biofilm development (Newton and Fray 2004; Loh et al. 2002). The biosurfactant (rhamnolipid) generated by *Pseudomonas* spp. modulates the quorum-sensing mechanism (intercellular communication). Biosurfactants are also said to alter microbe movement, contribute in signaling and differentiation, and help create biofilms (Ron and Rosenberg 2011). As a result, these green surfactants are critical factors for microorganisms to form a good relationship with plant roots and also boost plant growth. Furthermore, rhizobacteria produce biosurfactants that increase the bioavailability of hydrophobic compounds that could be used as nutrients. Soil bacteria create biosurfactants, which offer wettability and aid in the appropriate dispersion of chemical fertilizers in the soil, promoting plant growth (Dusane et al. 2010).

POTENTIAL OF BIOSURFACTANTS IN PESTICIDE INDUSTRIES

Fungicides, insecticides, and herbicides all require surfactants as adjuvants. Synthetic surfactants are being employed in the pesticide industry as emulsifying, dispersing, spreading, and wetting agents to improve pesticide efficiency. Furthermore, because these surfactants have defensive qualities, they are used in insecticides in modern agriculture (Rostas and Blassmann 2009). Surfactants of many sorts, including amphoteric, anionic, cationic, and nonionic, are being employed in the pesticide production industry (Mulqueen 2003). As a result, surfactants are commonly employed in pesticide composition. It is crucial to note, however, that the surfactants used in pesticide formulations accumulate inside the soil, affecting the texture, color, and growth of the plant. These hazardous chemicals are also leached into groundwater from the soil (Blackwell 2000). Pesticide residues are known to last for years in soil, and they can also spread through the air and water. These can even be seen on the outside of fruits and vegetables (Street 1969). Synthetic surfactants are also regarded to be powerful organic contaminants in soil (Petrovic and Barcelo 2004). Given the negative effects of pesticides and surfactants used in pesticides, environmentally acceptable biosurfactants must be used to replace these hazardous surfactants in the multibillion-dollar pesticide industry, reducing contamination (Hopkinson et al. 1997). Another solution to this environmental problem could be research into soil bacteria that can use the chemical surfactant in farm soil as a carbon source. *Pseudomonas* sp. and *Burkholderia* sp. bacteria from rice fields have been found to break down surfactants, according to a study. Important agricultural products, such as pesticides made with the help of biosurfactants, can be widely employed on agricultural fields. Many companies can use a blend of biosurfactants in diverse combinations with polymers to generate good formulations for agricultural uses, and to reach this goal, agrochemical manufacturers must develop effective formulation technology. (Nishio et al. 2002)

They can be widely used in agriculture to improve the condition of agricultural soils through soil remediation. As a result, biosurfactants derived from environmental isolates have the potential to aid plant growth and other agricultural applications. Surfactants are employed in crop protection and also pesticide formulations in the amount of 0.2 million tons per year, according to statistics from 2004. Green surfactants (biosurfactants derived from microorganisms) have been proven in several studies to have advantages over synthetic surfactants (Renfro et al. 2014; Liu et al. 2016). Antimicrobial and anti-biofilm effects can be used in the food business to sanitize production equipment and prevent food spoilage, in addition to their applications in food manufacturing (Kiran et al. 2017). Food applications primarily provide consistency control, stable solubilization of components such as flavor oils, and fat stability in food items. Furthermore, marine fungi can provide chemicals such as massoia lactone, which has a pleasing odor and flavor (Luepongpattana et al. 2017). It could be used as a new natural resource for the molecule as an alternative for the currently dominant synthetic chemistry-based production pathway. Finally, food wastes could be employed as a starting point for the production of biosurfactants. A clever mix of these applications could finally permit the "cradle to cradle" approach demanded by a strictly circular bio-economy (de Araujo et al. 2014).

CONCLUSION

Surfactants are used in agriculture and the agrochemical industry in a variety of ways. However, biosurfactants, which are more environmentally friendly, are rarely used. The precise role of surfactants in supporting other systems as biological control agents is still unknown, and further research is needed. Such research will support the replacement of toxic chemical surfactants with green alternatives. To obtain a net economic gain from biosurfactant application in agriculture and other sectors, work on the production cost of green surfactants is required. The use of agricultural waste for biosurfactant overproduction also requires greater consideration. The chemical compositions of biosurfactants that have been described as effective biocontrol agents can be changed by altering the

manufacturing process. This method could lead to the production of green surfactants with great target specificity. The significant frequency of biosurfactants
and biosurfactant-producing bacteria in the rhizosphere is evidence of the rhizosphere's importance in sustainable agriculture. In the literature, biosurfactant
producers are mostly *Pseudomonas* and *Bacillus* species, showing that only a few
genera have been examined thus far. Functional metagenomics, for example, is a
cutting-edge method that could lead to the identification of new green surfactants.
To avoid the negative consequences of synthetic surfactants, which are widely
used in many commercial sectors, including the agrochemical industry, intensive
research on green surfactants is a top priority. As a result, it can be determined
that researchers from many domains such as molecular biology, biochemistry,
microbiology, and computational biology can contribute to the final product.

REFERENCES

Al-Mallah, Maha, Madeleine Goutx, Gilbert Mille, and Jean-Claude Bertrand.
"Production of emulsifying agents during growth of a marine *Alteromonas* in sea
water with eicosane as carbon source, a solid hydrocarbon." *Oil and Chemical
Pollution* 6, no. 4 (1990): 289–305.
Bee, H., M.Y. Khan, and R.Z. Sayyed. "Microbial surfactants and their significance in
agriculture." In *Plant Growth Promoting Rhizobacteria (PGPR): Prospects for
Sustainable Agriculture*, pp. 205–215. Singapore: Springer, 2019.
Blackwell, P.S. "Management of water repellency in Australia and risks associated with
preferential flow, pesticide concentration and leaching." *Journal of Hydrology* 231–
232 (2000): 384–395.
Boyette, C.D., H.L. Walker, and H.K. Abbas. "Biological control of kudzu (*Pueraria
lobata*); with an isolate of *Myrothecium verrucaria*." *Biocontrol Science and
Technology* 12 (2002): 75–82.
Bustamante, M., N. Duran, and M.C. Diez. "Biosurfactants are useful tools for the bio-
remediation of contaminated soil: A review." *Journal of Soil Science and Plant
Nutrition* 12, no. 4 (2012): 667–687.
Castro, M.J.L., C. Ojeda, and A.F. Cirelli. "Surfactants in agriculture." *Green Materials
for Energy, Products and Depollution* (2013): 287–334.
Das, K., S. Das, M.K. Sinha, M. Das, M. Bolem, and N. Pal. "Cyanobacterial biosurfac-
tants in the bioremediation of oil industries effluents." In *Microbial Surfactants* (pp.
183–196). Boca Raton, FL: CRC Press, 2022.
Das, P., S. Mukherjee, and R. Sen. "Antimicrobial potential of a lipopeptide biosurfactant
derived from a marine *Bacillus circulans*." *Journal of Applied Microbiology* 104,
no. 6 (2008): 1675–1684.
Das, P., S. Mukherjee, and R. Sen. "Antiadhesive action of a marine microbial surfactant." *Colloids and Surfaces B: Biointerfaces* 71, no. 2 (2009): 183–186.
Davis, D.A., H.C. Lynch, and J. Varley. "The production of surfactin in batch culture by
Bacillus subtilis ATCC 21332 is strongly influenced by the conditions of nitrogen
metabolism." *Enzyme and Microbial Technology* 25, no. 3–5 (1999): 322–329.
de Araujo, L.V., D.M. Guimarães Freire, and M. Nitschke. *Perspectives on Using
Biosurfactants in Food Industry*. Boca Raton, FL: CRC Press Taylor & Francis
Group, 2014.

Desai, J.D., and I.M. Banat. "Microbial production of surfactants and their commercial potential." *Microbiology and Molecular Biology Reviews* 61, no. 1 (1997): 47–64.

Dusane, D., P. Rahman, S. Zinjarde, V. Venugopalan, R. McLean, and M. Weber. "Quorum sensing; implication on rhamnolipid biosurfactant production." *Biotechnology and Genetic Engineering Reviews* 27 (2010): 159–184.

Gronwald, J.W., K.L. Plaisance, D.A. Ide, and D.L. Wyse. "Assessment of *Pseudomonas syringae* pv. tagetis as a biocontrol agent for Canada thistle." *Weed Science* 50 (2002): 397–404.

Guo, P., W. Xu, S. Tang, B. Cao, D. Wei, M. Zhang, J. Lin, and W. Li. "Isolation and characterization of a biosurfactant producing strain *Planococcus* sp. XW-1 from the cold marine environment." *International Journal of Environmental Research and Public Health* 19, no. 2 (2022): 782.

Hopkinson, M.J., H.M. Collins, and G.R. Goss. "Pesticide formulations and application systems: ASTM Committee E-35 on Pesticides." *ASTM International* 17, no. 1328 (1997): 1–331.

Jazzar, C., and E.A. Hammad. "The efficacy of enhanced aqueous extracts of melia aze-darach leaves and fruits integrated with the *Camptotylus reuteri* releases against the sweet potato whitefly nymphs." *Bulletin of Insectology* 56 (2003): 269–275.

Joshi, S.J., Y.M. Al-Wahaibi, S.N. Al-Bahry, A.E. Elshafie, A.S. Al-Bemani, A. Al-Bahri, and M.S. Al-Mandhari. "Production, characterization, and application of *Bacillus licheniformis* W16 biosurfactant in enhancing oil recovery." *Frontiers in Microbiology* 7 (2016): 1853.

Kang, S.W., Y.B. Kim, J.D. Shin, and E.K. Kim. "Enhanced biodegradation of hydrocar-bons in soil by microbial biosurfactant, sophorolipid." *Applied Biochemistry and Biotechnology* 160 (2010): 780–790.

Khopade, A., R. Biao, X. Liu, K. Mahadik, L. Zhang, and C. Kokare. "Production and stability studies of the biosurfactant isolated from marine *Nocardiopsis* sp. B4." *Desalination* 285 (2012a): 198–204.

Khopade, Abhijit, Biao Ren, Xiang-Yang Liu, Kakasaheb Mahadik, Lixin Zhang, and Chandrakant Kokare. "Production and characterization of biosurfactant from marine *Streptomyces* species B3." *Journal of Colloid and Interface Science* 367, no. 1 (2012b): 311–318.

Kim, P.I., H. Bai, D. Bai, H. Chae, S. Chung, Y. Kim, R. Park, and Y.T. Chi. "Purification and characterization of a lipopeptide produced by *Bacillus thuringiensis* CMB26." *Journal of Applied Microbiology* 97 (2004): 942–949.

Kindall, H.W., and D. Pimentel. "Constraints on the expansion of the global food supply." *Ambio* 23 (1994): 198–205.

Kiran, G.S., S. Priyadharsini, A. Sajayan, G.B. Priyadharsini, N. Poulose, and J. Selvin. "Production of lipopeptide biosurfactant by a marine *Nesterenkonia* sp. and its application in food industry." *Frontiers in Microbiology* 8 (2017): 1138.

Kitamoto, D., H. Isoda, and T. Nakahara. "Functions and potential applications of gly-colipid biosurfactants—From energy-saving materials to gene delivery carriers." *Journal of Bioscience and Bioengineering* 94, no. 3 (2002): 187–201.

Kubicki, Sonja, Alexander Bollinger, Nadine Katzke, Karl-Erich Jaeger, Anita Loeschcke, and Stephan Thies. "Marine biosurfactants: Biosynthesis, structural diversity and biotechnological applications." *Marine Drugs* 17, no. 7 (2019): 408.

Liu, H., B. Shao, X. Long, Y. Yao, and Q. Meng. "Foliar penetration enhanced by biosurfactant rhamnolipid." *Colloids and Surfaces B: Biointerfaces* 145 (2016): 548–554.

Loh, J., E.A. Pierson, L.S. Pierson, G. Stacey, and A. Chatterjee. "Quorum sensing in plant-associated bacteria." *Current Opinion in Plant Biology* 5 (2002): 1–5.

Luepongpattana, S., J. Thaniyavarn, and M. Morikawa. "Production of massoia lactone by *Aureobasidium pullulans* YTP6–14 isolated from the Gulf of Thailand and its fragrant biosurfactant properties." *Journal of Applied Microbiology* 123 (2017): 1488–1497.

Maneerat, S. "Biosurfactants from marine microorganisms." *Songklanakarin Journal of Science and Technology* 27, no. 6 (2005): 1263–1272.

Margesin, R., and F. Schinner. "Potential of halotolerant and halophilic microorganisms for biotechnology." *Extremophiles* 5, no. 2 (2001): 73–83.

Mnif, I., and D. Ghribi. "Glycolipid biosurfactants: Main properties and potential applications in agriculture and food industry." *Journal of the Science of Food and Agriculture* 96, no. 13 (2016): 4310–4320.

Mnif, S., M. Chamkha, and S. Sayadi. "Isolation and characterization of *Halomonas* sp. strain C2SS100, a hydrocarbon-degrading bacterium under hypersaline conditions." *Journal of Applied Microbiology* 107, no. 3 (2009): 785–794.

Moldes, A.B., R. Paradelo, D. Rubinos, R. Devesa-Rey, J.M. Cruz, and M.T. Barral. "*Ex situ* treatment of hydrocarbon-contaminated soil using biosurfactants from *Lactobacillus pentosus*." *Journal of Agricultural and Food Chemistry* 59 (2011): 9443–9447.

Mulqueen, P. "Recent advances in agrochemical formulations." *Advances in Colloid and Interface Science* 106 (2003): 83–107.

Newton, J.A., and R.G. Fray. "Integration of environmental and hostderived signals with quorum sensing during plant–microbe interactions." *Cellular Microbiology* 6 (2004): 213–224.

Nihorimbere, V., M.M. Ongena, M. Smargiassi, and P. Thonart. "Beneficial effect of the rhizosphere microbial community for plant growth and health." *Biotechnology, Agronomy and Society and Environment* 15 (2011): 327–337.

Nishio, E., Y. Ichiki, H. Tamura, S. Morita, K. Watanabe, and H. Yoshikawa. "Isolation of bacterial strains that produce the endocrine disruptor, octylphenol diethoxylates, in paddy fields." *Bioscience, Biotechnology, and Biochemistry* 66 (2002): 1792–1798.

Petrovic, M., and D. Barcelo. "Analysis and fate of surfactants in sludge and sludge-amended soil." *Trends in Analytical Chemistry* 23 (2004): 10–11.

Ranjan, S., and S. Sow. "Biosurfactants as a biological tool to increase agricultural productivity." *Agriculture Food and Newsletter* 3, no. 4 (2021): 274–277.

Ray, M., V. Kumar, C. Banerjee, P. Gupta, S. Singh, and A. Singh. "Investigation of biosurfactants produced by three indigenous bacterial strains, their growth kinetics and their anthracene and fluorene tolerance." *Ecotoxicology and Environmental Safety* 208 (2021): 111621.

Reddy, M. Srikanth, B. Naresh, T. Leela, M. Prashanthi, N. Ch Madhusudhan, G. Dhanasri, and Prathibha Devi. "Biodegradation of phenanthrene with biosurfactant production by a new strain of *Brevibacillus* sp." *Bioresource Technology* 101, no. 20 (2010): 7980–7983.

Renfro, T.D., W. Xie, G. Yang, and G. Chen. "Rhamnolipid surface thermodynamic properties and transport in agricultural soil." *Colloids and Surfaces B: Biointerfaces* 115 (2014): 317–322.

Rodrigues, L., I.M. Banat, J. Teixeira, and R. Oliveira. "Biosurfactants: Potential applications in medicine." *Journal of Antimicrobial Chemotherapy* 57 (2006): 609–618.

Ron, E.Z., and E. Rosenberg. "Natural roles in biosurfactants." *Environmental Microbiology* 3 (2011): 229–236.

Rostas, M., and K. Blassmann. "Insects had it first: Surfactants as a defense against predators." *Proceedings of the Royal Society B* 276 (2009): 633–638.

Santos, Danyelle Khadydja F., Raquel D. Rufino, Juliana M. Luna, Valdemir A. Santos, and Leonie A. Sarubbo. "Biosurfactants: Multifunctional biomolecules of the 21st century." *International Journal of Molecular Sciences* 17, no. 3 (2016): 401.

Satpute, S.K., A.G. Banpurkar, P.K. Dhakephalkar, I.M. Banat, and B.A. Chopade. "Methods for investigating biosurfactants and bioemulsifiers: A review." *Critical Reviews in Biotechnology* 30, no. 2 (2010a): 127–144.

Shakeri, F., H. Babavalian, M. Ali Amoozegar, Z. Ahmadzadeh, S. Zuhuriyanizadi, and M.P. Afsharian. "Production and application of biosurfactants in biotechnology." *Biointerface Research in Applied Chemistry* 11 (2020): 10446–10460.

Singh, A., J.D. Van Hamme, and O.P. Ward. "Surfactants in microbiology and biotechnology: Part 2: Application aspects." *Biotechnology Advances* 25 (2007): 99–121.

Street, J.C. "Methods of removal of pesticides residues." *Canadian Medical Association Journal* 100 (1969): 154–160.

Thavasi, R., S. Jayalakshmi, T. Balasubramanian, and Ibrahim M. Banat. "Biosurfactant production by *Corynebacterium kutscheri* from waste motor lubricant oil and peanut oil cake." *Letters in Applied Microbiology* 45, no. 6 (2007): 686–691.

Tripathi, L., V.U. Irorere, R. Marchant, and I.M. Banat. "Marine derived biosurfactants: A vast potential future resource." *Biotechnology Letters* 40, no. 11 (2018): 1441–1457.

Vilela, W.F.D., S.G. Fonseca, F. Fantinatti-Garboggini, V.M. Oliveira, and M. Nitschke. "Production and properties of a surface-active lipopeptide produced by a new marine *Brevibacterium luteolum* strain." *Applied Biochemistry and Biotechnology* 174, no. 6 (2014): 2245–2256.

Voulgaridou, G.-P., T. Mantso, I. Anestopoulos, A. Klavaris, C. Katzastra, D.-E. Kiousi, M. Mantela et al. "Toxicity profiling of biosurfactants produced by novel marine bacterial strains." *International Journal of Molecular Sciences* 22, no. 5 (2021): 2383.

Zhu, Zhiwen, Baiyu Zhang, Bing Chen, Qinghong Cai, and Weiyun Lin. "Biosurfactant production by marine-originated bacteria *Bacillus subtilis* and its application for crude oil removal." *Water, Air, & Soil Pollution* 227, no. 9 (2016): 1–14.

12 Cosmetic Application of Surfactants from Marine Microbes

Tirth Bhatt, Avani Bhimani, Asmita Detroja,
Dhruv Gevariya and Gaurav Sanghvi

CONTENTS

DOI: 10.1201/9781003307464-12

INTRODUCTION

Cosmetic practices are challenging to abolish because of their pervasiveness. The word "cosmetics" originates from the Greek word "kosmeticos," which signifies embellishment (Khan and Alam 2019). The science and practice of cosmetology are said to have begun in the ancient world in nations such as Egypt and India, although the oldest records of cosmetic components and their use date back to the Indus Valley civilization between 2500 and 1500 BC (Duncan 1952). Cosmetics were fiercely opposed and seen as immoral in the world throughout the Victorian era. Historically, the Greeks, Egyptians, and Romans utilized a range of cosmetics containing mercury and lead. There was a famous historical notion that eye cosmetics could banish evil spirits and improve eyesight (Angeloglou 1973).

India's cosmetics and personal care sector is one of the fastest-growing consumer goods categories, with enormous prospects for foreign industries. In the projected timeslot (2016–2022), the global cosmetics sector is predicted to have a cumulative yearly prosperity rate of 4.3%, extending to $429.8 billion by 2022 (Martins et al. 2014). Cosmetics are not discouraged or outlawed in most of today's industrialized countries, and their consumption is prevalent. Various governmental authorities worldwide regulate the manufacture and sale of cosmetic items. There may be different regulatory systems in existence; however, they all aim to ensure that cosmetic items are safe and correctly labeled (Corinaldesi 2015). These constraints have evolved so that they are relatively stringent in advanced industrialized countries. The United States and the European Union are primary consumers of cosmetic items. India's cosmetics business is expanding at a 15-percentage-point annual rate. The cosmetics industry is mature and responsible enough to ensure the quality and safety of its products. Cosmetic regulations are complex and time consuming for pre-approval and marketing permission in India. Compared to the European Union and the United States, India's regulatory system is different. The Indian cosmetics business is developing at a 13–18% quicker rate than the US or European markets. According to a United Business Media (UBM) Limited India survey, India's cosmetics market would expand by 25% to $20 billion by 2025. The Indian beauty and personal care market (BPCM) divides cosmetic products into different categories: (1) color cosmetics, (2) haircare, (3) hand care, (4) face care, and (5) body care (Danovaro et al. 2014).

Cosmetics are classified as any product or composition that comes in contact with the various outermost components of the body layer like the epidermis

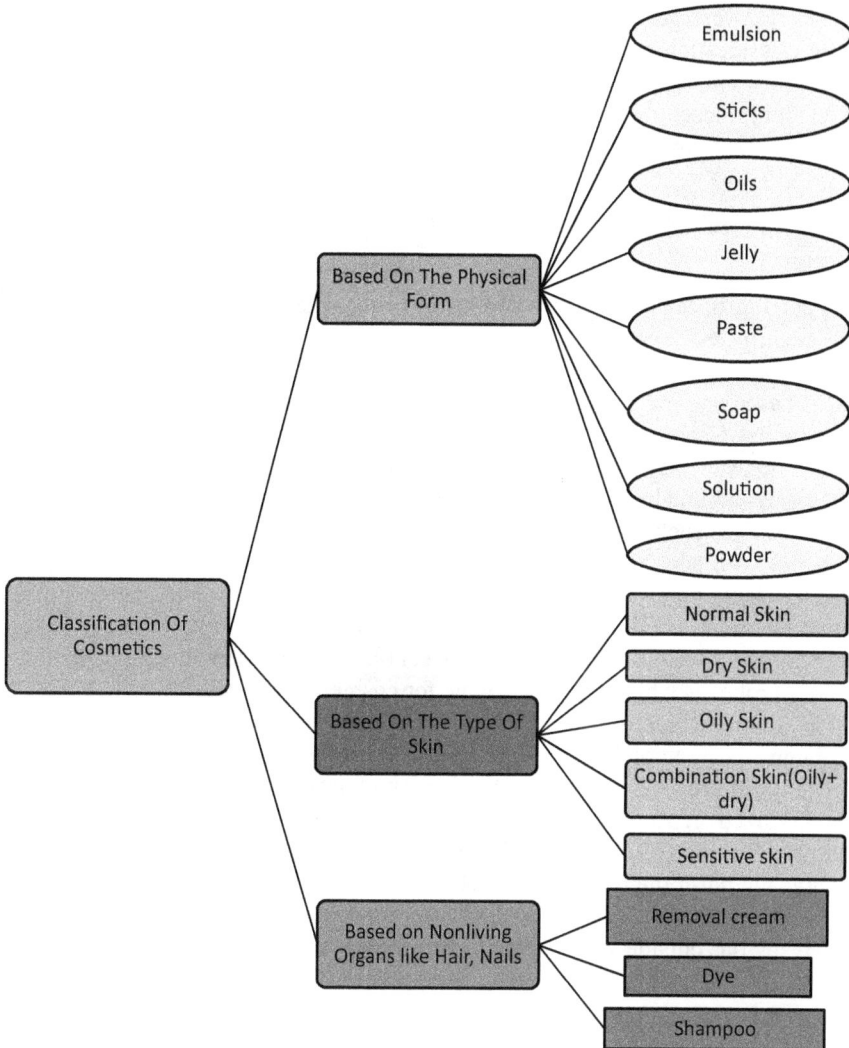

FIGURE 12.1 Classification of cosmetics.

layer, nails, hairs, mucous membranes, teeth, and so on. Polymers may be located in a range of hair products, such as conditioner, hair dye, shampoo, fixing gels, and moisturizing masks; in skincare products, including soaps, sunscreen, moisturizing lotions, for nail care, skincare, and aroma (Duncan 1952). Each of those products has a distinct purpose and usability and differences in formulation, production process, and physiochemical qualities that need a broad spectrum of polymers. Polymers are a significant primary adhesive type used in cosmetic synthesis, as they are required in the manufacturing process.

They are categorized as natural, synthetic, or semi-synthetic macromolecules. The classification is based on the many repeating chain monomer units. This systemic complexity is used in cosmetic preparations to enhance various roles, including thickening agents, rheological property adjusters, foam destabilizers, fixatives, and emulsifiers (Angeloglou 1973).

SYNTHETIC COSMETIC POLYMERS

Synthetic cosmetics are interesting as an active ingredient in cosmetics manufacturing because they can be customized for specific demands. They are frequently less expensive than natural cosmetic polymers. They all have a broad scale of consistency and a long service life. Polyacrylamides, acrylic acid-based polymers, and silicon are the most often seen synthetic polymers in cosmetics. Silicon products have been used in the cosmetics market in recent years (Gawade et al. 2020). Silicon and its compounds are employed as stabilizing agents and in activities such as emulsification or as a thickening material in cosmetics such as shampoo, lotions, and deodorants. Polyethylene glycols (PEGs) with non-ionic or anionic derivatives are broadly applied in cosmetics as emulsifiers, emollients, and penetration enhancers. They are found in bath items, skincare, shaving, makeup, skin cleansing products, shampoo, and deodorants. Because of their properties, aliphatic polyesters are gaining favor for microbeads which are sustainable for cosmetics. However, on the negative side, synthetic components harm skin, such as the use of silicon and PEGs. Silicon, for example, covers skin pores and impairs the nutrition-absorbing channel. It creates pimples on the skin and hurts the environment. PEGs are synthesized petrochemicals and contain impurities such as 1,4-dioane and ethylene oxide (Fruijtier and Polloth 2005). These compounds are carcinogenic and harm the respiratory system. These are just two cases, but the bulk of synthetic chemicals included in cosmetic preparations either react with the body or the outside environment, with adverse implications for the human body.

SEMI-SYNTHETIC COSMETIC POLYMERS

In this group, mainly cellulose ethers and esters are prominent replacements of cellulose with other mechanical and physicochemical characteristics. These derivatives display surface activity, solution viscosity, heat, and thermostable film qualities and show biodegradation, oxidation, and hydrolysis (Germenshaus et al. 2014). Hydrophilic ether derivatives of cellulose include ethylcellulose, methylcellulose, hydroxypropyl methylcellulose, hydroxyethylcellulose, and sodium carboxymethyl cellulose (Germenshaus et al. 2014). Hydrophobic esters of cellulose include hydroxypropyl methylcellulose phthalate, cellulose acetate butyrate, cellulose acetate, cellulose acetate trimelitate, and cellulose acetate phthalate. These polymers are frequently utilized as thickening agents, gelling agents, and bioadhesives in cosmetics such as shampoos, creams, gels, and lotions. They are less susceptible to microbial contamination than natural gelling agents like starch, agar, sodium alginate, and gelatin (Ammala 2012). These compounds are also a combination of natural and synthetic substances. Adopting semi-synthetic substances is quite expensive, and they may degrade more slowly in the environment than natural goods.

NATURAL COSMETIC POLYMERS

Natural polymers are heavily exploited in cosmetic usage because they are safe and biocompatible, which consumers find quite appealing. They are eco-friendly and appropriate for a wide range of applications, including skin makeup, hair care, and stabilizers (Klein and Povernov 2020). Polysaccharides, starch, guar gum, alginate, pectin agar, gelatin, carrageenan, and so on are among the most frequently exploited natural polymers. Starch is a naturally present polysaccharide that can be applied in two forms: granule starch and soluble starch. When soluble starch is boiled during extraction, it becomes humidity resistant and as result gives smooth hair and skin (Serzan and Atkins 2021). The hydrogen bonds in granule starch are unbreakable, resulting in smooth, moisturized skin and hair with reduced greasiness. Because of these variable amylose concentrations, various starch sources, such as potato, corn, and cassava root, all contribute distinct properties to cosmetic formulations. The mixing of starch with other organic polymers like chitosan to promote antioxidant production and other skin-related effects has been used. As listed, many natural compounds are used for making cosmetics. Most of the time, plant or animal extracts are applied as cosmetic polymers. The majority of the time, plant extractions are used for starch (Semenzato et al. 2014). At the same time, compounds such as tallow, which is a characteristic constituent in the formulation of lipstick, eye makeup, and other cosmetics, are created from animal fat. However, to minimize the expenses of this process, microorganisms are being used to create the molecules needed to make cosmetics.

In most cases, the name "microorganisms" emphasizes the size of the organisms. *Microorganisms* are too small to be studied with the naked eye. Almost any organism less than 0.1 mm fits within this group. Microorganisms are present in two of the three domains of life (Archaea and Bacteria). Microbes have a significant effect on the atmosphere and our daily lives. They can affect an ecosystem by causing disease in animals, humans, and plants. On the other hand, they are used to produce compounds such as vaccines, antibiotics, enzymes, yogurt, food, and other products like detergents. Microorganisms are everywhere on Earth. Microbes are typically isolated from three environments: soil, air, and water. The air and soil sectors are being explored in greater depth than water. Water is further classed as freshwater (river, pond, lake) and marine (ocean) (Sjollem and Weekamp 1989). Because of the challenging habitat, marine diversity is understudied compared to other diversities. There are microbiological diversities in the deep ocean that offer distinct features to help society thrive.

MARINE MICROBES

The oceans enclose various environmental conditions and habitats, containing a wide range of microbial biodiversity. Several unique marine microbes can produce multiple biologically active compounds for biological adaptations. Bacteria, algae, protista, and fungi are marine species with greatest capacity for creating bioactive substances that can be exploited for a number of diverse applications, such as cosmetics (Rocha et al. 2011). Marine organisms have some combinations that highlight the aquatic habitats. These compounds are superior to others in their probable use in beauty and cosmetic sectors. Microbial carotenoids; mycosporin; and its comparable amino acids, chitosan, fatty acids, and other substances may supply a trustworthy and quicker alternative to certain other biological substances employed in photoprotective, skin-whitening, and antiaging treatments (Corindesi et al. 2017).

ALGAE

Microalgae are eukaryotic and prokaryotic photosynthetic microorganisms. Prokaryotic algae are categorized into two classes (Prochlorophyta and Cyanophyta). Eukaryotic algae are categorized into various divisions (Chrysophyta, Bacillariophyta, Chlorophyta, Rhodophyta, Phaeophyta) (Imhoff et al. 2011). They simply grow in fresh or saltwater and extremely salty habitats. Cyanobacteria can also grow in saltwater, freshwater, and marine conditions. Prokaryotic organisms are capable of photosynthesis (Jin et al. 2016). Today, the majority of cyanobacteria and microalgae production is geared toward high-value applications (Andersen and Bux 2013). Proteins, pigments, vitamins, fatty acids, minerals, and polysaccharides are all found in algal biomass and are of significant importance in the preparation of natural products in food and cosmetics.

Components produced by microalgae and cyanobacteria may be of relevance to the beauty industry (Tamagnini et al. 2002). Certain microalgae produce active components, which are presently employed in beauty and cosmetic industry products.

BACTERIA

Bacteria are distributed in all marine ecosystems. Many species of bacteria are used for biotechnological applications. Marine bacteria produce many compounds with different activities like moisturizing, antiaging, photoprotective, antimicrobial, and antioxidant. Some antioxidants are alkaloids, peptides, lipids, proteins, mycosporine-like amino acids, isoprenoids, and glycosides. The microorganism can use a broad variety of organic molecules as a kind of energy and carbon essential to develop whenever there are insoluble carbon sources such as hydrocarbons. Microorganisms produce some compounds (biosurfactants) that help in the diffusion process to take these compounds inside the cell. Ionic surfactant-producing bacteria emulsify the CxHy compound in the growth medium. Sophorolipids generated by various *Trolopsis* spp. and rhamnolipids synthesized by various *Pseudomonas* spp. (Burger et al. 1963) are examples of this type of biosurfactant. Other microbes can influence the shape of their cell walls by generating lipopolysaccharide or nonionic surfactants in their cell wall. *Arthrobacter* spp., *Rhodococcus* spp., and different *Mycobacterium* spp. (Guerra et al. 1986) are examples of this group, which produce nonionic trehalose corynomycolates. *Bacillus subtilis* produces lipoproteins such as surfactin and subtilisin, while *Acinetobacter* spp. produces lipopolysaccharides such as emulsan (Cooper et al. 1981).

FUNGI

There is much research done on the production of biosurfactants from bacterial species. Only a few fungi are known to produce biosurfactants when compared to bacteria. Some fungi, like *Trichosporon ashii*, *Candida lipolytica*, *Candida ishiwadae*, and *Candida bombicola*, have been researched (Casas et al. 1997). The majority of these are grown to produce biosurfactants from minimal raw components (Cortes 2011). Sophorolipids are one of several kinds of biosurfactants produced by these strains. When cultivated on n-alkanes, *Candida lipolytica* generates lipopolysaccharides that are connected to the cell wall (Rufino et al. 2006).

BIOSURFACTANTS: BIOLOGICALLY DERIVED INDUSTRIALLY IMPORTANT MOLECULES

Biosurfactants are surface-active biomolecules synthesized by living organisms that have many uses. These biomolecules have some unique properties. Biosurfactants have been utilized in a range of industries due to their distinctive functional properties, including petroleum, metallurgy, petrochemicals, organic chemicals, mining, beverages, agrochemicals, pharmaceuticals, fertilizers, cosmetics, and many more. They can be employed as emulsifiers and demulsifiers, foaming agents, wetting agents, functional food components, spreading agents, and detergents, among other things. Biosurfactants have the ability to decrease surface tension, which allows them to play an important role in the bioremediation of crude and oil recovery. Biosurfactants provide three important functions: (1) raising the surface area of hydrophobic substrates, (2) expanding the availability of water insoluble substrates by solubilization, and (3) controlling the adhesion and expulsion of microbes from surfaces (Borowitzka 1995).

PROPERTIES

Microbial surfactants are distinguished by their surface activity, pH tolerance, temperature, ionic strength, biodegradability, low toxicity, antimicrobial activity, and emulsifying and demulsifying ability. Biosurfactants have extensive substrate obtain ability, which made them ideal for various uses (Das et al. 2009).

SURFACE AND INTERFACE ACTIVITY

Surfactants are substances that diminish surface and interfacial tension. The organism *B. subtilis* produces a biosurfactant (surfactin) that reduces water surface tension (Cooper et al. 1981). Biosurfactants are more effective and efficient than chemical surfactants. Their critical micelle concentration (CMC) is less, implying that less surfactant is required for maximal surface tension reduction.

TEMPERATURE AND PH TOLERANCE

The surface activity of most biosurfactants is resistive to ambient factors like pH and temperature. In recent decades, the synthesis of biosurfactants from extremophiles has received interest due to their potential commercial value (McInerney et al. 1990). *Bacillus licheniformis* was resistant to temperatures as high like 50°C and pH ranges from 4.5–9.0.

Another biosurfactant produced by *Arthrobacter protophormiae* was found stable at pH (2 to 12) and thermostable (30–100°C). Because industrial processes encompass extreme pH, pressure, and temperature, it is required to extract new microbial products that can work under these circumstances (Singh and Camiotra 2004).

BIODEGRADABILITY

Biosurfactants formed by bacteria are biodegradable and appropriate for environmental applications such as biosorption/bioremediation (Mohan et al. 2006). Synthetic surfactants might cause harm, so biodegradable surfactants are the best choice for the future (Mulligan et al. 2001).

LOW TOXICITY

Biosurfactants are frequently low-toxicity or non-toxic compounds that may be used in food, beauty, and medicines. They are non-mutagenic. Sophorolipids generated by *Candida bombicola* have a low toxicity profile, making them valuable in the food business (Poremba et al. 1991).

ANTI-ADHESIVE AGENTS

The role of biosurfactants in microbial adhesion has been extensively studied, and biosurfactant adsorption to solid surfaces may be an effective strategy for reducing microbial adhesion by pathogenic microorganisms not only in the biomedical field but also in cosmetics and the food industry (Hood and Zottola 1995). Some microorganisms produce biofilm, which can be used as a biosurfactant in the cosmetics industry. Biofilm is a set of organisms that grows on any surface. The first step of biofilm development is bacterial adherence to the surface, which is influenced by different factors such as the different microorganisms, electrical charges and hydrophobicity, climatic circumstances, and the potential of microbes to yield secreted polymers, which support cells anchoring to substrates. Biosurfactants would be used to change the hydrophobicity of the character, which impacts bacterial adhesion to the surface (Zottola and Sasahara 1994). The surfactant is derived from *Streptococcus thermophilus*, which reduces the growth of other extremophiles or thermophilic strains of *Streptococcus* over steel, which causes fouling. Likewise, a biosurfactant derived from *Pseudomonas fluorescens* prevented *Listeria monocytogenes* from adhering to metal substrate (Ciriglino and Carman 1985).

EMULSION BREAKING AND FORMING

Biosurfactants are either emulsified or de-emulsified. Emulsion is basically a heterogeneous system consisting of one immiscible fluid spread in different forms of fluid droplets with a diameter larger than 0.1 mm. Emulsions are categorized into two types: water-in-oil (w/o) and oil-in-water (o/w) emulsions (Kosaric 2010). These both have low stability, which can be modified by the addition of some biosurfactants, and it can be retained as a strong emulsion for an extended time. Liposan is a hydrophilic emulsifier compound produced by *Candida lipolytica* that has been exploited to emulsify edible oils via covering oil droplets and generating durable emulsions. These liposans are widely utilized in the food and cosmetics industries to create oil/water emulsions (Cooper et al. 1981).

CLASSIFICATION OF BIOSURFACTANTS

Biosurfactants are generally classed depending on their molecular weight. Higher molecular weight biosurfactants are better emulsifiers compared to common molecular weight substances. Biosurfactants with smaller molecular weights are more effective at minimizing contact and interfacial stresses (Simhi et al. 2000). Based on their chemical composition, each group is classified into multiple groups (Sen et al. 2020). Polysaccharides such as lipoproteins, protein, lipopolysaccharide, and fatty acid complexes are examples of larger molecular weight biosurfactants (Vecino et al. 2017). Lower molecular weight biosurfactants include glycolipids, lipopeptides, phospholipids, glycopeptides, and so on (Stipcevic et al. 2005). Biosurfactants are additionally classed according to their chemical composition and microbiological origin.

GLYCOLIPIDS

Glycolipids are the most recommended class of biosurfactants in the pharmaceutical, food, and beauty industries (Choi and Maibach 2005). Glycolipids are made up of a hydrophobic lipid tail coupled to a sugar molecule via a glycosidic or covalent bond (Saravanan and Vijaykumar 2015). Glycolipids are classified as trehalose, rhamnose, cellobiose, and sophorose, lipomannosyl-mannitols, mannosylerythritol lipids, lipoarabinomannan, and lipomannans. The features and sources of the various glycolipids are addressed further in the following (Benincasa 2007).

SOPHOROLIPIDS

These are one of the types of glycolipids. They are generated by yeasts. They consist of dimeric carbohydrate sophorose glycosidically attached to a long-chain hydroxyl fatty acid. Several applications favor sophorolipids, which are often a mixture of at least 6–9 different hydrophobic sophorolipids and lactone of the sophorolipid (Borowitzka 1995).

TREHALOLIPIDS

Most *Nocardia, Mycobacterium, Corynebacterium, Arthrobacter,* and *Rhodococcus erythropolis* species produce trehalolipids (Asselinau and Asselineau 1978).

RHAMNOLIPIDS

These are one example of glycolipids that have one or more rhamnose molecules coupled with one or two hydroxydecanoic acids. This is the most well-researched biosurfactant and is the naturally available glycolipid synthesized by *Pseudomonas aeruginosa* (Edward and Hayashi 1965).

LIPOPEPTIDES AND LIPOPROTEINS

A lipopeptide is just a molecule containing a peptide that is lipid-bound (Gallot and Douy 1986). Lipoproteins are surface-active biopolymers. Lipoproteins are biopolymers with surface activity (Kitamoto et al. 2002). The biosurfactant cyclic lipopeptide (CLP) is persistent across a broad pH range (7.0–12.0) (Corinaldesi et al. 2017). Even in extreme temperatures, the biosurfactants do not loss their fascinating surface-active properties (Anderson and Bux 2014). Lipopeptides have been proposed for use in cosmetics as emulsifiers and anti-wrinkle agents (Hajfarajollah et al. 2014).

SURFACTIN

An example of a cyclic lipopeptide is surfactin. It is made up of seven amino acid rings that are linked together by a lactone connection with a fatty acid chain. Preliminary studies on the physicochemical features of *Bacillus subtilis* surfactin reveal that it can reduce the surface interfacial and surface tension of water. Surfactin was also reported to have antiviral action against retrovirus and herpes (Borowitzka 1995).

FATTY ACIDS, PHOSPHOLIPIDS, AND NEUTRAL LIPIDS

During development in a culture medium containing n-alkanes, certain bacteria and yeast strains can generate a considerable number of phospholipids and fatty acid biosurfactants. The alkyl branch and OH group are present in this structure; corynomucolic acid is an example of a fatty acid utilized as a surfactant. The lipophilic and hydrophilic balance of fatty acids is largely governed by the length of the hydrocarbon chains (Abella 2006).

CORYNOMYCOLIC ACIDS

Some bacterial species, such as *Nocardia erythropolis* and *Corynebacterium lepus*, may manufacture a compound of fatty acids with alkyl and hydroxyl groups. Corynomycolic acid, a complex fatty acid, is a potent biosurfactant (Simhi et al. 2000).

Corynomycolic acids (R1--CH (OH)--CH (R2)--COOH) generated by *Corynebacterium lepus* have excellent surfactant ability and may efficiently reduce the surface tension of an aqueous solution. Corynomycolic acids, like 2-hydroxy fatty acids, exhibit properties that are somewhat insensitive to ionic strength and pH; they yield outcomes in pH ranges from 2 to 10 (Asselineau and Asselineau 1978).

POLYMERIC BIOSURFACTANTS

Emulsan, lipomannan, liposan, alasan, and other polysaccharide-protein complexes are widely investigated polymeric biosurfactants. Even at low concentrations, emulsan is a powerful emulsifying agent for hydrocarbons in water (0.001–0.01%). *Candida lipolytica* synthesizes liposan, a biosurfactant. It is an extrinsic hydrophilic emulsifier composed of 83% carbohydrates and 17% protein. In the cosmetic and food sectors, liposan is used as an emulsifying agent (Borowitzka 1995).

LIPOSAN

Candida lipolytica generates liposan, an extracellular, water-soluble emulsifier. It is made up of 17% protein and 83% carbohydrate, with the latter consisting of a heteropolysaccharide made up of galacturonic acid, glucose, galactose, and galactosamine (Zohdy 2021). *Candida lipolytica* manufactures this biosurfactant in the final step of fermentation utilizing hexadecane as a carbon substrate. Liposan has largely been used to stabilize O/W emulsions, which include a spectrum of vegetable oils (Saravanan and Vijaykumar 2015).

APPLICATION OF MARINE BIOSURFACTANTS

Biosurfactants have physical and biological properties, which allow them to be utilized as detergents, foaming agents, wetting agents, emulsifiers, and solubilizers.

Furthermore, the chemical industries are fast recognizing the relevance of these bio-based molecules as drivers of a bio-based economy. An increasing number of patents explaining various applications of these chemicals attest to this industrial interest. There are various scientific areas where applications are under study (Doye et al. 2017).

ANTI-AGING

There are extrinsic and intrinsic phases of aging that the human body goes through. In intrinsic aging, fibers and collagen become wider, more cohesive, and looser and result in brittle skin and, eventually, wrinkling and sagging. Extrinsic aging occurs as a result of environmental factors like smoke, UV rays, and pollution that produce free radicals that react with the skin cells, and those chemical changes cause aging. Moisturizing, protection, cleanliness, and prevention are all critical components of efficient cosmetic products. Research shows that biosurfactants can be employed as replacements for chemical surfactants in beauty care products. These will be possible if they can deliver better outcomes in their composition and have a market rate that makes them advisable. But bulk production and the restricted structural flexibility of microbial biosurfactants remain a barrier, even though a few have been marketed. Specific antioxidant-containing therapies, such as anti-aging face lotions and gels, can help minimize or slow the effects of skin aging (Morita et al. 2009).

TOOTHPASTE

Sophorolipid (SLP) biosurfactants have been created and commercialized as a strong element in cosmetics for the body and skin. SLPs are derived from the marine actinobacterium *Nocardiopsis* VITSISB. It is used to replace sodium lauryl sulfate (SLS), which is usually employed as a biosurfactant in commercial toothpaste, in the cosmetic makeup of toothpaste. Several experiments, including the cleaning and foaming test, spreadability test, brine shrimp hatchability (BSH) test, and abrasiveness test were applied to examine the quality of this biosurfactant toothpaste. According to the findings, biosurfactants are more efficient and less dangerous than chemical surfactants (Kulakovskaya et al. 2003). *Candida bombicola* ATCC 22214 produces sophorolipids, glycolipid biosurfactants commonly found in toothpaste. It also produces detergents, emulsifiers, solubilizers, foaming agents, and wetting agents. SLP demonstrates less cytotoxicity against human fibroblasts and keratinocytes, which is one of the required qualities for cosmetic application (Rincon et al. 2018).

Furthermore, SLP promotes fibroblast metabolism and collagen biosynthesis in the skin's dermis, acting as a restructurer, repairer, and toner. SLP promotes leptin synthesis in adipocytes, reducing subcutaneous fat excess, making it beneficial in treating cellulitis. Dermal fibroblasts stimulate elastase activity, which causes the skin to age and wrinkles to appear. SLP reduces free radical generation by decreasing elastase activity (Gudi et al. 2015). SLP functions as a macrophage activator, healing agent, depigmenting agent, and desquamating agent in the treatment of brown spots by partially blocking melanogenesis. Furthermore, its bactericides and bacteriostatic properties are used to control dandruff, treat acne, and as an active ingredient in deodorants (Toren et al. 2001).

SUNSCREEN PRODUCTS

Natural light exposure can cause skin damage attributable to ultraviolet (UV) radiation. Skin protection from sun exposure is vital to prevent immediate symptoms such as sunburn and to lessen the possibility of acquiring skin cancer, particularly in persons with fair skin. Sunscreen agents are used to protect the skin (Gudi et al. 2015). Sunscreen products are available in several formulations such as cream, lotion, and wipes and integrated into different products like moisturizers. The most-used biosurfactant is surfactin. As a sunscreen agent, biosurfactant from the agro-industrial stream is utilized to increase the defensive influence of mica minerals against UV rays. The sunscreen protection factor (SPF) of numerous biological composites is based on various mica minerals combined with a extract of the biosurfactant derived from the maize industry. This biosurfactant is synthesized by the organism *Bacillus subtilis*, which is exploited in the maize business (Toren et al. 2001).

FACE WASH AND SHAMPOO

Rhamnolipids are naturally occurring molecules with surface-active (surfactant) qualities. A natural fermentation process manufactures rhamnolipids from a renewable source—vegetable oil. It is mild when used to cleanse hair and skin. The rhamnolipid biosurfactant can be used in place of standard surfactants like sodium lauryl sulfate. Rhamnolipids comprise one or more rhamnose sugar groups connected to one or more fatty acid long chains. The sugar groups are hydrophilic, while the fatty acids assist the surfactants in grasping onto oils and other nonpolar molecules. Rhamnolipids are friendly on the skin, have a reduced carbon impact, function well in hard or soft water, and offer lovely foam. Rhamnolipids can be utilized as a mixture or selected for specific attributes by altering the line of the carbon chain and the number of rhamnose units (Morita et al. 2010). Rhamnolipids were initially identified in the 1950s and 1960s. *Pseudomonas aeruginosa*, Gram-negative, pathogenic bacteria, is employed by most producers to make rhamnolipids. According to industry players, rhamnolipids cost 10–30 times more than typical surfactants, depending on purity and volume. In a recent study, this bio-surfactant is produced by the organism *Pseudomonas aeruginosa* and used to make a hair product like shampoo containing 2% rhamnolipids mixed in water. After three days, the antibacterial action of the bio-surfactant kept the hair odorless and shiny (Brown 1991).

SKIN PROTECTIVE CREAM

Sophorolipids are used on the skin to protect, moisturize, and lubricate it. Water routinely evaporates from the skin's deeper layers, a phenomenon called transepidermal water loss (TEWL). Because dry skin is inflexible, any increase in skin water content improves skin quality. It can retain moisture as a biosurfactant. *Torulopsis bombicolu* sophorolipids interact with alkylene oxides to generate a long-chain of alkyl-sophorolipids. All these chemically changed components have been discovered to supplement the organic moisturizing components (Ambrico et al. 2019). The hydrophilic lipophilic balance value of the oleyl-sophorolipid was 7–8, indicating exceptional compatibility and great skin-moisturizing characteristics. Sophorolipids form a sophorose residue, a disaccharide molecule composed of two glucose residues connected by the 1,2′ bond, and fatty acid as an aglycone. It is acetylated on the 6′ sites of sophorose residue. Either terminal or subterminal, one hydroxylated fatty acid is glycosidically linked to the sophorose molecule. One or more unsaturated links can be observed in the hydroxy fatty acid residue. Fatty acid carboxylic groups are either free or internally esterified. Sophorolipids can arise as lactones in both mono and dimer forms (Smith and Alexander 2005).

BIOSURFACTANTS AS A SKIN SURFACE MOISTURIZER

Shampoos and other individual washing products are intended to make brief contact with skin and hair. However, their contact with the skin and its nearby cells throughout this time period may compromise the structural integrity of the stratum corneum, solubilize intracellular lipids, and denature proteins (Behera et al. 2017). Microbial biosurfactants, which are selected as chemical surfactant replacementsand to deliver great skin surface moisturization (Morris et al. 2019). *Candida* spp. models have unique applicability in this area. For example, when the dry and cracked surface of an in vitro skin model pretreated with sodium dodecyl sulfate (SDS) was re-treated with 10% MEL-A glycolipid biosurfactants after one day of incubation, cell viability increased to approximately 90% (Louhrith and Kalnayavattanakul 2009). Ceramide, an epidermal lipid, stimulates the development of the skin barrier and consequently protects epidermal hydration. Some research suggests that the reduction of ceramides in the stratum corneum leads to the genesis of skin illnesses such like psoriasis, atopic dermatitis, and eczema. Ceramides, whether natural or synthetic, improve skin surface roughness but are incredibly expensive to make (Vecino et al. 2017). As a result, monosylerythrotol lipids (MELs) with equivalent properties provide an acceptable replacement at lower production costs. Monoerythritol-4-phosphates have been demonstrated to have water-retention, moisturizing, rough skin–improving, and skin cell repair properties (Sen et al. 2020). MELs are often utilized in beauty products due to their ability to improve moisture-holding capacity in the stratum corneum and mend split hair (Stipcevic et al. 2005). Aquaporins (AQPs) are a protein family that produces water channels inside the cellular membranes of plants, mammals, and microbes (Morita et al. 2009a). In mammals, there are 13 AQPs (0–12) (Stipcevic et al. 2005). This plasma membrane allows moisture and other small solutes like glycerol and urea to pass through the skin epidermis, altering skin variables like wetness (Choi and Maibach 2005). The most abundant and well-studied aquaporin in human skin is AQP-3. In contact with water, it transfers inert solutes such as glycerol and urea, maintaining water content in the epidermis and the transport of tiny solutes (Sethi et al. 2016). Analyzing the connection of both age and illness to skin dryness and AQP-3 expression revealed that a decrease in AQP-3 formation at the mRNA and protein stages influences skin dryness (Paulino et al. 2016).

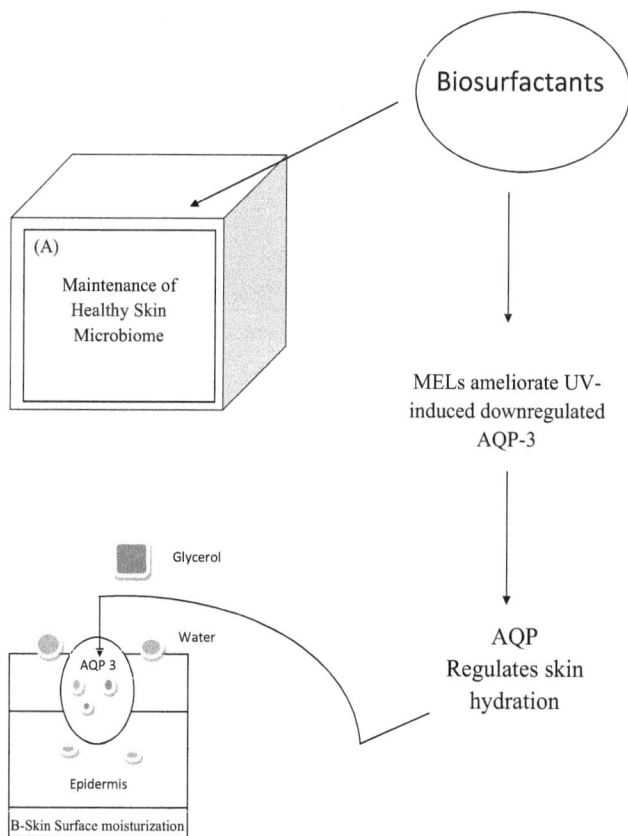

FIGURE 12.2 Potential benefits of microbial glycolipid and lipopeptide biosurfactants on human skin and its microbiome. (a) Maintenance of healthy skin microbiome; (b) skin surface moisturization.

Source: Figure referenced from Portilla et al. (2015).

AQP-3 = aquaporin 3; MELs = mannosylerythritol lipids. Figure referenced from (108,109,110); permission from Taylor & Francis for (31), 2017.

LIMITATIONS

Biosurfactant production on a large scale may be costly. This problem, however, could be solved by combining the process with the conversion of disposal substrates while countering their environmentally harmful effects and balancing the total costs (Pirri et al. 2009).

Pure biosurfactants are challenging to collect, which is especially important in pharmaceutical, food, and beauty care applications. This is due to the fact that downstream processing of diluted broths may necessitate multiple sequential steps.

Bacterial strains that produce a lot of biosurfactant are uncommon, and those organisms that do exist have very low productivity. Furthermore, selective optimized media must be practiced for the sample (Hu et al. 2021).

The control of biosurfactant production is poorly understood; it appears to reflect "secondary metabolite" regulatory oversight.

Strong foam formation impedes an increase in production yield (Avramora et al. 2008).

TABLE 12.1
Biosurfactant Types and Properties

Bio-Surfactant Type	Biosurfactant Name	Name of the Microorganism	Properties	References
Glycolipid	Rhamnolipid	*Pseudomonas aeruginosa*	Foaming, antibacterial, antiwrinkle, and anti-aging	(Piljac and Piljac 2007)
	Xylolipid	*Lactooccuslactis*	Antibacterial	(Saravanakumari and Mani 2010)
	Ployollipids	*Rhodotorulaglutinis*	Antifoaming agent, antibacterial agent	(Bonte 2011)
	Flocculosin	*Pseudozyma flocculosa*	Antifungal and antimicrobial	(Ekaterina et al. 2014)
Lipoproteins	Surfactin	*Bacillus subtilis*	Anti-aging, skin penetration agent, sunscreen agent, and emulsifier	(Jenn et al. 2014)
	Arthrofactin	*Arthrobacter* sp.	Treatment of acne, moisturizer	(Luci 2016)
	Iturin	*Bacillus subtilis*	Antifungal and antibacterial	(Khem et al. 2014)
	Fengycin	*Bacillus subtilis*	Antibacterial	(Bonte 2011)
	Pumilacidin	*Bacillus pumilus*	Antibacterial	(Sameh and Wallah 2017)
	Gramicidin	*Brevibacillus brevis*	Antibiotic	(Yamamoto et al. 2012)

Bio-Surfactant Type	Biosurfactant Name	Name of the Microorganism	Properties	References
Fatty acids, Phospholipids, and neutral lipids	Fatty acids	*Corynebacterium lupus*	Cleansing agent, surfactant, opacifying, emulsifying agent	(Brown 1991)
	Oleic acid	*Issatchenkia orientalis*	Emulsifying agent and emollient	(Bae et al. 2019)
	Phospholipids	*Acidithiobacillus thiooxidans*	Emulsifier, liposome former, solubilizer, and wetting agent	(Van Hoogevest et al. 2013)
Polymeric bio-surfactants	Emulsan	*Acinetobacter calcoaceticus*	Emulsifier	(Ambrico et al. 2019)
	Liposan	*Candida lipolytica*	Emulsifiers and stabilizers	(Cirigliano and Carman 1985)
	Alasan	*Acinetobacter calcoaceticus*	Emulsifier	(Eugene and Eliora 1997)
	Biodispersan	*Acinetobacter calcoaceticus*	Emulsifier and stabilizer	(Portilla et al. 2015)
	Bioemulsan	*Gordinia* sp.	Emulsifier, stabilizer, antioxidant	(Vecino et al. 2017)

CONCLUSION

As the aquatic ecosystem represents an encouraging source of new bioactive substances, there has been an increase in enthusiasm in the research and exploration of marine microbes as great possible producers of new compounds for application in various areas in recent years. Marine microbes can produce a variety of biosurfactants that could be useful in cosmetics and cosmeceuticals. Biosurfactants have several benefits versus chemically derived surfactants, including improved recyclability, increased environmental friendliness, and, in certain circumstances, higher foaming abilities and preserved effectiveness even at extreme pH and temperatures. Although marine biosurfactants have received less attention than terrestrial areas, they have several qualities that make them beneficial and potent for a variety of medicinal implementations as options to current drugs. However, because it is challenging to extract and grow these marine microbes, the majority of the marine microbial world is still uncharted. But with the fast discovery rate in science and people starting to explore marine resources as their current research field, biosurfactants may play an important role in a brighter future.

REFERENCES

Abella, M. 2006. Evaluation of anti-wrinkle efficacy of adenosine-containing products using the FOITS technique. *International Journal of Cosmetic Science*, *28*(6): 447–451. https://doi.org/10.1111/j.1467-2494.2006.00349.x.

Ambrico, A., Trupo, M. & Magarelli, R. A. 2019. Influence of phenotypic dissociation in *Bacillus subtilis* Strain ET-1 on iturin a production. *Current Microbiology*, *76*(12): 1487–1494. https://doi.org/10.1007/s00284-019-01764-y.

Ammala, A. 2012. Biodegradable polymers as encapsulation materials for cosmetics and personal care markets. *International Journal of Cosmetic Science*, *35*(2): 113–124. https://doi.org/10.1111/ics.12017.

Andersen, R. A. & Bux, F. 2013. Biotechnological applications of microalgae. Biodiesel and value-added products. CRC Press, Boca Raton, Florida, 239 pp. Hardcover $135.96; eBook $118.97. *Journal of Phycology*, *50*(6): 1155–1157. https://doi.org/10.1111/jpy.12233.

Angeloglou, M. 1973. A history of make-up. *Studio Vista*, *11*(1): 123–124. doi:10.1016/0015-626473)90079-5.

Asselineau, C. & Asselineau, J. 1978. Trehalose-containing glycolipids. *Progress in the Chemistry of Fats and other Lipids*, *16*: 59–99. https://doi.org/10.1016/0079-683278)90037-x.

Avramova, T., Sotirova, A., Galabova, D. & Karpenko, E. 2008. Effect of triton X-100 and rhamnolipid PS-17 on the mineralization of phenanthrene by *Pseudomonas* sp. cells. *International Biodeterioration & Biodegradation*, *62*(4): 415–420. https://doi.org/10.1016/j.ibiod.2008.03.008.

Bae, I. H., Lee, S. H., OH, S., Choi, H., Marinho, P. A., Yoo, J. W., Ko, J. Y., Lee, E. S., Lee, T. R., Lee, C. S. & Kim, D. Y. 2019. Mannosylerythritol lipids ameliorate ultraviolet A-induced aquaporin-3 downregulation by suppressing c-Jun N-terminal kinase phosphorylation in cultured human keratinocytes. *The Korean Journal of Physiology & Pharmacology*, *23*(2): 113. https://doi.org/10.4196/kjpp.2019.23.2.113.

Behera, A., Choudhury, S. B. & Routaray, M. 2017. A categorical constructional of minimal model matter. *International Journal of Science and Technology*, *1*(1): 48–63. https://doi.org/10.20319/mijst.2016.s11.4863.

Benincasa, M. 2007. Rhamnolipid produced from agroindustrial wastes enhances hydrocarbon biodegradation in contaminated soil. *Current Microbiology*, *54*(6): 445–449. https://doi.org/10.1007/s00284-006-0610-8.

Bonté, F. 2011. Skin moisturization mechanisms. *Annales Pharmaceutiques Françaises*, *69*(3): 135–141. https://doi.org/10.1016/j.pharma.2011.01.004.

Borowitzka, M. A. 1995. Microalgae as sources of pharmaceuticals and other biologically active compounds. *Journal of Applied Phycology*, *7*(1): 3–15. https://doi.org/10.1007/bf00003544.

Brown, M. 1991. Biosurfactants for cosmetic applications. *International Journal of Cosmetic Science*, *13*(2): 61–64. https://doi.org/10.1111/j.1467-2494.1991.tb00549.x.

Burger, M. M., Glaser, L. & Burton, R. M. 1963. The enzymatic synthesis of a rhamnose-containing glycolipid by extracts of *Pseudomonas aeruginosa*. *Journal of Biological Chemistry*, *238*(8): 2595–2602. https://doi.org/10.1016/s0021-925818)67872-x.

Casas, J., García De Lara, S. & García-Ochoa, F. 1997. Optimization of a synthetic medium for *Candida bombicola* growth using factorial design of experiments. *Enzyme and Microbial Technology*, *21*(3): 221–229. https://doi.org/10.1016/s0141-022997)00038-0.

Choi, M. J. & Maibach, H. I. 2005. Role of ceramides in barrier function of healthy and diseased skin. *American Journal of Clinical Dermatology*, *6*(4): 215–223. https://doi.org/10.2165/00128071-200506040-00002.

Cirigliano, M. C. & Carman, G. M. 1985. Purification and characterization of liposan, a bioemulsifier from *Candida lipolytica*. *Applied and Environmental Microbiology*, *50*(4): 846–850. https://doi.org/10.1128/aem.50.4.846-850.1985.

Cooper, D. G., Macdonald, C. R., Duff, S. J. B. & Kosaric, N. 1981. Enhanced production of surfactin from *Bacillus subtilis* by continuous product removal and metal cation additions. *Applied and Environmental Microbiology*, *42*(3): 408–412. https://doi. org/10.1128/aem.42.3.408-412.1981.

Corinaldesi, C. 2015. New perspectives in benthic deep-sea microbial ecology. *Frontiers in Marine Science*, 2. https://doi.org/10.3389/fmars.2015.00017.

Corinaldesi, C., Barone, G., Marcellini, F., Dell'Anno, A. & Danovaro, R. 2017. Marine microbial-derived molecules and their potential use in cosmeceutical and cosmetic products. *Marine Drugs*, *15*(4): 118. https://doi.org/10.3390/md15040118.

Cortes-Sánchez, Alejandro. 2011. Production of glycolipids with antimicrobial activity by *Ustilago maydis* FBD12 in submerged culture. *African Journal of Microbiology Research*, *5*(17): 2512–2523. https://doi.org/10.5897/ajmr10.814.

Danovaro, R., Snelgrove, P. V. & Tyler, P. 2014. Challenging the paradigms of deep-sea ecology. *Trends in Ecology & Evolution*, *29*(8): 465–475. https://doi.org/10.1016/j. tree.2014.06.

Das, P., Mukherjee, S. & Sen, R. 2009. Biosurfactant of marine origin exhibiting heavy metal remediation properties. *Bioresource Technology*, *100*(20): 4887–4890. https://doi.org/10.1016/j.biortech.2009.05.028.

Doye, P., Mena, T. & Das, N. 2017. Formulation and bio-availability parameters of pharmaceutical suspension. *International Journal of Current Pharmaceutical Research*, *9*(3): 8. https://doi.org/10.22159/ijcpr.2017.v9i3.18892.

Duncan, W. 1952. The library shelf. *Aircraft Engineering and Aerospace Technology*, *24*(12): 381. https://doi.org/10.1108/eb032243.

Edwards, J. R. & Hayashi, J. A. 1965. Structure of a rhamnolipid from *Pseudomonas aeruginosa*. *Archives of Biochemistry and Biophysics*, *111*(2): 415–421. https://doi. org/10.1016/0003-98616590204-3.

Fahim, S. & Hussein, W. 2017. Antibacterial potentials of surfactins against multi drug resistant bacteria. *Research Journal of Pharmaceutical, Biological and Chemical Sciences. RJPBCS 8*(3): 1076–1083.

Fruijtier-Pölloth, C. 2005. Safety assessment on polyethylene glycols PEGs and their derivatives as used in cosmetic products. *Toxicology*, *214*(1–2): 1–38. https://doi. org/10.1016/j.tox.2005.06.001.

Gallot, B. & Douy, A. 1986. *U.S. Patent No. 4,600,526*. Washington, DC: U.S. Patent and Trademark Office.

Gawade, R. P., Chinke, S. L. & Alegaonkar, P. S. (2020). Polymers in cosmetics. In *Polymer Science and Innovative Applications* (pp. 545–565). Amsterdam: Elsevier.

Germershaus, O., Lühmann, T., Rybak, J. C., Ritzer, J. & Meinel, L. 2014. Application of natural and semi-synthetic polymers for the delivery of sensitive drugs. *International Materials Reviews*, *60*(2): 101–131. https://doi.org/10.1179/1743280414y.0000000045.

Gudiña, E. J., Fernandes, E. C., Rodrigues, A. I., Teixeira, J. A. & Rodrigues, L. R. 2015. Biosurfactant production by *Bacillus subtilis* using corn steep liquor as culture medium. *Frontiers in Microbiology*, 6. https://doi.org/10.3389/fmicb.2015.00059.

Guerra-Santos, L. K., Käppeli, O. & Fiechter, A. 1986. Dependence of *Pseudomonas aeruginosa* continuous culture biosurfactant production on nutritional and environmental factors. *Applied Microbiology and Biotechnology*, *24*(6). https://doi. org/10.1007/bf00250320.

Hajfarajollah, H., Mokhtarani, B. & Noghabi, K. A. 2014. Newly antibacterial and antiad-hesive lipopeptide biosurfactant secreted by a probiotic strain. *Propionibacterium freudenreichii. Applied Biochemistry and Biotechnology, 174*(8): 2725–2740. https://doi.org/10.1007/s12010-014-1221-7.

Hood, S. & Zottola, E. 1995. Biofilms in food processing. *Food Control, 6*(1): 9–18. https://doi.org/10.1016/0956-71359591449-u

Hu, J., Sohn, A., George, J., Malik, R. & Ilonzo, N. 2021. Transcarotid artery revas-cularization and advances in vascular surgery for carotid artery disease. *Surgical Technology International, 39*. https://doi.org/10.52198/21.sti.39.cv1484

Imhoff, J. F., Labes, A. & Wiese, J. 2011. Bio-mining the microbial treasures of the ocean: New natural products. *Biotechnology Advances, 29*(5): 468–482. https://doi.org/10.1016/j.biotechadv.2011.03.001

Jenn-Kan, L. U., Hsin-Mei, W. A. N. G. & Xuan-Rui, X. U. 2014. Applications of surfac-tin in cosmetic products. *Patent No. US 2016/0030322 A1*, United States.

Jin, L., Quan, C., Hou, X. & Fan, S. 2016. Potential pharmacological resources: Natural bioactive compounds from marine-derived fungi. *Marine Drugs, 14*(4): 76. https://doi.org/10.3390/md14040076

Khan, A. D. & Alam, M. N. 2019. Cosmetics and their associated adverse effects: A review. *Journal of Applied Pharmaceutical Sciences and Research*: 1–6. https://doi.org/10.31069/japsr.v2i1.1

Kitamoto, D., Isoda, H. & Nakahara, T. 2002. Functions and potential applications of glycolipid biosurfactants—from energy-saving materials to gene delivery carriers. *Journal of Bioscience and Bioengineering, 94*(3): 187–201. https://doi.org/10.1016/s1389-17230280149-9.

Klein, M. & Poverenov, E. 2020. Natural biopolymer-based hydrogels for use in food and agriculture. *Journal of the Science of Food and Agriculture, 100*(6): 2337–2347. https://doi.org/10.1002/jsfa.10274.

Kosaric, N. 2010. ChemInform abstract: Biosurfactants. *ChemInform, 22*(12). https://doi.org/10.1002/chin.199112362.

Kulakovskaya, E. & Kulakovskaya, T. 2014. *In Extracellular Glycolipids of Yeasts. Biotechnology and Biology of Trichoderma Richard Bélanger, Yali Cheng, Caroline Labbé, David Menally. 2007.* Antimicrobial Molecule. Patent No. WO2004007514A1.

Kulakovskaya, T., Kulakovskaya, E. & Golubev, W. 2003. ATP leakage from yeast cells treated by extracellular glycolipids of *Pseudozyma fusiformata. FEMS Yeast Research, 3*(4): 401–404. https://doi.org/10.1016/s1567-13560200202-7.

Lourith, N. & Kanlayavattanakul, M. 2009. Natural surfactants used in cosmetics: Glycolipids. *International Journal of Cosmetic Science, 31*(4): 255–261. https://doi.org/10.1111/j.1468-2494.2009.00493.x.

Martins, A., Vieira, H., Gaspar, H. & Santos, S. 2014. Marketed marine natural prod-ucts in the pharmaceutical and cosmeceutical industries: Tips for success. *Marine Drugs, 12*(2): 1066–1101. https://doi.org/10.3390/md12021066.

McInerney, M. J., Javaheri, M. & Nagle, D. P. 1990. Properties of the biosurfactant pro-duced by *Bacillus licheniformis* strain JF-2. *Journal of Industrial Microbiology, 5*(2–3): 95–101. https://doi.org/10.1007/bf01573858.

Meena, K. R., Saha, D. & Kumar, R. 2014. Isolation and partial characterization of iturin like lipopeptides a bio-control agent from a *Bacillus subtilis* strain. *International Journal of Current Microbiology and Applied Sciences, 3*(10): 121–126.

Mohan, P. K., Nakhla, G. & Yanful, E. K. 2006. Biokinetics of biodegradation of sur-factants under aerobic, anoxic and anaerobic conditions. *Water Research, 40*(3): 533–540. https://doi.org/10.1016/j.watres.2005.11.030.

Morita, T., Fukuoka, T., Konishi, M., Imura, T., Yamamoto, S., Kitagawa, M., Sogabe, A. & Kitamoto, D. 2009a. Production of a novel glycolipid biosurfactant, mannosylmannitol lipid, by *Pseudozyma parantarctica* and its interfacial properties. *Applied Microbiology and Biotechnology*, 83(6): 1017–1025. https://doi.org/10.1007/s00253-009-1945-4.

Morita, T., Kitagawa, M., Suzuki, M., Yamamoto, S., Sogabe, A., Yanagidani, S., Imura, T., Fukuoka, T. & Kitamoto, D. 2009b. A yeast glycolipid biosurfactant, mannosylerythritol lipid, shows potential moisturizing activity toward cultured human skin cells: The recovery effect of MEL-A on the SDS-damaged human skin cells. *Journal of Oleo Science*, 58(12): 639–642. https://doi.org/10.5650/jos.58.639.

Morita, T., Kitagawa, M., Yamamoto, S., Sogabe, A., Imura, T., Fukuoka, T. & Kitamoto, D. 2010. Glycolipid biosurfactants, mannosylerythritol lipids, repair the damaged hair. *Journal of Oleo Science*, 59(5): 267–272. https://doi.org/10.5650/jos.59.267.

Morris, S. A. V., Thompson, R. T., Glenn, R. W., Ananthapadmanabhan, K. P. & Kasting, G. B. 2019. Mechanisms of anionic surfactant penetration into human skin: Investigating monomer, micelle and submicellar aggregate penetration theories. *International Journal of Cosmetic Science*, 41(1): 55–66. https://doi.org/10.1111/ics.12511.

Mulligan, C., Yong, R. & Gibbs, B. 2001. Remediation technologies for metal-contaminated soils and groundwater: An evaluation. *Engineering Geology*, 60(1–4): 193–207. https://doi.org/10.1016/s0013-79520000101-0.

Paulino, B. N., Pessôa, M. G., Mano, M. C. R., Molina, G., Neri-Numa, I. A. & Pastore, G. M. 2016. Current status in biotechnological production and applications of glycolipid biosurfactants. *Applied Microbiology and Biotechnology*, 100(24): 10265–10293. https://doi.org/10.1007/s00253-016-7980-z.

Piljac, T. & Piljac, G. 2007. Use of Rhamnolipids as cosmetics. *Patent No. EP, 1056462*, B1.

Pirri, G., Giuliani, A., Nicoletto, S., Pizzuto, L. & Rinaldi, A. 2009. Lipopeptides as anti-infectives: A practical perspective. *Open Life Sciences*, 4(3): 258–273. https://doi.org/10.2478/s11535-009-0031-3.

Poremba, K., Gunkel, W., Lang, S. & Wagner, F. 1991. Toxicity testing of synthetic and biogenic surfactants on marine microorganisms. *Environmental Toxicology & Water Quality*, 6(2): 157–163. https://doi.org/10.1002/tox.2530060205.

Portilla Rivera, O. M., Arzate Martínez, G., Jarquín Enríquez, L., Vázquez Landaverde, P. A. & Domínguez González, J. M. 2015. Lactic acid and biosurfactants production from residual cellulose films. *Applied Biochemistry and Biotechnology*, 177(5): 1099–1114. https://doi.org/10.1007/s12010-015-1799-4.

Rincón-Fontán, M., Rodríguez-López, L., Vecino, X., Cruz, J. & Moldes, A. 2018. Design and characterization of greener sunscreen formulations based on mica powder and a biosurfactant extract. *Powder Technology*, 327: 442–448. https://doi.org/10.1016/j.powtec.2017.12.093.

Rocha, C. A., Pedregosa, A. M. & Laborda, F. 2011. Biosurfactant-mediated biodegradation of straight and methyl-branched alkanes by *Pseudomonas aeruginosa* ATCC 55925. *AMB Express*, 1(1): 9. https://doi.org/10.1186/2191-0855-1-9.

Rosenberg, R. & Ron, E. Z. 1997. Bioemulsans: Microbial polymeric emulsifiers. *Current Opinion in Biotechnology*, 6: 313–316.

Rufino, R. D., Sarubbo, L. A. & Campos-Takaki, G. M. 2006. Enhancement of stability of biosurfactant produced by *Candida lipolytica* using industrial residue as substrate. *World Journal of Microbiology and Biotechnology*, 23(5): 729–734. https://doi.org/10.1007/s11274-006-9278-2.

Saravanakumari, P. & Mani, K. 2010. Structural characterization of a novel xylolipid biosurfactant from *Lactococcus lactis* and analysis of antibacterial activity against multi-drug resistant pathogens. *Bioresource Technology, 101*(22): 8851–8854. https://doi.org/10.1016/j.biortech.2010.06.104.

Saravanan, V. & Vijayakuma, S. 2015. Biosurfactants-types, sources and Applications. *Research Journal of Microbiology, 10*(5): 181–192. https://doi.org/10.3923/jm. 2015.181.192.

Semenzato, A., Costantini, A. & Baratto, G. 2014. Green polymers in personal care products: Rheological properties of tamarind seed polysaccharide. *Cosmetics, 2*(1): 1–10. https://doi.org/10.3390/cosmetics2010001.

Sen, S., Borah, S. N., Kandimalla, R., Bora, A. & Deka, S. 2020. Sophorolipid biosurfactant can control cutaneous dermatophytosis caused by *Trichophyton mentagrophytes*. *Frontiers in Microbiology, 11*. https://doi.org/10.3389/fmicb.2020.00329.

Serzan, M. & Atkins, M. B. 2021. Lymphocyte-activation Gene 3 and programmed cell death 1 checkpoint inhibition as a novel combination in advanced melanoma. *Oncology & Haematology, 17*(2): 58. https://doi.org/10.17925/ohr.2021.17.2.58.

Sethi, A., Kaur, T., Malhotra, S. & Gambhir, M. 2016. Moisturizers: The slippery road. *Indian Journal of Dermatology, 61*(3): 279. https://doi.org/10.4103/0019-5154.182427.

Simhi, E., Mei, H. C., Ron, E. Z., Rosenberg, E. & Busscher, H. J. 2000. Effect of the adhesive antibiotic TA on adhesion and initial growth of *E. coli* on silicone rubber. *FEMS Microbiology Letters, 19*(21): 97–100. https://doi.org/10.1111/j.1574-6968.2000. tb09365.x.

Singh, P. & Cameotra, S. S. 2004. Potential applications of microbial surfactants in biomedical sciences. *Trends in Biotechnology, 22*(3): 142–146. https://doi.org/10.1016/j. tibtech.2004.01.010.

Sjollem, J., Busscher, H. J. & Weerkamp, A. H. 1989. Experimental approaches for studying adhesion of microorganisms to solid substrata: Applications and mass transport. *Journal of Microbiological Methods, 9*(2): 79–90. https://doi.org/10.1016/0167-70128990058-4.

Smith, C. & Alexander, B. 2005. The relative cytotoxicity of personal care preservative systems in Balb/C 3T3 clone A31 embryonic mouse cells and the effect of selected preservative systems upon the toxicity of a standard rinse-off formulation. *Toxicology in Vitro, 19*(7): 963–969. https://doi.org/10.1016/j.tiv.2005.06.014.

Stipcevic, T., Piljac, T. & Isseroff, R. 2005. Di-rhamnolipid from displays differential effects on human keratinocyte and fibroblast cultures. *Journal of Dermatological Science, 40*(2): 141–143. https://doi.org/10.1016/j.jdermsci.2005.08.005.

Tamagnini, P., Axelsson, R., Lindberg, P., Oxelfelt, F., Wünschiers, R. & Lindblad, P. 2002. Hydrogenases and hydrogen metabolism of cyanobacteria. *Microbiology and Molecular Biology Reviews, 66*(1): 1–20. https://doi.org/10.1128/mmbr.66.1.1-20.2002.

Toren, A., Navon-Venezia, S., Ron, E. Z. & Rosenberg, E. 2001. Emulsifying activities of purified alasan proteins from *Acinetobacter radioresistens* KA53. *Applied and Environmental Microbiology, 67*(3): 1102–1106. https://doi.org/10.1128/ aem.67.3.1102-1106.2001.

Tournier-Couturier, L. 2016. Arthrofactin for the treatment of acne. *Patent No. WO 2018/114720 Al Luiz Santos*, Lucie Tournier-Couturier Nakako Shibagaki. 2016. Cosmetic use of arthrofactin. Patent no. WO 2018/115522 Al.

Van Hoogevest, P., Prusseit, B. & Wajda, R. 2013. Phospholipids: Natural functional ingredients and actives for cosmetic products. *SOFW-Journal, 139*(8): 9–13.

Vecino, X., Cruz, J. M., Moldes, A. B. & Rodrigues, L. R. 2017. Biosurfactants in cosmetic formulations: Trends and challenges. *Critical Reviews in Biotechnology, 37*(7): 911–923. https://doi.org/10.1080/07388551.2016.1269053.

Yamamoto, S., Morita, T., Fukuoka, T., Imura, T., Yanagidani, S., Sogabe, A., Kitamoto, D. & Kitagawa, M. 2012. The moisturizing effects of glycolipid biosurfactants, mannosylerythritol lipids, on human skin. *Journal of Oleo Science*, *61*(7): 407–412. https://doi.org/10.5650/jos.61.407.

Zohdy, S. 2021. Advances in parasitology. Volume 108. Edited by David Rollinson and Russell Stothard. Academic Press. Amsterdam The Netherlands and Boston Massachusetts: Elsevier. $196.35. ix + 229 p.; ill.; no index. ISBN: 978-0-12-820750-5. 2020. *The Quarterly Review of Biology*, *96*(4): 312–313. https://doi.org/10.1086/717357.

Zottola, E. A. & Sasahara, K. C. 1994. Microbial biofilms in the food processing industry—Should they be a concern? *International Journal of Food Microbiology*, *23*(2): 125–148. https://doi.org/10.1016/0168-16059490047-7.

13 Marine Biosurfactants Combined with Nanomaterials for Potential Oil Spill Remediation

Deviany Deviany[a] and Siti Khodijah Chaerun[b,c]*

a Department of Chemical Engineering, Institut Teknologi Sumatera, Lampung Selatan, Lampung 35365, Indonesia

b Department of Metallurgical Engineering, Faculty of Mining and Petroleum Engineering, Institut Teknologi Bandung, Ganesha 10, Bandung 40132, West Java, Indonesia

c Geomicrobiology-Biomining & Biocorrosion Laboratory, Microbial Culture Collection Laboratory, Biosciences and Biotechnology Research Center (BBRC), Institut Teknologi Bandung, Ganesha 10, Bandung 40132, West Java, Indonesia

* Corresponding author: Tel: +62 – 22 – 2502239; Fax: +62 – 22 – 2504209.

E-mail addresses: skchaerun@gmail.com; skchaerun@itb.ac.id (S.K. Chaerun)

CONTENTS

INTRODUCTION

Microorganisms synthesize secondary metabolites in the form of bioactive compounds such as biosurfactants and exopolysaccharides (EPSs; high molecular

DOI: 10.1201/9781003307464-13

weight biosurfactants) during their late growth phase and as a response to environmental stress factors such as pH changes, temperature changes, or the presence of xenobiotics (Tripathi et al. 2018). Biosurfactants have been the subject of significant research over the last decade due to their distinct advantages over synthetic chemical surfactants, particularly their biocompatibility and biodegradability. Biosurfactants, like their chemical counterparts, are surface-active molecules with amphiphilic properties. Each molecule has hydrophilic and hydrophobic components, allowing lowered interfacial tension between two surfaces.

Biosurfactants can be classified into numerous categories based on their molecular structures, including glycolipids, lipopeptides, phospholipids, lipoproteins, fatty acids, and polymeric biosurfactants. Glycolipids include rhamnolipids (produced by *Pseudomonas* sp.), sophorolipids (produced by *Candida* sp.), and trehalose lipids (produced by *Mycobacterium* sp. and *Rhodococcus erythropolis*). Glycolipids, extensively investigated, are biosurfactants consisting of carbohydrates and fatty acids. Lipopeptides are cyclic biosurfactants composed of hydrophilic peptide sequences and a hydrophobic fatty acid chain. Surfactin is a lipopeptide biosurfactant generated by *Bacillus* sp.

Marine-derived biosurfactants have a more comprehensive range of structural, biological, and functional properties than terrestrial and freshwater-derived biosurfactants, offering up an unlimited number of options in diverse industrial applications (Satpute et al. 2010). In contrast to the frequently reported biosurfactant producers, biosurfactants of marine origin are typically obtained from non-pathogenic bacteria. Certain biosurfactants derived from marine organisms are obtained from extremophiles and hence exhibit enhanced activity in adverse conditions, such as lower pH and increased salinity or temperature. Other marine biosurfactants were initially researched using microorganisms found in ecosystems saturated with pollutants/contaminants as a result of anthropogenic activities such as crude oil spillage.

This chapter focuses on the potential for biosurfactants produced by marine microorganisms in combination with nanomaterials to improve the remediation of marine oil spills. Several recent studies on using marine origin biosurfactants for crude oil spill mitigation established biosurfactants as a potential chemical dispersants alternative. The subsequent discussion was the implementation of nanotechnology using magnetic nanoparticle sorbents to remove and recover crude oil and nanomaterials functioning as Pickering emulsifiers to speed up bioremediation by improving bioavailability. In addition, various research investigations utilizing biosurfactant-modified nanomaterials for improved oil-in-seawater emulsification were highlighted.

MARINE-DERIVED BIOSURFACTANTS FOR ENHANCED OIL SPILL REMEDIATION

The accidental release of crude oil into the marine environment is one example of an ecological disaster caused by human activity that has long-term detrimental effects on a diverse array of aquatic organisms and indirectly on non-aquatic animals. The thick and viscous crude oil enters marine mammals' and fishes' respiratory organs and sticks to the furs and feathers of other aquatic species. Oil spills into bodies of water are frequently caused by shipwrecks and groundings, undersea pipeline ruptures, and the inevitability of drilling rig leaks. Spillage of crude oil containing hydrocarbons would result in the formation of an oil slick on the surface, depriving aquatic life of necessary oxygen and sunshine.

The well-documented *Exxon Valdez* oil spill occurred in 1989, when the name-sake supertanker ran aground off Alaska's once-pristine Prince William Sound coast, causing long-term environmental damage (Prabowo and Dong 2019; Barron et al. 2020). Another major ecological disaster regarded to be one of the largest accidental oil spills is the explosion of the Deepwater Horizon drilling rig in the northern Gulf of Mexico in 2010; even a decade later, the spill's effects on marine life may still be seen (Beyer et al. 2016; Kujawinski et al. 2020). Unfortunately, despite ostensibly significant efforts to prevent oil spill disasters, there have already been multiple reports of inadvertent crude oil spills into the environment in early 2022. In late January, a heavy crude oil pipeline in Ecuador collapsed due to a land-slide caused by severe rains, causing severe damage to the protected Amazon rain-forest and contaminating water resources (Oxford Analytica 2022). After Tonga's volcano erupted, Peru had the worst oil spill in its history. Thousands of barrels of crude oil spilled from a refinery into the ocean after the volcano burst, causing a lot of damage to marine protected areas and wildlife (Mega 2022).

Remediation strategies can be classed into chemical, physical/mechanical, biological, and thermal treatment in the case of oil spill catastrophe mitigation, as shown in Figure 13.1. Chemical remediation primarily uses emulsifiers and dispersants to break up giant oil slicks into smaller droplets, enhancing the bio-availability of the hydrocarbons for further microbial degradation (Tian et al. 2016). However, these petroleum-based compounds were ultimately discovered to be just as, if not more, detrimental to marine life (Tian et al. 2016). Oil slicks can be trapped with booms or skimmers for physical/mechanical remediation, then thermal treatment with controlled ignition or *in situ* burning (Dhaka and

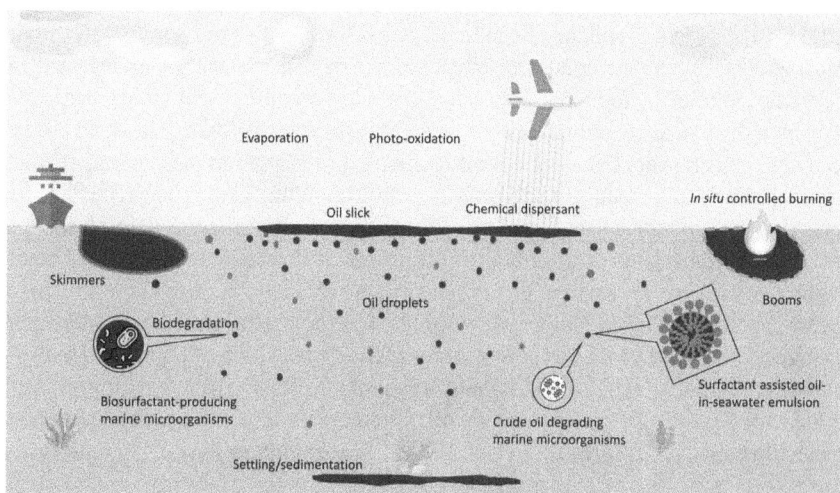

FIGURE 13.1   Strategies for marine oil spill remediation.

Chattopadhyay 2021). The effectiveness of this physical treatment is highly dependent on weather conditions when strong winds and surface waves significantly impair the performance of booms and skimmers. A promising technology that has emerged in recent years is the use of modified nanoparticles to replace conventional sorbents. They provide improved oil removal or recovery while also being easy to reuse. Natural sorbents derived from agricultural waste and human hair remain the preferred physical treatment due to their environmental and economic benefits (Dhaka and Chattopadhyay 2021).

Dispersants comprise one or more surfactants facilitating the solubilization of surface oil slick into accessible hydrocarbons for microorganisms. Despite their widespread use as an effective oil spill mitigation technique, the detrimental impacts on aquatic life have motivated a search for environmentally benign alternatives to chemical dispersants. Marine-derived biosurfactants have the potential to be a viable alternative to chemical dispersants, providing the same oil dispersing and solubilizing capabilities without the chemicals' harmful impacts on aquatic life. The bioremediation strategy, which utilizes microorganisms to remove pollutants/contaminants, has gained interest due to its ecologically beneficial nature in producing non-toxic end products such as carbon dioxide, water, fatty acids, and cell biomass (Xue et al. 2015). Microorganisms in the marine environments, particularly those that live in areas contaminated with hydrocarbons, utilize these hydrophobic chemicals as carbon and energy sources. Oil degradation by hydrocarbon-degrading marine microorganisms is influenced by interactions with various abiotic (nutrients, physicochemical characteristics) and biotic (non-hydrocarbon-degrading microorganisms, predators) components (Head et al. 2006). The degradation efficiency rate of crude oil by marine microorganisms might vary due to the complexity of crude oil compositions, which include aliphatic (alkanes, cycloalkanes, terpenes, etc.) and aromatic hydrocarbons (benzene and its derivatives, toluene, PAH, etc.) as well as the extrinsic marine environmental factors such as pH, salinity, and temperature (Xue et al. 2015). Biosurfactant-enhanced crude oil remediation is a promising strategy for marine oil spill management since many studies have demonstrated an increase in degradation rate due to increased dispersion of hydrocarbons available for utilization by a diverse array of microorganisms (McGenity et al. 2012; Xue et al. 2015).

Numerous research studies have documented the isolation of bacterial strains from a crude oil-contaminated marine environment. These bacterial strains include the genera *Cycloclasticus*, *Alcanivorax*, *Oleispira*, *Oleiphilus*, and *Thalassolituus* (Head et al. 2006). This group of bacteria is called hydrocarbonoclastic bacteria because of its exceptional capacity to degrade and utilize hydrocarbons as carbon and energy sources. *Cycloclasticus* and *Alcanivorax* are two genera that are particularly important in the degradation of petroleum hydrocarbons. *Cycloclasticus* is principally responsible for the polycyclic aromatic hydrocarbons (PAHs) found in crude oil, whereas *Alcanivorax* prefers linear and branching alkanes (Harayama et al. 2004). *Marinobacter hydrocarbonoclasticus*, *Desulfococcus oleovorans*, and *Mycobacterium vanbaalenii* were abundant in a marsh area contaminated with crude oil from the Deepwater Horizon oil spill, suggesting that these bacteria were also capable of metabolizing petroleum hydrocarbons (Atlas et al. 2015). Given the hydrophobicity of crude oil, it was believed that the majority of hydrocarbonoclastic bacteria generate surface-active biomolecules to aid in solubilizing hydrocarbon compounds and increasing their cellular intake. *Alcanivorax borkumensis* synthesizes biosurfactants of the glycolipid type structurally similar to rhamnolipids (Yakimov et al. 1998; Kubicki et al. 2019). Another species of *Alcanivorax*, *A. dieselolei*, was cultivated in media containing only diesel oil for carbon and energy and produced a biosurfactant classified as a proline lipid (Qiao and Shao 2010). *Marinobacter* strains isolated from harbor sediments have been shown to create biosurfactants of the phospholipopeptide class that are relatively stable and capable of dispersing crude oil in artificial seawater (Raddadi et al. 2017). The recovered chemicals were non-toxic to *Vibrio fischeri*, a non-pathogenic marine bacteria susceptible to many harmful toxins. Another study found that a glycolipid generated by a strain of *Marinobacter hydrocarbonoclasticus* was more effective in solubilizing crude oil than the chemical surfactant Tween 80 (Zenati et al. 2018). *Marinobacter*, a ubiquitous marine genus, was also identified in isolates collected from coastal and offshore sites for a study aimed at identifying alternative producers of rhamnolipids, a well-characterized class of biosurfactant produced primarily by opportunistic pathogenic bacteria *Pseudomonas aeruginosa*. Given the non-pathogenicity of *Marinobacter*, one of the collected bacterial strains was capable of producing biosurfactant, the chemical structure of which was investigated and proven to be a mixture of rhamnolipids, indicating the possibility for commercial production and industrial applications (Tripathi et al. 2019). Other hydrocarbonoclastic bacteria have not been documented to generate surface-active compounds, implying that either non-hydrocarbonoclastic bacteria found in crude oil-contaminated marine sites provide biosurfactants or that such compounds have not yet been discovered. A marine bacterium *Halomonas pacifica*, recently discovered from contaminated saltwater near a fishing port, could metabolize naphthalene and produce a lipopeptide biosurfactant (Cheffi et al. 2020). The biosurfactant, as mentioned, was capable of recovering crude oil from sand pack columns and demonstrating stability throughout a wide range of temperature, pH, and salinity conditions, indicating a potential future application for marine oil remediation. A specific *Planococcus* strain associated with a marine sponge developed a biosurfactant identified as a glycolipid, which was examined for pilot-scale production through submerged fermentation (Hema et al. 2019). The strain was capable of degrading up to 80% of the crude oil's different hydrocarbons. *Acinetobacter calcoaceticus*, a cold-water bacterial strain native to the North Atlantic, was examined for its cost-effective biosurfactant synthesis utilizing waste cooking oil as the sole carbon source (Moshtagh et al. 2021). Additionally, this biosurfactant can potentially be used for oil spill cleanup in the harsh marine environment because of its ability to withstand a variety of temperatures, pH, and salinity. The findings of many research investigating the biosurfactants produced by marine microorganisms and their potential for oil spill remediation are summarized in Table 13.1.

In a recent study, four different biosurfactants were utilized alone or in combination to disperse light crude oil and weathered crude oil compared to the

TABLE 13.1

Examples of Biosurfactants Produced by Marine Microorganisms and Their Potential for Oil Spill Remediation

Microorganism	Biosurfactant	Experiments	References
Marinobacter sp.	Phospholipo-peptide	Isolates of *Marinobacter* sp. from harbor sediments produced biosurfactants/bioemulsifiers after being grown in a soybean oil medium, which displayed low toxicity towards *Vibrio fischeri*. The partially purified biosurfactant dispersed light crude oil in artificial seawater.	Raddadi et al. (2017)
Marinobacter hydrocarbonoclasticus	Glycolipid	*Marinobacter hydrocarbonoclasticus* strain was isolated from a fishing harbor contaminated with diesel and lubricant oil. The produced biosurfactant was tested for short-term acute toxicity with *Artemia nauplii*. Sterile seawater with crude oil was used for the solubilization test against Tween 80.	Zenati et al. (2018)
Marinobacter sp.	Glycolipid (mixture of rhamnolipids)	Marine bacteria isolated from surface seawater were identified as the genus *Marinobacter*, grown in the pH and salt concentration-optimized medium supplemented with rapeseed oil. The biosurfactant-producing bacterial strain was tested with the *Galleria mellonella* infection model.	Tripathi et al. (2019)
Planococcus sp.	Glycolipid	*Planococcus* sp. from marine sponges was screened for biosurfactant production with hemolytic activity, emulsification activity, drop collapsing test, and oil displacement test. The bacterial adhesion to the hydrocarbons assay was performed to confirm the cell hydrophobicity. Degradation activity was screened on an agar plate containing crude oil.	Hema et al. (2019)
Halomonas pacifica	Lipopeptide	The bacterium *Halomonas pacifica*, isolated from the hydrocarbon-contaminated seawater near the fishing harbor, was able to degrade naphthalene. The generated biosurfactants were evaluated for surface tension and oil displacement.	Cheffi et al. (2020)
Acinetobacter calcoaceticus		Biosurfactant was produced by the bacterium *Acinetobacter calcoaceticus* isolated from an oil-contaminated sample of an offshore oil and gas site in Atlantic, Canada. Waste cooking oil was used as the sole carbon for biosurfactant production. The surface tension, emulsification index, and critical micelle concentration were measured.	Moshtagh et al. (2021)

commercial oil dispersant Corexit 9500A (Cai et al. 2021). The biosurfactants employed in the study included surfactins (generated by *Bacillus subtilis* isolated from crude oil-contaminated coastal sediment), trehalose lipids (produced by *Rhodococcus erythropolis*), rhamnolipids (produced by *Pseudomonas aeruginosa*), and exmulsins (a complex mixture primarily composed of lipopeptides, generated by *Exiguobacterium* sp.). With an emulsification-surface activity balance similar to Corexit 9500A, the study showed surfactin and trehalose lipids as promising commercial applications. Meanwhile, when rhamnolipids and surfactins were combined, they demonstrated superior dispersion performance compared to when employed alone, implying a possible balance of surface activity and emulsification for dispersant formulation (Cai et al. 2021).

As previously stated and evidenced by the findings of various studies cited previously, biosurfactants, particularly marine-derived biosurfactants, aid in the degradation of crude oil in the marine ecosystem. Biosurfactant commercialization is difficult due to the high cost of culturing biosurfactant-producing microorganisms and the complex downstream recovery processing (Patel et al. 2019). Additionally, there have been studies examining the addition of biosurfactants to bioremediation systems. Some results demonstrate no significant increase to a substantial increase in effectiveness depending on the species. Other results show oil degradation inhibition due to biosurfactant deposits, leading to contact limitation between hydrocarbon-degrading microorganisms and substrates (Ławniczak et al. 2013).

NANOMATERIALS FOR MARINE OIL SPILL REMEDIATION

Recent years have seen the emergence of nanotechnology as a possible treatment option for accelerating the microbial degradation of petroleum hydrocarbons. Nanotechnology-based oil remediation provides a sustainable and effective alternative to conventional oil spill remediation techniques, which are impacted by the weather and environmental conditions at contaminated sites (Pete et al. 2021). Nanotechnology is the application of materials that can be engineered or functionalized at the molecular level to obtain desired physical and chemical characteristics for specific purposes. These modified nanomaterials have the potential to be used as magnetic sorbents to remove and recover oil droplets. Additionally, nanomaterials can operate as emulsifiers, dispersing oil slicks and increasing their bioavailability, resulting in enhanced microbial degradation of petroleum hydrocarbons.

Furthermore, various nanomaterials have been studied as carriers for immobilizing microbial cells. Magnetic nanoparticle-based sorbents can be used to adsorb, absorb, or combine both to recover or remove oil. On the other hand, nanoparticles have a higher surface area-to-volume ratio than conventional sorbents, providing superior sorption capacity and oil retention. The adsorption efficacy is determined by the appropriate dispersion of nanoparticles in polluted water and the type of surface coatings (Simonsen et al. 2018). Magnetic nanoparticles can be modified to exhibit remarkable selectivity for oil uptake by enhancing their hydrophobicity, inhibiting water sorption, and therefore resulting in sorbent sinking (Qiao et al. 2019). Functionalized iron oxide-based magnetic nanoparticle sorbents were produced to boost sorption capacity, stability, and wettability while enhancing the sorbents' reusability and recyclability (Wu et al. 2008). One of the earliest studies on the use of magnetic nanoparticles for oil recovery described the use of magnetic collagen nanocomposites to absorb premium and used motor oil in oil-water mixtures, demonstrating selective oil absorption and magnetic tracking capability (Thanikaivelan et al. 2012). A study revealed that iron oxide nanoparticles coated with hydrophilic polyvinylpyrrolidone could recover nearly 100% of the oil under optimized conditions (Palchoudhury and Lead 2014). The polyvinylpyrrolidone coatings provided environmental stability to the water-soluble particles and were simple to produce. A recent work employed yeast biomass from the ethanol industry and magnetic nanoparticles to create a yeast magnetic bionanocomposite for oil removal, demonstrating that oil properties, temperature, contact time, and material interaction affected oil uptake (Debs et al. 2019). Numerous studies have established the efficacy and reusability of functionalized magnetic nanoparticles as excellent sorbents for treating marine oil spills. Before releasing them into oil spill-contaminated marine environments, additional research on their toxicity and potential environmental impacts is required. According to some studies, magnetic nanoparticles are biocompatible and have low toxicity; however, others indicate that their toxicity may vary depending on the type of coating (Qiao et al. 2019; Simonsen et al. 2018). Apart from biocompatibility concerns, there are several other factors to consider when utilizing functionalized iron oxide nanoparticles, including the expense of preparation/production, the intricate manufacturing process, and the feasibility of using them in marine environments (Pete et al. 2021).

Pickering emulsions, a concept first discovered by Ramsden and Pickering, are emulsions that are stabilized by the adsorption of solid particles, and nanomaterials can be used as emulsifiers (Pickering 1907). Solid particles such as chitosan, silica, clay, carbon, nanoparticles, and other materials form stable emulsions, providing an alternative to conventional chemical dispersants known to harm marine organisms when used to disperse oil slicks following a crude oil spill. Qi et al. (2018) synthesized polymer-coated nanoparticles to stabilize emulsions of heavy crude oil that could be destabilized by lowering the pH, resulting in the oil separation from the aqueous phase. The surface-active, pH-responsive polymer-modified silica nanoparticles may be used in reversible emulsification techniques for oil recovery and removal. One example of solid particles being used as emulsion stabilizers is the creation of chitosan nanoparticles to remove crude oil from polluted sediments (Saliu et al. 2021). By comparison, chitosan nanoparticles were as effective as a commercial surfactant mixture in removing crude oil from artificially contaminated sand. CO_2 also enhanced reversible emulsifications. Chitosan is chosen because it is derived from chitin, a plentiful natural polymer, which ensures its safety for use in marine oil spill remediation.

As previously discussed, using nanomaterials for the remediation of marine oil spills may prove to be an effective and sustainable approach to treating oil spills. The following are some of the examples: (1) biocompatible, low-toxicity materials in place of conventional oil spill mitigation, which is frequently costly and ineffective in inclement weather; (2) effective removal and recovery of oil due to the reusability of functionalized iron oxide nanoparticles; and (3) good dispersion activity and stable emulsification of oil in water via Pickering emulsions, promoting increased microbial degradation. Naturally, some limitations must be addressed, including the difficulty of selecting an appropriate method while taking into account the diverse indigenous microbial communities and diverse environmental factors, safety concerns regarding the introduction of nanomaterials into the environment, and effective separation techniques for nanomaterials from seawaters (Pete et al. 2021). Given the restrictions described, Pickering emulsification can be further enhanced by using naturally degradable nanoparticles such as chitosan or lignin that are abundant or commercially available, such as silica and clays.

BIOSURFACTANT-MODIFIED NANOMATERIALS AND THEIR POTENTIAL APPLICATION FOR MARINE OIL SPILL REMEDIATION

Using chemical dispersants to emulsify oil slicks after crude oil spills into the marine environment has been widely acknowledged as the most effective method. The crude oil bioremediation process is accelerated when hydrocarbon-degrading microorganisms consume the distributed oil droplets. Because of concerns about the potential toxicity of synthetic chemical dispersants to marine organisms, there have been suggestions to replace them with biosurfactants, which have been shown to be less toxic and more biocompatible and biodegradable (Judson et al. 2010). However, a recent study of the aquatic toxicity of commonly used oil spill dispersants revealed that they posed a lesser danger to the majority of marine species in field conditions when properly diluted and applied (Bejarano 2018).

As previously discussed, biosurfactants and Pickering emulsifiers may be utilized in place of environmentally benign dispersants in oil spill management. The benefits of solid particles for Pickering emulsifiers include stabilizing emulsions and preventing the coalescence of oil droplets. The solid particles can potentially be regarded as functionalized nanomaterials to improve biodegradation efficiency. Surfactants can be added to modify the surface properties of solid particles, facilitating the creation of a stable oil-in-seawater emulsion. On the other hand, the high salinity of saltwater may impede the production of Pickering emulsions.

Halloysite nanotubes, nanoclays, and functionalized nanoparticles are nanomaterials that can be used as solid particles to induce Pickering emulsions. Halloysite nanotubes are clay minerals with hollow tubular structures and diameters ranging from 500 to 1000 nm formed by weathering volcanic rocks. Surfactants derived from microorganisms stabilize created emulsions without the detrimental consequences of petroleum-based surfactants. In the food and pharmaceutical industries, halloysite nanotubes are frequently utilized. These nanotubes are typically loaded with surfactants, such as Tween 80-modified soybean lecithin. When evaluated for the cleanup of a crude oil spill, Span 80 demonstrated superior dispersion effectiveness compared to bared nanotubes, demonstrating that surfactants play a role in the process (Nyankson et al. 2015). Another study that used halloysite nanotubes that had been functionalized with iron oxide nanoparticles and encapsulated with surfactants reported that magnetic responsiveness led to fewer droplets coalescence and improved stabilization of oil-in-water emulsions, which, when combined with the released surfactant, resulted in droplets size smaller than 20 um (micron) (Owoseni et al. 2016). A Fe (III)-polyphenolic framework was created to cover the halloysite nanotubes to avoid the early release of surfactants placed inside the nanotubes. This framework could be dismantled at an acidic pH (Ojo et al. 2019). Figure 13.2 depicts the potential combination of biosurfactants and nanomaterials for improved oil slick emulsification.

Silica nanoparticles combined with caprylamidopropyl betaine, a zwitterionic surfactant containing two electrically neutral functional groups, exhibited a synergistic interaction for the formation and stabilization of emulsions in artificial seawater with a high salt concentration (Worthen et al. 2014). Another study used *in situ* rhamnolipid-modified silica nanoparticles to create an emulsion of tetradecane and crude oil in artificial seawater and observed improved stability with smaller oil droplet sizes (Pi et al. 2015). The synergistic combination of silica nanoparticles with rhamnolipid decreased coalescence and increased the stability of the emulsions. The high ionic strength of artificial seawater led to significant flocculation, when utilized individually, and silica nanoparticles were unable to stabilize the emulsions. Given the modest quantity of rhamnolipid necessary to stabilize the Pickering emulsion and its biodegradability, the Pickering emulsion is a safe and cost-effective option for removing marine oil spills.

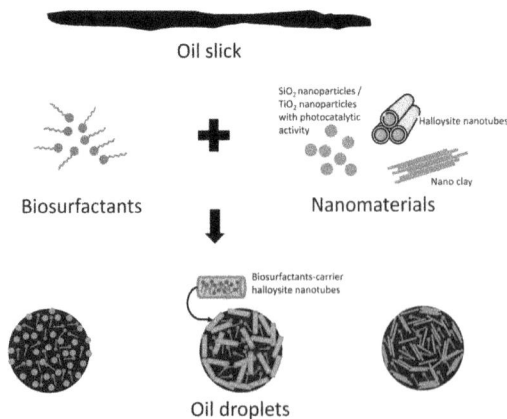

FIGURE 13.2 A combination of biosurfactants and nanomaterials for enhanced emulsification of oil slick.

Apart from silica nanoparticles, clay is another example of an environmentally benign solid particle for Pickering emulsifiers of oil in seawater. Palygorskite is a naturally occurring clay mineral composed of nanoscale hydrous magnesium aluminum silicate with a tubular microstructure. When employed alone, silica nanoparticles and palygorskite clay are ineffective emulsifiers in artificial seawater. Rhamnolipids effectively modify the surface characteristics of palygorskite and lower the interfacial tension of oil-artificial seawater. Rhamnolipid-modified palygorskite generated a robust coating, thus preventing oil droplet coalescence and encouraging prolonged emulsion stability. The synergistic interactions of palygorskite clay and rhamnolipid produced smaller droplets and increased the stability of tetradecane-in-artificial seawater emulsions (Chen et al. 2019).

Rhamnolipid-modified TiO_2 nanoparticles showed remarkable oil dispersion and promoted oil photodegradation in artificial seawater (Shi et al. 2021). TiO_2 nanoparticles were modified *in situ* with low concentrations of rhamnolipids. As a result, the modified nanoparticles demonstrated excellent diesel and crude oil dispersibility in artificial seawater with high ionic strength. Additionally, after 24 hours of UV light irradiation, the rhamnolipid-modified TiO_2 nanoparticles preserved their photocatalytic capabilities and exhibited diesel and crude oil photodegradation.

ACKNOWLEDGMENTS

This work was supported by the 2021 Research Program Grant of BBRC, Institute for Research and Community Services, Institut Teknologi Bandung, Indonesia.

REFERENCES

Atlas, Ronald M., Donald M. Stoeckel, Seth A. Faith, Angela Minard-Smith, Jonathan R. Thorn, and Mark J. Benotti. "Oil biodegradation and oil-degrading microbial populations in marsh sediments impacted by oil from the Deepwater Horizon well blowout." *Environmental Science & Technology* 49, no. 14 (2015): 8356–8366.

Barron, Mace G., Deborah N. Vivian, Ron A. Heintz, and Un Hyuk Yim. "Long-term ecological impacts from oil spills: comparison of *Exxon Valdez*, *Hebei Spirit*, and Deepwater Horizon." *Environmental Science & Technology* 54, no. 11 (2020): 6456–6467.

Bejarano, Adriana C. "Critical review and analysis of aquatic toxicity data on oil spill dispersants." *Environmental Toxicology and Chemistry* 37, no. 12 (2018): 2989–3001.

Beyer, Jonny, Hilde C. Trannum, Torgeir Bakke, Peter V. Hodson, and Tracy K. Collier. "Environmental effects of the Deepwater Horizon oil spill: a review." *Marine Pollution Bulletin* 110, no. 1 (2016): 28–51.

Cai, Qinhong, Zhiwen Zhu, Bing Chen, Kenneth Lee, Timothy J. Nedwed, Charles Greer, and Baiyu Zhang. "A cross-comparison of biosurfactants as marine oil spill dispersants: governing factors, synergetic effects and fates." *Journal of Hazardous Materials* 416 (2021): 126122.

Cheffi, Meriam, Dorra Hentati, Alif Chebbi, Najla Mhiri, Sami Sayadi, Ana Maria Marqués, and Mohamed Chamkha. "Isolation and characterization of a newly naphthalene-degrading Halomonas pacifica, strain Cnaph3: biodegradation and biosurfactant production studies." *3 Biotech* 10, no. 3 (2020): 1–15.

Chen, Dafan, Aiqin Wang, Yiming Li, Yajie Hou, and Zhining Wang. "Biosurfactant-modified palygorskite clay as solid-stabilizers for effective oil spill dispersion." *Chemosphere* 226 (2019): 1–7.

Debs, Karina B., Débora S. Cardona, Heron D. T. da Silva, Nashaat N. Nassar, Elma N. V. M. Carrilho, Paula S. Haddad, and Geórgia Labuto. "Oil spill cleanup employing magnetite nanoparticles and yeast-based magnetic bionanocomposite." *Journal of Environmental Management* 230 (2019): 405–412.

Dhaka, Abhinav, and Pradipta Chattopadhyay. "A review on physical remediation techniques for treatment of marine oil spills." *Journal of Environmental Management* 288 (2021): 112428.

Harayama, Shigeaki, Yuki Kasai, and Akihiro Hara. "Microbial communities in oil-contaminated seawater." *Current Opinion in Biotechnology* 15, no. 3 (2004): 205–214.

Head, Ian M., D. Martin Jones, and Wilfred F. M. Röling. "Marine microorganisms make a meal of oil." *Nature Reviews Microbiology* 4, no. 3 (2006): 173–182.

Hema, T., G. Seghal Kiran, Arya Sajayyan, Amrudha Ravendran, G. Gowtham Raj, and Joseph Selvin. "Response surface optimization of a glycolipid biosurfactant produced by a sponge associated marine bacterium *Planococcus* sp. MMD26." *Biocatalysis and Agricultural Biotechnology* 18 (2019): 101071.

Judson, Richard S., Matthew T. Martin, David M. Reif, Keith A. Houck, Thomas B. Knudsen, Daniel M. Rotroff, Menghang Xia et al. "Analysis of eight oil spill dispersants using rapid, in vitro tests for endocrine and other biological activity." *Environmental Science & Technology* 44, no. 15 (2010): 5979–5985.

Kubicki, Sonja, Alexander Bollinger, Nadine Katzke, Karl-Erich Jaeger, Anita Loeschcke, and Stephan Thies. "Marine biosurfactants: biosynthesis, structural diversity and biotechnological applications." *Marine Drugs* 17, no. 7 (2019): 408.

Kujawinski, Elizabeth B., Christopher M. Reddy, Ryan P. Rodgers, J. Cameron Thrash, David L. Valentine, and Helen K. White. "The first decade of scientific insights from the Deepwater Horizon oil release." *Nature Reviews Earth & Environment* 1, no. 5 (2020): 237–250.

Ławniczak, Łukasz, Roman Marecik, and Łukasz Chrzanowski. "Contributions of biosurfactants to natural or induced bioremediation." *Applied Microbiology and Biotechnology* 97, no. 6 (2013): 2327–2339.

McGenity, Terry J., Benjamin D. Folwell, Boyd A. McKew, and Gbemisola O. Sanni. "Marine crude-oil biodegradation: a central role for interspecies interactions." *Aquatic Biosystems* 8, no. 1 (2012): 1–19.

Mega, Emiliano Rodríguez. "Unprecedented oil spill catches researchers in Peru off guard." *Nature* (2022). https://doi.org/10.1038/d41586-022-00333-x

Moshtagh, Bahareh, Kelly Hawboldt, and Baiyu Zhang. "Biosurfactant production by native marine bacteria (*Acinetobacter calcoaceticus* P1-1A) using waste carbon sources: Impact of process conditions." *The Canadian Journal of Chemical Engineering* 99, no. 11 (2021): 2386–2397.

Nyankson, Emmanuel, Owoseni Olasehinde, Vijay T. John, and Ram B. Gupta. "Surfactant-loaded halloysite clay nanotube dispersants for crude oil spill remediation." *Industrial & Engineering Chemistry Research* 54, no. 38 (2015): 9328–9341.

Ojo, Olakunle Francis, Azeem Farinmade, James Trout, Marzhana Omarova, Jibao He, Vijay John, Diane A. Blake et al. "Stoppers and skins on clay nanotubes help stabilize oil-in-water emulsions and modulate the release of encapsulated surfactants." *ACS Applied Nano Materials* 2, no. 6 (2019): 3490–3500.

Owoseni, Olasehinde, Emmanuel Nyankson, Yueheng Zhang, Daniel J. Adams, Jibao He, Leonard Spinu, Gary L. McPherson, Arijit Bose, Ram B. Gupta, and Vijay T. John. "Interfacial adsorption and surfactant release characteristics of magnetically functionalized halloysite nanotubes for responsive emulsions." *Journal of Colloid and Interface Science* 463 (2016): 288–298.

Oxford Analytica. "Ecuador spill will boost criticism of Lasso oil plans", *Expert Briefings* (2022). https://doi.org/10.1108/OXAN-ES267056

Palchoudhury, Soubantika, and Jamie R. Lead. "A facile and cost-effective method for separation of oil—water mixtures using polymer-coated iron oxide nanoparticles." *Environmental Science & Technology* 48, no. 24 (2014): 14558–14563.

Patel, Seema, Ahmad Homaei, Sangram Patil, and Achlesh Daverey. "Microbial biosurfactants for oil spill remediation: pitfalls and potentials." *Applied Microbiology and Biotechnology* 103, no. 1 (2019): 27–37. https://doi.org/10.1007/s00253-018-9434-2

Pete, Amber J., Bhuvnesh Bharti, and Michael G. Benton. "Nano-enhanced bioremediation for oil spills: a review." *ACS ES&T Engineering* 1, no. 6 (2021): 928–946.

Pi, Guilu, Lili Mao, Mutai Bao, Yiming Li, Haiyue Gong, and Jianrui Zhang. "Preparation of oil-in-seawater emulsions based on environmentally benign nanoparticles and biosurfactant for oil spill remediation." *ACS Sustainable Chemistry & Engineering* 3, no. 11 (2015): 2686–2693.

Pickering, Spencer Umfreville. "Cxcvi.—emulsions." *Journal of the Chemical Society, Transactions* 91 (1907): 2001–2021.

Prabowo, Aditya Rio, and Dong Myung Bae. "Environmental risk of maritime territory subjected to accidental phenomena: Correlation of oil spill and ship grounding in the *Exxon Valdez*'s case." *Results in Engineering* 4 (2019): 100035.

Qi, Luqing, Chen Song, Tianxiao Wang, Qilin Li, George J. Hirasaki, and Rafael Verduzco. "Polymer-coated nanoparticles for reversible emulsification and recovery of heavy oil." *Langmuir* 34, no. 22 (2018): 6522–6528.

Qiao, Kaili, Weijun Tian, Jie Bai, Liang Wang, Jing Zhao, Zhaoyang Du, and Xiaoxi Gong. "Application of magnetic adsorbents based on iron oxide nanoparticles for oil spill remediation: a review." *Journal of the Taiwan Institute of Chemical Engineers* 97 (2019): 227–236.

Qiao, N., and Z. Shao. "Isolation and characterization of a novel biosurfactant produced by hydrocarbon-degrading bacterium *Alcanivorax dieselolei* B-5." *Journal of Applied Microbiology* 108, no. 4 (2010): 1207–1216.

Raddadi, Noura, Lucia Giacomucci, Grazia Totaro, and Fabio Fava. "*Marinobacter* sp. from marine sediments produce highly stable surface-active agents for combatting marine oil spills." *Microbial Cell Factories* 16, no. 1 (2017): 1–13.

Saliu, Francesco, Edoardo Meucci, Claudio Allevi, Alessandra Savini, Iikpoemugh Elo Imiete, and Roberto Della Pergola. "Evaluation of chitosan aggregates as Pickering emulsifier for the remediation of marine sediments." *Chemosphere* 273 (2021): 129733.

Satpute, Surekha K., Ibrahim M. Banat, Prashant K. Dhakephalkar, Arun G. Banpurkar, and Balu A. Chopade. "Biosurfactants, bioemulsifiers and exopolysaccharides from marine microorganisms." *Biotechnology Advances* 28, no. 4 (2010): 436–450.

Shi, Zhixin, Yiming Li, Limei Dong, Yihao Guan, and Mutai Bao. "Deep remediation of oil spill based on the dispersion and photocatalytic degradation of biosurfactant-modified TiO_2." *Chemosphere* 281 (2021): 130744.

Simonsen, Galina, Mikael Strand, and Gisle Øye. "Potential applications of magnetic nanoparticles within separation in the petroleum industry." *Journal of Petroleum Science and Engineering* 165 (2018): 488–495.

Thanikaivelan, Palanisamy, Narayanan T. Narayanan, Bhabendra K. Pradhan, and Pulickel M. Ajayan. "Collagen based magnetic nanocomposites for oil removal applications." *Scientific Reports* 2, no. 1 (2012): 1–7.

Tian, Wei, Jun Yao, Ruiping Liu, Mijia Zhu, Fei Wang, Xiaoying Wu, and Haijun Liu. "Effect of natural and synthetic surfactants on crude oil biodegradation by indigenous strains." *Ecotoxicology and Environmental Safety* 129 (2016): 171–179

Tripathi, Lakshmi, Victor U. Irorere, Roger Marchant, and Ibrahim M. Banat. "Marine derived biosurfactants: a vast potential future resource." *Biotechnology Letters* 40, no. 11 (2018): 1441–1457.

Tripathi, Lakshmi, Matthew S. Twigg, Aikaterini Zompra, Karina Salek, Victor U. Irorere, Tony Gutierrez, Georgios A. Spyroulias, Roger Marchant, and Ibrahim M. Banat. "Biosynthesis of rhamnolipid by a *Marinobacter* species expands the paradigm of biosurfactant synthesis to a new genus of the marine microflora." *Microbial Cell Factories* 18, no. 1 (2019): 1–12.

Worthen, Andrew J., Lynn M. Foster, Jiannan Dong, Jonathan A. Bollinger, Adam H. Peterman, Lucinda E. Pastora, Steven L. Bryant, Thomas M. Truskett, Christopher W. Bielawski, and Keith P. Johnston. "Synergistic formation and stabilization of oil-in-water emulsions by a weakly interacting mixture of zwitterionic surfactant and silica nanoparticles." *Langmuir* 30, no. 4 (2014): 984–994.

Wu, Wei, Quanguo He, and Changzhong Jiang. "Magnetic iron oxide nanoparticles: synthesis and surface functionalization strategies." *Nanoscale Research Letters* 3, no. 11 (2008): 397–415.

Xue, Jianliang, Yang Yu, Yu Bai, Liping Wang, and Yanan Wu. "Marine oil-degrading microorganisms and biodegradation process of petroleum hydrocarbon in marine environments: a review." *Current Microbiology* 71, no. 2 (2015): 220–228.

Yakimov, Michail M., Peter N. Golyshin, Siegmund Lang, Edward R. B. Moore, Wolf-Rainer Abraham, Heinrich Lünsdorf, and Kenneth N. Timmis. "*Alcanivorax borkumensis* gen. nov., sp. nov., a new, hydrocarbon-degrading and surfactant-producing marine bacterium." *International Journal of Systematic and Evolutionary Microbiology* 48, no. 2 (1998): 339–348.

Zenati, Billal, Alif Chebbi, Abdelmalek Badis, Kamel Eddouaouda, Hocine Boutoumi, Mohamed El Hattab, Dorra Hentati et al. "A non-toxic microbial surfactant from Marinobacter hydrocarbonoclasticus SdK644 for crude oil solubilization enhancement." *Ecotoxicology and Environmental Safety* 154 (2018): 100–107.

14 Marine Biosurfactants in Environmental Bioremediation
Scope and Applications

Swasti Dhagat and Satya Eswari Jujjavarapu

CONTENTS

ENVIRONMENTAL POLLUTION

For the past few decades, environmental pollution has become one of the major problems worldwide, affecting the health of plants, animals and humans. It is caused by inappropriate use and disposal of pollutants from industries, automobiles, domestic sewage and electronic waste containing heavy metals and radioactive compounds. These pollutants are increasing at an alarming rate and enter the food chain by contaminated water, air and soil, thereby threatening lives on Earth (Jan et al. 2016; Shackira et al. 2021; Patel et al. 2022).

A search for mitigation of environmental pollution has become necessary, with a sustainable method being the most suitable. Traditional methods of environmental mitigation are physical and chemical means. Physical methods for remediation of soil include capping, electrokinetic remediation, incineration, *in situ* grouting, *in situ* vitrification, stabilization/solidification, thermal desorption and vapor stripping, while remediation of water can be performed by air stripping/sparging and

DOI: 10.1201/9781003307464-14

incineration (Hamby 1996). However, the disadvantages of physical methods are that they are laborious and time consuming and hence are not economically viable (Dhaliwal et al. 2020). Chemically polluted soil can be treated using actinide chelators, immobilization of chemicals, critical fluid extraction, oxidation of contaminants, *in-situ* catalyzed peroxide remediation and photodegradation to recover uranium. The chemical treatments to remediate water consist of electron-beam irradiation, mercury extraction, radiocolloid treatment and sorption to organo-oxides (Hamby 1996). These methods have some shortcomings, such as being complex and requiring high energy for fusion. Also, the efficiency of these methods is low, they do not provide a permanent solution and they affect the biodiversity of treatment area (Dhaliwal et al. 2020). All of these techniques transform pollutants from one form to another instead of completely removing them from the site. On the other hand, remediation of pollution using biological sources is a highly promising technique to mitigate environmental pollution. Also termed green remediation, this technique utilizes biological sources, such as microorganisms and plants, to reduce air, water and soil pollution and hence preserve the area's biodiversity along with being eco-friendly and cost effective (Shackira et al. 2021).

ENVIRONMENTAL BIOREMEDIATION

Bioremediation is the process of employing biological agents to detoxify, transform or degrade pollutants. These biological agents include bacteria, fungi, algae and even plants in some cases. Due to the abundance of microbial fauna that can degrade pollutants, this process becomes effective, economical and environmentally friendly. Therefore, this method of bioremediation is growing exponentially compared to physical and chemical methods. The microorganisms used in bioremediation may be indigenous, that is, present at the site of contamination, or non-indigenous wherein the microorganisms are isolated from elsewhere and brought to the site of contamination. Bioremediation is a complex process that is regulated by many factors, such as the bioavailability of nutrients from pollutants to microbial population, the microbial species in the process, pH, temperature and moisture of the contaminated site and presence/absence of oxygen and nutrients. The concentration of contaminants also plays a vital role in determining the growth of microorganisms. Higher concentrations might be toxic to the microorganisms, while low concentrations might not be able to induce degradation pathways in microorganisms (Adams et al. 2015).

The process of bioremediation can be classified as *in situ* or *ex situ*. *In situ* bioremediation focuses on treating pollutants at the contaminated site itself, whereas in *ex situ* bioremediation, the pollutants are removed from the contaminated site for treatment (Boopathy 2000). As *in situ* remediation does not involve transport of pollutants, it is cost effective and does not disturb the ecology of the affected area. But this method is not desirable to remediate deep soil. The different types of *in situ* treatment are bioventing, *in situ* biodegradation, biosparging and bioaugmentation. *Ex situ* bioremediation methods include landfarming, composting, biopiles and bioreactors (Vidali 2001).

BIOREMEDIATION TECHNIQUES

Bioventing is used to treat soil contaminated with fuel residues and petrochemi-cal wastes, pesticides and volatile organic compounds (VOCs) as well as organic constituents present in water. This process increases the activity of indigenous microorganisms, which stimulates the *in situ* degradation of hydrocarbons aero-bically in the unsaturated zone. Air or oxygen is fed into the unsaturated zone along with nutrients, if required, through vertical injection wells. The most important requirement of this process is uniform distribution of air in the satu-rated soil. This is achieved by maintaining slower air injection rates, injection intervals and injection wells at the treatment site. The oxygen or air flow rate should also be monitored for sustenance of microorganisms. However, this pro-cess is time consuming and is dependent on temperature, pH and moisture con-tent of the contaminated soil (Speight 2019; Yadav, Singh et al. 2021; Chatterjee, Kumari et al. 2022). Biosparging also involves injecting air or oxygen into the contaminated zone to stimulate the metabolic activity of aerobic microorgan-isms. However, in this technique, air and nutrients (if required) are injected into saturated zone to aid degradation of contaminants (Turgeon 2009).

Bioaugmentation is the process of *in situ* bioremediation wherein indigenous or non-indigenous microorganisms are added at the site of pollution to accelerate degradation of pollutants. Introducing microbial fauna at the contaminated site increases the gene pool and genetic diversity of the site. On the other hand, in biostimulation, rate-limiting nutrients, such as nitrogen, oxygen, phosphorous and electron donors, are added to the contaminated site to stimulate the growth of pollutant-degrading microorganisms. This is the most efficient technique for bioremediation, especially for degradation of hydrocarbons and is cost effective and environmentally friendly. Both of these processes are affected by tempera-ture, pH, moisture, rate of aeration, organic matter, nutrient content and soil type of the contaminated area (Goswami, Chakraborty et al. 2018).

In case of biopiles, the contaminated soil is dug from its original site and piled up in a separate area for treatment. The contaminant pile is aerated through a series of plastic pipes with or without a vacuum pump. This helps in the growth of aerobic microorganisms inside the pile, which degrade the pollutants. The vapors released from the biopile are then collected by biological air filters or activated carbon to reduce emissions to the atmosphere. This technique requires less space and hence is a suitable option when space is a constraint. Also, it is a preferred method for odorous and volatile contaminants. Windrow or com-posting is similar to biopiles wherein the contaminants are heaped at a treat-ment site. Organic material is added to the contaminants to enhance microbial growth resulting in increased temperature of the pile. Aeration, in this technique, is provided by periodic turning of windrows by windrow turners. To improve soil characteristics, nutrients can be added at the time of turning (FLI 2022).

Landfarming is the simplest and cheapest form of bioremediation. In it, con-taminated soil is dug and spread into layers at the treatment site. The process of bioremediation can be increased by turning the beds or layers periodically and adding nutrients, but the disadvantage of this technique is that it requires large land area (FLI 2022). The contaminated soil can also be treated using bioreactors in either batch, fed-batch or continuous processes. This technique ensures optimal microbial growth due to controlled temperature, pH, substrate concentration, aeration and agitation and inoculum concentration. This reduces the time of bioremediation and abiotic losses (Sharma 2020).

MECHANISM OF ENVIRONMENTAL BIOREMEDIATION BY MICROORGANISMS

Bioremediation depends upon the microorganisms present at the site of contamination and the conditions of the contaminated site. In the process of bioremediation, microbial enzymes degrade pollutants either aerobically or anaerobically and modify the structure and toxicity of these pollutants (Supaphol et al. 2006). The key enzymes used for degradation of pollutants are oxidases, hydroxylases, hydrolases, dehydrogenases, dehalogenases and mono- and dioxygenases (Mohapatra et al. 2022). The aerobic route for bioremediation is followed inside the cytoplasmic membrane or on the surface of the cell. Here, the pollutant is taken into the cell and microbial oxygenases oxidize it by incorporating a molecule of oxygen (Cerniglia 1997; McDonald et al. 2006). This leads to formation of an alcohol, which is further oxidized to aldehyde and carboxylic acid. The final molecule is similar to fatty acid, and hence it is degraded by β-oxidation to yield acetyl-CoA. The degradation of polyaromatic hydrocarbons requires two oxygenase enzymes, dioxygenases in bacteria and monooxygenases in fungi. The dioxygenases help in catalysis of molecular oxygen to the pollutants. This leads to the formation of an intermediate, cis-dihydrodiol, which is further oxidized and ruptures the aromatic ring of the hydrocarbon into muconic acid (Cerniglia 1997). In the case of fungi, pollutants are oxidized by cytochrome P-450 and catalyzed by monooxygenases. A furan oxide is formed, leading to rearrangement of the pollutant to a trans dihydrodiol or hydrolyzed phenol. The phenol can be transformed to glycosides or sulfates by lignin-modifying enzymes (Romero et al. 2002; Dhagat and Jujjavarapu 2022). The aerobic process renders microbial cells maximum amount of energy (Field et al. 1995). In these processes, oxygen is the final electron acceptor, and it is also involved in activation of substrates (pollutants) through oxygenation reactions (Díaz 2004).

Microorganisms follow an anaerobic route of pollutant degradation when the contaminant is not present superficially. In this process, the electron acceptors are CH_2, Fe^{2+}, NO^{3-} and SO^{4-} (Akpor et al. 2007). The degradation of pollutants anaerobically is slower than the aerobic process, and the growth of microorganisms in this process is also slower (Peters et al. 2007). Many polluted areas, such as submerged soils and water bodies, are anoxic; that is, they do not contain dissolved oxygen. Bioremediation, in these areas, is by strict anaerobes or facultative microbes with CO_2, Fe(III), sulfate, nitrate, chlorate and so on as electron acceptors (Widdel and Rabus 2001; Gibson and Harwood 2002; Lovley 2003).

Heavy metals cannot be degraded completely but can only be transformed from one state to another by microorganisms. There are various methods by which bacteria protect themselves from heavy metals, namely adsorption, uptake, oxidation, methylation and reduction. In anaerobic respiration, these heavy metals are utilized as terminal electron acceptors by bacteria. In some cases, bacteria may also reduce metals as a part of their metal resistance apart from respiration (Zhu et al. 2008; Sayel et al. 2012). Methylation of metals by microorganisms makes them more volatile and hence is an important factor in bioremediation of heavy metals (Takeuchi and Sugio 2006; De, Ramaiah et al. 2008). Sulfate-reducing bacteria can also metabolize heavy metals and precipitate them as insoluble sulfides (White et al. 1998).

Nonetheless, the process of bioremediation is associated with certain challenges. Bioremediation approaches generally depend on the conditions of the polluted site and hence might not be effective in all locations. Initially, processes are tested at lab scale, which might not be able to work successfully in the field. Also, the mechanisms by which microorganisms grow and are active in the polluted sites are still not clear. All these bottlenecks limit the use of bioremediation for environmental clean-up (Lovley 2003; Malla et al. 2018).

ROLE OF MARINE-DERIVED BIOSURFACTANTS IN BIOREMEDIATION

The major problem associated with the efficiency of bioremediation is the bioavailability of pollutants. The process of bioremediation can be divided into several steps. First, the pollutants get sorbed in the soil matrix, and a non-aqueous phase is formed. The pollutant interacts with organic matter and is biologically transformed into simpler forms. Finally, the contaminant ages naturally. Bioavailability of pollutants is affected by desorption, diffusion and dissolution. Hence, all these processes limit bioavailability of pollutants and decrease the efficiency of bioremediation. Bioremediation of soil becomes difficult as oil forms films or droplets on soil particles and makes degradation by microorganisms difficult (Aulwar and Awasthi 2016).

A potential method that can improve the efficiency of bioremediation is the application of biosurfactants. Biosurfactants help in distribution of contaminants in the aqueous phase and increase their bioavailability. Biosurfactants decrease surface tension at the interface of the hydrocarbon contaminant and water, thereby increasing the area of contact at the hydrocarbon–water interface. This enhances the solubilization rate of hydrocarbons and hence increases the bioavailability of hydrocarbons for microbial degradation. Biosurfactants also increase the efficiency of bioremediation by interacting with microbial cell surfaces. This increases the hydrophobicity of microbial cells and allows the association of microbial cells with hydrophobic pollutants (Ron and Rosenberg 2002; Pacwa-Płociniczak et al. 2011; Aulwar and Awasthi 2016).

Biosurfactants from many marine microorganisms have been explored for their bioremediation potential. Marine γ-proteobacteria secrete amphiphilic biosurfactants that solubilize aromatic hydrocarbons (Tripathi et al. 2018). Production of biosurfactants by *Pseudomonas putida* has been shown to increase attachment of the microbial cell to polycyclic aromatic hydrocarbons (Rodrigues et al. 2005). The bioavailability of PAHs was increased by a biosurfactant produced by *Alcanivorax borkumensis*. This enhanced the biodegradation of PAHs by other marine microorganisms (McKew et al. 2007). Species of *Halomonas*, *Myroides* and *Marinobacter* and *Yarrowia lipolytica* also produce biosurfactants that can be used as potential hydrocarbon degraders (Tripathi et al. 2018). Glycolipid produced by *Halomonas* sp. has demonstrated its potency to recover crude oil. Also, glycolipid produced at low temperatures has also been shown to degrade organic compounds. Hence, it can be used to extract residual oil trapped in reservoirs in extreme environments as well as in bioremediation (Pepi et al. 2005; Dhasayan et al. 2014). Lipopeptide produced by marine *Achromobacter* sp. had a wide range of emulsifying activity and hence can be used to emulsify hydrocarbons (Deng et al. 2016). In addition to this, lipopeptide produced by *Brevibacterium luteolum* is capable of removing crude oil from sand, exhibiting its role in bioremediation (Vilela et al. 2014). L-ornithine lipid produced by *Myroides* sp. has been shown to emulsify crude oil. The emulsification ability of this biosurfactant was stable over a wide range of temperature, pH and salinity, with surface activities higher than chemical surfactants and surfactin (Maneerat et al. 2006). Bioemulsifiers produced by *Marinobacter* sp. dispersed crude oil in artificial marine water and were stable in extreme conditions (Raddadi et al. 2017), while rhamnolipid produced by *Nocardiopsis* sp. degraded engine oil under saline conditions (Roy et al. 2015). Similarly, a biosurfactant produced by *Marinobacter hydrocarbonoclasticus* and *Candida lipolytica* and a bioemulsifier (yansan) produced by *Yarrowia lipolytica* were able to emulsify hydrocarbons and demonstrated their potential application in bioremediation (Amaral et al. 2006; Santos et al. 2017; Zenati et al. 2018).

APPLICATIONS OF BIOSURFACTANTS IN ENVIRONMENTAL BIOREMEDIATION

PETROLEUM INDUSTRY, REMEDIATION OF OIL SPILLS AND MICROBIAL-ENHANCED OIL RECOVERY

Petroleum industries have been one of the major consumers of biosurfactants, as biosurfactants increase removal of oil by increasing the solubility of petroleum components in contaminated water and soil (Mulligan 2005; Kapadia and Yagnik 2013). Biosurfactants take up hydrocarbons in three major ways: emulsification, solubilization and facilitated transport (Mohanty et al. 2013). Biosurfactants reduce interfacial tension between oil and water by increasing solubility of non-aqueous phase liquids. This increases the concentration of the aqueous surfactant and decreases interfacial tension. The monomers of the mixture emulsify and form micelles. This enhances recovery of oil (Banat et al. 2000; Abdolhamid et al. 2009). They enable mass transfer of petroleum hydrocarbons from liquid sludge to bacterial cells. They also change the hydrophobicity of bacterial cells, which increases the attachment of petroleum hydrocarbons to cells (Yuste et al. 2000; Sponza and Gok 2011). Consequently, petroleum hydrocarbons become easily available to petrophilic bacteria and are utilized as substrates by bacteria, thereby cleaning up the oil spill (Figure 14.1).

Other than remediation of oil spills, biosurfactants are also used to reduce degradation of hydrocarbons due to fouling and biocorrosion in oil reservoirs (Perfumo et al. 2010). They also reduce the viscosity of heavy oil, thus enabling flow through pipelines. They have the potential to clean oil storage tanks and stabilize fuel-water-oil emulsions (Jain et al. 2012; Almansoory et al. 2015).

The addition of biosurfactant mixtures alone might be helpful in enhancing biodegradation of contaminants rich in hydrocarbons present in the environment. This can also be achieved by direct addition of bacterial cells to the contaminated sites, also termed *in situ* bioremediation (Figure 14.2). For this, biosurfactant-producing microorganisms are usually provided with inexpensive substrates, such as molasses and inorganic nutrients *in situ*, which enhance growth and bio-surfactant production by microorganisms (Reis et al. 2013). The conditions of oil reservoirs are generally harsh, with high temperature, pressure and salinity and low oxygen, and microorganisms should also be able to survive under these conditions. Hence, aerobic and anaerobic microorganisms that can withstand high pressure and salinity are suitable options (Usman et al. 2016). The use of biosur-factants for remediation of oil spills has been tested in closed systems, but their efficacy still needs to be evaluated in open systems (Banat et al. 2000).

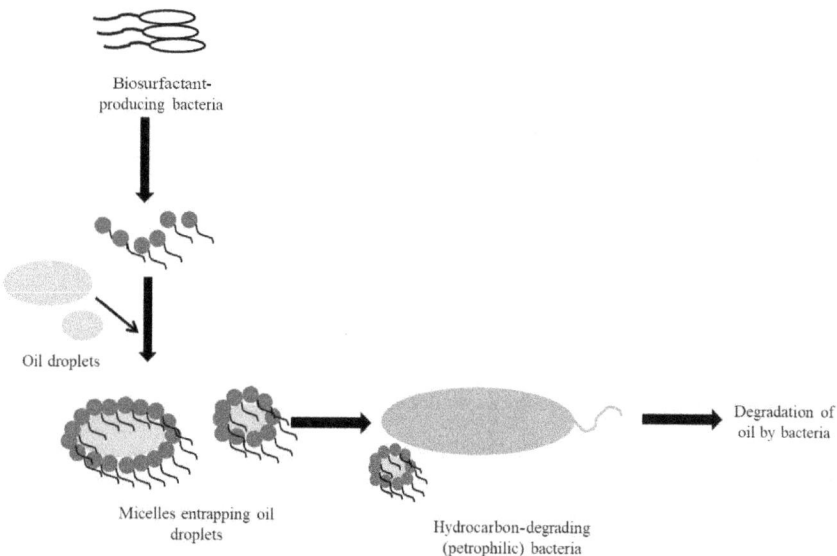

FIGURE 14.1 Bioremediation of oil spills by biosurfactants.

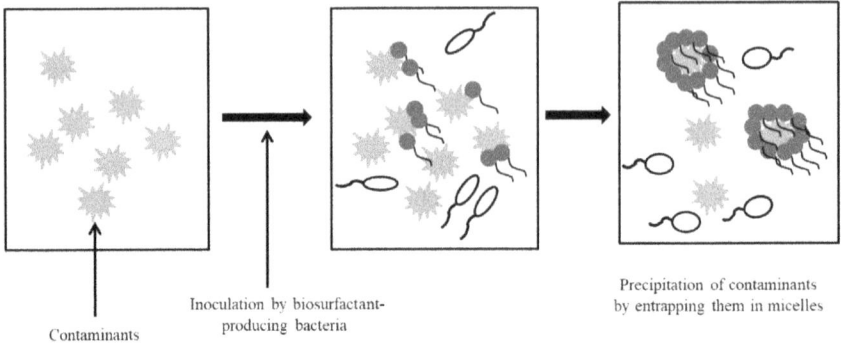

Contaminants

Inoculation by biosurfactant-producing bacteria

Precipitation of contaminants by entrapping them in micelles

FIGURE 14.2   Role of biosurfactants in degradation of contaminants.

Another application of biosurfactants in the petroleum industry is the recovery of oil by microorganisms, or microbial-enhanced oil recovery (MEOR). The oil from reservoirs is recovered by mechanical (primary) and physical (secondary) methods (Banat et al. 2000; Sen 2008). The oil remaining after two recovery procedures can be recovered by a tertiary method. The tertiary process involves the use of microorganisms and their metabolites to enhance recovery of oil from reservoirs. The metabolites include acids, enzymes, biomass, biopolymers, biosurfactants, gases and solvents. The residual oils in reservoirs are present in areas that are difficult to access, as oil is trapped in areas by capillary pressure (Sen 2008). Biosurfactants reduce interfacial tension between oil/water and oil/rock, which reduces capillary forces and prevents transfer of oil to deep pores (Figure 14.3). Hence, biosurfactants also find suitable application in microbial-enhanced recovery of oil.

A strain of *Bacillus pumilus* has demonstrated its potential to degrade 80.44% of total petroleum hydrocarbons (TPHs) within one month of incubation (Patowary et al. 2015). A study reported the ability of two biosurfactant-producing strains, *Achromobacter* sp. and *Ochrobactrum* sp., in degrading heavy oils efficiently (Primeia et al. 2020). *Pseudomonas aeruginosa* PG1 was also shown to degrade 81.8% of TPH five weeks after inoculation (Patowary et al. 2017). *Bacillus subtilis* A1 degraded crude oil with an efficiency of 87% in seven days (Parthipan et al. 2017).

DEGRADATION OF HYDROCARBONS AND POLYAROMATIC HYDROCARBONS

Polynuclear or polycyclic aromatic hydrocarbons are contaminants in soil that have mutagenic and carcinogenic effects. These hydrocarbons are produced during refining of petroleum, preservation of wood and coke production (Park et al. 1990). PAHs contain multiple aromatic rings. The degradability of compounds becomes difficult with an increase in the number of rings. This is because a high number of rings decreases the volatility and solubility of the compound, which

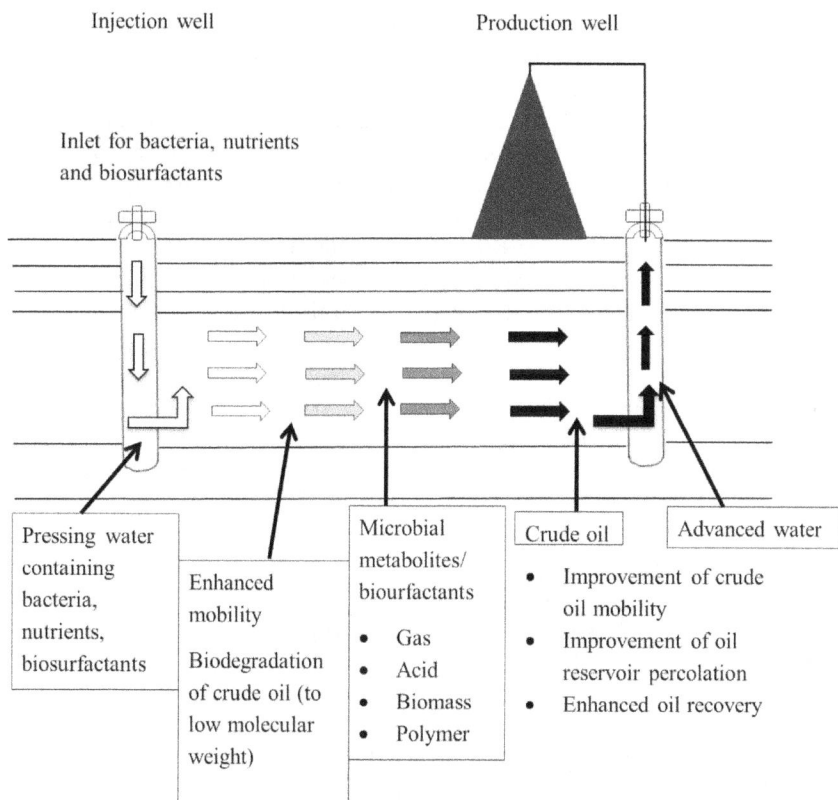

FIGURE 14.3 Recovery of residual oil from reservoirs by biosurfactant-producing microorganisms (microbial-enhanced oil recovery).

increases sorption. The degradation of PAHs occurs one ring at a time, which is like single-ring aromatics. Rhamnolipids have been shown to degrade naphthalene by increasing the solubility of naphthalene and using it as a carbon source (Vipulanandan and Ren 2000). Biosurfactants have also been shown to biodegrade phenanthrene (Providenti et al. 1995; Burd and Ward 1996; García-Junco et al. 2001), gasoline-contaminated soil (Rahman et al. 2002) and pentachlorophenol (Straube et al. 2003).

Halogenated aromatic compounds are also one of the xenobiotics that negatively affect the soil and environment. A few examples of chlorinated hydrocarbons are pentachlorophenol, polychlorinated biphenyls, plasticizers and pesticides, such as 2,4-dichlorophenoxyacetic acid, 2,4,5-trichlorophenoxyacetic acid and dichlorodiphenyltrichloroethane. The mechanism and rate of biodegradation of these halogenated aliphatic compounds depend upon the type of compound and number and position of halogens. Biosurfactants are capable of mineralizing polychlorinated biphenyls (Robinson et al. 1996); 4,4' chlorobiphenyl; Arochlor 1242 (Fiebig et al. 1997); and pesticides, such as atrazine, coumaphos and trifluralin (Mata-Sandoval et al. 2000). Pesticides are bound tightly in micelles by biosurfactants. This enables slow movement of the pesticide to the aqueous phase and helps in its microbial uptake (Mulligan 2005).

A few studies have been conducted for degradation of hydrocarbons by biosurfactants. A biosurfactant produced by marine *Pseudomonas aeruginosa* has been shown to degrade hexadecane, heptadecane, octadecane and nonadecane within a month of incubation, along with its ability to break tetradecane, 2-methylnaphthale and pristine (Zhuang et al. 2002). *Bacillus pumilus* has also been shown to degrade polyaromatic hydrocarbons present in crude oil (Patowary et al. 2015). A consortium of microorganisms dominated by *Pseudomonas aeruginosa*, *Bacillus megaterium*, *B. stratosphericus* and *B. subtilis* enhanced PAH degradation by 86.5% in the presence of biosurfactants after 45 days of incubation (Bezza and Chirwa 2016), and 63%, 88% and 98% degradation of chrysene, pyrene and fluoranthene, respectively, were observed by biosurfactants of *Pseudomonas aeruginosa* MS-1 and *Acinetobacter* sp. (Chirwa et al. 2021).

BIOREMEDIATION OF HEAVY METALS AND METALLOIDS

Heavy metals have been accumulated in soil and water bodies as a result of mining activities and industries such as paint, metal pipes, arms and ammunition, batteries and so on (Singh and Cameotra 2004) and result in toxification (Mao et al. 2016). The most prominent heavy metals acting as pollutants are arsenic, lead, copper, zinc, mercury, cadmium and nickel. These heavy metals have been linked to many congenital defects and damage to organs. Heavy metals are not easily degraded in the environment but can be transformed to less toxic forms (Santona et al. 2006). Conventionally heavy metals are removed by treating contaminated soil with organic and inorganic acids, metal-chelating agents and chemical surfactants. But these methods do not remove metal ions from soil completely (Singh and Cameotra 2004; Dahrazma and Mulligan 2007; Das et al. 2009). Other remediation techniques, such as landfilling and thermal treatment, require large amounts of space and hence are not preferred (Das et al. 2009; Wang and Mulligan 2009). Biosurfactants having good metal-complexing ability and stability in polluted environments and can help in remediation of heavy metal–contaminated surfaces (Singh and Cameotra 2004). Also, biosurfactants can be produced from cheap agricultural wastes and are stable over a wide range of pH, temperature and saline concentrations.

Biosurfactants tend to extract heavy metals by ion exchange, counter ion binding, electrostatic interaction and precipitation-dissolution. When biosurfactant complexes with non-ionic metals, the activity of metal in the solution phase decreases, which promotes desorption. Reducing interfacial tension allows biosurfactants to bind directly to sorbed metals and helps in the accumulation of metals at the solid-solution interface (Singh and Cameotra 2004). Metals have a high affinity for ionic biosurfactants and hence form complexes and are desorbed from the matrix (Fulke et al. 2020). Cationic surfactants with positively charged functional groups decrease the association of metal with surfaces, while anionic surfactants increase this association by either precipitating the complex or enhancing the sorption of metal-surfactant (Christofi and Ivshina 2002). This is because the negative charge of the anionic biosurfactant reacts with cationic metal and forms an ionic bond that is stronger than the bond of metal with soil (Açıkel 2011).

The mechanism of biosurfactant activity in metal-contaminated soil is given in Figure 14.4. There are four steps by which heavy metals trapped in soil are removed by biosurfactant solution. First, when heavy metals are released into the soil, they get adsorbed onto the surface of soil particles. After the addition of biosurfactant solution, it assimilates between the soil and heavy metal and forms a complex at the interface of soil and metal. The metal-biosurfactant complex gets desorbed from the soil surface, and the metal gets trapped into micelles formed by biosurfactant through electrostatic interaction. The polar head of micelles binds to the metal and makes it more soluble in the aqueous phase. The metals are recovered by flushing, whereas the biosurfactant is recovered by membrane separation (Frazer 2000; Aşçı et al. 2007; Ibrahim et al. 2016; Guan et al. 2017). The metals recovered from this process can be used again, reducing synthesis of heavy metals and ore mining (Ravindran et al. 2020). As soil particles have negative charges on their surfaces, cationic biosurfactants are easily adsorbed to the soil surface. Also, biosurfactants contain a carboxylic functional group that acts as an organic ligand and increases the extraction of heavy metals from soil (Tang et al. 2015).

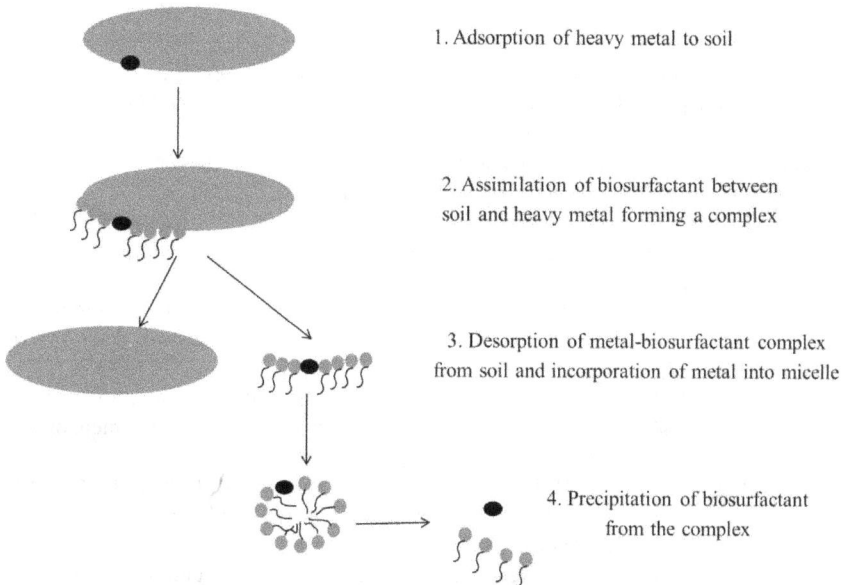

1. Adsorption of heavy metal to soil

2. Assimilation of biosurfactant between soil and heavy metal forming a complex

3. Desorption of metal-biosurfactant complex from soil and incorporation of metal into micelle

4. Precipitation of biosurfactant from the complex

FIGURE 14.4 Mechanism of biosurfactant activity in metal-contaminated soil.

Various studies have reported the role of biosurfactants in degradation of heavy metals. The external addition of rhamnolipid to *Burkholderia* sp. culture medium reduced the toxicity of Cd(II) (Sandrin et al. 2000). The efficiency of heavy metal removal by biosurfactants has also been reported for chromium, copper, cadmium and zinc, with success rates of 90–100% (Hong et al. 2002; Guan, Yuan et al. 2017). *Bacillus* sp. MTCC 5514 produced a biosurfactant that increased its tolerance to hexavalent chromium (Gnanamani et al. 2010). Also, the production of biosurfactant by marine actinobacter was found to increase with the presence of metal ions such as $MgSO_4$, $MnCl_2$, $CuSO_4$ and $FeCl_3$ (Kiran et al. 2010).

CONCLUSION

Biosurfactants produced by marine microorganisms can tolerate extreme conditions of temperature, pH and salinity and hence find many applications in environmental bioremediation. Biosurfactants help in distribution of pollutants, thereby increasing their bioavailability for microbial degradation. They have been used in bioremediation of oil spills, degradation of polyaromatic hydrocarbons, removal of heavy metals and microbial-enhanced oil recovery. However, the problem associated with the use of biosurfactants in environmental bioremediation is their large-scale production. Also, the efficiency of biosurfactants for the degradation of pollutants has only been tested in laboratories, and hence its testing in fields is required. Once the solutions to these problems have been established, biosurfactants have promising scope in bioremediation applications.

REFERENCES

Abdolhamid, H. R., M. A. S. AL-Baghdadi and A. K. E. Hinshiri. 2009. Evaluation of bio-surfactants enhancement on bioremediation process efficiency for crude oil contaminated soil at oilfield: strategic study. *Strategic Study. Ovidius University Annals of Chemistry.* 20: 25–30.

Açıkel, Y. S. 2011. Use of biosurfactants in the removal of heavy metal ions from soils. pp. 183–223. *Biomanagement of Metal-Contaminated Soils.* Springer.

Adams, G. O., P. T. Fufeyin, S. E. Okoro and I. Ehinomen. 2015. Bioremediation, biostimulation and bioaugmention: a review. *International Journal of Environmental Bioremediation & Biodegradation.* 3: 28–39.

Akpor, O., O. Igbinosa and O. Igbinosa. 2007. Studies on the effect of petroleum hydrocarbon on the microbial and physico-chemicals characteristics of soil. *African Journal of Biotechnology.* 6.

Almansoory, A. F., H. A. Hasan, M. Idris, S. R. S. Abdullah and N. Anuar. 2015. Potential application of a biosurfactant in phytoremediation technology for treatment of gasoline-contaminated soil. *Ecological Engineering.* 84: 113–120.

Amaral, P., J. Da Silva, B. M. Lehocky, A. Barros-Timmons, M. Coelho, I. Marrucho and J. Coutinho. 2006. Production and characterization of a bioemulsifier from *Yarrowia lipolytica. Process Biochemistry.* 41: 1894–1898.

Aşçı, Y., M. Nurbaş and Y. S. Açıkel. 2007. Sorption of Cd (II) onto kaolin as a soil component and desorption of Cd (II) from kaolin using rhamnolipid biosurfactant. *Journal of Hazardous Materials.* 139: 50–56.

Aulwar, U. and R. Awasthi. 2016. Production of biosurfactant and their role in bioremediation. *Journal of Ecosystem & Ecography*. 6: 202.

Banat, I. M., R. S. Makkar and S. S. Cameotra. 2000. Potential commercial applications of microbial surfactants. *Applied Microbiology and Biotechnology*. 53: 495–508.

Bezza, F. A. and E. M. N. Chirwa. 2016. Biosurfactant-enhanced bioremediation of aged polycyclic aromatic hydrocarbons (PAHs) in creosote contaminated soil. *Chemosphere*. 144: 635–644.

Boopathy, R. 2000. Factors limiting bioremediation technologies. *Bioresource Technology*. 74: 63–67.

Burd, G. and O. P. Ward. 1996. Bacterial degradation of polycyclic aromatic hydrocarbons on agar plates: the role of biosurfactants. *Biotechnology Techniques*. 10: 371–374.

Cerniglia, C. E. 1997. Fungal metabolism of polycyclic aromatic hydrocarbons: past, present and future applications in bioremediation. *Journal of Industrial Microbiology and Biotechnology*. 19: 324–333.

Chatterjee, S., S. Kumari, S. Rath and S. Das. 2022. Prospects and scope of microbial bioremediation for the restoration of the contaminated sites. pp. 3–31. *Microbial Biodegradation and Bioremediation*. Elsevier,

Chirwa, E. M. N., T. B. Lutsinge-Nembudani, O. M. Fayemiwo and F. A. Bezza. 2021. Biosurfactant assisted degradation of high molecular weight polycyclic aromatic hydrocarbons by mixed cultures from a car service oil dump from Pretoria Central Business District (South Africa). *Journal of Cleaner Production*. 290: 125183.

Christofi, N. and I. Ivshina. 2002. Microbial surfactants and their use in field studies of soil remediation. *Journal of Applied Microbiology*. 93: 915–929.

Dahrazma, B. and C. N. Mulligan. 2007. Investigation of the removal of heavy metals from sediments using rhamnolipid in a continuous flow configuration. *Chemosphere*. 69: 705–711.

Das, P., S. Mukherjee and R. Sen. 2009. Biosurfactant of marine origin exhibiting heavy metal remediation properties. *Bioresource Technology*. 100: 4887–4890.

De, J., N. Ramaiah and L. Vardanyan. 2008. Detoxification of toxic heavy metals by marine bacteria highly resistant to mercury. *Marine Biotechnology*. 10: 471–477.

Deng, M. C., J. Li, Y. H. Hong, X. M. Xu, W. X. Chen, J. P. Yuan, J. Peng, M. Yi and J. H. Wang. 2016. Characterization of a novel biosurfactant produced by marine hydrocarbon-degrading bacterium *Achromobacter* sp. HZ 01. *Journal of Applied Microbiology*. 120: 889–899.

Dhagat, S. and S. E. Jujjavarapu. 2022. Utility of lignin-modifying enzymes: a green technology for organic compound mycodegradation. *Journal of Chemical Technology & Biotechnology*. 97: 343–358.

Dhaliwal, S. S., J. Singh, P. K. Taneja and A. Mandal. 2020. Remediation techniques for removal of heavy metals from the soil contaminated through different sources: a review. *Environmental Science and Pollution Research*. 27: 1319–1333.

Dhasayan, A., G. S. Kiran and J. Selvin. 2014. Production and characterisation of glycolipid biosurfactant by *Halomonas* sp. MB-30 for potential application in enhanced oil recovery. *Applied Biochemistry and Biotechnology*. 174: 2571–2584.

Díaz, E. 2004. Bacterial degradation of aromatic pollutants: a paradigm of metabolic versatility. *International Microbiology: The Official Journal of the Spanish Society for Microbiology*. 7: 173–180.

Fiebig, R., D. Schulze, J.-C. Chung and S.-T. Lee. 1997. Biodegradation of polychlorinated biphenyls (PCBs) in the presence of a bioemulsifier produced on sunflower oil. *Biodegradation*. 8: 67–75.

Field, J. A., A. J. Stams, M. Kato and G. Schraa. 1995. Enhanced biodegradation of aromatic pollutants in cocultures of anaerobic and aerobic bacterial consortia. *Antonie van Leeuwenhoek.* 67: 47–77.

FLI, V. 2022. *Ex-Situ Bioremediation.* www.vertasefli.co.uk/our-solutions/expertise/ex-situ-bioremediation.

Frazer, L. 2000. Lipid lather removes metals. *Environmental Health Perspectives.* 108: A320–A323.

Fulke, A. B., A. Kotian and M. D. Giripunje. 2020. Marine microbial response to heavy metals: mechanism, implications and future prospect. *Bulletin of Environmental Contamination and Toxicology.* 105: 182–197.

García-Junco, M., E. De Olmedo and J. J. Ortega-Calvo. 2001. Bioavailability of solid and non-aqueous phase liquid (NAPL)-dissolved phenanthrene to the biosurfactant-producing bacterium *Pseudomonas aeruginosa* 19SJ. *Environmental Microbiology.* 3: 561–569.

Gibson, J. and C. S. Harwood. 2002. Metabolic diversity in aromatic compound utilization by anaerobic microbes. *Annual Reviews in Microbiology.* 56: 345–369.

Gnanamani, A., V. Kavitha, N. Radhakrishnan, G. S. Rajakumar, G. Sekaran and A. Mandal. 2010. Microbial products (biosurfactant and extracellular chromate reductase) of marine microorganism are the potential agents reduce the oxidative stress induced by toxic heavy metals. *Colloids and Surfaces B: Biointerfaces.* 79: 334–339.

Goswami, M., P. Chakraborty, K. Mukherjee, G. Mitra, P. Bhattacharyya, S. Dey and P. Tribedi. 2018. Bioaugmentation and biostimulation: a potential strategy for environmental remediation. *Journal of Microbiology & Experimentation.* 6: 223–231.

Guan, R., X. Yuan, Z. Wu, H. Wang, L. Jiang, Y. Li and G. Zeng. 2017. Functionality of surfactants in waste-activated sludge treatment: a review. *Science of the Total Environment.* 609: 1433–1442.

Hamby, D. 1996. Site remediation techniques supporting environmental restoration activities—a review. *Science of the Total Environment.* 191: 203–224.

Hong, K.-J., S. Tokunaga and T. Kajiuchi. 2002. Evaluation of remediation process with plant-derived biosurfactant for recovery of heavy metals from contaminated soils. *Chemosphere.* 49: 379–387.

Ibrahim, W. M., A. F. Hassan and Y. A. Azab. 2016. Biosorption of toxic heavy metals from aqueous solution by *Ulva lactuca* activated carbon. *Egyptian Journal of Basic and Applied Sciences.* 3: 241–249.

Jain, R. M., K. Mody, A. Mishra and B. Jha. 2012. Isolation and structural characterization of biosurfactant produced by an alkaliphilic bacterium *Cronobacter sakazakii* isolated from oil contaminated wastewater. *Carbohydrate Polymers.* 87: 2320–2326.

Jan, S., B. Rashid, M. Azooz, M. A. Hossain and P. Ahmad. 2016. Genetic strategies for advancing phytoremediation potential in plants: a recent update. *Plant Metal Interaction.* 431–454.

Kapadia, S. G. and B. Yagnik. 2013. Current trend and potential for microbial biosurfactants. *South Asian Journal of Experimental Biology.* 4: 1–8.

Kiran, G. S., T. A. Thomas and J. Selvin. 2010. Production of a new glycolipid biosurfactant from marine *Nocardiopsis lucentensis* MSA04 in solid-state cultivation. *Colloids and Surfaces B: Biointerfaces.* 78: 8–16.

Lew, D. 2022. *Ex Situ Bioremediation Technologies.* www.drdarrinlew.us/contaminated-soil/ex-situ-bioremediation-technologies.html.

Lovley, D. R. 2003. Cleaning up with genomics: applying molecular biology to bioremediation. *Nature Reviews Microbiology.* 1: 35–44.

Malla, M. A., A. Dubey, S. Yadav, A. Kumar, A. Hashem and E. F. Abd-Allah. 2018. Understanding and designing the strategies for the microbe-mediated remediation of environmental contaminants using omics approaches. *Frontiers in Microbiology.* 9: 1132.

Maneerat, S., T. Bamba, K. Harada, A. Kobayashi, H. Yamada and F. Kawai. 2006. A novel crude oil emulsifier excreted in the culture supernatant of a marine bacterium, *Myroides* sp. strain SM1. *Applied Microbiology and Biotechnology.* 70: 254–259.

Mao, X., Z. Yu, Z. Ding, T. Huang, J. Ma, G. Zhang, J. Li and H. Gao. 2016. Sources and potential health risk of gas phase PAHs in Hexi Corridor, Northwest China. *Environmental Science and Pollution Research.* 23: 2603–2612.

Mata-Sandoval, J. C., J. Karns and A. Torrents. 2000. Effect of rhamnolipids produced by *Pseudomonas aeruginosa* UG2 on the solubilization of pesticides. *Environmental Science & Technology.* 34: 4923–4930.

McDonald, I. R., C. B. Miguez, G. Rogge, D. Bourque, K. D. Wendlandt, D. Groleau and J. C. Murrell. 2006. Diversity of soluble methane monooxygenase-containing methanotrophs isolated from polluted environments. *FEMS Microbiology Letters.* 255: 225–232.

McKew, B. A., F. Coulon, M. M. Yakimov, R. Denaro, M. Genovese, C. J. Smith, A. M. Osborn, K. N. Timmis and T. J. McGenity. 2007. Efficacy of intervention strategies for bioremediation of crude oil in marine systems and effects on indigenous hydro-carbonoclastic bacteria. *Environmental Microbiology.* 9: 1562–1571.

Mohanty, S., J. Jasmine and S. Mukherji. 2013. Practical considerations and chal-lenges involved in surfactant enhanced bioremediation of oil. *BioMed Research International.* 2013.

Mohapatra, B., T. Dhamale, B. K. Saha and P. S. Phale. 2022. Microbial degradation of aromatic pollutants: metabolic routes, pathway diversity, and strategies for biore-mediation. pp. 365–394. *Microbial Biodegradation and Bioremediation.* Elsevier,

Mulligan, C. N. 2005. Environmental applications for biosurfactants. *Environmental Pollution.* 133: 183–198.

Pacwa-Płociniczak, M., G. A. Płaza, Z. Piotrowska-Seget and S. S. Cameotra. 2011. Environmental applications of biosurfactants: recent advances. *International Journal of Molecular Sciences.* 12: 633–654.

Park, K. S., R. C. Sims and R. R. Dupont. 1990. Transformation of PAHs in soil systems. *Journal of Environmental Engineering.* 116: 632–640.

Parthipan, P., E. Preetham, L. L. Machuca, P. K. Rahman, K. Murugan and A. Rajasekar. 2017. Biosurfactant and degradative enzymes mediated crude oil degradation by bacterium *Bacillus subtilis* A1. *Frontiers in Microbiology.* 8: 193.

Patel, H., S. Shakhreliya, R. Maurya, V. C. Pandey, N. Gohil, G. Bhattacharjee, K. J. Alzahrani and V. Singh. 2022. CRISPR-assisted strategies for futuristic phytore-mediation. pp. 203–220. *Assisted Phytoremediation.* Elsevier.

Patowary, K., R. Patowary, M. C. Kalita and S. Deka. 2017. Characterization of biosur-factant produced during degradation of hydrocarbons using crude oil as sole source of carbon. *Frontiers in Microbiology.* 8: 279.

Patowary, K., R. R. Saikia, M. C. Kalita and S. Deka. 2015. Degradation of polyaromatic hydrocarbons employing biosurfactant-producing *Bacillus pumilus* KS2. *Annals of Microbiology.* 65: 225–234.

Pepi, M., A. Cesàro, G. Liut and F. Baldi. 2005. An Antarctic psychrotrophic bacterium *Halomonas* sp. ANT-3b, growing on n-hexadecane, produces a new emulsyfying glycolipid. *FEMS Microbiology Ecology.* 53: 157–166.

Perfumo, A., T. Smyth, R. Marchant and I. Banat. 2010. Production and roles of bio-surfactants and bioemulsifiers in accessing hydrophobic substrates. pp. 1501–1512. *Handbook of Hydrocarbon and Lipid Microbiology*. Springer.

Peters, F., Y. Shinoda, M. J. McInerney and M. Boll. 2007. Cyclohexa-1, 5-diene-1-car-bonyl-coenzyme A (CoA) hydratases of *Geobacter metallireducens* and *Syntrophus aciditrophicus*: evidence for a common benzoyl-CoA degradation pathway in facultative and strict anaerobes. *Journal of Bacteriology*. 189: 1055–1060.

Primeia, S., C. Inoue and M.-F. Chien. 2020. Potential of biosurfactants' production on degrading heavy oil by bacterial consortia obtained from tsunami-induced oil-spilled beach areas in Miyagi, Japan. *Journal of Marine Science and Engineering*. 8: 577.

Providenti, M. A., C. A. Flemming, H. Lee and J. T. Trevors. 1995. Effect of addition of rhamnolipid biosurfactants or rhamnolipid-producing *Pseudomonas aeruginosa* on phenanthrene mineralization in soil slurries. *FEMS Microbiology Ecology*. 17: 15–26.

Raddadi, N., L. Giacomucci, G. Totaro and F. Fava. 2017. *Marinobacter* sp. from marine sediments produce highly stable surface-active agents for combatting marine oil spills. *Microbial Cell Factories*. 16: 1–13.

Rahman, K., I. Banat, J. Thahira, T. Thayumanavan and P. Lakshmanaperumalsamy. 2002. Bioremediation of gasoline contaminated soil by a bacterial consortium amended with poultry litter, coir pith and rhamnolipid biosurfactant. *Bioresource Technology*. 81: 25–32.

Ravindran, A., A. Sajayan, G. B. Priyadharshini, J. Selvin and G. S. Kiran. 2020. Revealing the efficacy of thermostable biosurfactant in heavy metal bioremediation and surface treatment in vegetables. *Frontiers in Microbiology*. 11: 222.

Reis, R., G. Pacheco, A. Pereira and D. Freire. 2013. Biosurfactants: production and applications. *Biodegradation-life of Science*: 31–61.

Robinson, K. G., M. M. Ghosh and Z. Shi. 1996. Mineralization enhancement of non-aqueous phase and soilbound PCB using biosurfactant. *Water Science and Technology*. 34: 303–309.

Rodrigues, A. C., S. Wuertz, A. G. Brito and L. F. Melo. 2005. Fluorene and phenanthrene uptake by *Pseudomonas putida* ATCC 17514: kinetics and physiological aspects. *Biotechnology and Bioengineering*. 90: 281–289.

Romero, M. C., M. L. Salvioli, M. C. Cazau and A. Arambarri. 2002. Pyrene degradation by yeasts and filamentous fungi. *Environmental Pollution*. 117: 159–163.

Ron, E. Z. and E. Rosenberg. 2002. Biosurfactants and oil bioremediation. *Current Opinion in Biotechnology*. 13: 249–252.

Roy, S., S. Chandni, I. Das, L. Karthik, G. Kumar and K. V. Bhaskara Rao. 2015. Aquatic model for engine oil degradation by rhamnolipid producing *Nocardiopsis* VITSISB. *3 Biotech*. 5: 153–164.

Sandrin, T. R., A. M. Chech and R. M. Maier. 2000. A rhamnolipid biosurfactant reduces cadmium toxicity during naphthalene biodegradation. *Applied and Environmental Microbiology*. 66: 4585–4588.

Santona, L., P. Castaldi and P. Melis. 2006. Evaluation of the interaction mechanisms between red muds and heavy metals. *Journal of Hazardous Materials*. 136: 324–329.

Santos, D. K., A. H. Resende, D. G. de Almeida, R. D. C. F. Soares da Silva, R. D. Rufino, J. M. Luna, I. M. Banat and L. A. Sarubbo. 2017. *Candida lipolytica* UCP0988 biosurfactant: potential as a bioremediation agent and in formulating a commercial related product. *Frontiers in Microbiology*. 8: 767.

Sayel, H., W. Bahafid, N. Tahri Joutey, K. Derraz, K. Fikri Benbrahim, S. Ibnsouda Koraichi and N. El Ghachtouli. 2012. Cr (VI) reduction by *Enterococcus gallinarum* isolated from tannery waste-contaminated soil. *Annals of Microbiology.* 62: 1269–1277.

Sen, R. 2008. Biotechnology in petroleum recovery: the microbial EOR. *Progress in Energy and Combustion Science.* 34: 714–724.

Shackira, A., K. Jazeel and J. T. Puthur. 2021. Phycoremediation and phytoremediation: promising tools of green remediation. pp. 273–293. *Sustainable Environmental Clean-up.* Elsevier,

Sharma, I. 2020. Bioremediation techniques for polluted environment: concept, advantages, limitations, and prospects. *Trace Metals in the Environment-New Approaches and Recent Advances.* IntechOpen,

Singh, P. and S. S. Cameotra. 2004. Enhancement of metal bioremediation by use of microbial surfactants. *Biochemical and Biophysical Research Communications.* 319: 291–297.

Speight, J. G. 2019. *Natural Water Remediation: Chemistry and Technology.* Butterworth-Heinemann,

Sponza, D. T. and O. Gok. 2011. Effects of sludge retention time (SRT) and biosurfactant on the removal of polyaromatic compounds and toxicity. *Journal of Hazardous Materials.* 197: 404–416.

Straube, W., C. Nestler, L. Hansen, D. Ringleberg, P. Pritchard and J. Jones-Meehan. 2003. Remediation of polyaromatic hydrocarbons (PAHs) through landfarming with biostimulation and bioaugmentation. *Acta Biotechnologica.* 23: 179–196.

Supaphol, S., S. Panichsakpatana, S. Trakulnaleamsai, N. Tungkananuruk, P. Roughjanajirapa and A. G. O'Donnell. 2006. The selection of mixed microbial inocula in environmental biotechnology: example using petroleum contaminated tropical soils. *Journal of Microbiological Methods.* 65: 432–441.

Takeuchi, F. and T. Sugio. 2006. Volatilization and recovery of mercury from mercury-polluted soils and wastewaters using mercury-resistant *Acidithiobacillus ferrooxidans* strains SUG 2-2 and MON-1. *Environmental Sciences: An International Journal of Environmental Physiology and Toxicology.* 13: 305–316.

Tang, Z., L. Zhang, Q. Huang, Y. Yang, Z. Nie, J. Cheng, J. Yang, Y. Wang and M. Chai. 2015. Contamination and risk of heavy metals in soils and sediments from a typical plastic waste recycling area in North China. *Ecotoxicology and Environmental Safety.* 122: 343–351.

Tripathi, L., V. U. Irorere, R. Marchant and I. M. Banat. 2018. Marine derived biosurfactants: a vast potential future resource. *Biotechnology Letters.* 40: 1441–1457.

Turgeon, M. 2009. *Fact Sheet: Biosparging.* https://gost.tpsgc-pwgsc.gc.ca/tfs.aspx?ID=4&lang=eng.

Usman, M. M., A. Dadrasnia, K. T. Lim, A. F. Mahmud and S. Ismail. 2016. Application of biosurfactants in environmental biotechnology; remediation of oil and heavy metal. *AIMS Bioengineering.* 3: 289–304.

Vidali, M. 2001. Bioremediation. an overview. *Pure and Applied Chemistry.* 73: 1163–1172.

Vilela, W., S. Fonseca, F. Fantinatti-Garboggini, V. Oliveira and M. Nitschke. 2014. Production and properties of a surface-active lipopeptide produced by a new marine *Brevibacterium luteolum* strain. *Applied Biochemistry and Biotechnology.* 174: 2245–2256.

Vipulanandan, C. and X. Ren. 2000. Enhanced solubility and biodegradation of naphthalene with biosurfactant. *Journal of Environmental Engineering.* 126: 629–634.

Wang, S. and C. N. Mulligan. 2009. Rhamnolipid biosurfactant-enhanced soil flushing for the removal of arsenic and heavy metals from mine tailings. *Process Biochemistry.* 44: 296–301.

White, C., A. K. Shaman and G. M. Gadd. 1998. An integrated microbial process for the bioremediation of soil contaminated with toxic metals. *Nature Biotechnology.* 16: 572–575.

Widdel, F. and R. Rabus. 2001. Anaerobic biodegradation of saturated and aromatic hydrocarbons. *Current Opinion in Biotechnology.* 12: 259–276.

Yadav, M., G. Singh and R. Jadeja. 2021. Bioremediation of organic pollutants: a sustainable green approach. pp. 131–147. *Sustainable Environmental Clean-up.* Elsevier.

Yuste, L., M. E. Corbella, M. J. Turiégano, U. Karlson, A. Puyet and F. Rojo. 2000. Characterization of bacterial strains able to grow on high molecular mass residues from crude oil processing. *FEMS Microbiology Ecology.* 32: 69–75.

Zenati, B., A. Chebbi, A. Badis, K. Eddouaouda, H. Boutoumi, M. El Hattab, D. Hentati, M. Chelbi, S. Sayadi and M. Chamkha. 2018. A non-toxic microbial surfactant from *Marinobacter hydrocarbonoclasticus* SdK644 for crude oil solubilization enhancement. *Ecotoxicology and Environmental Safety.* 154: 100–107.

Zhu, W., L. Chai, Z. Ma, Y. Wang, H. Xiao and K. Zhao. 2008. Anaerobic reduction of hexavalent chromium by bacterial cells of *Achromobacter* sp. strain Ch1. *Microbiological Research.* 163: 616–623.

Zhuang, W.-Q., J.-H. Tay, A. Maszenan and S. Tay. 2002. *Bacillus naphthovorans* sp. nov. from oil-contaminated tropical marine sediments and its role in naphthalene biodegradation. *Applied Microbiology and Biotechnology.* 58: 547–554.

15 Biosurfactants in Oil Spill Cleanup

E. Kardena, Q. Helmy, and Sukandar

CONTENTS

BRIEF HISTORY OF MAJOR OIL SPILL ACCIDENTS

Crude oil can also be called petroleum, which comes from the Greek words *petrus* (stone) and *oleum* or *oleus* (oil). It is a dark brown viscous liquid composed of several complex hydrocarbon compounds. As we know, oil and natural gas are some of the most-needed energy sources in various fields, according to their respective needs, for industry and transportation, including in various fields of household activities. In the process of exploration and exploitation of crude oil, it has the potential to cause damage or environmental pollution. Sources of petroleum pollution can come from leaks in distribution pipes and spills during the production, refinery, and transportation processes. In addition, waste treatment that does not meet quality standards for disposal into the environment is a source of pollution in the soil, air, and aquatic ecosystems.

Oil pollution at sea is caused by a variety of sources, including natural seeps, tanker accidents, ship operations, and offshore structures.

  a. Natural Seeps

    Natural seeps from seafloor rocks account for 40–50% of all oil released into the oceans. On a global scale, this equates to approximately 600,000

DOI: 10.1201/9781003307464-15

tons per year. While natural seeps are the single most common source of oil spills, they are regarded as less dangerous because ecosystems have adapted to such regular releases. Ocean bacteria, for example, have evolved to digest oil molecules near natural oil seeps (Burgherr 2007).

b. Tanker Transportation

The world's oil production is estimated to be between 3.5 and 4.5 billion tons per year, with half of it shipped by sea (IEA 2020). After loading the cargo oil, the tanker also transports ballast water (a ship stability system that uses a water loading and unloading mechanism), which is usually stored in a slop tank. After the unloading process is completed, the remaining oil cargo in the tank, as well as dirty ballast water, is channeled into the slop tank until the unloading port. The empty cargo tank is cleaned with a waterjet; this tank cleaning procedure is designed to keep the tank filled with new ballast water for the next voyage. The effluent, which is a mixture of water and oil, is pumped into the slop tank, resulting in a mixture of oil and water in the slop tank. Before the ship sets sail, a portion of the water in the slop tank must be pumped into a waste storage tank at the terminal or into the sea and replaced with new ballast water. It is undeniable that the waste water pumped into the sea still contains oil, resulting in pollution of the sea where tankers load and unload.

c. Tanker Accidents

Tanker accidents can occur due to hull leaks, running aground, explosions, fires, or collisions. Several major accidents have occurred around the world, including the leak of the Atlantic Empress tanker in the waters of Tobago on July 19, 1979, which spilled 287,000 tons of oil into the sea. The case of the ABT Summer ship fire on May 28, 1991, in the waters of Angola, which spilled about 260,000 tons of oil, is an example of a major oil tanker incident.

d. Offshore Infrastructure Incidents

Well drilling is one of the activities involved in the exploration and development of oil and gas. This operation is the most hazardous and poses the greatest risk in the entire process of oil and gas exploration and extraction. The drilling of wells is a step farther in the hunt for oil and gas. This is a method of establishing the presence or absence of oil or gas reserves by drilling holes gradually to a particular depth based on the findings of studies and seismic data analysis of subsurface conditions. Drilling oil and gas wells is considered a high-risk project with high costs. The occurrence of wild bursts, fire, and explosion is one of the most serious threats in drilling operations (Necci et al. 2019). Figure 15.1A shows the

FIGURE 15.1A. The explosion and following fire caused the Deepwater Horizon Platform to sink, resulting in a large-scale oil spill in the Gulf of Mexico. Courtesy: National Science Foundation (public domain).

uncontrollable explosion that occurred on the Deepwater Horizon Platform and caused the largest oil spill in US waters in 2010 and Figure 15.1B shows the damage to the subsea oil pipeline caused by the ship's anchoring activity, which culminated in an oil spill.

e. Unloading Terminals at Mid-Sea

The process of loading and unloading tankers takes place not only at ports but also in the middle of the sea. The loading and unloading process at these marine terminals is fraught with dangers such as broken pipes, leaks, and accidents caused by human error.

f. Bilges and Fuel Storage Tanks

When sailing normally or in bad weather, all ships require a ballast process. Because the ship's ballast tanks are usually used to load cargo, an empty fuel tank is usually used to carry additional ballast water. When the weather is bad, ballast water is pumped into the sea and mixed with oil. In addition to ballast water, bilge water, which

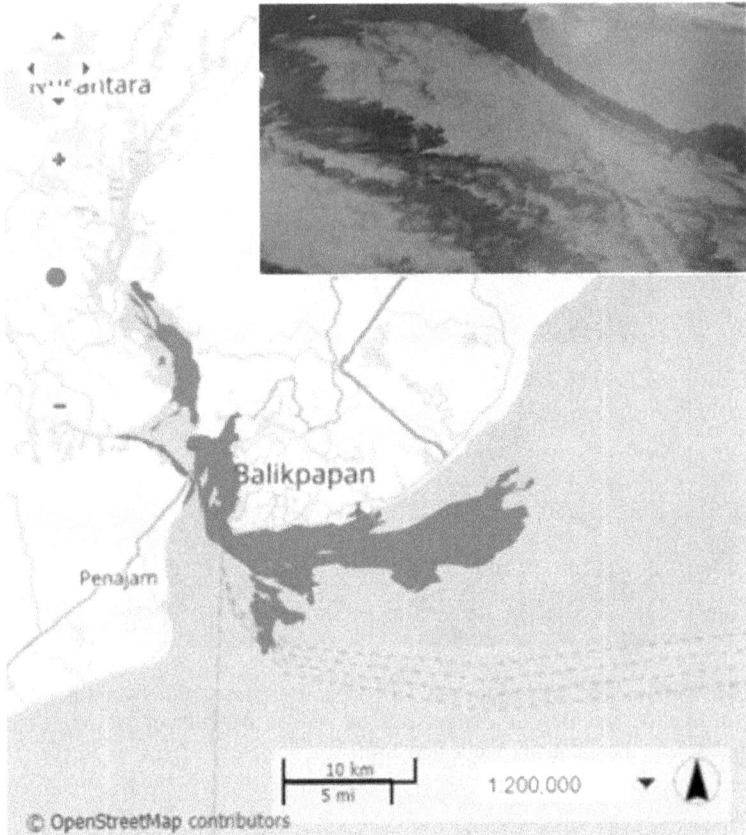

FIGURE 15.1B. Aerial view of the Balikpapan Bay oil spill, Indonesia, on March 30, 2018, caused by the MV Ever Judger anchoring activity fracturing the subsea oil pipeline. Source: National Institute of Aeronautics and Space, Indonesia 2022.

is mixed with oil, is pumped out. Bilge is a waste channel for waste water, oil, and lubricants produced by machine processes. International rules require that bilge water discharge enter the oil and water separator before being pumped into the sea, but in reality, a lot of illegal bilge effluent that does not comply with international regulations is dumped into the sea.

g. Ship Repair and Maintenance (Docking)

All ships must be repaired and maintained on a regular basis, including tank and hull cleaning. To avoid explosions and fires, all remaining fuel in the tank must be emptied during the docking process. According to regulations, all shipyards must have waste storage tanks; however, many shipyards lack this facility, so waste oil is pumped directly into the sea. It was estimated that approximately 30,000 tons of oil were wasted into the sea as a result of this docking process in 1981 (Clark 2001).

h. Shipwrecks

The ship scrapping process (cutting the hull to become scrap metal) is primarily carried out in India and Southeast Asia, including Indonesia. As a result of this process, a large amount of metal and other content, including oil, is dumped into the sea. It is estimated that approximately 1500 tons of oil are released into the sea each year as a result of this process, which harms the local environment (Clark 2001).

Damage to marine ecosystems is frequently caused by crude oil leaks in offshore mining. Leaks and explosions of offshore platforms and structures, subsea pipelines, and tanker operations can all result in oil spills. Based on the number of cases of oil leaks or spills in the ocean, special attention from the relevant authorities is required to prevent, overcome, and provide appropriate solutions to the problem of oil leaks in the ocean. This is done to forestall the destruction of marine ecosystems, which will have an impact on the degradation of marine waters, as well as other effects that are detrimental to aspects of life as a result of oil leaks or spills. Many countries overexploit the sea and its resources, with little regard for the preservation of the sea and its resources. Furthermore, actions to preserve and protect the marine environment are frequently overlooked and underutilized. Such actions are harmful not only to the territory of the country involved in the incident but also to the territory of other neighboring countries. Furthermore, this condition initiates a dispute between the state or party suspected of causing the damage or pollution and other countries affected by the damage or pollution. An oil spill will have a long-term impact on ecosystems and biological resources. In addition to fish, coral reefs, seaweed, mangroves, minerals, and other marine natural resources, it has an impact on the economy of fishermen, who will have difficulty finding fish in the sea. Oil spills in the sea have the characteristics of being soluble in sea water, floating on the sea surface, and sinking and settling as black deposits in sand sediments and coastal rocks. Floating oil darkens the color of the water's surface and can interfere with organisms that live on the surface. Furthermore, oil can inhibit the entry of sunlight into the sea, depleting oxygen in the sea and interfering with the photosynthesis process of aquatic plants in the sea.

For a long time, oil spill disasters have been a major source of concern in the marine world. They are both commercial and environmental disasters. As a result of a ship or oil rig accident, the ocean water becomes contaminated with liquid petroleum hydrocarbons, causing environmental damage for decades to come. Table 15.1 shows a summary of the disaster based on the amount of oil spilled into the environment.

TABLE 15.1
Major Oil Spills in History

No.	Oil Spill Disaster	Estimated Spillage, Barrels (US, million)	Ref.
1	1991 Persian Gulf: The spill (intentionally dumping oil into the Persian Gulf), considered an act of environmental terrorism, was a heated political move between Iraq and Kuwait that had ramifications for the larger Gulf War.	±10.9–11	Michel et al. 2005 Joyner and Kirkhope 1992
2	2010 Deepwater Horizon: The Deepwater Horizon is a semi-submersible offshore oil rig designed to operate in depths of up to 8000 feet. An explosion and fire occurred while drilling an exploratory well about 41 miles off the coast of Louisiana, USA, releasing approximately 4.9 million barrels of crude into the Gulf of Mexico.	±4.9	Makocha et al. 2019
3	1979 IXTOC I: A blowout occurred in the Bahia de Campeche, 600 miles south of Texas in the Gulf of Mexico, due to a loss of drilling mud circulation. The oil and gas blowing out of the well ignited, catching fire on the platform. The burning platform collapsed into the wellhead area, making immediate attempts to control the blowout impossible.	±3.5	Patton et al. 1981
4	1979 *Atlantic Empress*: During a tropical storm, two oil supertankers (*Atlantic Empress* and *Aegean Captain*) collided in the Caribbean Sea off the coast of Tobago, leaking approximately 88.3 million gallons of crude oil.	±2.1	Horn and Neal 1981
5	1992 Fergana Valley: A new oil well in Uzbekistan was damaged and began to leak a large amount of crude oil, estimated to be up to 60,000 barrels per day for two months.	±2.0	USEPA 1992
6	1983 Nowruz Oil Field: An oil tanker collided with the Nowruz Field platform in the Persian Gulf, resulting in a massive oil spill. The leaking of oil was the result of damage to the well caused by the accident. During the seven months following the accident, it is estimated that approximately 80 million gallons of oil—roughly 1500 barrels per day—flowed into the Persian Gulf.	±1.9	Pashaei et al. 2015

No.	Oil Spill Disaster	Estimated Spillage, Barrels (US, million)	Ref.
7	1991 *ABT Summer*: On May 28, 1991, around 1300 kilometers from the Angolan coast, the *ABT Summer* ship was rocked by an unexplained detonation, causing it to catch fire. The flames that engulfed the tanker ship raged uncontrollably for about three days before sinking to the depths of the sea, during which time over 260,000 metric tons of the laden oil cargo began to spill and spread onto the water surface.	±1.8	Galierikova and Materna 2020
8	1983 *Castillo de Bellver*: The oil tanker *Castillo de Bellver* caught fire off the coast of Saldanha Bay, about 70 miles northwest of Cape Town, South Africa, releasing approximately 252,000 tonnes of light crude oil. After drifting offshore, the burning vessel was abandoned and broke up.	±1.7	Moldan and Dehrman 1989
9	1978 *Amoco Cadiz*: On March 16, 1978, the oil tanker *Amoco Cadiz* ran aground on Portsall Rocks, 2 kilometers off the coast of Brittany, France, and eventually split in three and sank, resulting in a 232,000-tonne oil spill.	±1.5	Marchand 1980
10	1991 *M/C Haven*: On April 11, 1991, an explosion aboard the supertanker *M/C Haven* resulted in the loss of roughly 145,500 metric tons of heavy Iranian crude oil in Genoa, Italy, in the industrialized northern Ligurian Sea coastal region.	±0.99	Martinelli et al. 1995
11	1988 *Odyssey*: On 10 November 1988, while heading from Scotland to Canada, the oil tanker *Odyssey* got caught in a storm 700 miles off the Canadian coast. It sank in the North Atlantic off the coast of Canada due to an explosion, releasing its cargo into the sea.	±0.9	Rogowska and Namiesnik 2010
12	1989 *Exxon Valdez* The oil supertanker collided with Bligh Reef in Prince William Sound, Alaska, spilling 10.8 million US gallons of crude oil.	±0.25	Shigenaka 2014

OIL SPILLS IN AQUATIC ENVIRONMENTS: THEIR BEHAVIOR AND EFFECTS

When an oil spill occurs in the marine environment, the oil undergoes a series of physical and chemical changes (weathering). Some of these changes result in the loss of a portion of the oil from the sea surface, while others result in the presence of a portion of the oil material at sea level. Although the spilled oil will eventually be decomposed/assimilated by the marine environment, the time it takes depends on the initial physical and chemical properties of the oil as well as the natural oil weathering process. Some of the main factors that affect changes in oil properties, according to Baker et al. (1990), are:

1. Chemical characteristics and oil composition;
2. Physical characteristics of the oil, specifically specific gravity, viscosity, and boiling point;
3. Meteorological conditions (photooxidation), oceanographic conditions, and air temperature; and
4. Seawater characteristics (pH, specific gravity, current, temperature, presence of microorganisms, nutrients, dissolved oxygen, and suspended solids).

Each oil has unique physical and chemical properties. These properties influence how oil spreads and degrades, the danger it poses to aquatic and human life, and the likelihood that it will endanger natural and human-made resources. The rate at which an oil spill spreads determines its environmental impact. Most oils tend to spread horizontally on top of the water, forming a smooth and slippery surface known as a slick. Surface tension, specific gravity, and viscosity are all factors that influence an oil spill's ability to spread.

- Specific gravity is a measure of a substance's density in comparison to the density of water. Most oils float on top of water because they are lighter than water. The specific gravity of an oil spill, on the other hand, can rise if the lighter substances in the oil evaporate. Heavier oils, vegetable oils, and animal fats may sink and form tarballs, or they may interact with rocks or sediments on the water's bottom.
- Surface tension is a measure of the attraction between a liquid's surface molecules. The higher the surface tension of the oil, the more likely it will remain in place. If the oil's surface tension is low, it will spread even without the assistance of wind and water currents. Because rising temperatures reduce the surface tension of a liquid, oil is more likely to spread in warmer waters than in very cold waters.
- The viscosity of a liquid is a measure of its resistance to flow. The higher the viscosity of the oil, the more likely it is to stick to one spot.

Spreading, evaporation, dispersion, emulsification, dissolution, sedimentation, and oxidation are the physicochemical processes responsible for the transformation of petroleum hydrocarbons. Figure 15.2 shows the interacting processes involved in changing the nature of oil. Pollutants from the type of crude oil in the waters frequently become environmental issues, posing a regional threat to the investment climate. The impact of waste in the form of oil spills, in particular, has a significant negative impact on the coastal and marine environment,

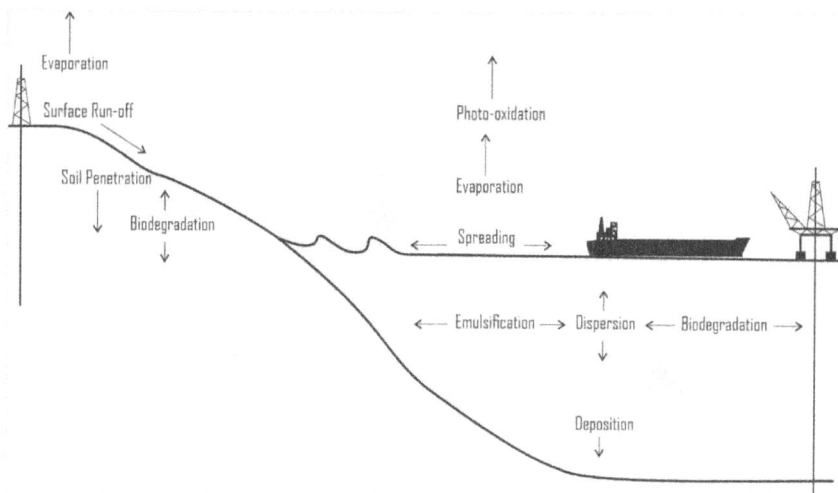

FIGURE 15.2   The fate of oil spilled.

particularly through direct contact with aquatic organisms, direct impacts on fishery activities including marine tourism, and indirect impacts through environmental disturbances.

Natural processes are constantly at work in aquatic environments. These can lessen the severity of an oil spill and hasten the recovery of an impacted area. Weathering, emulsification, dispersion, evaporation, oxidation, and biodegradation are examples of natural processes (Figure 15.3).

- Weathering refers to the chemical and physical changes that cause spilled oil to degrade and become heavier than water. Natural dispersion can occur as a result of wave action, which breaks a slick into droplets that are then distributed vertically throughout the water column. These droplets can also create a secondary slick or thin film on the water's surface.
- Emulsification is the process by which emulsions (mixtures of small droplets of oil and water) are formed. Emulsions are formed as a result of wave action and have significantly hampered weathering and cleanup processes. There are two types of emulsions: water-in-oil and oil-in-water. Water-in-oil emulsions, also known as "oil mousse" (Figure 15.4), form when water becomes trapped inside viscous oil due to strong wave action. Oil mousse emulsions can last for months or even years in the environment. Oil and water emulsions cause oil to sink and disappear from the surface, giving the appearance that it is no longer a threat to the environment.

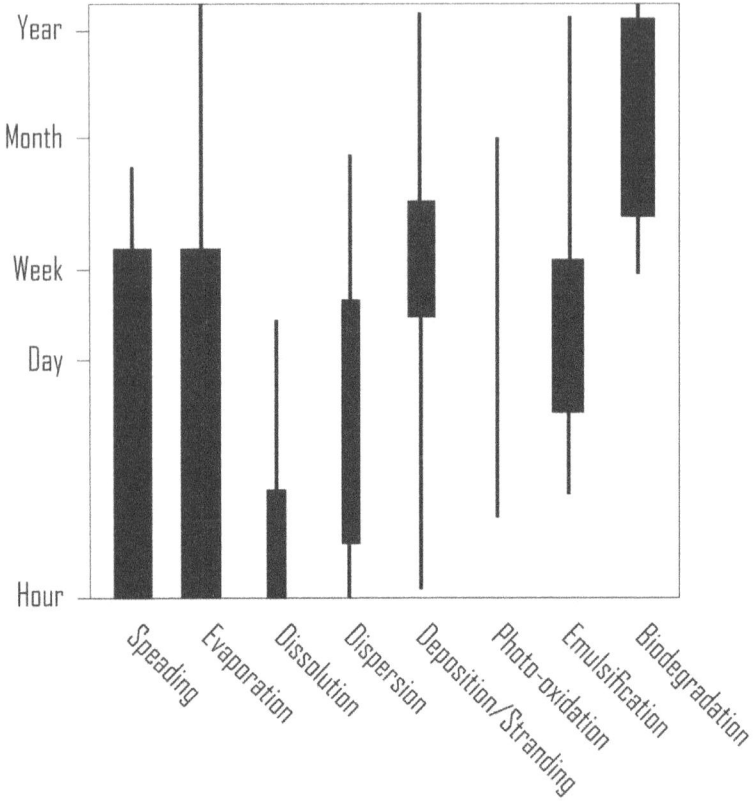

FIGURE 15.3 The basic physical/chemical processes involved in an oil spill at sea level. Source: Redrawn from USNRC 2003; Young et al. 2011.

- Natural oil dispersion consists of surface oil breakage, conversion to subsurface oil droplets, and transfer and dispersion of these droplets in the water column. The dispersion process should take into account the size distribution of oil droplets, the stability of dispersed oil in the water column, and re-entrainment to the surface. Oil is dispersed from the water's surface by dispersion. Natural dispersion is simply the mixing of oil droplets with water as it enters the sea body. As a result, the properties of the oil droplets are identical to those of the spilled oil. Oil drops dispersed in the water column are mostly exposed to fate processes, and their chemical properties change over time.
- Evaporation occurs when the lighter or more volatile substances in the oil mixture evaporate and leave the water's surface. The heavier components of the oil are left behind by this process, and they may weather further or sink to the ocean floor. Lighter refined product spills, such as kerosene and gasoline, contain a high proportion of flammable

FIGURE 15.4 A thick layer of weathered oil from the explosion of the Deepwater Horizon oil spill floats over the Gulf of Mexico, forming a water-in-oil emulsion known as oil mousse. Courtesy: National Science Foundation 2022.

components known as light ends. These may evaporate in a matter of hours, causing little harm to the aquatic environment. Heavier oils, vegetable oils, and animal fats produce a thicker, viscous residue. These oils are less prone to evaporation.

- Oxidation occurs when oil comes into contact with water and oxygen reacts with the oil hydrocarbons to form water-soluble compounds. This process primarily affects oil slicks at their edges. Thick slicks may only partially oxidize, resulting in the formation of tarballs. These dense, sticky black spheres may linger in the environment for a long time after a spill, washing up on shorelines.
- Biodegradation occurs when microorganisms like bacteria consume oil hydrocarbons. A diverse range of microorganisms is required for significant oil reduction. Nutrients such as nitrogen and phosphorus are sometimes added to water to encourage microorganisms to grow and reproduce in order to sustain biodegradation. Warm-water environments are ideal for biodegradation.

These natural processes differ in freshwater and marine environments. Because water movement is restricted in these habitats, freshwater environmental impacts can be more severe. Oil tends to pool in standing bodies of water and can remain in the environment for extended periods of time. Oil tends to collect on plants and grasses growing on the banks of flowing streams and rivers. Oil can also interact with sediment at the bottom of freshwater bodies, affecting organisms that live in or feed on sediment. The short-term effect of oil pollution is that hydrocarbon molecules damage the cell membranes of marine biota, resulting in the release of cell fluid and the absorption of these materials into the cells. Various types of shrimp and fish will smell like oil, resulting in a reduction in quality. Oil can directly kill fish by depriving them of oxygen, poisoning them with carbon monoxide, and poisoning them with toxic substances. Young marine life is affected by the long-term effects of oil pollution. Marine biota can absorb and consume oil, and some will accumulate in fat and protein compounds. The nature of this accumulation can be passed down the food chain from one organism to the next. Table 15.2 shows some of the damaging effects that oil spills have on ecosystems.

Pollutants in the form of oil spills, in particular, have a significant negative impact on the coastal and marine environment, particularly through direct contact with aquatic organisms, direct impacts on fishing activities including marine tourism, and indirect impacts through environmental disturbances. The lethal impact of oil spills in offshore waters is frequently caused by tanker accidents, offshore activities, or natural seepage of petroleum from the seabed (oil seep); however, no reports of industrial activities on land have been reported to dispose of oily waste far into the oceans. In the case of oil spills in open water, the concentration of oil below the slick is usually very low, and the maximum will be in the range of 0.1 ppm, so that oil spills in offshore waters do not cause mass death of organisms, particularly fish. The issue is that the majority of oil spills occur in coastal or inshore waters. It was reported that during the *Amono Cadiz* tanker accident in 1978 in British and French waters, populations of *Pleurenectes platessa* and *Solea vulgaris* fish died in large numbers. The risk of mass mortality will be even higher for fish in ponds or cages, as well as shellfish species with limited migration ability to avoid spills (Davis et al. 1984). Sublethal impacts will be more accurate if proven in the laboratory, as opposed to lethal effects, which can be easily quantified in the field. The concentration of oil in the water was found to affect the reproduction and behavior of fish and shellfish in laboratory tests. Egg hatchability, survival rate, number of defective larvae, and shell closure (in shellfish) were all significantly affected by relatively low concentrations (0.1 ppm). Many types of shrimp and crabs develop a keen olfactory system to direct their numerous activities; as a result, exposure to toxic materials causes shrimp and crabs to experience behavioral disruptions such as in the ability to search, eat, and mate.

The direct impact of an oil spill on aquaculture, for example, is tainting. Tainting can occur in caged fish and ponds, as well as shellfish that lack the ability to move away from contaminants, rendering them unfit for sale. Organisms contaminated with oil will produce unpleasant odors and tastes, as well as changing color in their tissues. Fish with high fat content are more likely to become contaminated than fish with lean muscle. The odor and taste of oil in organisms will

TABLE 15.2
Impact of Oil Spills on Marine Life

No.	Biota	Response	Ref.
1	Mangrove	Small mangroves (less than 1 m in height) are defoliated and start dying; the aerial root community is lost in 15–30 days. Defoliation and death of medium (3 m in height) mangroves; tissue damage to aerial roots observed over a 30-day, 1-year period. Death of larger (>3 m in height) mangroves; loss of oiled aerial roots and regrowth of new (sometimes deformed) ones; recolonization of oil-damaged areas by new seedlings observed over a period of 1–5 years. Reduced litter fall, reproduction, and seedling survival; death or reduced growth of young trees colonizing oiled site is observed during 1- to 10-year observation period. After more than 10 years of observation, complete recovery of mangrove growth was observed.	Lewis 1983
2	Mangrove	The first effect, 0–1 year: Seedlings and saplings die; no structural changes are visible. Damage to the tree structure, 1–4 years: There is a high mortality rate, and the oil impact can be measured in terms of major structural changes. Stabilization, 4–9 years: No or few additional structural parameters are changed; sapling growth is observed. Recovery time, >9 years: It is possible to measure improvements in structural tree parameters; however, the ecosystem may not fully recover to its original state.	Lamparelli et al. 1997
3	Marshes	*Spartina alterniflora* impacted by light crude oil requires a recovery time of 8 months to 5 years.	Hoff 1995
4	Birds	Observations of oiled birds were reported during the Deep Water Horizon spill. There were nearly 10,000 bird observations, with 2085 classified as visibly oiled alive and 2303 classified as visibly oiled dead.	Barron 2011
5	Oysters	One year after the Erika oil spill, severe immunological changes were discovered in an oil-impacted area. Because the oysters at that location contained higher levels of PAH than those at the other sites, it was hypothesized that chronic contamination, possibly caused by oil trapped in the sediments, had caused immunotoxicity. Furthermore, moderate variations in some hemolymph parameters observed in the unaffected area strongly suggested that natural environmental factors could have caused physiological stress.	Auffret et al. 2004
6	Parasitoid wasps	Parasitoid abundance was lower in one of the sites affected by the recent 2014 oil spill, but not in the site affected by the 1975 oil spill. In some sampling site/year combinations, oil-polluted trees supported less parasitoid diversity than unpolluted trees; however, such negative effects were inconsistent, and pollution explained only a small proportion of the variation in parasitoid community composition.	Moleer et al. 2020

(continued)

TABLE 15.2
Impact of Oil Spills on Marine Life (Continued)

No.	Biota	Response	Ref.
7	Marine birds	Pathological methods were used to examine a total of 2465 seabirds (*Uria aalge*, *Alca torda*, and *Fratercula arctica*) that beached in the northwestern part of Spain following the "Prestige" oil spill on November 19, 2002. Birds were divided into three groups: dead birds with oil covering their bodies (group 1) or uncovered bodies (group 2), and birds recovered alive but dying after being treated at a rescue center (group 3). Severe dehydration and emaciation were the most visible gross lesions. Microscopically, hemosiderin deposits associated with cachexia and/or hemolytic anemia were found in the intestines of birds harboring oil. Severe aspergillosis and ventriculus ulcers were found only in group 3 birds, most likely as a result of stress associated with attempted rehabilitation at the rescue center.	Balseiro et al. 2005
8	Marine mammals	Hematological injury, immune function and organ weight modulation, genotoxicity, eye irritation, neurotoxicity, lung disease, adrenal dysfunction, metabolic and clinical abnormalities related to pelage oiling, behavioral impacts, decreased reproductive success, mortality, and population-level declines are common endpoints observed in conventional crude oil exposures and oil spills.	Ruberg et al. 2021a
9	Seaside sparrows	The Deepwater Horizon oil spill in 2010 released an estimated 4.9 million barrels of oil into the Gulf of Mexico, causing coastal ecosystems to suffer. Seaside sparrows (*Ammospiza maritima*), a year-round resident of Gulf Coast salt marshes, were exposed to oil and their nests were monitored in Plaquemines Parish, Louisiana, USA, from 2012 to 2017 to assess potential impacts on the nesting biology of seaside sparrows. During the study, the majority of nests failed (76% of known-fate nests, $N = 252$ nests, 3521 exposure-days), and predation was the leading cause of nest failure (91% of failed nests).	Hart et al. 2021
10	Bacterial communities	Following the 2010 Deepwater Horizon oil spill, coastal salt marshes along the northern Gulf of Mexico shoreline received varying types and amounts of weathered oil residues. Oil replaced native natural organic matter (NOM) originating from *Spartina alterniflora* and marine phytoplankton in the marshes between May and September 2010, according to hydrocarbon biomarker indices calculated. The major class- and order-level shifts among the phyla were observed in all of the marshes studied. Proteobacteria, Firmicutes, Bacteroidetes, and Actinobacteria were discovered during the first 4 months, but another community shift occurred during peak oiling in 2011. After 2 years, hydrocarbon levels had decreased and bacterial communities had become more diverse, with Alphaproteobacteria (Rhizobiales), Chloroflexi (Dehalococcoidia), and Planctomycetes dominating.	Engel et al. 2017

No.	Biota	Response	Ref.
11	Turtles and marine iguanas	Crude oil can modify skin function, energy metabolism, immune responses, diving patterns, and respiration in adult turtles. Crude oil can decrease embryonic survival, alter incubation time, and cause hatching deformities in exposed eggs. Ingestion of low levels of petroleum can increase corticosterone levels in marine iguanas and wipe out gut bacteria.	Ruberg et al. 2021b
12	Fish	Haddock and cod are both sensitive to very low concentrations of oil (10 µg/L tPAH) that cause toxicity; however, haddock is more affected by oil droplets than cod. This is due to the addition of an adhesive membrane covering the primary egg envelop of haddock eggs, as well as the distinct structure of the haddock chorion, which allows oil droplets to interact and adhere to it. Cod, on the other hand, is more sensitive to the water-soluble fraction of PAHs. PAHs were discovered in the internal embryos of haddock and cod samples. This was due to the binding of oil to the chorion, which exposed and increased PAH uptake by the embryo, resulting in extreme and heightened toxicological responses such as deformation and cardio-toxicity in haddock.	Yuewen and Adzigbli 2018
13	Plankton	Copepods, euphausiids, and mysids, for example, assimilate hydrocarbons directly from seawater and by ingesting oil droplets and oil-contaminated food. Ingestion of oil by these organisms frequently results in death, while survivors frequently exhibit developmental and reproductive abnormalities. Because of their high lipophilicity, plankton can accumulate aromatic hydrocarbons. According to the researchers, marine plankton is extremely sensitive to the petroleum fraction, with lethal concentrations as low as mg/L. The toxicity of ten polycyclic aromatic hydrocarbons associated with the Prestige fuel oil spill on adult copepods (*Oithona davisae*) was discovered to have narcotic effects on these organisms.	Saadoun 2015

disappear by depuration processes depending on the type of waste, species, and optimal living conditions for these species (Baker et al. 1990). For a small spillage (e.g., 50 tons), the impact on aquaculture activities will be significant; aside from the cultured organisms that will be directly affected, some cultivation equipment, such as nets and rigging, will no longer be usable. Furthermore, stock can be harmed if seawater is taken in to meet stock requirements. Impact on coastal and marine ecosystems (mangroves, river deltas, estuaries, seagrass beds, and coral reefs) plays critical ecological, economic, and socio-cultural roles. Ecologically, the ecosystem serves as a breeding area, providing habitat and food for adult organisms as well as supporting food networks (for example, nutrient input from dead leaves) for the ecosystem or other habitats nearby. The pressure from pollutant entry will affect the designation of these systems, and the vulnerability of these ecosystems is very high, in addition to natural attenuation (dispersion and dilution) in several ecosystems such as mangroves, estuaries, and seagrass beds being relatively slow (taking years of recovery time) (Sulistyono 2013).

OIL SPILL REMEDIATION TECHNOLOGY

Crude oil that has a chemical component or composition at the time of a spill in the sea will cause changes both chemically and physically. Among these changes, there will be dispersion, formation of layers, evaporation, polymerization, the occurrence of emulsions, and the formation of clots. The changes that occur are driven by wind movement, current waves, and surface tension. The content of hydrocarbons in water is volatile, and when dissolved in water they will form a thick layer known as oil mousse. When an oil spill occurs on water, it is critical that the spill be contained as soon as possible in order to minimize danger and potential damage to people, property, and natural resources. Containment equipment is used to keep oil from spreading and to allow for its recovery, removal, or dispersal. Because of the lower density of oil, it is much easier to clean up an oil spill. However, it is difficult to clean up a spill if the oil is denser than the water, forming a layer along the bottom rather than the surface. Floating barriers known as booms are the most common type of equipment used to control the spread of oil. Containment, recovery, and removal/cleanup are the three major steps in controlling oil spills.

CONTAINMENT

Containment, or the prevention of oil spills through the use of mechanical equipment, is the first treatment by localizing the oil spill through the use of barrier buoys (oil booms). Because it is easier to clean up oil if it is all in one place, containment booms act as a fence to keep the oil from spreading or floating away. Booms float on the surface and are made up of three parts: a "freeboard," or part that rises above the water surface and contains the oil and keeps it from splashing over the top; a "skirt" that rides below the surface and keeps the oil from being pushed under the booms and escaping; and some kind of cable or chain that connects, strengthens, and stabilizes the boom (Figure 15.5). Connected boom sections are placed around the oil spill until it is completely surrounded and contained. Oil booms are the most effective tool for dealing with oil spills in the water as soon as they occur. Because oil has a lower specific gravity than water, it will float to the surface if it spills into it. Oil that has spilled into the water must be contained immediately so that it does not spread and be carried away by water currents or waves in the ocean. The longer it takes to respond to a localized oil spill, the larger the spill's area. Following the localization of the oil, various equipment, such as an oil skimmer, oil absorbent, and oil vacuum, can be used to recover the oil spill until the oil on the water's surface can be recovered. Booms are classified into several types. Fence booms have a high freeboard and a flat flotation device, making them ineffective in rough water where wave and wind action can twist the boom. Round or "curtain" booms have a more circular

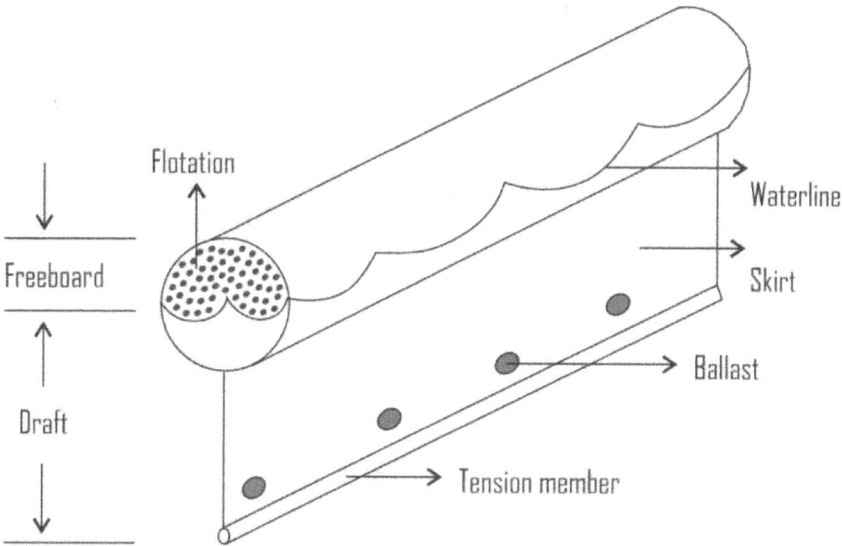

FIGURE 15.5 Schematic of typical oil boom component.

flotation device and an uninterrupted skirt. They are more difficult to clean and store than fence booms, but they perform well in rough water. Non-rigid inflatable booms come in a variety of shapes and sizes. They're simple to clean and store, and they hold up well in rough seas. However, they are more expensive and more difficult to use and easily puncture and deflate. The sea conditions have a large impact on all boom types; the higher the waves swell, the less effective the booms become (USEPA 1999).

Depending on the water and weather conditions, there are several configurations of localizing an oil spill with an oil boom that can be done when an oil spill occurs (Figure 15.6), including:

- U configuration: using two floating units (ship/boat).
- U configuration (single or double sweep): employs a single unit of floating means outfitted with a sweeping arm.
- J configuration: using two floating units (ship/boat).

After successfully localizing the oil with an oil boom, a vacuum apparatus or a pump is used to physically remove the oil that has accumulated in the boom.

RECOVERY

After successfully localizing the oil with an oil boom, oil recovery is performed with an oil skimmer, which is then transferred by pumping device to a "reservoir"

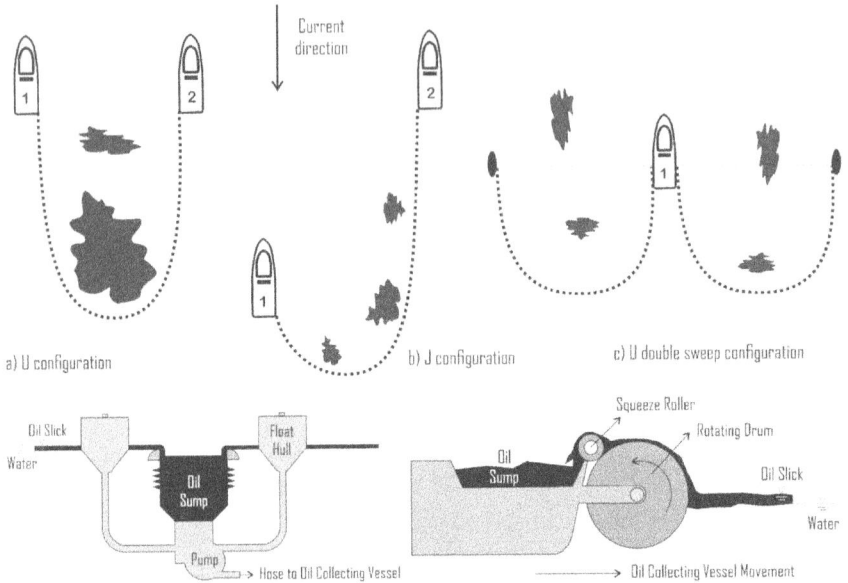

FIGURE 15.6 Configuration of localizing oil spill with oil boom (U, J, and U double sweep configuration).

receiving facility in the form of a tank or balloon. Large amounts of oil can be extracted, though residual oiling will remain after the majority of the heavy oil has been collected. Low-pressure flushing is typically used to help move oil to collection points where other removal equipment, such as vacuums or boom/skimmer collectors, is operating (Figures 15.7 and 15.8). Vacuum removal combined with low-pressure flushing can also be effective. One disadvantage of this method is that it works best in waters with low hydrodynamics (currents, tides, waves, etc.) and less extreme weather. This method is also difficult to implement at the port because it can interfere with the activities of ships entering and leaving the port. Furthermore, the most common method for collecting oil spills in surface water is to use adsorbent. Adsorption methods are mostly considered because of their simplicity and economic viability. At the moment, the majority of adsorbents are organic synthetics (such as polypropylene and polyurethane), inorganic mineral products (such as zeolite and silica gel), and organic naturals (such as sawdust, coconut husks, nutshells, and other natural cellulose/fibers). Sorbents are materials that absorb liquids. They can be used to recover oil via absorption, adsorption, or both at the same time. The adsorbent attracts oil to its surface but does not allow it to penetrate into the material's pore space, whereas an absorbent allows oil to penetrate into the material's pore space. To be effective in fighting oil spills, the sorbent must be both oleophilic and hydrophobic. While sorbents can be used as the sole cleaning method in the case of a small-scale oil spill, they are typically used to remove the last traces of oil or in areas where skimmers cannot reach. The sorbent must be removed from the water after being used to recover oil and properly disposed of on land or cleaned for reuse. Any oil extracted from the sorbent material must be properly disposed of or recycled. Table 15.3 summarizes materials that can be used as crude oil spill sorbents.

TABLE 15.3
Low-Cost Materials That Can Be Used as Sorbents for Crude Oil Spills

No.	Material	Response	Ref.
1	Groundnut shells	According to the findings of the intra-diffusion model, both intra-diffusion and external mass transfer are rate-determining stages of the adsorption process. According to the study's findings, activated ground shell carbon is a good mop-up and low-cost alternative medium for oil-splattered surfaces.	Uzoije et al. 2011
2	Wood particles	As crude oil sorbent materials, hardwood and semi-hardwood are used. For all observations, the maximum uptake of the oil by both adsorbents occurred at 180 and 240 minutes. For all results, smaller particle sizes resulted in the greatest uptake. This study suggests that more research be conducted on the use of modified smaller particle sizes of semi-hardwood sawdust in order to increase its potential as a low-cost adsorbent in oil spill cleanup.	Ikenyiri and Ukpaka 2016
3	Meshed corncobs	Meshed corncobs, particularly those with particle sizes smaller than 80 micrometers, demonstrated a high affinity for crude oil adsorption. For the temperature range considered, the maximum adsorption was observed at 15°C. The results demonstrated that crude oil adsorption can be accomplished using this readily available and biodegradable waste material. Meshed corncob adsorbent is convincingly oleophilic or hydrophobic due to its high affinity for oil and low water pick-up.	Olufemi et al. 2014
4	Rice husks	The sorbents' efficiency for removing oil compounds from water was in the following order: black rice husk ash > raw rice husk > white rice husk ash. The maximum adsorption capacity of black rice husk ash for the adsorption of engine, spent, and crude oils, respectively, was 2000, 1250, and 1000 mg g^1. The amount of black rice husk ash required for total removal of engine, spent, and crude oils from aqueous solutions was two, three, and six times the oil concentration, respectively.	Razavi et al. 2015
5	Mango shells	More than 90% crude oil removal was observed with 2 g of 80 μm mango shell–activated carbon at 0.025 g/L initial crude oil concentration.	Olufemi and Otolorin 2017
6	Activated carbon	Activated carbon (AC) was dip coated with a copolymer of polydopamine (PDA) and cationic polyethylenimine (PEI) and used to treat crude oil spills in synthetic seawater. The addition of 10% PDA-PEI to the AC increased its dispersion efficiency to 61%, with a surface sorption efficiency of 30%, resulting in a total treatment efficiency of 91% and a dispersion-sorption balance of 0.82.	Giwa and Taher 2020

(continued)

TABLE 15.3
Low-Cost Materials That Can Be Used as Sorbents for Crude Oil Spills
(Continued)

No.	Material	Response	Ref.
7	Coconut shells	Using coconut shell–based activated carbon and iron oxide nanoparticles, simple chemical methods are used to create an efficient magnetic adsorbent nanocomposite material. The composite material has a high oil retention capacity (up to 6 g/g) and fast kinetics, and it can be recovered with the adsorbed oil using an external magnet. After being recovered by heat treatment or solvent extraction, the adsorbent material can be reused. As a result, the magnetic nanocomposite is demonstrated to be an efficient and recyclable potential candidate for magnetic separation of oil spills.	Raj and Joy 2015
8	Coconut husks	Coconut husk (CH) was chemically activated with zinc chloride and then pyrolyzed at a different combination of temperature-retention times of 400–800°C and 30–60 minutes to produce unactivated and activated coconut husk–derived biochar (CHB and ACHB), respectively, while acetic anhydride was used to produce acetylated-coconut husk (ACCH). The maximum monolayer sorption capacities for raw CH, ACCH, CHB_{800-60}, and $ACHB_{800-60}$, respectively, were 12.11, 15.06, 16.10, and 16.84 g/g; thus, the performance of the sorbents was in the following order: $ACHB_{800-60}$ > CHB_{800-60} > ACCH > raw CH.	Agarry et al. 2020
9	Coconut husk composite	The use of stearic acid–grafted coconut husk composite at low pH resulted in high adsorption efficiency, as the surfaces were predominantly positive at 0.05 and 0.10 mol/dm^3. Furthermore, the composite demonstrated good adsorption potential, with 96.4% crude oil removal after 50 minutes at 0.2 $g/100\ cm^3$. The adsorption isotherm data were well fitted to the Langmuir isotherm model, as expected given the conjugated material's structural homogeneity, with a maximum adsorption capacity of 69.86 mg/g. The pseudo-second-order kinetic model adequately matched the experimental kinetic data (R^2 0.99). With ethanol, crude oil was eluted from the composite and regenerated to the original pattern. The regenerated material had the same functionality as the original and was used in several adsorption-elution-reuse cycles.	Asadu et al. 2021
10	Corn silk fiber	Corn silk absorbs 8.6 and 9.4 g/g of Tapis and Arabian crude oil in water, respectively. Because of more hydrophobic properties and acetylation of the hydroxyl groups in the molecular structure of corn silk, this capacity could be improved by the acetylation process. The best acetylation conditions are a 3% catalyst for 6 hours of reflux at 120°C, with maximum weight percent gain and oil sorption capacity of 14.02 and 16.68 g/g for Tapis and Arabian crude oil, respectively. Tapis and Arabian oil had the best contact times of 30 and 40 minutes, respectively. The acetylated corn silk can be reused up to five times.	Asadpour et al. 2015

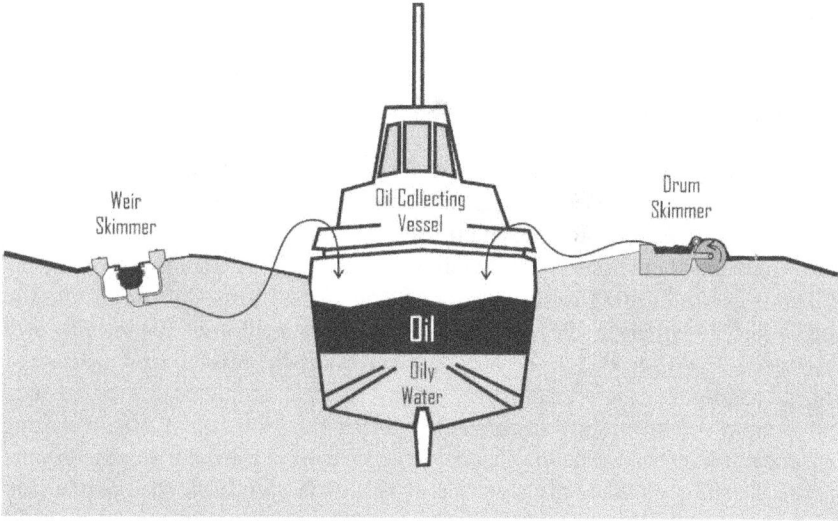

FIGURE 15.7   Schematic of typical weir skimmer (upper left), drum skimmer (upper right), and oil collecting vessel in operation (below).

FIGURE 15.8   Oil spill response crews work to collect oil on May 8, 2010, around the location where the Deepwater Horizon oil platform caught fire and sank. At least 193 vessels are assisting in the oil spill recovery after the platform sank. (Source: U.S. Coast Guard 2010, photo by Petty Officer 3rd Class Casey J. Ranel, under creative commons license.)

REMOVAL OR CLEANUP

In Situ Burning

Any response to an oil spill must consider oil containment, recovery, disposal, and the logistics of quickly delivering adequate response equipment to the spill site. The use of burning as an oil spill response method is appealing, particularly in remote areas. Because the oil is gasified during combustion, the need for physical collection, storage, and transport of recovered product is reduced to a few percent of the original spill volume that remains as residue after burning. Oil spill *in situ* burning is not a novel concept. The Torrey Canyon incident in the United Kingdom in 1967 was the first major oil spill in which burning was attempted. However, due to the oil's emulsification, the results were unsuccessful, discouraging others from trying. Many research studies and experimental burns on *in situ* burning were conducted during the 1970s and 1980s, including one successful burn during the *Exxon Valdez* oil spill in 1989, but the results were varied. Burning oil spills in place usually results in a visible smoke plume containing soot and other combustion products. The public is concerned about the effects of intentional burning of large crude oil spills due to a lack of knowledge about the extent of the area affected by the smoke plume produced by burning crude oil spills and the possibility of undesirable combustion products carried in the plume. Unanswered questions about the safety of personnel and equipment from the heat and thermal radiation produced by large fires have also hampered the use of burning to clean up oil spills. Burning, once considered a last resort, is now one of the first response methods considered by authorities in the event of a spill (Evans et al. 2001; Walton and Jason 1999).

With devices as simple as an oil-soaked sorbent pad, thick, fresh slicks can be ignited very quickly. *In situ* burning can remove oil from the water's surface very efficiently and quickly. For thick slicks, removal efficiencies can easily exceed 90%. With a fire area of only about 10,000 m^2 or a circle of about 100 m in diameter, removal rates of 2000 m^3/hr are possible (Figure 15.9). The use of a towed fire containment boom to capture, thicken, and isolate a portion of a spill, followed by ignition, is far less complex than mechanical recovery, transfer, storage, treatment, and disposal operations. Three elements must be present in order to burn oil spilled on water: fuel, oxygen, and a source of ignition. The oil must be heated to a temperature high enough to vaporize enough hydrocarbons to support combustion in the air above the slick. It is the vapors of hydrocarbons above the slick that burn, not the liquid itself. Oil's flashpoint and fire point are two properties that are frequently used to determine its ignitability. The flashpoint is the temperature at which the slick produces enough vapors to ignite. The fire point is a few degrees above the flashpoint when the oil is warm enough to supply vapors at a rate sufficient to support continuous burning. The rising column of combustion gases carries the majority of the heat from a burning oil slick away, but a small percentage (about 1 to 3%) radiates from the flame back to the slick's surface. This heat is used to partially vaporize the liquid hydrocarbons that rise to mix with the air above the slick and burn; a small amount of this heat

FIGURE 15.9   Vessels of opportunity pull oil into a fire boom in a controlled burn of the Deepwater Horizon oil spill with a second controlled burn visible in the distance on June 17, 2010. (Source: US Coast Guard 2010, photo by Chief Petty Officer Bob Laura, under creative commons license.)

is transferred into the slick and eventually to the underlying water. Once ignited, a burning thick oil slick reaches a steady state in which the vaporization rate sustains the combustion reaction, which radiates the necessary heat back to the slick surface to keep the vaporization going (Fingas 2016; Buist et al. 1999).

Dispersing Agents

Oil dispersants are commonly used to clean up oil spills in marine environments. They are chemical compounds, such as detergents, that are made up of surfactants that have been dissolved in one or more solvents. The use of dispersants reduces the surface tension between water and oil, allowing currents and waves to break the oil down into smaller particles. The breakdown of oil into smaller particles increases the contact surface area between the oil and biological agents, allowing the biodegradation process to proceed more easily. As long as surfactants are present in these waters, oil particles that have been broken down do not stick to sediments, aquatic life, or shorelines. Dispersants are well suited to a variety of environmental conditions. Dispersants can be used to handle oil in unfavorable environmental conditions such as choppy seas, strong winds, or oil spills in difficult-to-isolate waters. Dispersants are typically sprayed on the surface of seawater (Figure 15.10) affected by oil spills to allow the dispersants to diffuse into the oil and seawater layers. When other methods, such as oil containment and removal, are insufficient, oil dispersants are quickly used. The consequences of the toxicity of oil spill dispersants alone

FIGURE 15.10 Dispersants were applied from a boat to combat an oil spill approaching the shoreline in Balikpapan Bay, Indonesia, in March 2018.

or in the presence of oil, on the other hand, must be assessed. In general, undispersed oil is the most dangerous to shorelines and surface-dwelling organisms. The majority of dispersed oil, however, remains in the water column, where it primarily threatens pelagic and benthic organisms. Several studies have been conducted to compare the toxicity of oil spill dispersants when used alone or in conjunction with oil. Analyses of tests performed on a variety of aquatic life species revealed that crustaceans are more sensitive to oil dispersant exposure than fish. Oil dispersant exposure is most dangerous to species with the least amount of protective shell or external tissue. It has been demonstrated that the use of oil dispersants increases fish exposure and uptake of PAHs. This is especially true for fish that live in the water column of coastal areas, the ocean, and lakes. Following the application of chemical dispersants to the surface slicks, concentrations of low and high molecular weight polyaromatic hydrocarbons were found to be higher in the water column. Burridge and Shir (1995) investigated the toxic effects of the oil-dispersed chemical dispersants corexit 7664, corexit 8667, corexit 9500, and corexit 9527 on algae growth. Toxic effects were measured using the half-maximal effective concentration (EC_{50}) test method over a 48-hour period. The EC_{50} test is a statistical method for determining the substrate concentration that produces a specific effect (change in behavior) in 50% of the population. The toxic effects of using chemical dispersants in marine oil spill applications are summarized in Table 15.4. To classify toxicants based on their acute toxicity to aquatic organisms, the USEPA employs a five-step toxicity category scale, with an LC_{50} value of more than 100 mg/L considered practically non-toxic, between 10 and 100 mg/L slightly toxic, between 1 and 10 mg/L moderately toxic, between 0.1 and 1 mg/L highly toxic, and less than 0.1 mg/L very highly toxic, while the toxicity categories for terrestrial organisms are shown in Table 15.5 (USEPA 2022).

TABLE 15.4
The Toxicity of Chemical Dispersants Used in Marine Oil Spill Applications

Dispersant	Test Species	Toxicity Result	Ref.
Corexit 9500	*Palaemon serenus*	LC_{50} 96-h: 83.1 ppm	Gulec and Holdway 2000
	Macquaria novemaculeata	LC_{50} 96-h: 19.8 ppm	Gulec and Holdway 2000
	Americamysis bahia	LC_{50} 48-h: 38–47 µl/L	Hemmer et al. 2011
	Menidia beryllina	LC_{50} 48-h: 122–138 µl/L	Hemmer et al. 2011
	Mysidopsis bahia	LC_{50} 48-h: 32.23 ppm	Lindgren et al. 2001
Corexit 9527	*Palaemon serenus*	LC_{50} 96-h: 49.4 ppm	Gulec and Holdway 2000
	Macquaria novemaculeata	LC_{50} 96-h: 14.3 ppm	Gulec and Holdway 2000
	Menidia beryllina	LC_{50} 96-h: 14.57 ppm	Lindgren et al. 2001
	Mysidopsis bahia	LC_{50} 48-h: 24.14 ppm	Lindgren et al. 2001
	Allorchestes compressa	LC_{50} 96-h: 3.03 ppm	Gulec and Holdway 1997
	Allorchestes compressa	EC_{50} 24-h: 33.8 ppm	Gulec and Holdway 1997
	Haliotis rufescens (embryos)	LC_{50} 96-h: 2.2 ppm	George-Ares and Clark 2000
	Crassostrea gigas (embryos)	LC_{50} 24-h: 3.1 ppm	George-Ares and Clark 2000
Dispersed oil using corexit 9500	*Phyllospora comosa*	EC_{50} 48-h: 340 µl/L	Burridge and Shir 1995
	Palaemon serenus	LC_{50} 96-h: 3.6 ppm	Gulec and Holdway 2000
	Macquaria novemaculeata	LC_{50} 96-h: 14.1 ppm	Gulec and Holdway 2000
	Americamysis bahia	LC_{50} 48-h: 4.9–67 µl/L	Hemmer et al. 2011
	Menidia beryllina	LC_{50} 48-h: 6.2–8.5 µl/L	Hemmer et al. 2011
	Mysidopsis bahia	LC50 48-h: 3.4 ppm	Lindgren et al. 2001
Dispersed oil using corexit 9527	*Phyllospora comosa*	EC_{50} 48-h: 380 µl/L	Burridge and Shir 1995
	Palaemon serenus	LC_{50} 96-h: 8.1 ppm	Gulec and Holdway 2000
	Macquaria novemaculeata	LC_{50} 96-h: 28.5 ppm	Gulec and Holdway 2000
	Menidia beryllina	LC_{50} 96-h: 4.49 ppm	Lindgren et al. 2001
	Mysidopsis bahia	LC_{50} 48-h: 6.60 ppm	Lindgren et al. 2001
	Allorchestes compressa	LC_{50} 96-h: 16.2 ppm	Gulec and Holdway 1997
	Allorchestes compressa	EC_{50} 24-h: 26.3 ppm	Gulec and Holdway 1997
Dispersit 1000	*Americamysis bahia*	LC_{50} 48-h: 1.9–2.2 µl/L	Hemmer et al. 2011
	Menidia beryllina	LC_{50} 96-h: 3.5 ppm	Lindgren et al. 2001
	Mysidopsis bahia	LC_{50} 48-h: 16.6 ppm	Lindgren et al. 2001
Dispersed oil using Dispersit 1000	*Menidia beryllina*	LC_{50} 96-h: 7.9 ppm	Lindgren et al. 2001
	Mysidopsis bahia	LC_{50} 48-h: 8.2 ppm	Lindgren et al. 2001

LC_{50}: lethal concentration, EC_{50}: effective concentration, IC_{50}: inhibitory concentration

TABLE 15.5

Categories of Toxicity for Aquatic and Terrestrial Organisms

No.	Toxicity Category	Aquatic Organisms LC_{50} (mg/L)	Terrestrial Organisms Oral dose LD50 (mg/kg-bw)	Dietary LC50 (mg/L)
1	Very highly toxic	<0.1	<10	<50
2	Highly toxic	>0.1–1	10–50	50–500
3	Moderately toxic	>1–10	51–500	501–1000
4	Slightly toxic	>10–100	501–2000	1001–5000
5	Practically nontoxic	>100	>2000	>5000

Source: USEPA 2022.

Oil exposure of aquatic receptors to physically and/or chemically dispersed oils can occur through one of four basic routes, each of which may contribute differently to the overall exposure of each biological resource:

1. Direct contact: Wildlife at the air–water interface (e.g., turtles, birds, and marine mammals) may come into direct skin contact with oil while swimming, surfacing, or diving through oil surfaces where their bodies are oil-coated. Oils on fur or hair can wreak havoc on their waterproofing and insulation properties, resulting in hypothermia and even death. Skin contact with liquid oil can have effects ranging from relatively harmless to more severe on sensitive tissues, particularly the eye membranes, depending on the amount.

2. Inhalation and aspiration: When breathing air, air-breathing wildlife (e.g., sea turtles, birds, and marine mammals), particularly those breathing above the air–water interface, may be exposed to volatile organic compounds and potentially aerosolized oil droplets on slippery surfaces caused by breaking waves, wind, and rain. Cetaceans (such as whales and dolphins) also inadvertently ingest liquid oil-containing seawater into their lungs. Oil inhalation and aspiration can cause respiratory tract irritation and serve as a source of hydrocarbon compounds in the bloodstream.

3. Absorption from water: Aquatic organisms can be exposed to oil and oil residues in the water column through absorption of bioavailable hydrocarbon compounds directly through the outer membrane, skin, or respiratory membranes exposed to dissolved concentrations.

4. Ingestion: Aquatic organisms may be exposed to oil or oil residues through the ingestion of water, sediment, and/or oil-containing food or water containing oil microdroplets. While some oil fractions are insoluble in digestive tract fluids, others are absorbed and transported into the bloodstream. The oil can irritate the digestive tract depending on the amount consumed.

BIOLOGICAL AGENTS IN OIL SPILL CLEANUP

Almost all oil and gas corporations, as well as countries experiencing oil spill situations, rely on dispersants to combat oil spills, purportedly to avoid spreading the oil-exposed areas, which could raise claims from the enormously affected communities. However, scientists continue to debate the efficacy of dispersants and the toxicity of dispersed oil mixtures. Numerous long-term environmental effect and toxicity tests have been conducted following dispersant application to ascertain the environmental impact. To avoid long-term damage from dispersants, it is vital to deal with oil spills using bio-based chemicals that are more environmentally friendly.

Biological agents are substances such as nutrients, enzymes, or microbes that accelerate the pace of natural biodegradation. Biodegradation is the process by which microorganisms such as bacteria, fungus, and yeast degrade complex molecules to get energy and nutrients. Oil biodegradation is a natural process that gradually eliminates oil from the environment over weeks, months, or years. However, removing spilt oil rapidly from beaches and wetlands may be necessary to avoid environmental damage to these delicate environments. Bioremediation technology can expedite the biodegradation process. Bioremediation is the process of introducing elements into the environment, such as fertilizers or microbes, that accelerate natural biodegradation. Additionally, bioremediation is frequently used when all mechanical means of oil recovery have been exhausted. For oil spill cleanup, two bioremediation techniques have been used: biostimulation and bioaugmentation.

Biostimulation is a process that involves the addition of nutrients such as phosphorus and nitrogen to a contaminated environment in order to accelerate the growth of oil-degrading bacteria. The scarcity of these essential nutrients typically limits the expansion of the indigenous microbial population. Native microbial populations can develop fast in the presence of nutrients, potentially increasing the rate of biodegradation.

Bioaugmentation is a process that increases the population of microbes in a contaminated environment if there aren't enough bacteria to digest the amount of pollution released into marine waterways. The type of microbe given to the polluted environment can be a single culture or a consortium with a mixture of various microorganisms with the potential to break down contaminants. In general, the application of a bacterial consortium is a frequently utilized strategy in decomposing crude oil or hydrocarbon spills, where the consortium is a variety of microbial populations in the form of a community that has mutually supporting interactions and helps one another. Several genera of bacteria, including *Achromobacter, Actinomycetes, Acinetobacter, Alcaligenes, Archrobacter, Aureobasidium, Bacillus, Candida, Coryneforms, Flavobacterium, Microbacterium, Micrococcus, Nocardia, Pseudomonas, Rhodotorula*, and *Sporobolomyces*, have been found in hydrocarbon-polluted settings and have the potential to break down hydrocarbons. In addition, some species of Fungi, including *Aspergillus, Fusarium, Gordonia, Mucor*, and *Penicillium* are also known as petroleum-degrading microorganisms. Occasionally, bacterial species that do not occur natively in an area are introduced into the local population. As with nourishment, seeding aims to expand the number of microorganisms capable of decomposing oil spills. This technique, however, is rarely essential, as hydrocarbon-degrading bacteria are widespread and non-native species frequently do not compete successfully with native microorganisms. Table 15.6 summarizes the bacteria reported by researchers to have the ability to degrade hydrocarbon compounds in the marine environment.

TABLE 15.6
Previous Studies on the Removal of Hydrocarbon Compounds by Marine Bacteria and Microalgae

Microbe	Culture Type	Contaminant	Removal Efficiency	Ref.
Pseudomonas sp., *Marinomonas* sp., *Oleispira* sp., *Cycloclasticus* sp., *Paraperlucidibaca* sp.	Mixed	Crude oil; 1%	71% in 4 months, 85% in 8 months	Nõlvak et al. 2021
Oleispira sp., *Thalassolituus* sp., *Marinobacter* sp., *Amphritea* sp. *Pseudomonas_D* sp.	Mixed	Diesel fuel; 0.1%	43% (Alkane), 70% (PAH) in 71 days	Murphy et al. 2021
Achromobacter sp., *Microbacterium testaceum,* *Aquicoccus* sp., *Frondibacter* sp., *Pseudomonas* sp., *Acinetobacter* sp., *Pseudooceanicola marinus*	Mixed	Crude oil; 2.5%	23–30% in 15 days	Tomasino et al. 2021
Bacillus licheniformis	Single	Diesel fuel; 40%	25.03% in 5 days	Purwanti et al. 2015
Bacillus subtilis	Single	Diesel fuel; 40%	5.42% in 5 days	
Bacillus cereus	Single	Diesel fuel; 40%	5.9% in 5 days	
Bacillus subtilis, *Bacillus jeotagali,* *Bacillus foraminis*	Mixed	Crude oil; 1%	95.5% in 20 days	Prabhakaran et al. 2014
Bacillus cereus *Bacillus* sp.	Mixed	Crude oil; 5%	67.2% in 5 weeks	Wardhani and Titah 2020
Pseudomonas sp. *Bacillus* sp.	Mixed	Crude oil; 5%	73.2% in 5 weeks	
Bacillus cereus *Pseudomonas* sp.	Mixed	Crude oil; 5%	89.1% in 5 weeks	
Bacillus sphaericus *Pseudomonas aeruginosa*	Mixed	Crude oil; 5%	78.5% in 5 weeks	
Pseudomonas sp.	Single	Crude oil; 1%	67% in 4 weeks	
Bacillus sp.	Single	Crude oil; 1%	61% in 4 weeks	
Bacillus cereus	Single	Crude oil; 1%	54% in 4 weeks	

Microbe	Culture Type	Contaminant	Removal Efficiency	Ref.
Micrococcus sp.	Single	Diesel fuel; 15%	61–80% in 48 hours	Titah et al.
Staphylococcus sp.	Single	Diesel fuel; 15%	41–60% in 48 hours	2018
Pseudomonas sp.	Single	Diesel fuel; 0.5%	49.93% in 20 days	Panda et al. 2013
Bacillus subtilis	Single	Crude oil; 2%	80% in 10 days	Sakthipriyaa et al. 2015
Acinetobacter sp. Y9, W3, F9 and *Gordonia* sp. X1	Mixed	Crude oil; 787 mg/L	98% in 24 hours	Luo et al. 2021
Rhodococcus erythropolis, Pseudomonas sp.	Mixed	Crude oil; 250mg/L	47% in 15 days	Perdigão et al. 2021
Haloferax sp., *Halobacterium* sp., *Halococcus* sp.		Crude oil; 0.2%	Crude oil (13–47%); n-octadecane (28–67%); phenanthrene (13–30%) in 3 weeks	Al-Mailem et al. 2010
Chlorella vulgaris BS1	Single	Crude oil; 115 mg/L	98% in 14 days	Das and Deka 2019
Chlorella vulgaris	Single	Crude oil; 20 g/L	94% (light fraction), 88% (heavy fraction) in 14 days	Kalhor et al. 2017
Anabaena variabilia, Lyngbya digueti, Oscillatoria pranceps, Phormidium mucicola, Westiellopsis prolific	Single	Waste petroleum oil; 7 mg/L	87%, 90%, 92%, 85%, 88% in 96 hours, respectively	Al-Hussieny et al. 2020
Phormidium sp.	Single	Hexadecane (0.3%), diesel oil (0.5%)	45%, 37% in 10 days, respectively	Ammar et al. 2018

In general, the process of crude oil biodegradation can be explained as follows:

1. A percentage of hydrocarbon-degrading microorganisms will naturally exist in an environment contaminated with crude oil (hydrocarbonoclastic).
2. Hydrocarbonoclastic bacteria use hydrocarbons for energy as well as carbon and hydrogen.
3. Degrading microbes synthesize biosurfactants, which dissolve hydrocarbons in the liquid phase, reduce surface tension, and increase degrading microorganisms' accessibility to oil droplets.
4. Microbes transport hydrocarbons into microbial cells by cell interactions with dissolved hydrocarbons in the aqueous phase, diffusion or active transport mechanisms, or cell interactions with bacterially emulsified hydrocarbon droplets.
5. The characteristics, quantity of cells, and enzymes found in microorganisms will influence the result of petroleum hydrocarbon breakdown.

In general, enzymes will oxidize or break down hydrocarbons to form aldehydes and simple acids that can enter metabolic processes such as the Krebs cycle and electron transport pathways.

6. The Krebs cycle generates ATP, NADH, FADH$_2$, and CO$_2$, while NADH and FADH$_2$ join the electron transport chain to generate H$_2$O and some ATP.

According to the described procedure, bacteria use petroleum hydrocarbon molecules as substrates and convert them into non-toxic compounds such as CO$_2$ and H$_2$O during the biodegradation process (Figure 15.11). In contrast to most other components of the environment, the primary mediators of bioremediation are degrader microbes and their products as they convert or mineralize contaminants, thereby decreasing their masses and toxicities. The efficacy of bioremediation is determined by three key factors: the availability of microorganisms, the accessibility of toxins, and the presence of a favorable environment. The bioremediation efficiency is determined by the microbial ability to breakdown these complex combinations and their rate-limiting kinetics. Contaminants in the environment are bioremediated because of their widespread distribution, poor bioavailability and durability in the marine environment, and potentially harmful effect on biota health (Kardena et al. 2015). The rate of absorption and metabolism (the intrinsic activity of the cell) and the rate of transfer to the cell regulate chemical conversion during biodegradation by microbial cells (mass transfer). These variables influence a chemical's so-called bioavailability. Hydrocarbons, for example, are hydrophobic molecules with limited water solubility; as a result, microorganisms have developed different ways to enhance the bioavailability of these compounds in order to use them as potential carbon and energy sources.

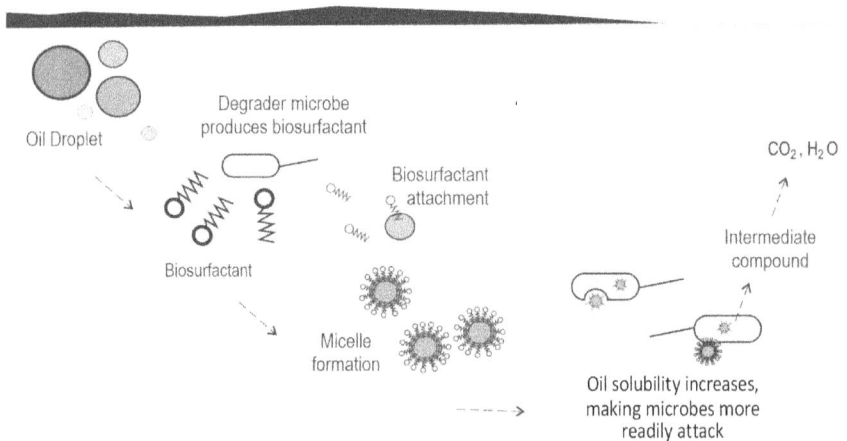

FIGURE 15.11 Schematic of the biodegradation process of an oil spill in the marine environment.

Biosurfactant-Enhanced Oil Spill Cleanup

The bioavailability of the pollutant is the most important factor influencing the efficiency of biological treatment. These naturally occurring processes, such as molecular sorption into solids, development of non-aqueous phases, interactions with organic matter, biotransformation, and contaminant aging or weathering, frequently result in restricted bioavailability, reducing the efficacy of biodegradation. Various studies and reports on biosurfactant synthesis by marine microflora, as well as its involvement in enhancing biodegradation of oil spills, have been comprehensively examined (Table 15.7), justifying the use of biosurfactants for oil spill bioremediation. The growing interest is due to the fact that biosurfactants can improve the solubilization of petroleum oil in seawater, increasing its solubility and, as a result, its bioavailability for microbial attack. Biosurfactants are involved in oil spill bioremediation in at least two ways: increasing the surface area of low-solubility to insoluble petroleum oil and enhancing its bioavailability. The addition of biosurfactants or *in situ* generation of biosurfactants can also improve the biodegradation of petroleum oil in seawater. It was discovered that the degrading time, especially the adaption time, of degrader microorganisms was shortened.

TABLE 15.7
Biosurfactant-Producing Marine Bacteria

Isolate	Results	Ref.
Pseudomonas sp.	Emulsification activity of isolate was recorded as 0.11 OD, and emulsification index was recorded as 66.6%. When salt concentrations were increased from 1 to 15%, emulsification activity gradually increased from 68 to 80%, and emulsification index was increased from 0.03 to 0.2 OD. The crude oil (10% v/v) biodegradation by the isolate was recorded as 94% after 15 days of incubation period.	Saravana and Amruta 2013
Phormidium sp.	When grown in actual seawater enhanced with nutrients and in the presence of hydrocarbons, marine cyanobacteria *Phormidium* sp. was able to remove hexadecane (45%) and diesel oil (37%). The partially purified surface-active compounds produced by *Phormidium* helped to improve this strain's hexadecane and diesel removal capabilities. In autotrophic conditions, the monospecific cyanobacterium *Phormidium* demonstrated degradative capacity on hydrocarbons with C10–C28 carbon atom numbers in axenic cultures.	Morales and Paniagua 2014
Bacillus salmalaya	Strain 139SI can significantly reduce the surface tension (ST) from 70.5 to 27 mN/m, with a critical micelle concentration of 0.4%. Moreover, the biosurfactant demonstrated high stability at different ranges of salinity, pH, and temperature and enhanced the biodegradation of lubricating oil (2% v/v) with 71.5% removal efficiency after 20 days.	Dadrasnia and Ismail 2015

(continued)

TABLE 15.7

Biosurfactant-Producing Marine Bacteria (Continued)

Isolate	Results	Ref.
Pantoea spp-1 and *Pantoea* spp-2	A definite decrease in surface tension from 69.9 and 41.2 mN.M^{-1}. After 24 hours of growth, the cell free extracts of both the cultures exhibited emulsifying activity with groundnut oil, coconut oil, and kerosene.	Gopalakrishnan et al. 2006
Bacillus megaterium, Corynebacterium kutscheri, Pseudomonas aeruginosa	Emulsification activity (D$_{610}$) on crude oil was 1.72, 1.69, and 1.85, respectively. Bacterial adhesion to hydrocarbons assay results revealed that *P. aeruginosa* showed high affinity (95.3%) with crude oil, followed by *C. kutscheri* (49.7%) and *B. megaterium* (40.2%).	Thavasi et al. 2010
Moorea bouillonii	Fatty acid amides extracted as columbamide D and serinolamide C exhibited biosurfactant activity with critical micelle concentrations of about 0.34 and 0.78 mM, respectively.	Mehjabin et al. 2020
Bacillus circulans	Emulsification activity (D$_{610}$) on crude oil was 1.2, and biosurfactant production yield was 0.7 mg/ml.	Harikrishnan and Jayalakshmi 2020

1. *Increasing the Surface Area of a Water-Insoluble Oil*—In an open environment, such as oil-polluted saltwater, the concentration of cells never reaches a high enough amount to properly emulsify the oil. One method to reconcile the available evidence with these theoretical considerations is to propose that emulsifying agents play a natural function in oil degradation but not in the formation of macroscopic emulsions in the bulk liquid. If emulsion happens at or near the cell surface and no mixing occurs at the microscopic level, each cluster of cells forms its own micro-environment and there is no overall cell-density dependence. Biosurfactants, being amphiphiles, have a proclivity to deposit at the oil/water interface. Through particular interactions that result in solubilization and micellization, biosurfactants may enhance the transfer of hydrophobic pollutants into the aqueous phase. Aside from interactions with contaminants, biosurfactants may have a direct impact on the efficiency of degrader bacteria. Biosurfactants have a high level of biological activity, particularly at the cellular membrane level. These changes may result in increased hydrophobicity, which is thought to be important in terms of biodegradation efficiency, or they may modify the permeability of cellular membranes, which could be helpful during biodegradation.

2. *Increasing Oil Bioavailability*—Desorption, diffusion, and dissolution all have an impact on the bioavailability of oil in seawater to degrading microorganisms. Biosurfactants are established to reduce tension at the hydrophobic-water interface in order to pseudosolubilize the hydrophobic molecule, hence enhancing mobility, bioavailability, and biodegradability. The low water solubility of large molecular weight hydrocarbons,

which promotes their sorption to surfaces and limits their availability to biodegrading bacteria, is one of the key reasons for their long-term persistence. Biodegradation is impeded when organic molecules are permanently attached to surfaces. Biosurfactants can promote growth on bound substrates by either desorbing them from the surface or enhancing their apparent water solubility. Biosurfactants exist as monomers when dissolved in water at very low concentrations. In such cases, the hydrophobic tail, unable to make hydrogen bonds, disrupts the water structure in its vicinity, increasing the system's free energy. At greater concentrations, when this effect is more pronounced, the free energy can be lowered by aggregation of the biosurfactant molecules into micelles, with the hydrophobic tails in the inner part of the cluster and the hydrophilic heads exposed to the bulk water phases. Additional quantities of biosurfactants in the water body will facilitate the establishment of more micelles at concentrations above the critical micelle concentrations (CMC) (Figure 15.12). As these molecules can partition into the central core of a micelle, the apparent solubility of hydrophobic organic compounds increases significantly, even exceeding their water solubility limit. The result of such a process is an increase in the mobilization and dispersion of organic molecules in solution. The reduction of interfacial tension between immiscible phases also contributes to this effect.

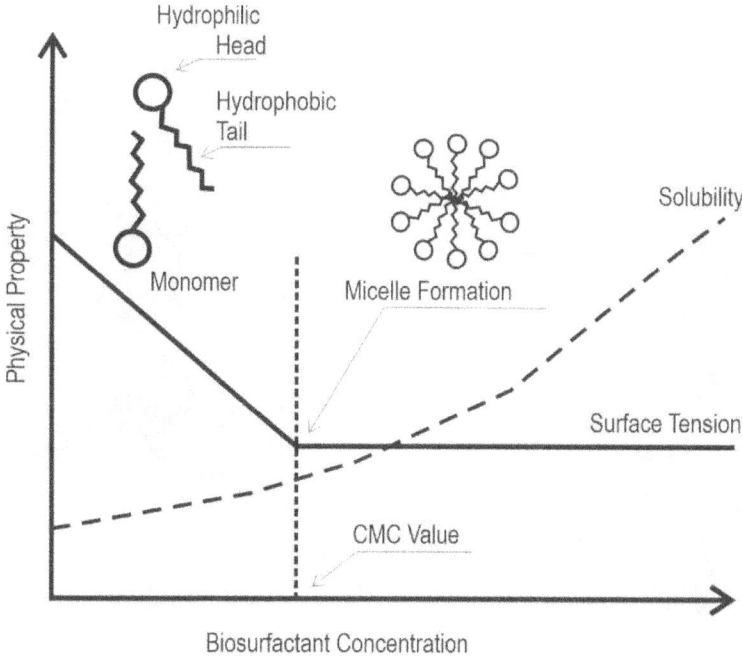

FIGURE 15.12 The value of surface tension as a function of biosurfactant concentration and the formation of micelles at the CMC value.

CASE STUDY: REMEDIATION OF MANGROVE SEDIMENT AFFECTED BY OIL SPILL

As Indonesia has an abundance of maritime resources, including oil and natural gas, numerous exploration activities take place in the sea. These conditions have become increasingly intense as the focus of upstream oil and gas activity has shifted from onshore fields to offshore and deep-sea areas. This activity poses threats to the marine environment, including oil spills caused by oil platform damage. An oil spill is a significant issue that could result in both short- and long-term economic and ecological consequences. On July 12, 2019, an oil leak was discovered near the Pertamina Hulu Energi-Offshore North West Java (PHE-ONWJ) YYA-1 oil platform. The oil well kick happened in the re-activated YYA-1, which is located beneath the PHE-ONWJ's offshore platform 2 km off the northern coast of Karawang, Java Sea. The first oil flow appeared in the YYA-1 well on July 16, 2019 (Figure 15.13). Oil spill management was carried out by localizing the distribution of oil in the sea through:

1. Primary containment of YYA-1, with the deployment of two layers of static oil boom to prevent further spread of oil in the sea. Oil spills that collect in a static oil boom are then picked up with a skimmer for further management,
2. Secondary containment by using a dynamic oil boom to anticipate if there is still oil escaping from the first containment due to bad weather or other factors,

FIGURE 15.13   Oil leakage observed from YYA-1 platform (upper left) causes damage to shoreline due to tarballs landing along the beach (upper right), polluting the sediment (lower left), and damaging mangrove roots (lower right).

3. Shoreline protection and oil cleanup by the emergency response team
for tarballs that landed on the coast, predicted by modeling and satellite
imagery. In this onshore treatment, sensitive areas and infrastructure
are protected, including mangrove forests, tourist beaches, settlements,
and fishing facilities.

The main source of pollutants is medium type crude oil, the characteristics of
which are shown in Table 15.8. Oil spills in the sea will be carried by the waves
until some of them reach the coast. While floating in the sea, some of the oil
spills will experience weathering and are rolled up by the waves, breaking up

TABLE 15.8
YYA-1 Crude Oil Characteristics

Composition	Result	Unit
API gravity @60°F	30.6	—
Pour point	40	°C
Kinematic viscosity @140°F	10.210	cSt
Kinematic viscosity @160°F	7.337	cSt
Kinematic viscosity @180°F	5.505	cSt
Ash content	0.055	% wt
Conradson carbon residue	2.248	% wt
Paraffin wax content	29.32	% wt
Asphaltenes	1.29	% wt
Total acid number	0.146	mg KOH/gr
Color	Pale brown–brownish black	
Fraction composition		
Saturated	%	48.47–73.05
Aromatic	%	16.01–21.42
NSO+polar	%	5.53–35.52
Antimony, Sb	<0.1	mg/dry kg
Arsenic, As	<0.1	mg/dry kg
Cadmium, Cd	0.03	mg/dry kg
Chromium, Cr	0.3	mg/dry kg
Cobalt, Co	0.3	mg/dry kg
Copper, Cu	0.2	mg/dry kg
Lead, Pb	<0.2	mg/dry kg
Manganese, Mn	6.0	mg/dry kg
Mercury, Hg	<0.001	mg/dry kg
Nickel, Ni	0.9	mg/dry kg
Selenium, Se	<0.1	mg/dry kg
Silver, Ag	<0.02	mg/dry kg
Tin, Sn	1.2	mg/dry kg
Vanadium, V	0.5	mg/dry kg
Zinc, Zn	<1	mg/dry kg

and forming lumps of oil called tarballs. The oil spill and tarball will land on the coast and adhere to the mangrove roots, but only the tarball will settle and be buried in the sand because it has a heavier density. The landing mechanism of an oil tarball can be seen in Figure 15.14.

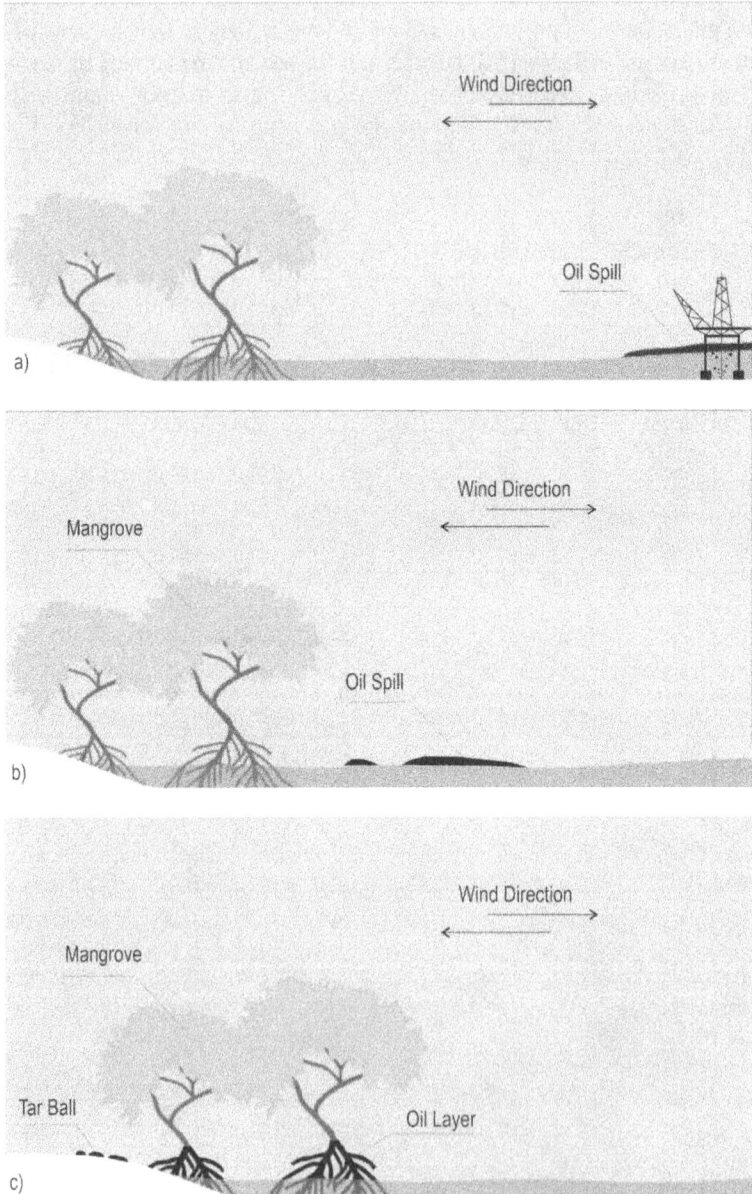

FIGURE 15.14   A conceptual site model illustrating the impact of the YYA-1 oil spill on land.

The scale of contamination of oil spill exposure is grouped based on the concentration of contamination, which is divided into three categories:

1. Heavy polluted category: concentration of long chain C_{10-36} petroleum hydrocarbon > 5000 mg/kg and/or short chain C_{6-9} > 325 mg/kg,
2. Medium polluted category: concentration of long chain C_{10-36} petroleum hydrocarbon > 1000–5000 mg/kg and/or short chain C_{6-9} > 100–325 mg/kg,
3. Lightly polluted category: concentration of long chain C_{10-36} petroleum hydrocarbon < 1000 mg/kg and/or short chain C_{6-9} < 100 mg/kg.

The determination of the pollution level category follows the Indonesian Government Regulation No. 101 of 2014 concerning Management of Hazardous and Toxic Waste, Appendix V. Recovery and remediation of contaminated land is declared successful if the concentration of long-chain petroleum hydrocarbons C_{10-36} < 1000 mg/kg and/or short-chain C_{6-9} <100 mg/kg, and other parameters of concern such as heavy metals are below the total concentration level C. The delineation of the area affected by the oil spill and the determination of the volume of contaminated land to be recovered are shown in Figure 15.15 and Table 15.9.

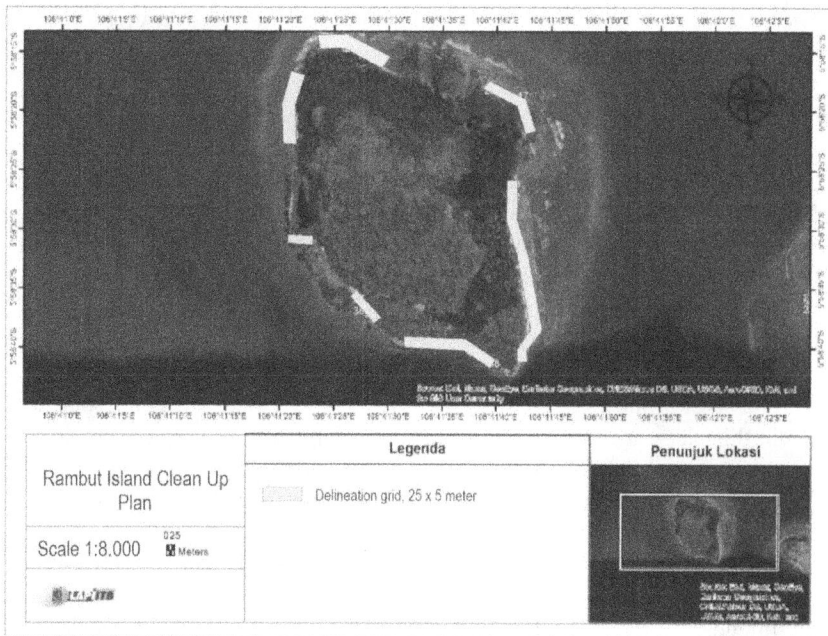

FIGURE 15.15 Delineation of the area affected by the oil spill on Rambut Island.

TABLE 15.9
Estimated Area and Volume of Contaminated Land Affected by YYA-1 Oil Spill in Untung Jawa and Rambut Island

Location	Area (m²)	Volume (m³)	Rocks, Coral (m²)
Untung Jawa Island	1,375	412.5	60
Rambut Island	1,225	720	400
Total	2,600	1,132.5	460

The selection of the most appropriate cleaning technique requires an evaluation of the magnitude of the impact and type of contamination, the nature of the contaminants, and the accessibility of the affected site. In determining priority actions, the demands of other environmental interests need to be considered. For example, contamination in densely populated residential areas/tourist areas/ other important infrastructures demands a fast and effective recovery method to remove oil, but the technique may not be compatible with environmental considerations; for example, the use of chemical dispersants will remove the oil quickly, but emulsified oil will be more widely dispersed and potentially toxic to other living things. This is in contrast to contamination in insensitive areas, which may require less aggressive techniques and slower oil removal but will be more environmentally friendly. In situations such as the previous, a balance must be struck between these two potentially conflicting interests for the overall or specific recovery method for each polluted site. The cleanup and restoration of oil-polluted land on Untung Jawa Island and Rambut Island is divided into three stages:

- Stage 1—Emergency Response Phase: Emergency response due to an oil pollution incident. The main activities carried out are the collection/retrieval of floating oil close to the shoreline, emulsified oil in tidal areas, beach sand, sediments, and oil puddles in residential areas. The response in this phase is intended so that oil contamination does not spread more widely to other areas.
- Stage 2—Recovery Activity Phase: Removal of oil stranded/adhered to beach materials such as sand, gravel, coral, coral reefs, mangrove roots, and other properties that are contaminated/sticky with oil. The response in this phase is aimed at restoring the land affected by oil pollution in a more comprehensive manner up to the determined level of success. At this stage, the process of delineating the distribution of land affected by contamination is carried out as the basis for the preparation of the method for the recovery plan for the contaminated land.
- Stage 3—Final Phase And Monitoring: The response in this phase is intended as a final polishing/cleaning of the remaining oil residue and monitoring the results of the recovery that was carried out in the previous phase. Depending on the situation at hand, the development of each

of these stages may not be necessary. In some cases, the entire recovery operation can be completed in one stage, while in others, Stages 1 and 2 may be combined activities, while after Stage 2 is complete, the remaining oil may best be allowed to degrade naturally. In each case, the first priority is to recover the oil that floats on the shore as quickly as possible to prevent it from moving to previously uncontaminated areas or clean land. The same is true for stranded oil accumulations that can change locations during the next high tide. The use of oil booms/barriers to hold oil is necessary while recovery is underway. After the oil contaminants have been collected, it is necessary to consider advanced recovery techniques or wait until all the oil remaining in the sea has come ashore. This aims to avoid cleaning the same oil-polluted area more than once because oil keeps coming from the waters with the waves.

The method of recovering contaminated areas from oil spills on Untung Jawa Island and Rambut Island is a combination of physical and biological techniques:

1. Manual cleanup

   The main activities carried out are the collection/retrieval of floating oil close to the shoreline, emulsified oil in tidal areas, tarballs stranded on the beach, contaminated sand and sediment, oil puddles in mangrove-growing media, and oil attached to roots and mangrove trunks (Figure 15.16).

2. Pressure washing using biosurfactants

   Pressurized water washing is ideal for cleaning oil-affected surfaces, starting from the surface of the rock, coral fragments on the beach, and sand to the oil that sticks to the roots and stems of mangroves (Figure 15.17). The addition of biosurfactants is used to help release oil that is bound and cannot be separated under normal conditions. Installation of oil booms/barriers is necessary so that the oil released due to the application of biosurfactants can be localized and prevent cross-contamination. The biosurfactants used are derived from producing bacteria that are produced on the mainland due to the limited culture propagation facilities onsite. The biosurfactant-producing isolates used were *Pseudomonas aeruginosa* PAU01, *Burkholderia* sp. PAU02, and *Azotobacter vinelandii* GNC01, as reported by Arsyah et al. (2018), Effendi et al. (2018), Devianto et al. (2020), and Leslie et al. (2021). In applications that require high volumes of biosurfactants at low prices, the use of growing media in the form of agricultural by-products becomes feasible (Helmy et al. 2011). Molasses is used as the sole carbon source in the production of biosurfactants and is harvested in the form of cell-free broth without further purification for later use in pressure-washing applications in Untung Jawa Island and Rambut Island.

FIGURE 15.16 Manual cleanup of stranded tarballs along shoreline of affected beach.

FIGURE 15.17 Pressure washing using biosurfactant on affected coral and rock.

3. *In situ* biostimulation and bioaugmentation

Biostimulation is a process in which the growth of onsite natural oil-degrading microbes is stimulated by the addition of nutrients or other growth-limiting co-substrates and/or habitat changes. In the case of Untung Jawa and Rambut Island remediation, biostimulation was carried out by adding nutrients to stimulate the native microbial population in order to accelerate the biodegradation process of the oil contaminants. On the shorelines of Untung Jawa and Rambut Island, especially the mangrove sediments/growing media were contaminated with oil. Nutrients are added in the form of urea fertilizer, NPK fertilizer, and trace minerals. The ratio of added nutrients C:N:P:K:trace minerals is 100:10:1:0.5:0.01. In addition to stimulating the growth of native microbes in the sediment, the provision of nutrition also helps the growth of mangroves affected by the oil spill. Bioaugmentation is a technique of spreading microbes when oil pollution occurs. The microbes that work to decompose the oil are not only one species but a mixed culture community that was pre-selected in the laboratory for its ability to degrade YYA-1 crude oil. Each type of microbe has its own ability to break down oil. The effectiveness of bacteria in degrading oil varies, depending on the type of bacteria, from 0–100%, including symbiotic bacteria that do not have a special ability to degrade petroleum oil but are able to stimulate petrophilic bacteria to adapt better to a new environment (Figure 15.18).

FIGURE 15.18 Injection of nutrient on sediment to stimulate the native microbial growth and to promote mangrove growth (upper), and augmentation of petrophilic bacteria (PETREA) to enhance the biodegradation of crude oil (lower).

FIGURE 15.19 *Ex situ* bioremediation (landfarming) of soil, sediment, and sand affected by oil spill.

4. *Ex situ* bioremediation

Landfarming was chosen as an *ex situ* bioremediation technique to degrade soil and sediment contaminated with oil spills and the remaining tarballs mixed with sand. The selection of the location for the temporary bioremediation facility was carried out with several needs in mind, including flat contours, unaffected by coastal tides, compacted bottom soil, and covered with a waterproof membrane. To avoid excessive water content due to rainwater, the temporary bioremediation facility is covered with a waterproof membrane (Figure 15.19).

REFERENCES

Agarry, S.E., Oghenejoboh, K.M., Oghenejoboh, E.O., Owabor, C.N., and Ogunleye, O.O. 2020. Adsorptive Remediation of Crude Oil Contaminated Marine Water Using Chemically and Thermally Modified Coconut (*Cocos nucifera*) Husks. *Journal of Environmental Treatment Techniques* 8(2):694–707.

Al-Hussieny, A.A., Imran, S.G., and Jabur, Z.A. 2020. The Use of Local Blue-Green Algae in the Bioremediation of Hydrocarbon Pollutants in Wastewater from Oil Refineries. *Plant Archives* 20(2):797–802.

Al-Mailem, D.M., Sorkhoh, N.A., Al-Awadhi, H., Eliyas, M., and Radwan, S.S. 2010. Biodegradation of Crude Oil and Pure Hydrocarbons by Extreme Halophilic Archaea from Hypersaline Coasts of the Arabian Gulf. *Extremophiles* 14:321–328. http://doi.org/10.1007/s00792-010-0312-9.

Ammar, S.H., Khadim, H.J., and Mohamed, A.I. 2018. Cultivation of *Nannochloropsis oculata* and *Isochrysis galbana* Microalgae in Produced Water for Bioremediation and Biomass Production. *Environmental Technology & Innovation* 10:132–142. http://doi.org/10.1016/j.eti.2018.02.002.

Arsyah, D.M., Kardena, E., and Helmy, Q. 2018. Characterization of Biosurfactant Produced by Petrofilic Bacteria Isolated from Hydrocarbon Impacted soil and Its Potential Application in Bioremediation. *IOP Conf. Series: Earth and Environmental Science* 106(2018):012101. http://doi.org/10.1088/1755-1315/106/1/012101.

Asadpour, R., Sapari, N.B., Isa, M.H., Kakooei, S., and Orji, K.U. 2015. Acetylation of Corn Silk and Its Application for Oil Sorption. *Fibers and Polymers* 16(9):1830–1835. http://doi.org/10.1007/s12221-015-4745-8.

Asadu, C.O., Anthony, E.C., Elijah, O.C., Ike, I.S., Onoghwarite, O.E., and Okwudili, U.E. 2021. Development of an Adsorbent for the Remediation of Crude Oil Polluted Water Using Stearic Acid Grafted Coconut Husk (*Cocos nucifera*) Composite. *Applied Surface Science Advances* 6:100179. http://doi.org/10.1016/j.apsadv.2021.100179.

Auffret, M., Duchemin, M., Rousseau, S., Boutet, I., Tanguy, A., Moraga, D., and Marhic, A. 2004. Monitoring of Immunotoxic Responses in Oysters Reared in Areas Contaminated by the "Erika" Oil Spill. *Aquatic Living Resources* 17:297–302. http://doi.org/10.1051/alr:2004035.

Baker, J.M., Clark, R.B., Kingston, P.F., and Jenkins, R.H. 1990. Natural Recovery of Cold Water Marine Environments after an Oil Spill. *Paper presented at the 13th AMOP Seminar*, Edmonton. Canada.

Balseiro, A., Espi, A., Marquez, I., Perez, V., Ferreras, M.C., Marin, J.F.G., and Prieto, J.M. 2005. Pathological Features in Marine Birds Affected by The Prestige's Oil Spill in the North of Spain. *Journal of Wildlife Diseases* 41(2):371–378. http://doi.org/10.7589/0090-3558-41.2.371.

Barron, M.G. 2011. Ecological Impacts of the Deepwater Horizon Oil Spill: Implications for Immunotoxicity. *Toxicologic Pathology* 40(2):315–320. http://doi.org/10.1177/0192623311428474.

Buist, I., McCourt, J., Potter, S., Ross, S., and Trudel, K. 1999. *In Situ* Burning. *Pure and Applied Chemistry* 71(1):43–65.

Burgherr, P. 2007. In-Depth Analysis of Accidental Oil Spills from Tankers in the Context of Global Spill Trends from All Sources. *Journal of Hazardous Materials* 140(1–2):245–256. http://doi.org/10.1016/j.jhazmat.2006.07.030.

Burridge, T.R., and dan Shir, M.-A. 1995. The Comparative Effects of Oil Dispersants and Oil/Dispersant Conjugates on Germination of the Marine Macroalga *Phyllospora comosa* (Fucales: Phaeophyta). *Marine Pollution Bulletin* 31:446–452.

Clark, R.B. 2001. *Marine Pollution*, 5th edition. Oxford: Oxford University Press.

Dadrasnia, A., and Ismail, S. 2015. Biosurfactant Production by *Bacillus salmalaya* for Lubricating Oil Solubilization and Biodegradation. *International Journal of Environmental Research and Public Health* 12:9848–9863. http://doi.org/10.3390/ijerph120809848.

Das, B., and Deka, S. 2019. A Cost-Effective and Environmentally Sustainable Process for Phycoremediation of Oil Field Formation Water for Its Safe Disposal and Reuse. *Scientific Reports* 9:15232. http://doi.org/10.1038/s41598-019-51806-5.

Davis, W.P., Hoss, D.E., Scott, G.I., and Sheridan, P.F. 1984. Fisheries Resource Impacts from Spills of Oil or Hazardous Substances. In J. Cairns and A.L. Buikema (Eds.), *Restoration of Habitats Impacted by Oil Spills*. Boston: Butterworth Publishers, 182 pp.

Devianto, L.A., Latunussa, C.E.L., Helmy, Q., and Kardena, E. 2020. Biosurfactants Production Using Glucose and Molasses as Carbon Sources by *Azotobacter vinelandii* and Soil Washing Application in Hydrocarbon-Contaminated Soil. *OP Conf. Series: Earth and Environmental Science* 475(2020):012075. http://doi.org/10.1088/1755-1315/475/1/012075.

Effendi, A.J., Kardena, E., and Helmy, Q. 2018. Biosurfactant-Enhanced Petroleum Oil Bioremediation. In V. Kumar, M. Kumar and R. Prasad (Eds.), *Microbial Action on Hydrocarbons*. Singapore: Springer. http://doi.org/10.1007/978-981-13-1840-5_7.

Engel, A.S., Liu, C., Paterson, A.T., Anderson, L.C., Turner, R.E., and Overton, E.B. 2017. Salt Marsh Bacterial Communities before and after the Deepwater Horizon Oil Spill. *Applied and Environmental Microbiology* 83(20):e00784–17. http://doi.org/10.1128/AEM.00784-17.

Evans, D.D., Mulholland, G.W., Baum, H.R., Walton, W.D., and McGrattan, K.B. 2001. *In Situ* Burning of Oil Spills. *J Res Natl Inst Stand Technol.* 106(1):231–278. http://doi.org/10.6028/jres.106.09.

Fingas, M. 2016. *In-Situ* Burning: An Update. In M. Fingas (Ed.), *Oil Spill Science and Technology*. Oxford, UK: Gulf Professional Publishing, pp. 483–676.

Galierikova, A., and Materna, M. 2020. World Seaborne Trade with Oil: One of Main Cause for Oil Spills?. *Transportation Research Procedia* 44(2020):297–304. http://doi.org/10.1016/j.trpro.2020.02.039.

George-Ares, A., and Clark, J.R. 2000. Aquatic Toxicity of Two Corexit Dispersants. *Chemosphere* 40(2000):897–906.

Giwa, A., and Taher, H. 2020. Dispersion-Sorption Balance (DSB) of Pickering Emulsions of Polydopamine-Polyethylenimine-Modified Activated Carbon for Oil Spill Treatment. *Journal of Environmental Chemical Engineering* 8(4):103950. http://doi.org/10.1016/j.jece.2020.103950.

Gopalakrishnan Kumar, A.S., Mody, K., and Jha, B. 2006. Biosurfactant Production by Marine Bacteria. *National Academy Science Letters* 29(3–4):95–101.

Gulec, I., and Holdway, D.A. 2000. Toxicity of Crude Oil and Dispersed Crude Oil to Ghost Shrimp *Palaemon serenus* and Larvae of Australian Bass *Maquaria novemaculeata*. *Environmental Toxicology* 15(2):91–98. http://doi.org/10.1002/(SICI)1522-7278.

Gulec, I., Holdway, D.A., and Melbourne, V. 1997. Toxicity of Dispersant, Oil, and Dispersed Oil to Two Marine Organisms. *International Oil Spill Conference Proceedings* 1997(1). http://doi.org/10.7901/2169-3358-1997-1-1010.

Harikrishnan, S., and Jayalakshmi, S. 2020. Marine Biosurfactant Production from *Bacillus circulans*. *Compendium of Research Insights of Life Science Students* 2(356):893–894.

Hart, M.E., Perez-Umphrey, A., Stouffer, P.C., Burns, C.B., Bonisoli-Alquati, A., Taylor, S.S., and Woltmann, S. 2021. Nest Survival of Seaside Sparrows (*Ammospiza maritima*) in the Wake of the Deepwater Horizon Oil Spill. *PLoS ONE* 16(10):e0259022. http://doi.org/10.1371/journal.pone.0259022.

Helmy, Q., Kardena, E., Funamizu, N., and Wisjnuprapto. 2011. Strategies Toward Commercial Scale of Biosurfactant Production as Potential Substitute for It's Chemically Counterparts. *International Journal of Biotechnology* 12(1/2):66–86.

Hemmer, M.J., Barron, M.G., and Greene, R.M. 2011. Comparative Toxicity of Eight Oil Dispersants, Louisiana Sweet Crude Oil (LSC), and Chemically Dispersed LSC to Two Aquatic Test Species. *Environ Toxicol Chem* 30(10):2244–2252. http://doi.org/10.1002/etc.619.

Hoff, R.Z. 1995. Responding to Oil Spills in Coastal Marshes: The Fine Line Between Help and Hindrance. *HAZMAT Report 96–2.* Seattle: Hazardous Materials Response and Assessment Division, NOAA, 16 pp.

Horn, S.A., and Neal, C.P. 1981. The Atlantic Empress Sinking-A Large Spill Without Environmental Disaster. *International Oil Spill Conference Proceedings* 1981(1):429–435. http://doi.org/10.7901/2169-3358-1981-1-429.

IEA. 2020. *World Oil Production by Region, 1971–2020*, The International Energy Agency, Paris. www.iea.org/data-and-statistics/charts/world-oil-production-by-region-1971-2020

Ikenyiri, P.N., and Ukpaka, C.P. 2016. Overview on the Effect of Particle Size on the Performance of Wood Based Adsorbent. *Journal of Chemical Engineering & Process Technology* 7(5):1–4. http://doi.org/10.4172/2157-7048.1000315.

Joyner, C.C., and Kirkhope, J.T. 1992. The Persian Gulf War Oil Spill: Reassessing the Law of Environmental Protection and the Law of Armed Conflict. *Case Western Reserve Journal of International Law* 24(1):29–62.

Kalhor, A.X., Movafeghi, A., Mohammadi-Nassab, A.D., Abedi, E., and Bahrami, A. 2017. Potential of the Green Alga *Chlorella vulgaris* for Biodegradation of Crude Oil Hydrocarbons. *Marine Pollution Bulletin* 123(1–2):286–290. http://doi.org/10.1016/j.marpolbul.2017.08.045.

Kardena, E., Helmy, Q., and Funamizu, N. 2015. Biosurfactant and Soil Bioremediation. In N. Kosaric and F.V. Sukan (Eds.), *Biosurfactant Production and Utilization—Processes, Technologies, and Economics.* Boca Raton: CRC Press.

Lamparelli, C.C., Rodeigues, F.O., and Orgler de Moura, D. 1997. Long-Term Assessment of an Oil Spill in a Mangrove Forest in Sao Paulo, Brazil. In B. Kjerfve, L. Drude de Lacerda, and W.H. Salif Diop (Eds.), *Mangrove Ecosystem Studies in Latin America and Africa.* Paris, France: UNESCO, pp. 191–203.

Leslie, M., Kardena, E., and Helmy, Q. 2021. Biosurfactant and Chemical Surfactant Effectiveness Test for Oil Spills Treatment in a Saline Environment. *IOP Conf. Series: Earth and Environmental Science* 896(2021):012041. http://doi.org/10.1088/1755-1315/896/1/012041.

Lewis, R.R. 1983. Impacts of Oil Spills on Mangrove Forests. In H.J. Teas (Eds.), *Biology and Ecology of Mangroves.* Task for Vegetation Science, Vol 8. Dordrecht, Netherlands: Springer, pp. 171–183.

Lindgren, C., Lager, H., and Fejes, J. 2001. *Oil Spill Dispersants: Risk Assessment for Swedish Waters.* Swedish Environmental Protection Agency. Stockholm, Sweden: IVL Publikationsservice.

Luo, Q., Hou, D., Jiang, D., and Chen, W. 2021. Bioremediation of Marine Oil Spills by Immobilized Oil-Degrading Bacteria and Nutrition Emulsion. *Biodegradation* 32(2):165–177. http://doi.org/10.1007/s10532-021-09930-5.

Makocha, I.R., Ete, T., and Saini, G. 2019. Deepwater Horizon Oil Spill: A Review. *International Journal of Technical Innovation in Modern Engineering & Science* 5(1):65–71.

Marchand, M.H. 1980. The Amoco Cadiz Oil Spill. Distribution and Evolution of Hydrocarbon Concentrations in Seawater and Marine Sediments. *Environment International* 4(5–6):421–429. http://doi.org/10.1016/0160-4120(80)90021-5.

Martinelli, M., Luise, A., Trome, E., Sauer, T.C., Neff, J.M., and Douglas, G.S. 1995. The M/C Haven Oil Spill: Environmental Assessment of Exposure Pathways and Resource Injury. *International Oil Spill Conference Proceedings* 1995(1):679–685. http://doi.org/10.7901/2169-3358-1995-1-679.

Mehjabin, J.J., Wei, L., Petitbois, J.G., Umezawa, T., Matsuda, F., Vairappan, C.S., Morikawa, M., and Okino, T. 2020. Biosurfactants from Marine Cyanobacteria Collected in Sabah, Malaysia. *Journal of Natural Products* 83(6):1925–1930. http://doi.org/10.1021/acs.jnatprod.0c00164.

Michel, J., Hayes, M.O., Getter, C., and Cotsapas, L. 2005. The Gulf War Oil Spill Twelve Years Later: Consequences of Eco-Terrorism. *International Oil Spill Conference Proceedings* 2005(1):957–961. http://doi.org/10.7901/2169-3358-2005-1-957.

Moldan, A., and Dehrman, A. 1989. Trends in Oil Spill Incidents in South African Coastal Waters. *Marine Pollution Bulletin* 20(11):565–567. http://doi.org/10.1016/0025-326X(89)90358-5.

Moller, D.M., Ferrante, M., Moller, G.M., Rozenberg, T., and Segoli, M. 2020. The Impact of Terrestrial Oil Pollution on Parasitoid Wasps Associated with Vachellia (Fabales: Fabaceae) Trees in a Desert Ecosystem, Israel. *Environmental Entomology* 49(6):1355–1362. http://doi.org/10.1093/ee/nvaa123.

Morales, A.R., and Paniagua, M.J. 2014. Bioremediation of Hexadecane and Diesel Oil Is Enhanced by Photosynthetically Produced Marine Biosurfactants. *Journal of Bioremediation & Biodegradation* S4:1–5. http://doi.org/10.4172/2155-6199. S4-005.

Murphy, S., Bautista, M.A., Cramm, M.A., and Hubert, C. 2021. Diesel and Crude Oil Biodegradation by Cold-Adapted Microbial Communities in the Labrador Sea. *Applied and Environmental Microbiology* 87(20):e00800–21. http://doi.org/10.1128/AEM.00800-21.

National Institute of Aeronautics and Space. 2022. *Oil Spill in East Kalimantan on 02 April 2018 Indonesia*. https://spbn.pusfatja.lapan.go.id/layers/geonode%3Aosd_20180402_0450_s1a. Accessed April 30th, 2022.

National Science Foundation. 2022. *Researchers Studied the Effects of the Oil Spill on Life in Gulf Waters and Sediments*. Photographed by Roy McKay, University of Georgia. https://nsf.gov/news/mmg/mmg_disp.jsp?med_id=72299&from= (public domain)

Necci, A., Tarantola, S., Vamanu, B., Krausmann, E., and Ponte, L. 2019. Lessons Learned from Offshore Oil and Gas Incidents in the Arctic and Other Ice-Prone Seas. *Ocean Engineering* 185:12–26. http://doi.org/10.1016/j.oceaneng.2019.05.021.

Nõlvak, H., Dang, N.P., Truu, M., Peeb, A., Tiirik, K., O'Sadnick, M., and Truu, J. 2021. Microbial Community Dynamics during Biodegradation of Crude Oil and Its Response to Biostimulation in Svalbard Seawater at Low Temperature. *Microorganisms* 9(12):2425. http://doi.org/10.3390/microorganisms9122425.

Olufemi, B.A., Jimoda, L.A., and Agbodike, N.F. 2014. Adsorption of Crude Oil Using Meshed Corncobs. *Asian Journal of Applied Science and Engineering* 3(1):63–75.

Olufemi, B.A., and Otolorin, F. 2017. Comparative Adsorption of Crude Oil Using Mango (*Mangifera indica*) Shell and Mango Shell Activated Carbon. *Environmental Engineering Research* 22(4):384–392. http://doi.org/10.4491/eer.2017.011.

Panda, S.K., Kar, R.N., and Panda, C.R. 2013. Isolation and Identification of Petroleum Hydrocarbon Degrading Microorganisms from Oil Contaminated Environment. *International Journal of Environmental Sciences* 3(5):1314–1321. http://doi.org/10.6088/ijes.2013030500001.

Pashaei, R., Gholizadeh, M., Iran, K.J., and Hanifi, A. 2015. The Effects of Oil Spills on Ecosystem at the Persian Gulf. *International Journal of Review in Life Sciences* 5(3):82–89. http://doi.org/10.13140/RG.2.1.2239.3684.

Patton, J.S., Rigler, M.W., Boehm, P.D., and Fiest, D.L. 1981. Ixtoc 1 Oil Spill: Flaking of Surface Mousse in the Gulf of Mexico. *Nature* 290(5803):235–238.

Perdigão, R., Almeida, C., Magalhães, C., Ramos, S., Carolas, A.L., Ferreira, B.S., Carvalho, M.F., and Mucha, A.P. 2021. Bioremediation of Petroleum Hydrocarbons in Seawater: Prospects of Using Lyophilized Native Hydrocarbon-Degrading Bacteria. *Microorganisms* 9(11):2285. http://doi.org/10.3390/microorganisms9112285.

Prabhakaran, P., Sureshbabu, A., Rajakumar, S., and Ayyasamy, P.M. 2014. Bioremediation of Crude Oil in Synthetic Mineral Salts Medium Enriched with Aerobic Bacterial Consortium. *International Journal of Innovative Research in Science, Engineering and Technology* 3(2):9236–9242.

Purwanti, I.F., Abdullah, S.R.S., Hamzah, A., Idris, M., Basri, H., Mukhlisin, M., and Latif, M.T. 2015. Biodegradation of Diesel by Bacteria Isolated from *Scirpus mucronatus* Rhizosphere in Diesel-Contaminated Sand. *Journal of Computational and Theoretical Nanoscience* 21(2):140–143. http://doi.org/10.1166/asl.2015.5843.

Raj, K.G., and Joy, P.A. 2015. Coconut Shell Based Activated Carbon–Iron Oxide Magnetic Nanocomposite for Fast and Efficient Removal of Oil Spills. *Journal of Environmental Chemical Engineering* 3(3):2068–2075. http://doi.org/10.1016/j.jece.2015.04.028.

Razavi, Z., Mirghaffari, N., and Rezaei, B. 2015. Performance Comparison of Raw and Thermal Modified Rice Husks for Decontamination of Oil Polluted Water. *CLEAN—Soil, Water, Water* 43(2):182–190. http://doi.org/10.1002/clen.201300753.

Rogowska, J., and Namiesnik, J. 2010. Environmental Implications of Oil Spills from Shipping Accidents. *Reviews of Environmental Contamination and Toxicology* 206:95–114. http://doi.org/10.1007/978-1-4419-6260-7_5.

Ruberg, E.J., Elliott, J.E., and Williams, T.D. 2021a. Review of Petroleum Toxicity and Identifying Common Endpoints for Future Research on Diluted Bitumen Toxicity in Marine Mammals. *Ecotoxicology* 30(4):537–551. http://doi.org/10.1007/s10646-021-02373-x.

Ruberg, E.J., Williams, T.D., and Elliott, J.E. 2021b. Review of Petroleum Toxicity in Marine Reptiles. *Ecotoxicology* 30(4):525–536. http://doi.org/10.1007/s10646-021-02359-9.

Saadoun, I.M. 2015. Impact of Oil Spills on Marine Life. In M.L. Larramendy and S. Soloneski (Eds.), *Emerging Pollutants in the Environment—Current and Further Implications*. London: IntechOpen. http://doi.org/10.5772/60455.

Sakthipriya, N., Doble, M., and Sangwai, J.S. 2015. Bioremediation of Coastal and Marine Pollution due to Crude Oil Using a Microorganism *Bacillus subtilis*. *Procedia Engineering* 116(2015):213–220. http://doi.org/10.1016/j.proeng.2015.08.284.

Saravana, K.P., and Amruta, S.R. 2013. Analysis of Biodegradation Pathway of Crude Oil by *Pseudomonas* sp. Isolated from Marine Water Sample. *Archives of Applied Science Research* 5(4):165–171.

Shigenaka, G. 2014. *Twenty-Five Years After the Exxon Valdez Oil Spill: NOAA's Scientific Support, Monitoring, and Research*. Seattle: NOAA Office of Response and Restoration, 78 pp.

Sulistyono, S. 2013. Dampak Tumpahan Minyak (Oil Spill) di Perairan Laut pada Kegiatan Industri Migas dan Metode Penanggulangannya (Impact of Oil Spills in Sea Waters on Oil and Gas Industry Activities and Their Countermeasures Methods). *Swara Patra* 3(1):49–57.

Thavasi, R., Jayalakshmi, S., and Banat, I.M. 2010. Biosurfactants from Marine Bacterial Isolates. In A. Mendez-Vilas (Ed.), *Current Research, Technology and Education Topics in Applied Microbiology and Microbial Biotechnology*. Badajoz, Spain: Formatex.

Titah, H.S., Pratikno, H., Moesriati, A., Imron, M.F., and Putera, R.I. 2018. Isolation and Screening of Diesel Degrading Bacteria from Ship Dismantling Facility at Tanjungjati, Madura, Indonesia. *Journal of Engineering and Technological Sciences* 50(1):99–109. http://doi.org/10.5614/j.eng.technol.sci.2018.50.1.7.

Tomasino, M.P., Aparício, M., Ribeiro, I., Santos, F., Caetano, M., Almeida, C., de Fátima Carvalho, M., and Mucha, A.P. 2021. Diversity and Hydrocarbon-Degrading Potential of Deep-Sea Microbial Community from the Mid-Atlantic Ridge, South of the Azores (North Atlantic Ocean). *Microorganisms* 9(11):2389. http://doi.org/10.3390/microorganisms9112389.

US Coast Guard. 2010. *Vessels of Opportunity Pull Oil into a Fire Boom in a Controlled Burn with a Second Controlled Burn Visible in the Distance June 17, 2010*. Photographed by Chief Petty Officer Bob Laura. https://nara.getarchive.net/media/office-of-the-administrator-lisa-p-jackson-various-images-bp-oil-spill-gulf-8aa51d (creative commons licences).

USEPA. 1992. *Uzbekistan Oil Well Release: EPA After-Action Report*. National Service Center for Environmental Publications (NSCEP).

USEPA. 1999. Understanding Oil Spills and Oil Spill Response. *Office of Emergency and Remedial Response, EPA 540-K-99–007*. United States Environmental Protection Agency.

USEPA. 2022. https://www3.epa.gov/pesticides/endanger/litstatus/effects/redleg-frog/naled/appendix-i.pdf. Accessed April 24th, 2022.

US National Research Council (USNRC). 2003. *Committee on Oil in the Sea: Inputs, Fates, and Effects. Oil in the Sea III: Inputs, Fates, and Effects*. Washington, DC: National Academies Press. http://doi.org/10.17226/10388.

Uzoije, A.P., Onunkwo-A, A., and Egwuonwu, N. 2011. Crude Oil Sorption onto Groundnut Shell Activated Carbon: Kinetic and Isotherm Studies. *Research Journal of Environmental and Earth Sciences* 3(5):555–563.

Walton, W.D., and Jason, N.H. 1999. *In situ* Burning of Oil Spills. *Workshop Proceeding November 2–4, 1998*, National Institute of Standards and Technology, New Orleans, LA.

Wardhani, W.K., and Titah, H.S. 2020. Studi Literatur Alternatif Penanganan Tumpahan Minyak Mentah Menggunakan *Bacillus subtilis* dan *Pseudomonas putida* (Studi Kasus: Tumpahan Minyak Mentah Sumur YYA-1). *Jurnal Teknik ITS* 9(2):97–102.

Young, J.W., Skewes, T.D., Lyne, V.D., Hook, S.E., Revill, A.T., Condie, S.A., Newman, S.J., Wakefield, C.B., and Molony, B.W. 2011. *A Review of the Fisheries Potentially Affected by the Montara Oil Spill off Northwest Australia and Potential Toxicological Effects. Montara Well Release Scientific Monitoring Programme Study S4B*. CSIRO Marine and Atmospheric Research. Tasmania. ISBN 9780643107281.

Yuewen, D., and Adzigbli, L. 2018. Assessing the Impact of Oil Spills on Marine Organisms. *Journal of Oceanography and Marine Research* 6(1):1–7. http://doi.org/10.4172/2572-3103.1000179.

16 Extensive Studies on Fermentative Production of Biosurfactants from Extremophilic Marine Microbes

Nabya Nehal, Sonal, Sapna Jaiswar and Priyanka Singh

CONTENTS

DOI: 10.1201/9781003307464-16

16.1 INTRODUCTION

Surfactants such as amphiphilic moieties reduce interfacial tension at phase interfaces, emulsify oil–water mixtures, and form both stable gels and foams. They display physiochemical properties like phase dispersion, solubilization, lubrication, and stabilizing and foaming capacity. Many microbes like fungi, bacteria, and yeast have shown efficiency for production of biosurfactants (BSs) with a variety of molecular structures. In the bacterial domain, *Pseudomonas aeruginosa*, *Bacillus subtilis*, and *Acinetobacter calcoaceticus* have been reported as excellent producers of BSs. Globally, BSs receive attention over chemical surfactants because of several advantages such as low toxicity, high bio-degradability, cost-effectiveness, and availability of raw materials (cassava flour, sunflower oil, waste refinery oil, etc.) (Das and Kumar, 2018). A global marketing survey showed that in 2017, the global market of BSs was valued at USD 32.120 million. A rising demand for usage of surfactants in cosmetic (personal care) (Papageorgiou et al., 2010; Kefala et al., 2011) products, cleaning products, and agriculture (organosilicon surfactants) products drives the market. It has been forecast that 2018 to 2026 will be the most active growth period for marketing. In 2018, the global market of BSs was valued at USD 4.70 billion and is expected to rise by USD 7.25 billion by 2026. Per the demand for amphoteric surfactants, the compound annual growth rate has been raised to around 7%, and for the forecast period between 2018 and 2023, the demand of surfactants is projected to rise from USD 3.52 billion to USD 4.94 billion (Prashanthi et al., 2017). BASF Cognis, Ecover (Belgium), Urumqi Unite Bio Technical Co. Ltd (China), MG Intobio Co. Ltd (South Korea), Saraya Co Ltd (Japan), AkzoNobel; (Netherlands), Mitsubishi Chemical Holdings (Japan), Croda International (UK), Evonik Industries (Germany), and Chemtura Corporation (US) are the top international suppliers of BSs.

BSs also drew attention in the market as a biological hydrocarbon degrader with several advantages over synthetic surfactants (sulfonates, carboxylates, and sulfate esters) and are widely used in several industrial sectors such as biomedical and pharmaceuticals, food, agriculture, and oil and petroleum industries (Saenz-Marta et al., 2015; Shete et al., 2006; Varvaresou and Iakovou, 2015).

Some microbial BSs have the ability to degrade plastic items efficiently by acting as wetting agents, emulsifiers, solubilizers, and antistatic agents (Kaur, 2012). BSs are extracellularly produced micellar compound (Kitamoto et al., 2002; Marchant and Banat, 2012a). BSs are secondary metabolites produced during the stationary phase by wide range of microbes. The large-scale of production of BSs has major limitations of high-cost purification strategy and pathogenicity of some biosurfactant-producing strains (Irorere et al., 2017). Therefore, the search for non-pathogenic strains remains important to improve production (Elshikh et al., 2017; Funston et al., 2016). Marine organisms are known to metabolically survive under extreme temperature, pressure, pH, and salinity conditions (Das et al., 2010; Thavasi et al., 2014). BSs produced from marine bacteria can facilitate hydrocarbon dispersion, emulsification, and biodegradation (Tripathi et al., 2018; Mapelli and Scoma, 2017; Das et al., 2010). BSs from psychrophilic marine microbes showed higher stability against freezing temperatures and are therefore, preferred in the formulation of laundry detergent to process washing at low temperatures for energy conservation (Perfumo et al., 2018; Marchant and Banat, 2012b). Currently, marine microbes are in great demand for the production of new biosurfactant molecules with distinctive properties (Tripathi et al., 2018; Irorere et al., 2017; Uzoigwe et al., 2015). In this study, the physiochemical properties and molecular-level characterization of different microbes for the production of diversified BSs are discussed. The mechanism of action of different BSs against environmental contaminants and their specific roles in various industrial sectors are subsequently discussed.

16.2   LITERATURE REVIEW

BSs are surface-active amphipathic compounds and have been found to be significantly effective as compared to synthetic surfactants (Datta, 2011). High growth and awareness of eco-friendly, less toxic emulsifiers and biodegradable and bio-based byproducts support the demand of microbial BSs in the global market (Volkering et al., 1997). Marketers have an opportunity to make their place in the market utilization of agro-industrial wastes and usage of BSs in unconventional markets. The market for BSs is mainly distributed in Asia Pacific, Europe, the Middle East, North America, Latin America, and Africa. In 2017, Europe had the largest market for BSs and is expected to lead the global market during the forecast period. The European market for BSs is largely driven by increasing consciousness among buyers about protection of the environment from hazardous toxic chemicals. Ecover (Belgium), Evonik, Biotensidon (Germany), and Jeneil Biotech (US) are some BS product-dealing operators. Ecover and BASF Cognis are the top BS suppliers who have ventured into the market. Meanwhile, it has been observed that rhamnolipids are in demand because of their functionality and purity level, which help increase interest (Gharaei-Fathabad, 2011; Varvaresou and Iakovou, 2015). BSs can serve as indispensable components, demulsifiers-emulsifiers, and foaming and wetting agents, as shown in Table 16.1. BSs have important output for versatile industrial applications (Makkar et al., 2011), including petrochemicals, organic chemicals, mining, petroleum, metallurgy (mainly bio-leaching), agrochemicals, foods, fertilizers, beverages, pharmaceuticals, and cosmetics (Vaz et al., 2012; Geys et al., 2014; Rebello et al., 2014). Their ability to reduce interfacial surface tension plays a beneficial role in bioremediation (heavy metals) and oil recovery (crude oils) (Volkering et al., 1997). The cost of production of BSs has been reduced by the fermentative approach, with low-cost media components using waste products from the agricultural, food, and oil industries like waste oil (olive and cooking) (Badrul et al., 2019). It is also important to select microbial organisms with the ability to optimize large-scale fermentation and high-yield production and recovery processes. Aquatic ecosystems are microbially rich ecosystems with the potential to produce BSs.

16.2.1 GENERAL CLASSIFICATION AND CHEMICAL NATURE OF BIOSURFACTANTS

In 1999, Rosenberg and Ron classified biologically synthesized surfactants on the basis of their origin, molecular mass, and chemical composition. BSs are

TABLE 16.1
List of Biosurfactants with Their Applications in Different Industrial Sectors

Biosurfactant	Micro-Organisms	Application	References
Lipoproteins	*Bacillus subtilis* *B. licheniformis*	Degradation of toxins (heavy metals) from contaminated water and soil sediments; antimicrobial activity against profound mycosis; chemo repellent	Ahimou et al., 2000
Glycolipids	*Torulopsis bombicola* *T. apicola* *T. petrophilum* *Mycobacterium* *Nocardia* *Corynebacterium* *Pseudomonas aeruginosa* *Burkholderia plantarii* *B.glumae* *B.thailandensis*	Bioremediation (degradation of heavy metals from sediments of soil)	Franzetti et al., 2010
Phospholipids	*Rhodococcus erythropolis* *Micrococcus* sp. *Acinetobacter* sp.	Enhance the tolerance capacity of bacteria to heavy metals	Aparna et al., 2011
Polymeric biosurfactants	*Saccharomyces cerevisiae* *Candida lipolytica*	Stabilizes hydrocarbon in water emulsion; bioemulsifier	Kaplan et al., 1987
Particulate biosurfactants	*Acinetobacter* sp. *Cyanobacteria*	Bio-degradation of hydrocarbons	Rosenberg and Ron, 1999

categorized into low molecular mass (LMM), with poor levels of interfacial tension, and high molecular mass (HMM), more efficient emulsion agents. In the global market, the main classes of low-mass surfactants are lipopeptides, glycolipids, and phospholipids, whereas high molecular mass BSs include polymeric and particulate surfactants (Mukherjee et al., 2006). Microbial surfactants (MSs) are versatile molecules with unique physiochemical properties like antibiotics, peptides, lipopeptide, fatty acids, glycolipids, phospholipids, and amino acids (Rosenberg et al., 1999). Nitschke et al. (2011) classified BSs into four categories according to their hydrophilic moiety:

- Ionic
- Non-ionic
- Amphoteric (peptides, saccharides)
- Others (organosilicon surfactant)

Ionic surfactants are subdivided into anionic and cationic surfactants. Anionic surfactants are negatively charged (head), increase cleaning efficiency, and tends toward positively charged particles and soil particles as well as carpet fibers. Anionic surfactants are prominently used to clean dirt, stains, and clay (Rosen et al., 1989). Sulfonic acid salts, ammonium laureth sulfate, carboxylic acid salts, alkyl benzene sulfonates, sodium lauryl sarcosinate, alcohol sulfates, α olefin sulfonate, and phosphoric acid are some example of anionic surfactants (O'lenick, 2011; Williams, 2007). Globally, anionic and cationic surfactants are the most widely used surfactants. Magnesium and calcium are used in surfactants as sequestrants because hard water ions may interact with anionic surfactants. The hardness of water can deactivate the reaction of surfactants. Cationic surfactants are positively charged (head); they are commonly used in softening of fabrics and detergents (Rhein, 2007). In laundry, cationic-based detergents improve the packing of negatively charged molecules at the water interface or for greasy stains (St Laurent et al., 2007). Non-ionic surfactants are the most promising and are most prominently used as a pesticide in agricultural sectors. They are less hard compared to other surfactants because they possess no electric charge. The use of amphoteric surfactants in various applications may increase the demand for surfactants in the global market for use in personal care (hair, home, and body care products). They have also been used in several industrial sectors for cleaning purposes, as oil field chemicals, and others (Yuan et al., 2014). Amine oxide, betaine (alkyl amido betaine, alkyl betaine and others), sultaines, amphoacetates, and amphopropionates are some examples of amphoteric surfactants. Organosilicon is a non-ionic surfactant used as an adjuvant pesticide. The hydrophobic domain is usually composed of fatty acids (saturated, unsaturated, linear, hydroxylated, or branched) (Desai et al., 1997).

These BSs have been also categorized into two major classes on the basis of their molecular weight.

16.2.2 Biosurfactants with Low Molecular Weight

16.2.2.1 Lipoproteins/Lipopeptides

Lipopeptides are a group of low molecular mass surfactants derived from amino acids. Lipopeptides and lipids are linked together with polypeptide bonds. These cyclic peptides are produced by various classes of micro-organisms, including *Lactobacillus*, *Pseudomonas*, *Streptomyces*, *Serratia*, and *Bacillus* (Santos et al., 2016). Most lipopeptide-producing bacteria show reduction of interfacial surface tension and have significant bioactivity, including antibiotics of polymyxins B, daptomycin, and gramicidins. Surfactin and lichensin are produced by distinct species of *Bacillus*. *Pseudomonas syringae*, a plant bacterium, produced syringomycin E, composed of a hydrophobic head (3-hydroxydecanoic acid) and hydrophilic tail (amino acid residues) (Sorensen et al., 1996).

16.2.2.2 Glycolipids

Glycolipid BSs are commonly considered sugar-containing lipids in which both moieties may be linked either with a glycosidic functional group, as in the sophorose, rhamnose, and cellobiose lipids, or via acylation, as in the acyl polyols, trehalose lipids, and sugar mycolates. Rhamnolipids, trehalose lipids, sophorolipid, and mannosyl erythritol lipids are some of the best-known classified glycolipids (Li et al., 1984).

Rhamnolipids were first discovered and isolated from *Pseudomonas* sp. in 1949. They are amphiphilic compounds (anionic molecules), with hydrophilic moieties (sugar heads) and lipophilic moieties (tails). They are composed of mono- or di-L-rhamnose units, commonly linked to two units of 3-hydroxy fatty acid moieties across glycosidic linkage (Figure 16.1). Because of this, they have varying congeners, with carbon atoms 8 and 16 (Li et al., 1984). The Critical Micelle Concentration (CMC) range of rhamnolipids is between 20 and 250mg/L, while the surface and interfacial tension may be between 25–30 and 1–4 mN/m (Jauregi et al., 2019).

Trehalose lipids (Figure 16.1) are non-reducing disaccharides in which two glucose molecules are coupled in an α, α-1,1-glycosidic linkage (Stick and Williams, 2010).

Sophorolipids consist of 17-hydroxyl oleic acids and two units of saccharide of sophorose (Figure 16.1). *Torulopsis bombicola* and *Starmerella bombicola* are reported to produce sophorolipids with the ability to reduce interfacial surface tension (Kubicki et al., 2019).

16.2.2.3 Phospholipids and Neutral Lipids

Phospholipids are ubiquitous components of all living microorganisms, as they form the major constituent of all cell membranes—both of the outer cell envelope and of many intracellular structures. Phospholipids form the bilayer membrane, applied for various uses such as dispersion, solubilization, and emulsification. Phosphatidylcholine was the first discovered phospholipid (Da Costa et al., 2003).

16.2.3 High Molecular Weight Biosurfactants

16.2.3.1 Polymeric Biosurfactants

Liposan, emulsan, alasan, polysaccharide-proteins, and lipomanan are some of the best-known polymeric BSs. Polymeric BSs are composed of fatty acids with three to four repeating sugar molecules, as shown in Figure 16.1. Liposan has the ability to emulsify oils (edible) and reduce surface tension (Rosenberg et al.,1999). Emulsan (MW = 1000 KD) is an acylatedpolysacharide complex, extracellularly synthesized by the Gram-negative bacterial strain *Acinetobacter*

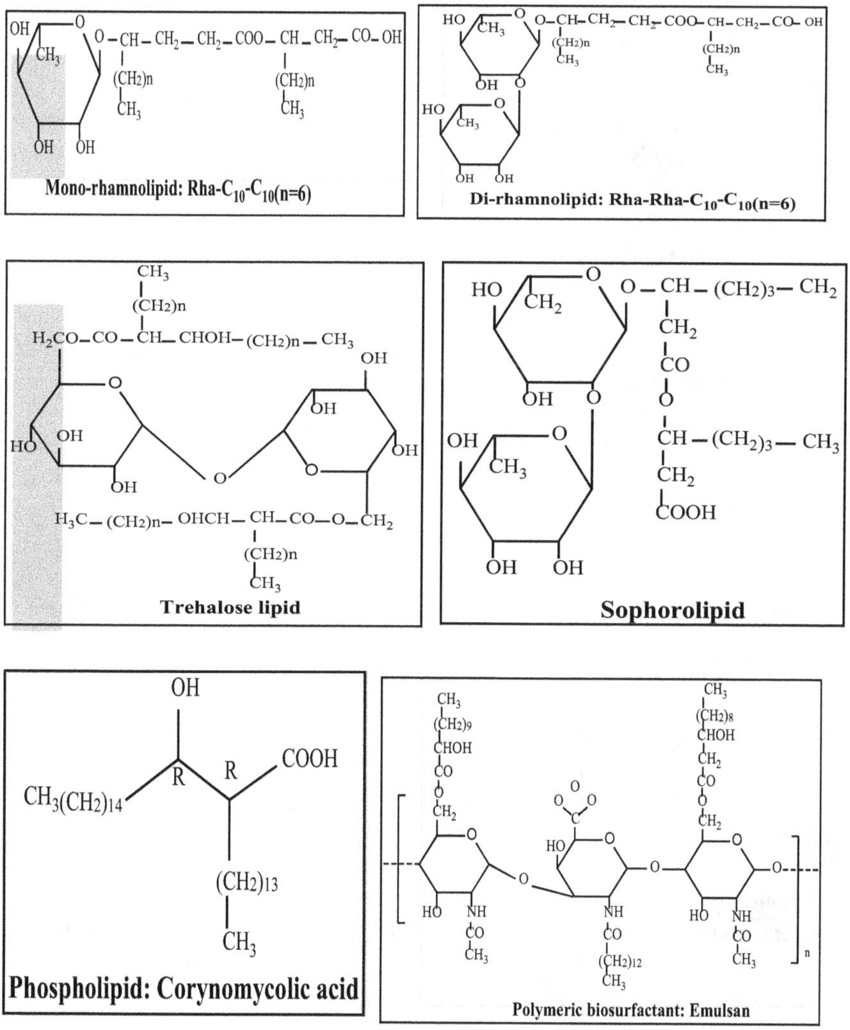

FIGURE 16.1 Structural composition of surface-active compounds of rhamnolipids, trehalose lipids, sophorolipids, phospholipids (corynomycolic acid), and polymeric surfactants (emulsan) (Desai et al., 1997).

calcoaceticus (Kim et al., 1997) and has been usually applied in biotechnological industries as a good emulsifier (Gorkovenko et al., 1999). First, emulsan appears on the surface of the cell wall and then binds to hydrocarbon moieties.

16.2.3.2 Particulate Biosurfactants

Microbial cells and vesicles (extracellular) are types of particulate BSs (Kappeli, 1979). Micro-emulsions are formed by partitioned hydrocarbon; microbial cells are responsible for uptake of hydrocarbons. In polymeric BSs, microbial cells act as BSs (Vijayakumar and Saravanan, 2015).

16.2.4 PHYSIOCHEMICAL PROPERTIES

The effectiveness and efficiency of interfacial and surface tension are keys to a good biosurfactant. Effectiveness is measured by interfacial and surface tension, whereas efficiency is related to CMC.

16.2.4.1 Surface and Interfacial Tension

Surface and interfacial tension are intermolecular. The upper layers of water are in tension (miscibility of water and oil), known as surface tension. A force applied between amphiphilic compounds reduces the tension between two immiscible liquids. The reduction rate of interfacial and surface tension helps to stabilize emulsions. Surfactants are amphiphilic molecules; they can reduce intermolecular forces that help in the replacement of oil/water, water/air, or liquid/liquid interfaces (Jahan et al., 2020). The surface tension of BSs follows the du Noüy ring method and is measured by tensiometer, whereas manually it has been based on the drop count principle using s stalagmometer.

16.2.4.2 Critical Micelle Concentration

BSs are amphiphilic compounds for bilayers, micelles, or vesicles. Surface tension and BS concentration are inversely proportional to each other. As the surface tension of a solution drops, the concentration of BS increases up to a critical level, and at that point, surfactants start forming micellar-like structures. Micelles are measured in nanometer; that is, the radius of micelles is very small. They are thermodynamically stable with a net charge of zero (equilibrium) (Jahan et al., 2020). The size and shape of micelles are directly proportional to the concentration and dilution, pressure, salt concentration, pH, and temperature of the biosurfactant. Covalent bond (Van der Waals force, hydrophobic and hydrogen bonding) interactions enhance the self-assembly tendency of BSs. The final concentration point of a surfactant where there is no more reduction in surface tension is the CMC. CMC can be calculated by plotting surface tension on the Y-axis and the concentration of the standard BS on the X-axis. The CMC ranges of rhamnolipids are reported to be between 1 and 200 mg/l (Nitschte et al., 2011). The CMC value of mono-rhamnolipids is lower than that of di-rhamnolipids (36%) because of a higher repulsion force (electrostatic) that occurs between hydrophilic moieties of di-rhamnolipids (30). The CMC values of *Starmerella bombicola* and *Candida batistae* are 95 and 138 mg/l, respectively. They are both sophorolipid-producing microbes and have different structure and fatty acid composition, due to which *Candida batistae* is highly hydrophobic and soluble in water (Konishi et al., 2008).

16.2.4.3 Emulsification

Emulsification is a process in which large droplets break into smaller ones. It is the mixture of an immiscible liquid such as oil in water or water in oil (where oil is a non-continuous phase and water is a medium, and vice versa). Emulsions are thermodynamically unstable. Coalescence (bumping of droplets tends to the formation of bigger droplets), soluble components (kerosene, crude oil), flocculation (an attractive force applied in between droplets), and creaming (a creamy layer of droplets occurs on the topmost surface of the emulsion) can affect the stability of an emulsion. Biosurfactants/surfactants can ensure an emulsion is stable. Emulsions can be characterized based on viscosity (stable emulsions are higher in comparison to unstable emulsions) and color (unstable emulsions are black, and stable emulsions are solid reddish). The size of emulsions varies from the nano to micron scale, and they are round or spherical (Chang, 2016). Fats can be emulsified either by homogenization or by adding an agent such as soap, detergent, or gum. Neethirajan and Jayas (2011) demonstrated that emulsification has a unique property that may lead to the formation of nano-capsulation, alginate, and chitosan used as a wall material on *Curcuma longa* and *Cymbopogan citratus*. It is not easy to ignite stable crude oil emulsions because water consumes a higher amount of energy for ignition, and the vaporization of an oil emulsion needs an additional energy source. The emulsification activity of natural surfactants for paraffin, toluene, lubricant oil, and palm oil is higher than that of toxic liquids (Tween 80). Toluene showed the best emulsification activity at 0.60 (Lima et al., 2009). Initially, the production of low molecular weight BSs was by bacterial strains (rhamnolipid from *Pseudomonas aeruginosa* and surfactin from *Bacillus subtilis*). Rhamnolipids exhibited high emulsifying activity (E24) against water and vegetable oil (Saharan et al., 2011), whereas surfactin showed high emulsifying activity against soya bean and kerosene oil.

16.2.5 Sources for Production of Biosurfactants

Most microbial surfactants are hydrocarbon degraders (Volkering et al., 1997; Willumsen et al., 1996). BSs have been extracellularly produced by *Pseudomonas aeruginosa*, *Rhodococcus erythropolis*, *Serratia rubidea*, *Arthrobacter* sp., *Mycobacterium* sp., *Nocardia erythropolis*, *Corynebacterium* sp., and *Pseudomonas chlororaphis* (Desai et al., 1997). Rhamnolipid BSs are produced by various species of *Pseudomonas* (Guerra-Santos et al., 1986; Burger et al., 1963), and different species of *Torulopsis* can produce sophorolipid BSs (Cooper and Paddock, 1984). Some fungi like *Aspergillus ustus*, *Candida* sp., and *Trichosporon ashii* are also reported to produce BSs (Alej et al., 2011; Onur et al., 2015; Konishi et al., 2008; Sarubbo et al., 2007). Among yeasts, *Candida lipolytica* produces a periplasmic glycolipid biosurfactant (Rufino et al., 2007). Some marine microbes isolated from hydrocarbon-polluted environments have been induced to produce BSs under extreme conditions (Table 16.2). These microbes showed the production of BSs with emulsification activities and facilitated the degradation of hydrocarbon to a greater extent (Thavasi et al., 2014). Microbes like *Alteromonas, Halomonas, Alcanivorax, Colwellia, Cycloclasticus,* and *Pseudoalteromonas* are being isolated for marine sites of oil spill for the production of highly efficient BSs under extreme environmental conditions (Mapelli et al., 2017; Tripathi et al., 2018). Some bacteria, like *Pseudomonas, Bacillus,* and *Acinetobacter* isolated from oil-contaminated waters also show secretion of BSs (Ortega-de la Rosa et al., 2018; Hentati et al., 2016; Gerard et al., 1997). The addition of inorganic components increased the populations of some marine bacteria like *A. borkumensis* by inducing them to consume hydrocarbons as a media component and energy source. The cell population density of *Alcanivorax* has increased up to 91% in zones of oil contaminant seawater within 10–15 days (Tripathi et al., 2018; Syutsubo et al., 2001, Roling et al., 2002; Roling et al., 2004).

16.2.6 MODE OF ACTION OF BIOSURFACTANTS

Solubilization, biodegradation, emulsification, or mobilization are some factors that help increase the degradation of hydrocarbons by BSs. Solubilization is a factor that can enhance the dispersion of hydrophobic components in the aqueous phase of surfactants. In this process, the solubilization of hydrocarbons directly depends on the concentration of BS (Pacwa-Plociniczak et al., 2011).

TABLE 16.2
List of Marine Microbes for Production of Different Types of Biosurfactants

Marine Microbes	Types of Biosurfactant	References
Brevibacterium casei MSA19, *Staphylococcus saprophyticus* SBPS15, *Serratia marcescens, Streptomyces* sp. B3, *Streptomyces* sp. MAB36, *Aspergillus ustus* MSF3, *Halomonas* sp., MB-30	Glycolipid	Kiran et al., 2010a, 2010b; Mani et al., 2016; Hamza et al., 2017; Dhasayan et al., 2014
Bacillus circulans DMS-2, *Bacillus licheniformis* NIOT-AMKV06, *Brevibacterium aureum* MSA13, *Brevibacterium luteolum, Nocardiopsis alba* MSA10, *Bacillus subtilis* SDNS, *bacillus mojavensis* B0621A, *Bacillus megaterium, Brevibacillus Laterosporus* PNG-276, *Achrobacter* sp. HZ01	Lipopeptide	Kiran et al., 2017; Deng et al., 2016; Vilela et al., 2014; Sivapathasekaran et al., 2010; Das et al., 2009a; Mukherjee et al., 2009; Ma and Hu, 2014; Dey et al., 2015; Balan et al., 2017; Desjardine et al., 2007
Nocardiopsis VITSISB (KC958579)	Rhamnolipid	Roy et al., 2015

Mobilization can reduce the concentrations of CMC and can be distinct in dispersion and displacement. In the displacement process, interfacial tension is reduced, due to which hydrocarbons are released from the medium (porous) (Santos et al., 2016). Interfacial tension is directly proportional to hydrocarbon, and the two liquid phase (aqueous and oil) reduction may help hydrocarbons reduce to overcome the forces of capillarity. There are no emulsion formations. Dispersion of hydrocarbons in the aqueous phase (as emulsions) is termed the dispersion process. Generally, emulsions are thermodynamically unstable, but because of kinetic restrictions, they may remain stable for some period of time. There is no co-relation between dispersion and displacement; they differ from each other in relation to interfacial surface tension, BS concentration, and emulsion formation (Bai et al., 1997). The mechanism of action of BSs on contaminated soil sediments is shown in Figure 16.2. Ionic strength and the pH of the solution are the main factors on which abolition of hydrophobic compounds depend and that can change the sorption of soil (surfactant) and aggregated micelles, which restrains the movement of the hydrocarbon by BS. Most of BSs have been proposed for the removal of oil-derived products from contaminated water and soil. Rhamnolipids have been successfully used in biotechnological decontamination processes (Santos et al., 2016). Most surfactants produced by microbial strains such as *Bacillus, Candida* (Bezerra et al., 2013), and *Pseudomonas* species (Vijayakumar and Saravanan, 2015) have been used in bioremediation of soil.

Some marine microbes produce different types of glycolipids and glycoproteins with superior emulsifying activity in saline conditions. These halophilic microbes play a vital role in the secretion of surfactants and emulsifiers in oil-contaminated saline environments (Pepi et al., 2005; Tan et al., 2011). Metagenomic approaches based on sequence analysis and functional characterization have been efficiently applied to screen biosurfactant-secreting microbial sources from a huge number of microbes in marine water (Kennedy et al., 2011). The functional characterization includes screening of metagenomics libraries constructed from marine ecosystems to explore genes encoding the secretion of BSs with *Escherichia coli* as a heterologous host. The sequence-based approach involves the identification of coding regions on the basis of homology sequencing in genomic databases (BLAST, MEGAN, KEGG).

16.2.6.1 Biosynthesis of L-Rhamnose

Rhamnolipid is one of the most studied members of class of glycolipids (Thakur et al., 2021). Variants microbial strains and nutritional sources (C- and N-sources) may affect the biosynthesis of the rhamnolipid deoxythymidine (Figure 16.3). Diphosphate-L-rhamnoses are the main precursors that can be synthesized by *Pseudomonas aeruginosa* (Dobler et al., 2016). Glucose is the main factor that can synthesize rhamnolipid moieties. In the gluconeogenesis pathway, fructose 6 phosphate can be directly synthesized by glucose-6-phosphate. Alg C plays an important role in the biosynthesis of alginate by the catalysis process. Alg C converts glucose-6-phosphate to glucose-1-phosphate, and the further four sequential enzymatic conversions generate dTDP-L-rhamnose by catalyzing the Rml A,

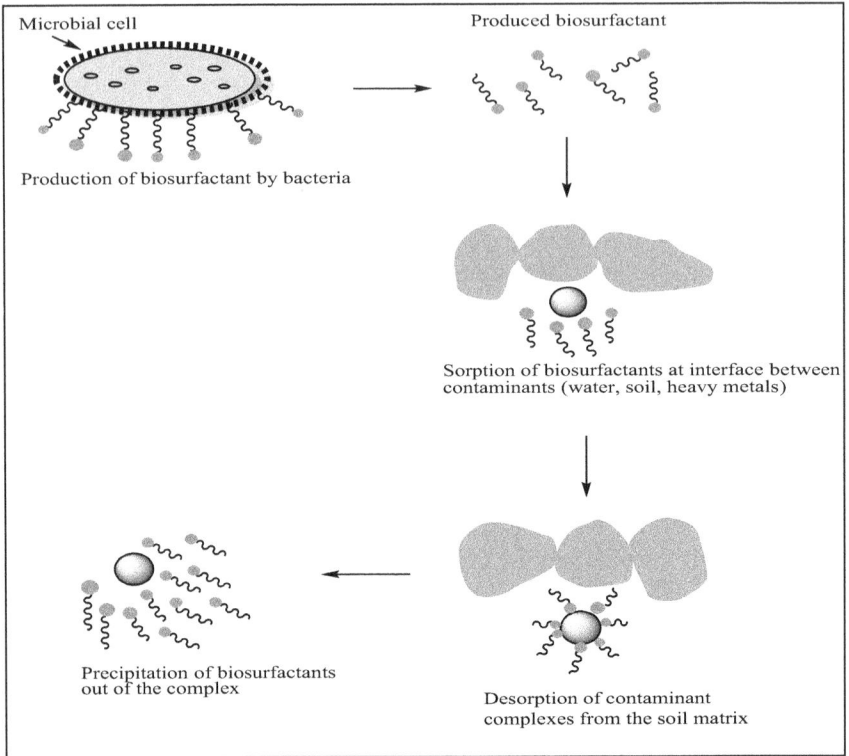

FIGURE 16.2 Mode of action of microbial biosurfactant against contaminants in soil sample.

B, C, and D enzymes. Rhamnosyltransferase (Rhl) enzyme acts as precursor to convert L-rhamnose to mono- and di-rhamnolipids, as shown in Figure 16.3. It has been found that *Pseudomonas aeruginosa* is regulated by a quorum-sensing (cell-cell communication) system. Rhl, Las, and P are quorum-sensing systems that have been identified in *P. aeruginosa* (Lee et al., 2015). First, Las (LasI and LasR) was discovered for the regulation and production of alkaline phosphatase, exotoxin A, and elastase.

16.2.6.2 Biosynthesis of Lipopeptides

Iturin, surfactin, and fengycin are some well-known lipopeptides. In 1968, sur-factin was first isolated from a *Bacillus subtilis* culture. It can be synthesized by seven modules. There are three main steps that can synthesize surfactin—initia-tion of synthesis, elongation, and cyclization of the peptide chain. In the initia-tion of surfactin, a carbon source synthesizes into malonyl CoA and amino acids (L-valine, L-leucine, and L-isoleucine) with the help of enzymes. In elongation and cyclization, four sequential conversions generate surfactin by catalyzing the

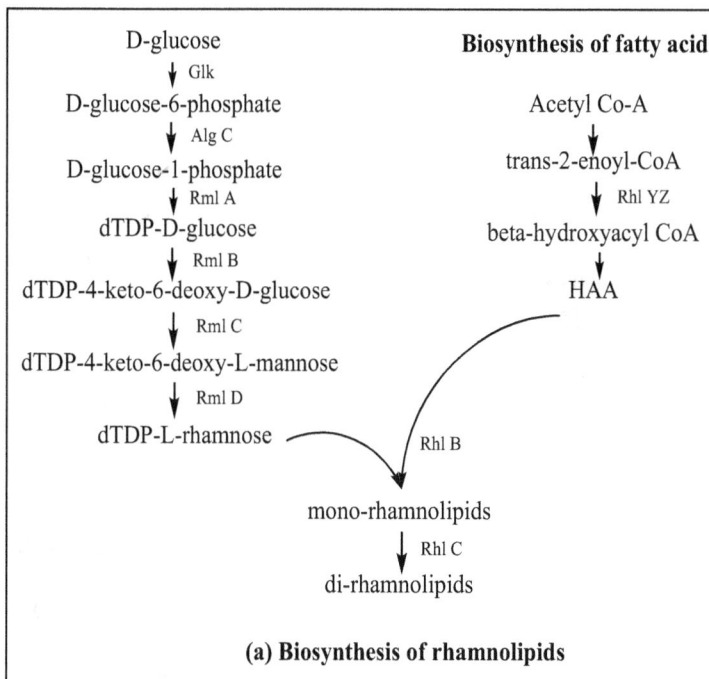

D-glucose   **Biosynthesis of fatty acid**

D-glucose
↓ Glk
D-glucose-6-phosphate   Acetyl Co-A
↓ Alg C   ↓
D-glucose-1-phosphate   trans-2-enoyl-CoA
↓ Rml A   ↓ Rhl YZ
dTDP-D-glucose   beta-hydroxyacyl CoA
↓ Rml B   ↓
dTDP-4-keto-6-deoxy-D-glucose   HAA
↓ Rml C
dTDP-4-keto-6-deoxy-L-mannose
↓ Rml D
dTDP-L-rhamnose ─── Rhl B

mono-rhamnolipids
↓ Rhl C
di-rhamnolipids

(a) Biosynthesis of rhamnolipids

FIGURE 16.3 Metabolic pathway for synthesis of rhamnolipid biosurfactant in microbes.

FabH and FabD enzymes, described in Figure 16.4. Modifications of the regulatory genes of SrfA and overexpression to help efflux surfactin are some strategies used to enhance the production of lipopeptides (Hu et al., 2013).

Enzymatic action of L-rhamnose biosynthesis pathways and surfactin biosynthesis pathways are listed in Table 16.3. phdABCD (pyruvate dehydrogenase), ilvGM (acetohydroxy acid synthase), ilvD (dihydroxy acid dehydrogenase), ilvC (acetohydroxyacidisomeroreductae), accABCD (acetyl CoA carboxylase), leuACDB (leuA, 2-isopropylmate synthase), ilvE (aminotransferase), FabD (malonyl CoA-ACP trans-acylase), and FabH (beta-ketoacyl-ACP synthases) are some enzymes that show action in regulation of surfactin.

16.2.7 CULTURAL FACTORS AFFECTING FERMENTATIVE PRODUCTION OF BIOSURFACTANTS

There are various optimization conditions that can affect the qualitative and quantitative nature of BSs. The production can be either spontaneous or induced. Salinity; pH; dilution rate; temperature; agitation speed; and variation in concentration of metal ions, nitrogen, and carbon are some bio-process optimization cultural conditions on which production is directly dependent (Desai and Banat, 1997). To reduce the costs of biosurfactant production, it is important to select

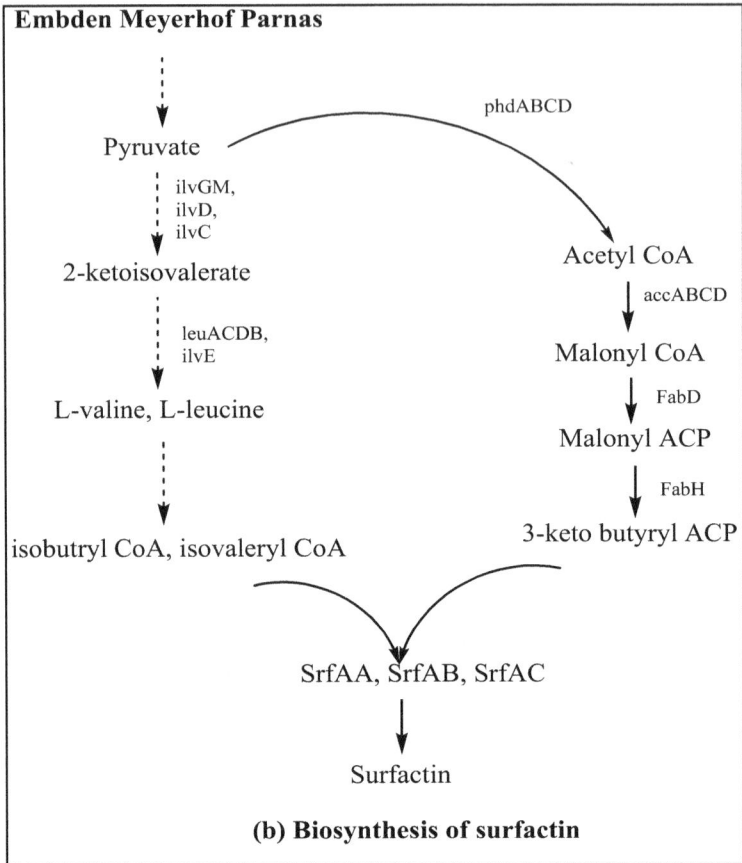

FIGURE 16.4 Schematic representation of biosynthetic pathways of (a) L-rhamnose, (b) surfactin.

microorganisms with the ability to optimize large-scale fermentation and high-yield production and recovery processes. Many bacteria have been reported to produce BSs, except species of *Acinetobacter*. Effective and efficient activity of interfacial and surface tension are keys to a good biosurfactant. Effectiveness is measured by interfacial and surface tension, whereas efficiency is related to CMC. Ranges of CMC of BSs are between 1 and 2000 mg/L, whereas surface and interfacial tensions (oil/water) are in between 1 and 30 N/m. Reduction of interfacial tension of n-hexadecane (hydrocarbon) from 40 to 1 mN/m and a water surface tension range from 72–35 N/m are signs of good production of BS. The fermentative production of BSs is affected by changes in the culture conditions of media components, pH, temperature, and fermentation time (Bhardwaj et al., 2013).

In biosurfactant production, microbes utilize carbon sources as energy for their growth. *Pseudomonas* spp. utilize water soluble C-sources such as glycerol and mannitol for the production of rhamnolipid. In 2001, Sarubbo et al. reported that *Yarrowia lipolytica* IA 1055 used glucose as C-source for BS production. Starch, glycerol, sodium acetate, n-hexadecane, n-alkanes, carbohydrates, and oleic acid are some different carbon sources. Various articles have reported that some agro-industrial wastes and oils (carrot peel, banana waste, lime and orange peel, molasses and whey, vegetable oil, waste frying oil, and vegetable oil byproducts) are cheap sources of carbon with maximum yield production of BSs. Sucrose and glucose are some carbon sources that can help in reduction of surface tension. Nitrogen sources play a critical role in BS production. They are one of the major nutritional constituents in media other than C-sources that can affect biomass growth. *Pseudomonas* strain 44T1 uses nitrate as a nitrogen source for BS production, and *Rhodococcus* strain ST uses paraffin and olive oil as substrates for growth. Peptone, urea, beef extract, yeast extract, sodium nitrate, ammonium nitrate, malt extracts, and a few essential amino acids are different nitrogen sources that have been used for BS production (Adamczak et al., 2000). Temperature variation also affects the production of rhamnolipids from *Pseudomonas aeruginosa*. On the basis of several studies, it was stated that 25–37°C was the optimum temperature for all types of microbial strains (Saharan et al., 2011). The parameters of cultures, like the pH of media, aeration, and agitation, also affect the fermentative production of BSs. The pH factor also affects the production of biosurfactant from *Yarrowia lipolytica* (Zinjarde et al., 2002), and optimum production has been observed at a media pH of 8. Agitation and aeration rate factors are crucial for uniform mixing of oxygen and media components through fermentation broth (Shaligram and Singhal, 2010). These factors can facilitate oxygen (air) transfer from two phases (gas to aqueous phase). Adamczak found that best surfactant production value (45.5 g/l) was obtained when dissolved oxygen (DO) concentration was maintained at 50% of saturation, and the flow rate should be around 1 vvm.

16.2.8 DIFFERENT DOWNSTREAM PROCESSING METHODOLOGIES FOR RECOVERY OF BIOSURFACTANTS

BSs can be recovered by different types of downstream techniques, as listed in Table 16.3. In the biotechnological area, downstream processes account for around 70–80% of total yield costs. The low cost effectiveness of production of BSs is due to the complicated recovery process. Mulligan and Gibbs (1990) reported that ultrafiltration membranes can be used to recover the amount of BS. Ultrafiltration membranes have an anisotropic structure and are used to filter macromolecules with a molecular weight range of 2000–50,000 (Mulligan and Gibbs, 1990). Some commonly used extraction methods are ethyl acetate, pentane, chloroform:methanol (2:1), acetic acid, and di-chloromethane; methanol, butanol, hexane, and many more solvents are used. There are disadvantages of using solvents in the recovery process such as chloroform that are reported as toxic chloro-organic compounds that can harm human health as well as the ecosystem. Apart from toxicity, they are expensive because of the use of a huge quantity of solvents for recovery methods. Other precipitation techniques used to recover BSs such as adsorption, centrifugation, and ammonium sulfate are listed in Table 16.3. Water solubilization, ionic charge, and cellular location are some methods that are used to recover BS (Helmy et al., 2011). In foam fractionation (solvent-free method), due to surface activity (high), there was continuous formation of foam in the produced BS. Moreover, due to their tiny size and structure, BSs do not readily undergo denaturation.

16.2.9 INDUSTRIAL APPLICATION OF BIOSURFACTANTS

BS are one of those important groups of chemical compounds with a wide range of industrial applications such as cosmetics, detergent, agricultural, food, pharmaceuticals, textile, paper, paint, and medical.

TABLE 16.3
Downstream Processing Techniques Employed for Bio-Separation of Different Types of Biosurfactants (Santos et al., 2016)

Biosurfactant	Recovery Process	Separating Factors	Advantages
Glycolipids (rhamnolipids), lipopeptide, mannosyl-erythritol lipids (MEL)	Adsorption (polystyrene resins, wood-activated carbon)	BSs are absorbed (polystyrene resins as well as activated carbon) and desorbed using organic solvents	Cost effective, high purity, reusable, recovered from continuous culture
Glycolipids	Centrifugation, ion-exchange chromatography (IEC), and ultrafiltration	Because of centrifugal force, insoluble BSs are precipitated out; in IEC, the charged BSs are attached to resins and elute out with a buffer; polymeric membranes can trap micelles attached above CMC	Reusable, fast recovery, high purity, one-step recovery (efficient in microbial surfactant recovery process)
Surfactin	Acid precipitation, foam fractionation	BSs are water immiscible at low pH, BSs can be partitioned by foam	Used to purify the product, cost effective, efficient in microbial surfactant recovery
Sophorolipid, trehalolipid, liposan	Organic solvent extraction	BSs are miscible in organic solvents	Reusable in nature, partial purification, used to enhance microbial surfactant recovery
Biodispersion, emulsan, lipopeptides	Precipitation by ammonium sulfate	Salting out (protein-rich and polymeric BS)	Shows effect in polymeric BS

16.2.9.1 Biomedical and Pharmaceutical Industry

Surfactants can destabilize the membranes (disrupting permeability and integrity) (Ortiz et al., 2009), and they have capability to affect the adhesive properties of microorganisms by modifying their characteristics and making them useful in biomedical industries. Lipopeptide BSs can show antimicrobial activity and lyse the bilayer membranes consisting of lipids. The mode of action occurs after oligomerization of lipopeptide in which a few of them are Ca^{2+}-dependent multimers and form pores in the cell membrane (Scott et al., 2007). These pores in cell membranes can cause trans-membrane influxes (ion), which result in cell apoptosis (death) and disruption of the cell membrane. There major properties may lead to use in biomedical and pharmaceutical industries, where lipopeptides have drawn attention because of resistant antibiotics against microbes (fungus and bacteria) (Mandal et al., 2006). Cationic surfactants are prominently used in biomedical industries because they are positively charged surfactants and have disinfecting properties and can be used in sanitizers. They have been broadly applied in household cleaning. *Bacillus polymyxa* soil bacteria can produce the first lipopeptide (Polymyxin A) enriched with antimicrobial function (Table 16.4). Similarly, microbes such as *Bacillus subtilis* can produce iturin, fengycin, surfactin, bacillomycins, and mycosubtilins, whereas *B. licheniformis* and *Bacillus pumilus* can produce lichenysin and pumilacidin, *Streptomyces roseosporus* can produce daptomycin and viscosin. Some lipopeptides have reached commercial antibiotic status, like caspofungin, daptomycin, anidulafungin, and micafungin. Daptomycin (cyclic nonribosomal, lipopeptide antibiotic) has been isolated from *S. roseosporus*. The mode of action of the daptomycin/Ca^{2+} complex interacts with head groups of membrane composed of phosphatidyl glycerol and causes a second conformational change that encourages oligomerization of the membrane that may lead to membrane perforation (Muraih et al., 2011). Non-competitive inhibitions of β-(1, 3)-D-glucan synthase of echinocandins are responsible for disruption of the mycelial cell wall (Schneider et al., 2010). The lack of a specific enzyme (essential component for cell wall) may lead to cell apoptosis and deterioration (Yao et al., 2012). They may lead to applications that include less toxicity and antifungal activity against some species of *Candida*. They can inhibit *Pneumocystis carinii* and *Aspergillus* species but were unable to show an inhibitory spectrum in *Cryptococcus neoformans* (Denning, 2002). Caspo-fungin was the first licensed echinocandin product.

16.2.9.2 Detergent Industry

BSs act as good cleaners and are used because of their antimicrobial application; and the lower use of solvents for the manufacture of detergents has brought them more attention. Biologically synthesized surfactants are used in detergent formulations because of advantages over other surfactants such as ecosystem protection, low toxicity, use of compounds, and high biodegradability (Kourmentza et al., 2017), as shown in Table 16.4. Consortia of glycolipids, nonionic surfactants, and composition of sophorolipids are used to manufacture detergents used for laundry purposes. Sophorolipid surfactants (low foam) are used in mixtures of soap and other cleaners (such as sprays, heavy-duty power cleaners, window sprays, wax cleaners, and car washes) produced by Ecover, with additional applications, such as low-foaming agents used for rough surfaces. Household cleaning products (detergents) are the most significant application sector for BSs. They have the best cleaning properties; hence, they are one of the most widely used surfactants in laundering, dishwashing liquids, and shampoos. These types of surfactants are particularly good at removing stain residues and making the texture of fabrics softer. Non-ionic surfactants are effective for soil suspension and oily soil cleaning. Surfactants used in flotation deinking cannot work in the way that surfactants probably work in wash deinking. Larger print particles with hydrophobic surfaces are particles that float, and many surfactants have been identified with improved flotation deinking quality (Prashanthi et al., 2017).

16.2.9.3   Food Industry

BSs are used as food additives in food industry. For raw material processing, BSs have been used as emulsifiers (Table 16.4). Derivatives of lecithin, fatty acid esters constituting sorbitan, glycerol/ethylene glycol, and ethoxylated derivatives of mono-glycerides, including a recently synthesized oligo-peptide, are used as emulsifiers worldwide. BSs help to improve organo-leptic properties and solubilization of flavor oils, retard staleness (such as in ice cream and bakery goods), and act as fat stabilizers during cooking of fats (Kosaric, 2001). Adding a specific amount of rhamnolipid BSs may improve the texture, volume, and stability of dough and conserve properties of ice cream and bakery products (Van Haesendonck et al., 2004). Rhamnolipids are in demand because they can improve the properties of croissants, frozen confectionery, and buttercream products. Recently, *Enterobacter cloaceae*, a marine strain that can isolate a bio-emulsifier, has been used as a viscosity enhancer agent in the food industry that may lower the pH in food products enriched with Vitamin C (citric acid) (Iyer et al., 2006). *Enterobacter sakazakii*, *Salmonella enteritidis*, and *Listeria monocytogenes* are examples of some pathogenic bacteria incriminated in outbreaks with ingestion of contaminated food.

16.2.9.4   Textile Industry

The textile finishing industry is one of the highest water-consuming industries. Therefore, there are some textile finishing areas where the use of BSs has been reported, such as detergents, emulsification, wetting, dispersing, and solubilization with characteristics of reducing environmental pollution (Table 16.4). (Mohan et al., 2006). In the textile industry, dye solubilization is one of the core problems. Poor solubility of dye in water causes non-homogenous dispersion of the dye in the fabric solid phase and accumulation of dye-privileged fiber surfaces (Quagliotto et al., 2006). The addition of surfactants has been claimed to enhance the dispersion and water solubilization of dyes to improve dye perforation into fibers (Montoneri et al., 2008).

16.2.10 ENHANCED OIL RECOVERY

Rhamnolipid was the first surfactant to show a microbial application. Microbial-enhanced oil recovery (MEOR) was the main application for which these kinds of surface-active compounds have been reported (Table 16.4). In particular, in the enhancement process of oil recovery, some lipo-peptide (lichenysin, surfactin, and emulsan) BSs have been reported in the microbial recovery process (Sen, 2008). Lately, the stability of different BSs produced by *Bacillus subtilis, P. aeruginosa*, and *Bacillus cereus* was determined at temperatures, salinity conditions, and pH, mimicking those found in oil reservoirs. One of the most promising experiments is displacement of oil performed in micromodels (made of glass) for BS production using Gram-positive bacteria (*B. subtilis*). In this, around 25% of oil was enhanced or recovered (Amani et al., 2010).

TABLE 16.4
Physiochemical Properties of Biosurfactants and Their Role in Different Industrial Sectors

Industries	Applications	Physiochemical Properties of Biosurfactants
Environment	Bioremediation (soil), cleaning (oil spill)	Lowering of interfacial tension, wetting, spreading, soil flushing, detergency, foaming, emulsification and dispersion of oils.
Mining	Remediation of soil, cleaning operations of heavy metal and flotation	Spreading, soil and sediments, wetting and foaming, degradation of metal ions from solutions, heavy metal sequestrants.
Petroleum	De-emulsification, oil recovery enhancement	Corrosion inhibition in fuel oils and equipment, dispersion of oils, detergency, spreadability and foaming, de-emulsification of oil emulsions.
Medicine	Pharmaceuticals, microbiological and therapeutics	Antimicrobial (anti-fungal and anti-bacterial), anti-viral and anti-adhesive agents, gene therapy, vaccines and immune-modulatory molecules.
Agriculture	Fertilizers, bio-control	Emulsification (in pesticides), elimination of plant pathogens, increased bioavailability of nutrients.
Textiles	Fiber preparation, printing and dyeing, finishing (textiles)	Emulsification (in finishing), wetting, dispersion and detergency, softening, solubilized in nature.
Cleaning	Detergents (washing)	Sanitizers and detergents for wetting, laundry, corrosion inhibition, spreading.
Cosmetics	Beauty and health products	Foaming agent, wetting agents, emulsification, antimicrobial agents, cleansers, solubilization.
Nanotechnology	Synthesis of nanoparticles	Stabilization and emulsification.
Food	Emulsification and de-emulsification	Detergency, thickener, foaming, control of consistency, wetting agent, solubilization of flavored oils.

16.2.10.1   Agriculture Sector

One of the best ways to enhance the solubility of bio-hazardous chemical compounds such as polycyclic aromatic hydrocarbons (PAHs) is to apply surfactants as mobilizing agents. Rhamnolipids show antimicrobial properties and can inhibit mycelium growth of *Bacillus cinerea* and spore formation. Moreover, combinations of rhamnolipids have potential to protect plants from diseases, such as protecting grapevines against gray mold disease (Varnier et al., 2009). Non-ionic surfactants have currently been suggested as potential adjuvants for improving sustainability and efficiency. Non-ionic surfactants are used for agricultural applications and have low potency in other industrial sectors. They are used as pesticide adjuvants (Table 16.4). Organo-silicone surfactants have nonionic division that may show effects on antagonistic and synergistic conditions between nutrient uptake and surfactants (Baratella et al., 2018).

16.2.10.2   Cosmetic Industry

Surfactants (chemical surfactants) have a wide range of properties (detergency, wetting, foaming, dispersing and solubilizing, emulsifying) that may be useful in the cosmetic industry (Table 16.4). Long-term exposure to these toxic chemicals in the environment may have adverse effects on humans and the ecosystem. Biologically synthesized surfactants are used for animal- and environmentally friendly (cosmetics) (Traversier et al., 2018). Currently, it has been observed that biosurfactant patents and products have been reviewed for broad application in the cosmetic and healthcare industries (Lourith et al., 2009; Banat et al., 2010; Morita et al., 2013). There are some healthcare products that use variant formulations of rhamnolipids, such as body spray (deodorants), antacids, insect repellents, hair products (anti-dandruff), acne pads, solutions (contact lens), toothpastes, and nail care products (Maier et al., 2000). Some topical gel formulations containing high amounts of rhamnolipids have been patented as anti-aging and anti-wrinkle products (Haba et al., 2003). Amide oxide showed excellent cleaning and foaming properties; therefore, it has been widely used in home cleaning products. Amphoteric surfactants have been used in cosmetic care products; they help to reduce skin irritation (Prashanthi *et al.*, 2017).

16.2.10.3   Environmental Remediation

Microbes secrete BSs for remediation of different derivatives of soil pollutants, and a list of some microbes is given in Table 16.4. *Corynebacterium simplex* and *Pseudomonas* strains were introduced into Bushnell and Hass mineral-rich media (magnesium sulfate, calcium chloride anhydrous, potassium dihydrogen phosphate, dipotassium hydrogen phosphate, ammonium nitrate, ferric chloride) supplemented with hydrocarbons (Ilori et al., 2008). Bioremediation is a waste remediation technique that involves biological remediation of xenobiotics from contaminated sites. This is one of the cheapest and most eco-friendly naturally occurring methods used to remove waste from soil sediments and groundwater, or it helps to reduce the toxicity of waste substances (Dzionek et al., 2016). Based on waste treatment, bioremediation is segmented into *in situ* and *ex situ*. *In situ* treatment can treat a contaminated site, whereas *ex situ* treatment removes contaminated material to be treated elsewhere (Ron and Rosenberg, 2002). Synthetic dyes, gasoline hydrocarbons, and spillage of oils in a marine system can play a major role in the exploitation of our aqua-system. Previously, it has been reported that gasoline hydrocarbons may harm the environment and human health. They are carcinogenic and toxic and can cause mutation in human bodies. Industrial organic pollutants are very tedious to remove because they are less soluble in water, with high interfacial tensions, and are bound with soil sediments and move toward underground water bodies. Around 90% of these hydrocarbons can transfer directly to humans as a source of potable water (Janbandhu and Fulekar, 2011). Continuous 28-day incubation of rhamnolipid-producing bacteria has shown the degradation of n-hexadecane, hepta-decane, n-octadecane, nanodecane, and other hydrocarbons. Microbial production of glycolipids and sophorose lipid surfactants can significantly degrade polycyclic aromatic hydrocarbons. When glycolipids and other BSs were added to contaminated sites, the rate of biodegradation of hazardous toxic chemicals (2,4-DCPIP, tetra-decane, naphthalene, toluene, and pristine) (Karlapudi et al., 2018) will increase and can remediate the environment in a month. The presence of hydrocarbons to microbes is one of the major limiting factors of bioremediation (Vijayakumar et and Saravanan, 2015). The biological degradation efficiency of hydrocarbon depends on the proportion of microbes present in the earth's crust and water. The physical properties of hydrocarbon directly affect the rate of degradation. The efficacy of bioremediation increases after adding surface-active molecules, that is, BSs. BSs may enhance the growth rate of hydrocarbon-degrader microbes by providing suitable substrates for their growth (Karanth et al., 1999).

16.3   CONCLUSION

Marine-derived BSs can effectively bioremediate a variety of hydrocarbons from the environment. These BSs could be used efficiently under extreme or harsh conditions for bioremediation and product formulation for different industrial sectors. Some marine bacteria have potent abilities to produce BSs and are therefore commercially preferred over mesophilic microbe-derived BSs for bioremediation of oil spills. Psychrophilic bacteria-derived BSs have a potent ability for degradation of crude oil at very low temperatures and are efficiently used for bioremediation of contaminated sites in cold regions. These BSs could be widely used for bioremediation purposes under extreme environmental conditions.

16.4   ACKNOWLEDGMENTS

The authors would like to thank the Department of Bioscience and Biotechnology of Banasthali Vidyapith, Rajasthan, India, for providing research support to carry out this research work.

16.5 REFERENCES

Adamczak, M. and Odzimierz Bednarski, W., 2000. Influence of medium composition and aeration on the synthesis of biosurfactants produced by *Candida antarctica*. *Biotechnology Letters*, 22(4), pp. 313–316.

Ahimou, F., Jacques, P. and Deleu, M., 2000. Surfactin and iturin A effects on *Bacillus subtilis* surface hydrophobicity. *Enzyme and Microbial Technology*, 27(10), pp. 749–754.

Alej, C.S., Humberto, H.S. and Maria, J.F., 2011. Production of glycolipids with antimicrobial activity by *Ustilago maydis* FBD12 in submerged culture. *African Journal of Microbiology Research*, 5(17), pp. 2512–2523.

Amani, H., Sarrafzadeh, M.H., Haghighi, M. and Mehrnia, M.R., 2010. Comparative study of biosurfactant producing bacteria in MEOR applications. *Journal of Petroleum Science and Engineering*, 75(1–2), pp. 209–214.

Aparna, A., Srinikethan, G. and Hedge, S., 2011. Effect of addition of biosurfactant produced by *Pseudomonas* ssp. on biodegradation of crude oil. *International Proceedings of Chemical, Biological & Environmental Engineering*, 6, pp. 71–75.

Bai, G., Brusseau, M.L. and Miller, R.M., 1997. Biosurfactant-enhanced removal of residual hydrocarbon from soil. *Journal of Contaminant Hydrology*, 25(1–2), pp. 157–170.

Balan, S.S., Kumar, C.G. and Jayalakshmi, S., 2017. Aneurinifactin, a new lipopeptide biosurfactant produced by a marine *Aneurinibacillus aneurinilyticus* SBP-11 isolated from Gulf of Mannar: Purification, characterization and its biological evaluation. *Microbiological Research*, 194, pp. 1–9. https://doi.org/10.1016/j.micres.2016.10.005

Banat, I.M., Franzetti, A., Gandolfi, I., Bestetti, G., Martinotti, M.G., Fracchia, L., Smyth, T.J. and Marchant, R., 2010. Microbial biosurfactants production, applications and future potential. *Applied Microbiology and Biotechnology*, 87(2), pp. 427–444.

Baratella, V. and Trinchera, A., 2018. Organosilicone surfactants as innovative irrigation adjuvants: Can they improve water use efficiency and nutrient uptake in crop production? *Agricultural Water Management*, 204, pp. 149–161.

Bezerra de Souza Sobrinho, H., de Luna, J.M., Rufino, R.D., Figueiredo Porto, A.L. and Sarubbo, L.A., 2013. Assessment of toxicity of a biosurfactant from *Candida sphaerica* UCP 0995 cultivated with industrial residues in a bioreactor. *Electronic Journal of Biotechnology*, 16(4), pp. 4–4.

Bhardwaj, G., Cameotra, S.S. and Chopra, H.K., 2013. Biosurfactants from fungi: A review. *Journal of Petroleum & Environmental Biotechnology*, 4(6), pp. 1–6.

Burger, M.M., Glaser, L. and Burton, R.M., 1963. The enzymatic synthesis of a rhamnose-containing glycolipid by extracts of *Pseudomonas aeruginosa*. *Journal of Biological Chemistry*, 238(8), pp. 2595–2602.

Chang, Qing. 2016. *Colloid and Interface Chemistry for Water Quality Control*. Elsevier Science.

Cooper, D.G. and Paddock, D.A., 1984. Production of a biosurfactant from *Torulopsis bombicola*. *Applied and Environmental Microbiology*, 47(1), pp. 173–176.

Da Costa, T.H.M. and Ito, M.K., 2003. Phospholipids physiology. In *Encyclopedia of Food Sciences and Nutrition, Ten-volume Set*, pp. 4523–4531. Elsevier Ltd

Das, A.J. and Kumar, R., 2018. Utilization of agro-industrial waste for biosurfactant production under submerged fermentation and its application in oil recovery from sand matrix. *Bioresource Technology*, 260, pp. 233–240.

Das, P., Mukherjee, S. and Sen, R., 2009a. Antiadhesive action of a marine microbial surfactant. *Colloids and Surfaces B, 71*, pp. 183–186.

Das, P., Mukherjee, S., Sivapathasekaran, C. and Sen, R., 2010. Microbial surfactants of marine origin: potentials and prospects. *Advances in Experimental Medicine and Biology, 672*, pp. 88–101.

Datta, S., 2011. *Optimization of Culture Conditions for Biosurfactant Production from Pseudomonas Aeruginosa OCD1.* Sciensage Publications.

Deng, M.C., Li, J., Hong, Y.H., Xu, X.M., Chen, W.X., Yuan, J.P., Peng, J., Yi, M. and Wang, J.H., 2016. Characterization of a novel biosurfactant produced by marine hydrocarbon-degrading bacterium *Achromobacter* sp. HZ01. *Journal of Applied Microbiology, 120*, pp. 889–899.

Denning, D.W., 2002. Echinocandins: A new class of antifungal. *Journal of Antimicrobial Chemotherapy, 49*(6), pp. 889–891.

Desai, J.D. and Banat, I.M., 1997. Microbial production of surfactants and their commercial potential. *Microbiology and Molecular Biology Reviews, 61*(1), pp. 47–64.

Desjardine, K., Pereira, A., Wright, H., Matainaho, T., Kelly, M. and Andersen, R.J. 2007. Tauramamide, a lipopeptide antibiotic produced in culture by *Brevibacillus laterosporus* isolated from a marine habitat: Structure elucidation and synthesis. *Journal of Natural Products, 70*, pp. 1850–1853.

Dey, G., Bharti, R., Dhanarajan, G., Das, S., Dey, K.K., Dumar, B.N.P., Sen, R. and Mandal, M., 2015. Marine lipopeptide Iturin a inhibits Akt mediated GSK3beta and FoxO3a signaling and triggers apoptosis in breast cancer. *Scientific Reports, 5*, p. 10316. https://doi.org/10.1038/srep10316

Dhasayan, A., Kiran, G.S. and Selvin, J., 2014. Production and characterisation of glyco-lipid biosurfactant by *Halomonas* sp. MB-30 for potential application in enhanced oil recovery. *Biotechnology and Applied Biochemistry, 174*, pp. 2571–2584.

Dobler, L., Vilela, L.F., Almeida, R.V. and Neves, B.C., 2016. Rhamnolipids in perspective: gene regulatory pathways, metabolic engineering, production and technological forecasting. *New Biotechnology, 33*(1), pp. 123–135.

Dzionek, A., Wojcieszyńska, D. and Guzik, U., 2016. Natural carriers in bioremediation: A review. *Electronic Journal of Biotechnology, 23*, pp. 28–36.

Elshikh, M., Funston, S., Chebbi, A., Ahmed, S., Marchant, R. and Banat, I.M., 2017. Rhamnolipids from non-pathogenic *Burkholderia thailandensis* E264: Physicochemical characterization, antimicrobial and antibiofilm efficacy against oral hygiene related pathogens. *New Biotechnology, 36*, pp. 26–36.

Franzetti, A., Gandolfi, I., Bestetti, G., Smyth, T.J. and Banat, I.M., 2010. Production and applications of trehalose lipid biosurfactants. *European Journal of Lipid Science and Technology, 112*(6), pp. 617–627.

Funston, S.J., Tsaousi, K., Rudden, M., Smyth, T.J., Stevenson, P.S., Marchant, R. and Banat, I.M., 2016. Characterising rhamnolipid production in *Burkholderia thailandensis* E264, a nonpathogenic producer. *Applied Microbiology and Biotechnology, 100*, pp. 7945–7956.

Gerard, J., Lloyd, R., Barsby, T., Haden, P., Kelly, M.T. and Andersen, R.J., 1997. Massetolides A-H, antimycobacterial cyclic depsipeptides produced by two Pseudomonads isolated from marine habitats. *Journal of Natural Products, 60*, pp. 223–229.

Geys, R., Soetaert, W. and Van Bogaert, I., 2014. Biotechnological opportunities in bio-surfactant production. *Current Opinion in Biotechnology, 30*, pp. 66–72.

Gharaei-Fathabad, E., 2011. Biosurfactants in pharmaceutical industry: A mini-review. *American Journal of Drug Discovery and Development, 1*(1), pp. 58–69.

Gorkovenko, A., Zhang, J., Gross, R.A. and Kaplan, D.L., 1999. Control of unsaturated fatty acid substituents in emulsans. *Carbohydrate Polymers, 39*(1), pp. 79–84.

Guerra-Santos, L.H., Käppeli, O. and Fiechter, A., 1986. Dependence of *Pseudomonas aeruginosa* continuous culture biosurfactant production on nutritional and environmental factors. *Applied Microbiology and Biotechnology, 24*(6), pp. 443–448.

Haba, E., Pinazo, A., Jauregui, O., Espuny, M.J., Infante, M.R. and Manresa, A., 2003. Physicochemical characterization and antimicrobial properties of rhamnolipids produced by *Pseudomonas aeruginosa* 47T2 NCBIM 40044. *Biotechnology and bioengineering, 81*(3), pp. 316–322.

Hamza, F., Satpute, S., Banpurkar, A., Kumar, A.R. and Zinjarde, S., 2017. Biosurfactant from a marine bacterium disrupts biofilms of pathogenic bacteria in a tropical aquaculture system. *FEMS Microbiology Ecology, 93*(11), p. 140.

Helmy, Q., Kardena, E., Funamizu, N. and Wisjnuprapto, 2011. Strategies toward commercial scale of biosurfactant production as potential substitute for it's chemically counterparts. *International Journal of Biotechnology, 12*(1–2), pp. 66–86.

Hentati, D., Chebbi, A., Loukil, S., Kchaou, S., Godon, J.J., Sayadi, S. and Chamkha, M., 2016. Biodegradation of fluoranthene by a newly isolated strain of *Bacillus stratosphericus* from Mediterranean seawater of the Sfax fishing harbour, Tunisia. *Environmental Science and Pollution Research, 23*, pp. 15088–15100.

Hu, R., Li, G., Jiang, Y., Zhang, Y., Zou, J.J., Wang, L. and Zhang, X., 2013. Silver–zwitterion organic–inorganic nanocomposite with antimicrobial and antiadhesive capabilities. *Langmuir, 29*(11), pp. 3773–3779.

Ilori, M.O., Adebusoye, S.A. and Ojo, A.C., 2008. Isolation and characterization of hydrocarbon-degrading and biosurfactant-producing yeast strains obtained from a polluted lagoon water. *World Journal of Microbiology and Biotechnology, 24*(11), pp. 2539–2545.

Irorere, V.U., Tripathi, L., Marchant, R., McClean, S. and Banat, I.M., 2017. Microbial rhamnolipid production: a critical reevaluation of published data and suggested future publication criteria. *Applied Microbiology and Biotechnology, 101*, pp. 3941–3951.

Iyer, A., Mody, K. and Jha, B., 2006. Emulsifying properties of a marine bacterial exopolysaccharide. *Enzyme and Microbial Technology, 38*(1–2), pp. 220–222.

Jahan, R., Bodratti, A.M., Tsianou, M. and Alexandridis, P., 2020. Biosurfactants, natural alternatives to synthetic surfactants: physicochemical properties and applications. *Advances in Colloid and Interface Science, 275*, p. 102061.

Janbandhu, A. and Fulekar, M.H., 2011. Biodegradation of phenanthrene using adapted microbial consortium isolated from petrochemical contaminated environment. *Journal of Hazardous Materials, 187*(1–3), pp. 333–340.

Jauregi, P. and Kourmentza, K., 2019. Membrane filtration of biosurfactants. In *Separation of Functional Molecules in Food by Membrane Technology*, pp. 79–112. Academic Press.

Kaplan, D., Christiaen, D. and Arad, S., 1987. Chelating properties of extracellular polysaccharides from Chlorella spp. *Applied and Environmental Microbiology, 53*(12), pp. 2953–2956.

Kappeli, O. and Finnerty, W.R., 1979. Partition of alkane by an extracellular vesicle derived from hexadecane-grown *Acinetobacter. Journal of Bacteriology, 140*(2), pp. 707–712.

Karanth, N.G.K., Deo, P.G. and Veenanadig, N.K., 1999. Microbial production of biosurfactants and their importance. *Current Science*, pp. 116–126.

Karlapudi, A.P., Venkateswarulu, T.C., Tammineedi, J., Kanumuri, L., Ravuru, B.K., Ramu Dirisala, V. and Kodali, V.P., 2018. Role of biosurfactants in bioremediation of oil pollution—a review. *Petroleum*, *4*(3), pp. 241–249.

Kaur Sekhon, K. (2012). Biosurfactant production and potential correlation with esterase activity. *Journal of Petroleum & Environmental Biotechnology*, 3.

Kefala, V., Kintziou, H., Protopapa, E., Varvaresou, A., Papageorgiou, S. and Raikou, V., 2011. Tyrosinase inhibitors from natural sources for potential use in the aesthetic and cosmetology practice. *Review of Clinical Pharmacology and Pharmacokinetics, International Edition, 25*(2), pp. 65–68.

Kennedy, J., O'Leary, N.D., Kiran, G.S., Morrissey, J.P., O'Gara, F., Selvin, J. and Dobson, A.D., 2011. Functional metagenomic strategies for the discovery of novel enzymes and biosurfactants with biotechnological applications from marine ecosystems. *Journal of Applied Microbiology, 111*, pp. 787–799.

Kim, P., Oh, D.K., Kim, S.Y. and Kim, J.H., 1997. Relationship between emulsifying activity and carbohydrate backbone structure of emulsan from *Acinetobacter calcoaceticus* RAG-1. *Biotechnology Letters, 19*(5), pp. 457–459.

Kiran, G.S., Priyadharsini, S., Sajayan, A., Priyadharsini, G.B., Poulose, N. and Selvin, J., 2017. Production of lipopeptide biosurfactant by a marine Nesterenkonia sp. and its application in food industry. *Frontiers in Microbiology, 8*, p. 1138.

Kiran, G.S., Sabarathnam, B. and Selvin, J., 2010a. Biofilm disruption potential of a glycolipid biosurfactant from marine *Brevibacterium casei*. *FEMS Immunology and Medical Microbiology, 59*, pp. 432–438.

Kiran, G.S., Sabu, A. and Selvin, J., 2010b. Synthesis of silver nanoparticles by glycolipid biosurfactant produced from marine *Brevibacterium casei* MSA19. *Journal of Biotechnology, 148*, pp. 221–225. https://doi.org/10.1016/j.jbiotec.2010.06. 012

Kitamoto, D., Isoda, H. and Nakahara, T., 2002. Functions and potential applications of glycolipid biosurfactants-from energy-saving materials to gene delivery carriers. *Journal of Bioscience and Bioengineering, 94*(3), pp. 187–201.

Konishi, M., Fukuoka, T., Morita, T., Imura, T. and Kitamoto, D., 2008. Production of new types of sophorolipids by *Candida batistae*. *Journal of Oleo Science, 57*(6), pp. 359–369.

Kosaric, N., 2001. Biosurfactants and their application for soil bioremediation. *Food Technology and Biotechnology, 39*(4), pp. 295–304.

Kourmentza, C., Freitas, F., Alves, V. and Reis, M.A., 2017. Microbial conversion of waste and surplus materials into high-value added products: The case of biosurfactants. In *Microbial Applications Vol. 1* (pp. 29–77). Springer.

Kubicki, S., Bollinger, A., Katzke, N., Jaeger, K.E., Loeschcke, A. and Thies, S., 2019. Marine biosurfactants: biosynthesis, structural diversity and biotechnological applications. *Marine Drugs, 17*(7), p. 408.

Lee, J. and Zhang, L., 2015. The hierarchy quorum sensing network in *Pseudomonas aeruginosa*. *Protein & Cell, 6*(1), pp. 26–41.

Li, Z.Y., Lang, S., Wagner, F., Witte, L. and Wray, V., 1984. Formation and identification of interfacial-active glycolipids from resting microbial cells. *Applied and Environmental Microbiology, 48*(3), pp. 610–617.

Lima, Á.S. and Alegre, R.M., 2009. Evaluation of emulsifier stability of biosurfactant produced by *Saccharomyces lipolytica* CCT-0913. *Brazilian archives of Biology and Technology, 52*, pp. 285–290.

Lourith, N. and Kanlayavattanakul, M., 2009. Natural surfactants used in cosmetics: glycolipids. *International Journal of Cosmetic Science, 31*(4), pp. 255–261.

Ma, Z. and Hu, J., 2014. Production and characterization of Iturinic lipopeptides as antifungal agents and biosurfactants produced by a marine *Pinctada martensii*-derived *Bacillus mojavensis* B0621A. *Biotechnology and Applied Biochemistry, 173*, pp. 705–715.

Maier, R.M. and Soberon-Chavez, G., 2000. *Pseudomonas aeruginosa* rhamnolipids: biosynthesis and potential applications. *Applied Microbiology and Biotechnology, 54*(5), pp. 625–633.

Makkar, R.S., Cameotra, S.S. and Banat, I.M., 2011. *Advances in Utilization of Renewable Substrates for Biosurfactant Production*. BioMed Central Ltd. BioMed Central Ltd. www.amb-express.com/content/1/1/5.

Mandal, D., Bolander, M.E., Mukhopadhyay, D., Sarkar, G. and Mukherjee, P., 2006. The use of microorganisms for the formation of metal nanoparticles and their application. *Applied Microbiology and Biotechnology, 69*(5), pp. 485–492.

Mani, P., Dineshkumar, G., Jayaseelan, T., Deepalakshmi, K., Ganesh Kumar, C. and Senthil Balan, S., 2016. Antimicrobial activities of a promising glycolipid biosurfactant from a novel marine *Staphylococcus saprophyticus* SBPS 15. *3 Biotech, 6*, p. 163.

Mapelli, F., Scoma, A., Michoud, G., Aulenta, F., Boon, N., Borin, S., Kalogerakis, N. and Daffonchio, D., 2017. Biotechnologies for marine oil spill cleanup: indissoluble ties with microorganisms. *Trends in Biotechnology, 35*, pp. 860–870.

Marchant, R. and Banat, I.M., 2012a. Microbial biosurfactants: challenges and opportunities for future exploitation. *Trends in Biotechnology, 30*, pp. 558–565.

Marchant, R. and Banat, I.M., 2012b. Biosurfactants: a sustainable replacement for chemical surfactants? *Biotechnology Letters, 34*(9), pp. 1597–1605.

Md Badrul Hisham, N.H., Ibrahim, M.F., Ramli, N. and Abd-Aziz, S., 2019. Production of biosurfactant produced from used cooking oil by *Bacillus* sp. HIP3 for heavy metals removal. *Molecules, 24*(14), p. 2617.

Mohan, P.K., Nakhla, G. and Yanful, E.K., 2006. Biokinetics of biodegradation of surfactants under aerobic, anoxic and anaerobic conditions. *Water Research, 40*(3), pp. 533–540.

Montoneri, E., Savarino, P., Bottigliengo, S., Musso, G., Boffa, V., Prevot, A.B., Fabbri, D. and Pramauro, E., 2008. Humic acid-like matter isolated from green urban wastes. Part II: performance in chemical and environmental technologies. *Bioresources, 3*(1), pp. 217–233.

Morita, T., Fukuoka, T., Imura, T. and Kitamoto, D., 2013. Production of mannosylerythritol lipids and their application in cosmetics. *Applied Microbiology and Biotechnology, 97*(11), pp. 4691–4700.

Mukherjee, S., Das, P. and Sen, R., 2006. Towards commercial production of microbial surfactants. *TRENDS in Biotechnology, 24*(11), pp. 509–515.

Mukherjee, S., Das, P., Sivapathasekaran, C. and Sen, R., 2009. Antimicrobial biosurfactants from marine Bacillus circulans: Extracellular synthesis and purification. *Letters in Applied Microbiology, 48*(3), pp. 281–288.

Mulligan, C.N. and Gibbs, B.F., 1990. Recovery of biosurfactants by ultrafiltration. *Journal of Chemical Technology & Biotechnology, 47*(1), pp. 23–29.

Muraih, J.K., Pearson, A., Silverman, J. and Palmer, M., 2011. Oligomerization of daptomycin on membranes. *Biochimica et Biophysica Acta (BBA)-Biomembranes, 1808*(4), pp. 1154–1160.

Neethirajan, S. and Jayas, D.S., 2011. Nanotechnology for the food and bioprocessing industries. *Food and Bioprocess Technology*, *4*(1), pp. 39–47.

Nitschke, M., Costa, S.G. and Contiero, J., 2011. Rhamnolipids and PHAs: recent reports on *Pseudomonas*-derived molecules of increasing industrial interest. *Process Biochemistry*, *46*(3), pp. 621–630.

O'Lenick, T., 2011. Anionic/cationic complexes in hair care. *Journal of Cosmetic Science*, *62*(2), pp. 209–228.

Onur, G., Yilmaz, F. and Icgen, B., 2015. Diesel oil degradation potential of a bacterium inhabiting petroleum hydrocarbon contaminated surface waters and characterization of its emulsification ability. *Journal of Surfactants and Detergents*, *18*(4), pp. 707–717.

Ortega-de la Rosa, N.D., Vázquez-Vázquez, J.L., Huerta-Ochoa, S., Gimeno, M. and Gutiérrez-Rojas, M., 2018. Stable bioemulsifiers are produced by Acinetobacter bouvetii UAM25 growing in different carbon sources. *Bioprocess and Biosystems Engineering*, *41*(6), pp. 859–869.

Ortiz, A., Teruel, J.A., Espuny, M.J., Marqués, A., Manresa, Á. and Aranda, F.J., 2009. Interactions of a bacterial biosurfactant trehalose lipid with phosphatidylserine membranes. *Chemistry and Physics of Lipids*, *158*(1), pp. 46–53.

Pacwa-Płociniczak, M., Płaza, G.A., Piotrowska-Seget, Z. and Cameotra, S.S., 2011. Environmental applications of biosurfactants: recent advances. *International Journal of Molecular Sciences*, *12*(1), pp. 633–654.

Papageorgiou, S., Varvaresou, A., Tsirivas, E. and Demetzos, C., 2010. New alternatives to cosmetics preservation. *Journal of Cosmetic Science*, *61*(2), p. 107.

Pepi, M., Cesaro, A., Liut, G. and Baldi, F., 2005. An antarctic psychrotrophic bacterium *Halomonas* sp. ANT-3b, growing on n-hexadecane, produces a new emulsyfying glycolipid. *FEMS Microbiology Ecology*, *53*, pp. 157–166.

Perfumo, A., Banat, I.M. and Marchant, R., 2018. Going green and cold: biosurfactants from low-temperature environments to biotechnology applications. *Trends in Biotechnology*, *36*, pp. 277–289.

Prashanthi, M., Sundaram, R., Jeyaseelan, A. and Kaliannan, T., 2017. *Bioremediation and Sustainable Technologies for Cleaner Environment*. Springer International Publishing.

Quagliotto, P., Montoneri, E., Tambone, F., Adani, F., Gobetto, R. and Viscardi, G., 2006. Chemicals from wastes: compost-derived humic acid-like matter as surfactant. *Environmental Science & Technology*, *40*(5), pp. 1686–1692.

Rebello, S., Asok, A.K., Mundayoor, S. and Jisha, M.S., 2014. Surfactants: toxicity, remediation and green surfactants. *Environmental Chemistry Letters*, *12*(2), pp. 275–287.

Rhein, L., 2007. Surfactant action on skin and hair-C.3: cleansing and skin reactivity mechanisms. In *Handbook for Cleaning/Decontamination of Surfaces*, pp. 305,IV–369,V. Elsevier B.V.

Roling, W.F., Milner, M.G., Jones, D.M., Fratepietro, F., Swannell, R.P., Daniel, F. and Head, I.M., 2004. Bacterial community dynamics and hydrocarbon degradation during a field-scale evaluation of bioremediation on a mudflat beach contaminated with buried oil. *Applied and Environmental Microbiology*, *70*, pp. 2603–2613.

Roling, W.F., Milner, M.G., Jones, D.M., Lee, K., Daniel, F., Swannell, R.J. and Head, I.M., 2002. Robust hydrocarbon degradation and dynamics of bacterial communities during nutrient-enhanced oil spill bioremediation. *Applied and Environmental Microbiology*, *68*, pp. 5537–5548.

Ron, E.Z. and Rosenberg, E., 2002. Biosurfactants and oil bioremediation. *Current Opinion in Biotechnology*, *13*(3), pp. 249–252.

Rosen, M.J., 1989. *Surface and Interfacial Phenomena*, 2nd ed. Wiely, p. 151.

Rosenberg, E. and Ron, E.Z., 1999. High-and low-molecular-mass microbial surfactants. *Applied Microbiology and Biotechnology*, *52*(2), pp. 154–162.

Roy, S., Chandni, S., Das, I., Karthik, L., Kumar, G. and Bhaskara Rao, K.V., 2015. Aquatic model for engine oil degradation by rhamnolipid producing *Nocardiopsis* VITSISB. *3 Biotech*, *5*, pp. 153–164. https://doi.org/10.1007/s13205-014-0199-8

Rufino, R.D., Sarubbo, L.A. and Campos-Takaki, G.M., 2007. Enhancement of stability of biosurfactant produced by *Candida lipolytica* using industrial residue as substrate. *World Journal of Microbiology and Biotechnology*, *23*(5), pp. 729–734.

Saenz-Marta, Claudia Isabel, Ballinas-Casarrubias, María de Lourdes, Rivera-Chavira, Blanca E. and Nevárez-Moorillón, Guadalupe Virginia. 2015. *Biosurfactants as Useful Tools in Bioremediation*. IntechOpen. www.intechopen.com/articles/show/title/biosurfactants-as-useful-tools-in-bioremediation.

Saharan, B.S., Sahu, R.K. and Sharma, D., 2011. A review on biosurfactants: fermentation, current developments and perspectives. *Genetic Engineering and Biotechnology Journal*, *2011*(1), pp. 1–14.

Santos, D.K.F., Rufino, R.D., Luna, J.M., Santos, V.A. and Sarubbo, L.A., 2016. Biosurfactants: multifunctional biomolecules of the 21st century. *International Journal of Molecular Sciences*, *17*(3), p. 401.

Sarubbo, L.A., Farias, C.B. and Campos-Takaki, G.M., 2007. Co-utilization of canola oil and glucose on the production of a surfactant by *Candida lipolytica*. *Current Microbiology*, *54*(1), pp. 68–73.

Schneider, T. and Sahl, H.G., 2010. Lipid II and other bactoprenol-bound cell wall precursors as drug targets. *Current Opinion in Investigational Drugs (London, England: 2000)*, *11*(2), pp. 157–164.

Scott, W.R., Baek, S.B., Jung, D., Hancock, R.E. and Straus, S.K., 2007. NMR structural studies of the antibiotic lipopeptide daptomycin in DHPC micelles. *Biochimica et Biophysica Acta (BBA)-Biomembranes*, *1768*(12), pp. 3116–3126.

Sen, R., 2008. Biotechnology in petroleum recovery: the microbial EOR. *Progress in Energy and Combustion Science*, *34*(6), pp. 714–724.

Shaligram, N.S. and Singhal, R.S., 2010. Surfactin—a review on biosynthesis, fermentation, purification and applications. *Food Technology and Biotechnology*, *48*(2), pp. 119–134.

Shete, A.M., Wadhawa, G., Banat, I.M. and Chopade, B.A. (2006). Reviews—Mapping of patents on bioemulsifier and biosurfactant: a review. *Journal of Scientific & Industrial Research*, *65*, p. 91.

Sivapathasekaran, C., Das, P., Mukherjee, S., Saravanakumar, J., Mandal, M. and Sen, R., 2010. Marine bacterium derived lipopeptides: characterization and cytotoxic activity against cancer cell lines. *International Journal of Peptide Research and Therapeutics*, *16*, pp. 215–222.

Sorensen, K.N., Kim, K.H. and Takemoto, J.Y., 1996. In vitro antifungal and fungicidal activities and erythrocyte toxicities of cyclic lipodepsinonapeptides produced by Pseudomonas syringae pv. syringae. *Antimicrobial Agents and Chemotherapy*, *40*(12), pp. 2710–2713.

St, L.J.B., de, B.F., De, C.K., Demeyere, H., Labeque, R., Lodewick, R. and van, L.L., 2007. Laundry cleaning of textiles-B.1.I. In *Handbook for Cleaning/Decontamination of Surfaces*, pp. 57–102. Elsevier B.V.

Stick, R.V. and Williams, S.J., 2010. *Carbohydrates: The Essential Molecules of Life.* Elsevier Science.

Syutsubo, K., Kishira, H. and Harayama, S., 2001. Development of specific oligonucleotide probes for the identification and *in situ* detection of hydrocarbon-degrading *Alcanivorax* strains. Environmental Microbiology, *3*, pp. 371–379.

Tan, D., Xue, Y.S., Aibaidula, G. and Chen, G.Q., 2011. Unsterile and continuous production of polyhydroxybutyrate by *Halomonas* TD01. *Bioresource Technology, 102*, pp. 8130–8136. https://doi.org/10.1016/j.biortech.2011.05.068

Thakur, P., Saini, N.K., Thakur, V.K., Gupta, V.K., Saini, R.V. and Saini, A.K., 2021. Rhamnolipid the Glycolipid Biosurfactant: emerging trends and promising strategies in the field of biotechnology and biomedicine. *Microbial Cell Factories, 20*(1), pp. 1–15.

Thavasi, R., Jayalakshmi, S. and Banat, I.M., 2014. Biosurfactants and bioemulsifiers from marine sources. In Mulligan, C.N., Sharma, S.K. and Mudhoo, A. (Eds.), *Biosurfactants,* pp. 125–146. Hardback—CRC Press. https://doi.org/10. 1201/b16383-6

Traversier, M., Thomas, G., Sandrine, M., Sylvie, M. and Eldra, D. 2018. Polar lipids in cosmetics: recent trends in extraction, separation, analysis and main applications. *Phytochemistry Reviews: Fundamentals and Perspectives of Natural Products Research, 17*(5), pp. 1179–1210.

Tripathi, L., Irorere, V.U., Marchant, R. and Banat, I.M., 2018. *Marine Derived Biosurfactants: A Vast Potential Future Resource.* Biotechnology Letters. https://doi.org/10.1007/s10529-018-2602-8

Uzoigwe, C., Burgess, J.G., Ennis, C.J. and Rahman, P.K., 2015. Bioemulsifiers are not biosurfactants and require different screening approaches. *Frontiers in Microbiology, 6*, p. 245. https://doi. org/10.3389/fmicb.2015.00245

Van Haesendonck, I.P.H. and Vanzeveren, E.C.A., 2004. Rhamnolipids in bakery products. *International Application Patent (PCT) WO*, pp. 2004–040984.

Varnier, A.L., Sanchez, L., Vatsa, P., Boudesocque, L., Garcia-Brugger, A.N.G.E.L.A., Rabenoelina, F., Sorokin, A., Renault, J.H., Kauffmann, S., Pugin, A. and Clément, C., 2009. Bacterial rhamnolipids are novel MAMPs conferring resistance to Botrytis cinerea in grapevine. *Plant, Cell & Environment, 32*(2), pp. 178–193.

Varvaresou, A. and Iakovou, K., 2015. Biosurfactants in cosmetics and biopharmaceuticals. *Letters in Applied Microbiology, 61*(3), pp. 214–223.

Vaz, D.A., Gudina, E.J., Alameda, E.J., Teixeira, J.A. and Rodrigues, L.R., 2012. Performance of a biosurfactant produced by a *Bacillus subtilis* strain isolated from crude oil samples as compared to commercial chemical surfactants. *Colloids and Surfaces B: Biointerfaces, 89*, pp. 167–174.

Vijayakumar, S. and Saravanan, V., 2015. Biosurfactants—types, sources and applications. *Research Journal of Microbiology, 10*(5), p. 181.

Vilela, W.F.D., Fonseca, S.G., Fantinatti-Garboggini, F., Oliveira, V.M. and Nitschke, M., 2014. Production and properties of a surface-active lipopeptide produced by a new marine Brevibacterium luteolum strain. *Applied Biochemistry and Biotechnology, 174*(6), pp. 2245–2256.

Volkering, F., Breure, A.M. and Rulkens, W.H., 1997. Microbiological aspects of surfactant use for biological soil remediation. *Biodegradation, 8*(6), pp. 401–417.

Williams, J.J., 2007. Formulation of carpet cleaners. In *Handbook for Cleaning/ Decontamination of Surfaces,* pp. 103–123. Elsevier Science BV.

Willumsen, P.A. and Karlson, U., 1996. Screening of bacteria, isolated from PAH-contaminated soils, for production of biosurfactants and bioemulsifiers. *Biodegradation, 7*(5), pp. 415–423.

Yao, J., Liu, H., Zhou, T., Chen, H., Miao, Z., Sheng, C. and Zhang, W., 2012. Total synthesis and structure–activity relationships of new echinocandin-like antifungal cyclolipohexapeptides. *European Journal of Medicinal Chemistry*, *50*, pp. 196–208.

Yuan, C.L., Xu, Z.Z., Fan, M.X., Liu, H.Y., Xie, Y.H. and Zhu, T., 2014. Study on characteristics and harm of surfactants. *Journal of Chemical and Pharmaceutical Research*, *6*(7), pp. 2233–2237.

Zinjarde, S.S. and Pant, A., 2002. Emulsifier from a tropical marine yeast, *Yarrowia lipolytica* NCIM 3589. *Journal of Basic Microbiology: An International Journal on Biochemistry, Physiology, Genetics, Morphology, and Ecology of Microorganisms*, *42*(1), pp. 67–73.

Index

For Product Safety Concerns and Information please contact our EU
representative GPSR@taylorandfrancis.com
Taylor & Francis Verlag GmbH, Kaufingerstraße 24, 80331 München, Germany

www.ingramcontent.com/pod-product-compliance
Lightning Source LLC
Chambersburg PA
CBHW060747220326
41598CB00022B/2358

9 781032 309668